Heat and Mass Transfer

Series editors

D. Mewes, Hannover, Germany
F. Mayinger, München, Germany

For further volumes:
http://www.springer.com/series/4247

Mihail-Dan Staicovici

Coabsorbent and Thermal Recovery Compression Heat Pumping Technologies

Springer

Mihail-Dan Staicovici
Clean Energy Research
S.C. Varia Energia S.R.L. & S.C.
Incorporate Power-absorption
Engineering S.R.L.
Bucharest
Romania

ISSN 1860-4846 ISSN 1860-4854 (electronic)
ISBN 978-3-642-54683-9 ISBN 978-3-642-54684-6 (eBook)
DOI 10.1007/978-3-642-54684-6
Springer Heidelberg New York Dordrecht London

Library of Congress Control Number: 2014933780

Printed on acid-free paper

Springer is part of Springer Science+Business Media (www.springer.com)

To my wife
To my children
To the memory of my parents

Preface

The made by the arm of man production of cooling and heating is amongst the energy processes of vital importance today in all domains of human activity, technology, food, health, social, research, etc. Also, the global population increase, the limited energy resources, and the accelerated global warming process, has forced people involved in the energy industry to find rapid solutions, much more effective in cooling and heating. Bearing these in mind, a recent research activity of the author of this book has brought to light new solutions, enjoying a high potential in primary energy saving, which can be applied immediately, without long-term research.

After an introductory part, providing the reader with the selected topic on thermodynamics, this book presents briefly the most relevant theoretical aspects of this research. Reference is first made to *the coabsorbent technology*. Although the coabsorbent technology bases on the classic absorption technology, it is more general. The coabsorbent technology can find effective solutions, practically for all kinds of heat pumping (cooling and/or heating) applications, met in industry, agriculture, district, household, etc., provided that two supplying sources with temperatures outdistanced by minimum (12–15) °C are available. It is mostly indicated that the coabsorbent technology be utilized in applications that recover free low-grade heat sources, coming of naturally, or from the industry technological processes. The book describes, in this respect, *the new coabsorbent cycles, of nontruncated, truncated, hybrid truncated and multi-effect type*, with applications in cooling (industrial, medium, and air conditioning) and heating (domestic warm water, house heating, and industrial). The operation in cogeneration of cooling and heating and in trigeneration of mechanical work, cooling and heating working modes of some coabsorbent cycles with special design is also presented. The first principle recommendation of improving cycle COP through internal sensible and latent heat recovery has been constantly taken into consideration in the cycle design. A new, important, and very useful tool of investigation for the absorption processes heat exchange, *the divided device method*, is introduced. This method has been applied extensively to absorption processes suffering latent–sensible or latent–latent heat transfer processes. In this respect, with its help, a thorough study of the generation–absorption (gax) recovery taking place at large intervals with temperature overlapping, and the results thereof, are given for the cooling and heating truncated cycles operating with ammonia–water and water–lithium bromide working combinations.

The theoretical presentation is completed by chapters giving the thermodynamical ideal limits of the cooling and heating coabsorbent cycles and showing the way the exergy efficiency of these cycles should be computed when the exergy balance is extended on boundaries comprising not only the processes at hand, but their different supplying sources of energy as well. Particular attention has also been paid to emphasize the second principle recommendations of improving the COP.

The second technology refers to cooling and/or heating production using the mechanical vapor compression. This technology is the most widespread in the world today, occupying the first place in what concerns the number of applications and the cooling and heating capacity. As compared to the classic cycles, this technology proposes as novelty, *the recovery of the discharge gas superheating*, with positive energy consequences. The discharge gas superheat recovery is converted into useful issues with the help of two methods. According to the first, the heat is converted into work which diminishes the compressor work input, and is termed *thermal-to-work recovery compression* (*TWRC*). According to the second, the heat is converted into useful cooling and/or heating effect, added to the cycle output effect via the coabsorbent technology, and is termed *thermal-to-thermal recovery compression* (*TTRC*). Other important aspects concerning the *TWRC* and *TTRC* methods, including the theoretical and ideal COP calculation, the choice of the recovery compression cycles, cycles structure solutions, the coabsorbent technology use in *TTRC* application, and the computation of the optimum intermediary pressures of multi-stage compression, are given as well. The methods are analyzed for single-, two-, and three-stage compression cooling and heating cycles, and the model effectiveness results are given.

Next, the book includes the author's own researches concerning fundamental aspects of the absorption processes, in completion of the coabsorbent technology. First, reference is made to *a non-equilibrium phenomenological theory of mass and heat transfer in physical and chemical interactions*. This theory postulates the existence of the *natural forces* governing the physical and chemical interactions and brings to light *the ideal point approaching effect* suffered by a natural force in the proximity of an ideal point (the denomination of the classic equilibrium point in the phenomenological approach). With the help of this new effect it was possible to explain phenomena which otherwise were difficult to be explained by the classic theory of mass and heat transfer (e.g. *the problem of the ammonia bubble absorption*, why an absorption process is *a mass phenomenon and not a surface one*, or *the heat pipe high heat transfer properties*). Based on the non-equilibrium phenomenological approach, a *two-point theory* (*TPT*) of mass and heat transfer is proposed, where *the equilibrium point* and *the ideal point* play an important role in the *non-coupled* and *the coupled mass and heat transfer*, respectively. Further on, *a new wording of the Laplace equation*, more general than the already known wording, and the variational numerical and analytical approach of the liquid capillary rise effect are presented. In the last part of the book, *the Marangoni convection basic mechanism explanation* is given. The Marangoni convection stimulation is a means of increasing the mass and heat transfer, generally, and particularly in the absorption interactions. Its true explanation is a consequence of the *le Chatelier principle* respect and is done with the help of the new Laplace equation and the *TPT*. In order to increase the mass and heat transfer of the pseudo-Marangoni cells, a *Marangoni—Gravity Forces Dimensionless Criterion*, created with the help of *TPT*, is applied and effective absorption–desorption mass and heat exchangers with horizontal free surfaces are proposed.

Mihail-Dan Staicovici

Contents

Nomenclature

A	Caloric equivalent of mechanical work, kcal $kgfm^{-1}$, anergy, kJ, absorber;
A	Species of reactant, reduced excess heat, chemical affinity, kJ kg^{-1}, area, m^2;
A(y)	Integration constant in Eq. (1.257);
$A_{i,j}$	Coefficients in Eq. (9.69);
$[A_i']$	Molar concentration of species A_i' of reactant, kmol $kmol^{-1}$;
$[A_i'']$	Molar concentration of species A_i'' of reactant, kmol $kmol^{-1}$;
AR	Aspect ratio, m m^{-1};
ARC	Absorption recovery cycle;
a	Integration constant in Eq. (1.251); symbol in Eq. (2.8);
a, a_1	Specific inlet and outlet absorber mass flow factors, kg kg^{-1};
a, b, c, d	Factors intervening in Eqs. (A9.1–A9.4);
acr	Adjacent concentration removal;
$a_1, \ldots, a_4$	Empiric coefficients of liquid volume, v^l, Eq. (1.285);
B(y)	Integration constant in Eq. (1.257), K^{-1};
$b = L * R^{-1}$	Constant in Eq. (1.251); symbol in Eq. (2.9), bubble lifetime;
$b_1, \ldots, b_3$	Empiric coefficients of liquid specific heat at constant pressure;
C	Heat capacity, kJ kg^{-1}, local thermal capability, kJ $kmol^{-1}$, velocity of sound, m s^{-1};
C	Absorption heat, kJ kg^{-1};
COP	Coefficient of performance;
$C_{Mg/mg}$	Marangoni—gravity forces dimensionless criterion;
c	Liquid phase mass fraction, kg kg^{-1};
c_p	Specific heat at constant pressure, kJ $kg^{-1}K^{-1}$;
c_v	Liquid specific heat at constant volume, kJ kg^{-1} K^{-1};
$c_1, \ldots, c_4$	Empiric coefficients of gas volume v^g, Eq. (1.287);
D	Desorber;
D, D_X	Mass diffusivity coefficient, m^2 s^{-1};
D_T	Heat diffusivity coefficient, m^2 s^{-1};
d	Inlet specific desorber mass flow factor, injection nozzle diameter, m, surfactant droplet half length, m;
d_r	Reversibility flaw, Eq. (9.183), kJ kg^{-1} K^{-1};

d, d_1	Specific inlet and outlet desorber mass flow factors, kg kg^{-1};
d_1, …, d_3	Empiric coefficients of gas specific heat at constant pressure, c_p^g, Eq. (1.228);
dT	Temperature pinch, °C;
E	Exergy, availability, kJ;
e	Thermodynamic function, Eq. (9.123), kJ kg^{-1} K^{-1};
e	Interface specific free energy, J m^{-2};
e_1, …, e_{16}	Coefficients of excess functions $f_1(T, p)$, …, $f_3(T, p)$, Eqs. (1.282–1.284);
e_{ht}	Heating temperature effectiveness;
F	Free energy of Helmholtz, kJ;
F	Force, N;
F_i	External force, N;
$f = F * G^{-1}$	Specific free energy of Helmholtz, kJ kg^{-1}, fresh air fraction;
f	Figure of merit showing the increase of a phenomenological function approaching an ideal state, as compared to its far equilibrium states;
f	Free energy potential, kJ kg^{-1};
f_{mixed}	Ratio of coupled and noncoupled natural forces of a mixed transfer, Eq. (9.131);
f_M	Mean y_M concentration cumulated absorbent flow, kg kg^{-1};
$f_1(T, p)$, …, $f_3(T, p)$	Functions of excess enthalpy, $\phi_E(T, p)$, kJ kg^{-1};
G, G_M	Mass, kg, generator, molar mass, kg $kmol^{-1}$;
GHE	Generator excess heat;
gax	Generator–absorber heat exchange;
g, g_1	Specific inlet and outlet generator mass flow factors, kg kg^{-1};
$g_{1,i}$	The *i*-th generator absorbent mass outlet flow, $i = 1, \ldots, n$;
H	Enthalpy, kJ; gas phase enthalpy, kJ kg^{-1}, discharge gas enthalpy, $kJkg^{-1}$;
$h = H*G^{-1}$	Specific enthalpy, kJ kg^{-1} ; liquid phase enthalpy, kJ kg^{-1}, Marangoni cell height, m;
h_r	Radiation heat exchange coefficient with the sky, $Wm^{-2}K^{-1}$;
h_t	Height of a truncation column, kg kg^{-1} K^{-1};
I	A function definition interval;
I, J, K, L	Cooling or heating effectiveness of subcycles connected in a cascade; L-known effectiveness of the $(i+1)$-th subcycle added to an i-th cascade in order to construct the $(i+1)$-th cascade;
i	Specific mass current in the liquid, mol s^{-1} m^{-2};
i.p.a.	Ideal point approaching effect;
$i_d(r)$	Relative desorption interval function, Eq. (3.101);
J	Mechanical equivalent of heat, $kcal^{-1}$ kgfm;
J_v, J_c, J_T	Velocity, mass, and heat fluxes, N m^{-2}, kg m^{-2} s^{-1} and kg K m^{-2} s^{-1}, respectively;

j_i, j_q, j_l	Interaction mass, heat and work currents, mol s^{-1}, kg s^{-1}, kW;
j_r	Reduced interface mass current;
$j'_{i,g}$	Interface mass current, mol s^{-1};
$j'_{i,g,a}$	Cumulated absorbed (interface) mass current, mol;
$j'_{i,g,a,s}$	Specific $j'_{i,g,a}$ or average absorption indicator, mol s^{-1};
$j''_{q,l,a}$	$j'_{i,g,a}$ coupled heat current, J;
$j''_{q,l,a,s}$	Specific $j''_{q,l,a}$, W;
K_g, K_l	Functions of the global resistance to mass transfer, kg m^{-2} s^{-1};
k	Adiabatic (isentropic) exponent, proportionality constant, global mass transfer coefficient, kg m^{-2} s^{-1};
k_{COP}	Constant in Eq. (3.100);
k_i	Stage i reduced heat source input;
$k_{i,j}$	$ARC_{i,j}$ reduced heat source input;
k_{mvc}	Constant in Eq. (3.113);
L	Marangoni cell characteristic length, m;
$L_{i,j}$	Phenomenological coefficient associated with coupled currents i and j, mol^2 K J^{-1} s^{-1}, kg K s m^{-2};
L, l	Mechanical work, kJ;
$L = r + l$	Accounts for phase changing and mixing heats, kJ kg^{-1};
l	Specific mechanical work, kJ kg^{-1}, liquid–liquid mixing heat, kJ kg^{-1};
l	Marangoni cell half length, m;
M	Number of mol, kmol, mixing point, molar mass, kg $kmol^{-1}$, Marangoni number;
$M(p_{r,e}, x_e)$	Match function in Eq. (9.34), given by Eq. (9.69), kg $kmol^{-1}$;
m	Mass, kg, specific mass flow rate, kg kg^{-1} s^{-1}, Chap. 8;
$\dot{m}$	Bubble feeding mass flow rate, kg s^{-1};
max(a, b)	Maximum value amongst a and b;
min(a, b)	Minimum value amongst a and b;
m_m	Molecular mass, kg;
mvc	Mechanical vapor compression;
N	Molar fraction, kmol $kmol^{-1}$, mass flux, kg m^{-2} s^{-1};
Nu	Nusselt number;
N_A	Avogadro's number of molecules per kmol;
n	Number of components, cycles, tabulation intervals, number of species or moles of a specie involved in a chemical reaction, number of bubbles;
$n'_{0,i} i = 1, \ldots, n'$	Initial number of moles of the reactants;
P	Control volume center;
Pe	Peclet number;
Pr	Prandtl number;
p	Pressure, partial pressure, kPa, bar, ata;
p_{low}, p_l	Low pressure, bar;
p_h	High pressure, bar;

Q	Heat, kJ;
$q = Q * G^{-1}$	Intensive heat, kJ kg^{-1} heat, kJ, heat flux, kJ m^{-2};
q	Specific heat current in the liquid, coupled with i, W m^{-2};
q,$\dot{q}_t$	Tube feeding mass flow rate, kg s^{-1};
$\dot{q}$	Heat flux, W m^{-2} K^{-1};
R	Gas constant, kJ kg^{-1} K^{-1}, universal gas constant, kJ $kmol^{-1}$ K^{-1}, Eqs. (1.260) and (1.261);
R	Resorber, bubble radius, m;
R′	Gas constant, kJ kg^{-1} K^{-1};
Re	Reynolds number;
RH	Relative humidity, %;
RHE	Resorber heat excess;
RII	Incomplete preheated absorbent;
R_M	Universal gas constant, kJ $kmol^{-1}$ K^{-1};
RRC	Rankine recovery cycle;
r, R, R_1, R_2	Radius, main surface radii, m;
r	Volumetric fraction, m^3 m^{-3}, liquid–gas phase change heat, kJ kg^{-1}, radial coordinate, m;
r, r_1	Specific inlet and outlet resorber mass flow factors, kg kg^{-1};
$r_{1,i}$	The i-th resorber absorbent mass outlet flow, i = 1, …, n;
r_i	Stage *i* reduced compression work input;
r_h, r_i, r_q	Ratios in Eqs. (9.139) (9.143) (9.144);
S	Entropy, kJ K^{-1}, coabsorbent cooling and heating cycle symmetry factor;
S	Specific heat exchange area, m^2 kg^{-1}, bubble area, m^2;
Sc	Schmidt number;
Sh	Sherwood number;
$\dot{S}$	Entropy source, W K^{-1};
$s = S * G^{-1}$	Specific entropy, kJ kg^{-1} K^{-1}, half-distance of infinite capillary walls, m;
s_j	Symmetry factor of the coabsorbent j-th stage, j = 1, …, n, either in cooling, Eq. (4.15), or in heating, Eq. (4.38);
ss(T_i)	Supplying source of temperature T_i;
T	Temperature, K, °C;
T_0	Temperature of infinite reservoir, K;
t	Temperature, K, °C, time, s;
T_M	y_M mean concentration absorbent temperature, K, °C;
T_{AO}, T_{DO}	Mixing temperatures of nontruncated cooling coabsorbent cycle, K, °C;
T_{GO}, T_{RO}	Mixing temperatures of nontruncated heating coabsorbent cycle, K, °C;
$T_{Rf,O}$	Resorber outlet temperature of the usefully heated fluid, °C;
TTRC	Thermal-to-thermal recovery compression;
TWRC	Work-to-thermal recovery compression;

U	Internal energy, kJ;
$u = U * G^{-1}$	Specific internal energy, kJ kg^{-1}, liquid velocity along x coordinate, m s^{-1};
V, V_M	Volume, m^3, molar volume, m^3 $kmol^{-1}$;
V	Desorbed or generated vapor mass flow, kg;
V_i	*i*-th generator generated vapor, $i = 1, \ldots, n$, kg kg^{-1};
$\dot{V}$	Temporal derivative of the total volume, m^3 s^{-1};
V	Bubble, tube volume, m^3;
v	Liquid velocity along y coordinate, m s^{-1};
v	Liquid velocity, m s^{-1};
$v = V * G^{-1}$	Specific volume, m^3 kg^{-1};
W	Content of moisture, kg moisture/kg dry air, mechanical work, kJ;
$\dot{W}, \dot{w}$	Temporal derivatives of specific mechanical work and mechanical work, kW kg^{-1}, kW;
w_t	Turbine mechanical work output, kW;
w_p	Pump mechanical work input, kW;
X, ($-X_r$)	Generalized thermodynamic natural force, kJ $kmol^{-1}$ K^{-1};
X(y)	Thermal capability, kJ;
X(T)	Thermal capability, kJ K^{-1};
$\bar{X}$	Thermal capability mean integral value, kJ kg^{-1} K^{-1};
x	Liquid phase mass or molar concentration, kg kg^{-1}, mol mol^{-1}, variable in Eq. (A1.1);
x	Specific absorbent mass flow factor of rectification column inlet, kg kg^{-1};
x	Real variable, m;
x, (z)	Coordinate along Marangoni cell length, m;
x^*	Liquid phase molar fraction in equilibrium with gas phase molar fraction y, respectively, mol mol^{-1};
Y	Gas phase mass fraction, kg kg^{-1};
y	Liquid or gas phase mass or molar fraction, kg kg^{-1}, mol mol^{-1};
y, y^*	Gas phase molar fraction and that in equilibrium with liquid phase molar fraction x, respectively, mol mol^{-1};
y	Coordinate along Marangoni cell high, m;
y	Interface real function of x, m;
y_{AO}, y_{DO}	Mixing concentrations of nontruncated cooling coabsorbent cycle, kg kg^{-1};
y_{GO}, y_{RO}	Mixing concentrations of nontruncated heating coabsorbent cycle, kg kg^{-1};
y_{AO}	Threshold concentration of nontruncated heating coabsorbent cycle, kg kg^{-1};
y_{RO}	Threshold concentration of nontruncated cooling coabsorbent cycle, kg kg^{-1};

y_M	Nontruncated cycle mean concentration, kg kg^{-1};
$y_{M,j}$	Coabsorbent mixer concentration of *j*-th stage;
z	Tubes number, coordinate in the natural force definition domain, space coordinate, m;

Greek Letters

ρ	Density, kg m^{-3};
α	Volume thermal isobar expansion coefficient, °C^{-1}, heat exchange coefficient, W m^{-2} K^{-1}, angle between two directions of a plane hosting Marangoni convection, rad;
η	Coefficient;
η_{work}	Work efficiency;
η_E	Exergy efficiency;
ε	Emittance, relative to Carnot efficiency, finite value;
ε_0	Compressor constructive characteristic;
σ	Stefan–Boltzmann constant, $Wm^{-1}K^{-4}$;
θ_{work}	Exergetical temperature factor or the heat-to-work Carnot efficiency factor;
$\xi, \xi \in I$	Variable value for which a function takes the mean value within *I*;
Φ	Free enthalpy of Gibbs, kJ;
$\varphi = \Phi G^{-1}$	Specific free enthalpy, kJ kg^{-1};
φ_E	Specific excess enthalpy, kJ kg^{-1};
Δ, δ	Variable increment, finite difference;
$\sum \Delta T_{Rf,i}$	Temperature increase of heated fluid in resorber;
μ	Chemical potential, kJ kg^{-1}, dynamic viscosity, N s m^{-2};
λ	Degree of chemical reaction progress, kmol $kmol^{-1}$; kg kg^{-1}, thermal conduction, W m^{-1} K^{-1};
Δ	Mass transfer driving force, according to classical interpretation, variation, duration;
ν_i', ν_i''	Stoichiometric coefficients of reactants, $i = 1, \ldots, n'$, and produces of reaction, $i = 1, \ldots, n''$, respectively, kmol, kg;
σ	Superficial tension, N m^{-1}, surface energy, J m^{-2};
σ'_{ik}	Second order viscous tensor, N m^{-2};
τ	Bubble lifetime, s, shear stress tensor, N m^{-2};
υ	Cinematic viscosity, m^2 s^{-1};
β	Thermal expansion coefficient, K^{-1};
$\Gamma_v, \Gamma_c, \Gamma_T$	Proportionality function in Eq. (11.15), Ns/m^2, kg/ms and kg/ms, respectively;
Π	Viscous tensor, N m^{-2};
γ	Mass expansion coefficient;
γ	Liquid weight, N m^{-3};
θ	Liquid/solid contact angle, deg.;

φ_v, φ_c, φ_T	Searched functions along direction x: velocity, m s^{-1}, mass fraction, kg kg^{-1}, and temperature, K;
ε	Absorption efficiency;

Subscripts

A	Thermodynamic state, absorber;
AC	Absorber–condenser;
AD	Absorber–desorber;
AI, AO	Absorber inlet and outlet;
ARC	Absorption recovery cycle;
a	Absorption, adiabatic, cumulated;
abs	Absorption;
atm	Atmospheric;
aux	Auxiliary;
B	Thermodynamic state;
b	Normal boiling point;
boiler	SRC steam generator;
C	Thermodynamic state, condenser, condensation, Carnot;
Carnot	Carnot cycle;
CpI	Compressor suction (inlet);
CpO	Compressor discharge (outlet);
c	Cooling, number of components, compression, critical, classic, kinetic;
cbs	Coabsorbent;
cc	Cooling cycle;
cd	Condensing;
cf	Cooling fractal;
ch	Heating cycle, chemical;
cg, cogen	Cogeneration;
compr	Compression;
cond	Condensation;
cp	Complete preheating;
cpl	Coupled
cs	Complete subcooling;
c_1, c_2	Two different components;
D	Thermodynamic state, desorber;
Db	Desorber brine;
DI, DO	Desorber inlet, outlet;
DII	Incomplete subcooled;
d	Direct, depleted;
$d_{1(2)}$	Differential heat for $m_{2(1)}$ = const.;
dh	Double heating;
dilat	Dilation;

dphlgm	Dephlegmation;
E	Exergy, evaporator;
e	Excess, external, electrical, equilibrium, specific surface energy when $[\sigma] = J * m^{-2}$;
e, w, s, n	East, west, south, north;
ef	Effective;
ep	Equilibrium point;
ex	Exergy;
f	Number of phases, fresh air fraction, fluid, interface specific force when $[\sigma] = N * m^{-1}$;
G	Constant mass, generator;
GI, GO	Generator inlet, outlet;
GE	Generator–evaporator;
GR	Generator–resorber;
GRZ	Gax recovery zone;
g	Generator, gas, gaseous phase, gravity;
gax	Generator–absorber heat exchange;
gen	Generator, generation, electrical generator;
H_2O	Water;
h	Heating, heat source, enthalpy;
hf	Heating fractal;
hcf	Hybrid cooling fractal;
hhf	Hybrid heating fractal;
hp	Heat pump;
hs	Heat source;
I	Inlet;
IJ	Cascade of I and J subcycles;
i	Current index, current number of components, initial, isentropic, indoor, isothermal, mass, indirect;
i	Before interaction;
ii	After interaction;
ics	Incomplete subcooled;
id	Ideal;
int	Intermediary;
irrev	Irreversible;
is	Isentropic;
icp	Incomplete preheating;
ip	Ideal point;
isothermal	Isothermal;
i, k	Indices referring to each compression stage number;
i, j, k	Variable indices;
j	Index referring to each ARC of a stage;
l	Liquid, liquid phase;
lg	Logarithmic;

lt	Long term;
M	Molar, mol, kmol, mixer, internal sink;
m	Molecule, mean, monothermal, mixing, melting;
mix	Mixture;
mech	Mechanical;
min, max	Minimum, maximum;
mvc	Mechanical vapor compression;
m_1, m_2	Constant mass of solute or solvate;
NH_3	Ammonia;
n	*i* and *j* indices maximum value, current time iteration;
nt	Nontruncated;
O	Outlet;
OTO	Operation temperature overlapping;
o	Outdoor;
p	Constant pressure, pump, polytropic, pure, potential;
pd	Preheating deficit;
ph, phenom	Phenomenological;
p_1, p_2	Two different phases;
q	Heat;
R	Resorber;
Rf	Resorber fluid;
Rf, O	Resorber heated fluid outlet temperature;
RI, RO	Resorber inlet, outlet;
ROf	Resorber cooling fluid outlet;
RII	Incomplete preheated absorbent prior to enter the resorber;
r	Reverse, number of independent chemical reactions, resorber, recovery, reduced, reversible;
rctf, rectif	Rectification;
rec	Received, recovered;
rel	Released, relative to;
rev	Reversible;
S, s	Constant entropy;
SRC, src	Steam Rankine cycle;
s	Source, sink source, isentropic, specific;
sc	Heat exchanger;
sd	Subcooling deficiency;
sh	Superheating;
sink	Sink source;
syst	System;
ss	Sink source;
st	Short term;
T	Constant temperature;
t	Tube;
tcf	Truncated cooling fractal;

thermal	Heat;
thf	Truncated heating fractal;
trgn, trg	Trigeneration;
turbine	Steam turbine
U	Constant internal energy;
u	Useful, unit;
V	Constant volume;
v	Vapor, vaporization, constant volume, number of independent variables (see rule of phases);
vt	Vapor tubes;
vol	Volumetric;
WB	Wet bulb;
w	Mechanical work, pump work;
w	Wall, warm, power;
w+q	Cumulated work and heat;
$[X]_Y$	Amount X calculated for amount Y = const.;
0	Reference state at 0 °C, or at ambient temperature, or initial, capillary head;
0i	Bare internal;
1,2	Two different thermodynamic states, mixture species;
12	Value over an integration interval 1–2;
1–2	Refering to the resorption process plotted in Fig. 4.5;
4–3	Refering to the generation process plotted in Fig. 4.5;
I	First principle;
II	Second principle;
σ	Surface;
p	Potential
g	Gravity

Superscripts

C, c	Complementary;
e	Excess;
l	Liquid;
m	Main;
g	Gas;
$'$	Currents getting in, gas phase;
$''$	Currents getting out.

Chapter 1
Introduction

This book is presenting in its first part two important heat pumping technologies (production of cooling and heating), introduced by the author lately. These are the coabsorbent technology (Chaps. 2–7) and the mechanical vapor compression refrigeration and heating COP increase, using the discharge gas superheat (Chap. 8). In the next chapters, the book is including author's own researches concerning a non-equilibrium phenomenological two-point theory of mass and heat transfer in physical and chemical interactions (Chap. 9), a new wording of the Laplace equation and the variational numerical and analytical approach of the liquid capillary rise effect (Chap. 10). In the book last part, Marangoni convection basic mechanism explanation, pseudo-Marangoni cells model and the absorption–desorption mass and heat exchangers model application are presented (Chap. 11). In this first introductory chapter, selected topic of thermodynamics is attached, in order to complete the subjects elaborated and facilitate the book reading.

1.1 First and Second Principles of Thermodynamics

First, a few introductory definitions are given. The univocally determination of a substance state is done with the help of the state parameters. A substance properties are intensive and extensive. The intensive properties are not depending on the quantity of substance of a system (e.g. temperature, pressure). The properties depending on the substance quantity are extensive properties (e.g. volume) and are noted by capitals. The specific extensive properties, i.e. those referred to the unit of a substance quantity, are intensive properties (e.g. specific volume, specific heat). The intensive properties describing the state of a substance (thermodynamic system) are named state thermodynamic parameters. The most used state parameters are the absolute pressure, p, the absolute temperature, T, and the specific volume $v = \frac{V}{G}$, or the density, $\rho = \frac{G}{V} = \frac{1}{v}$. Except the absolute temperature, the rest of intensive parameters are noted by small letters. When an external action is absent, the state

M.-D. Staicovici, *Coabsorbent and Thermal Recovery Compression Heat Pumping Technologies*, Heat and Mass Transfer, DOI: 10.1007/978-3-642-54684-6_1,

of a substance is completely determined if two independent state parameters were known, i.e. $p = p(T, v)$, $v = v(p, T)$ and $T = T(p, v)$. Each of these functions can be written in an implicit form as state equations, $E(p, v, T) = 0$. The relationship between the state parameters is graphically shown by state diagrams, which usually have the coordinates p–v, p–T, or v–T.

A thermodynamic system gathers material objects which are interacting both one another and with the ambient (sources). Modifying one of the state parameters within a domain of values will change the system state, which suffers a transformation. The equilibrium transformations cover successively and continuously states of equilibrium, characterized by the fact that all system parts have the same temperature and the same pressure. On the contrary, a system suffering a nonequilibrium transformation covers in its constitutive parts states which are not in equilibrium, characterized by different values of temperature, pressure, density, concentration, etc. All real transformations are more or less nonequilibrium transformations, the smaller the unfolding speed of the transformation, the closer the system to the equilibrium state. Except Chaps. 9 and 11, the close-to-equilibrium (quasistatic) transformations are considered in all book chapters. A state diagram cannot plot but only quasistatic transformations. Depending on the state parameter kept constant during its transformation, or on the process feature, a thermodynamic system can be of isothermal ($dT = 0$), isobar ($dp = 0$), isocor ($dv = 0$), isolated ($dG = 0$, $dq = 0$ and $dl = 0$), adiabatic ($dG \neq 0$, $dq = 0$ and $dl \neq 0$), closed ($dG = 0$, $dq \neq 0$ and $dl \neq 0$), open ($dG \neq 0$, $dq \neq 0$ and $dl \neq 0$), or global ($dG \neq 0$, $dq \neq 0$ and $dl \neq 0$, including also its sources which the system is connected to) type. The systems presented in this book are mostly of closed and global type. These have one or more component sub-systems of the types mentioned above. According to the thermodynamics convention, heats inputs/outputs and work outputs/inputs are considered positive/negative quantities. A stationary system is characterized by the time independence of its intensive parameters, while its extensive parameters are constant during a transformation. On the contrary, a non-stationary system is characterized by the time dependence of both intensive and extensive parameters during a transformation.

1.1.1 Ideal (Perfect) Gas Laws

The gases having pressures close to the ambient pressure, obey the following laws:

- *R. Boyle (1662) and E. Mariotte (1676) law*. During an isothermal transformation ($dT = 0$):

$$pV = const. \tag{1.1}$$

In Eq. (1.1), p and V hold for the gas pressure and volume, respectively.

- *G. Gay-Lussac (1802) law*. During an isobar transformation ($dp = 0$):

$$V = V_0(1 + \alpha t) \tag{1.2}$$

In Eq. (1.2), V_0, and V hold for the gas volume at temperatures 0 and t °C, respectively, and α is the volume thermal isobar expansion coefficient, $\alpha = const. = 1/273.15 = 0.003661$ °C^{-1} valid for all gases. A temperature value, T, can be defined by:

$$T = \frac{1}{\alpha} + t = 273.15 + t, \ ^\circ\mathrm{C} \tag{1.3}$$

belonging to the Kelvin temperature scale. Equation (1.3) shows the relationship between Celsius, $[t]$ = °C, and Kelvin, $[T]$ = K, temperature scales, which have the origins 0 and −273.15 °C, respectively. Writing Eq. (1.2) for two temperatures t_1 and t_2, the G. Gay-Lussac law results in:

$$\frac{V_2}{V_1} = \frac{T_2}{T_1} = const. \tag{1.4}$$

- *Charles law*. During an isocore transformation ($dv = 0$):

$$p = p_0(1 + \alpha t) \tag{1.5}$$

  and

$$\frac{p_2}{p_1} = \frac{T_2}{T_1} = const. \tag{1.6}$$

- *Avogadro (1811) law*. Equal volumes of different ideal gases having the same temperature and the same pressure contain the same number of molecules, N_m. A mol/kmol is defined as the substance amount G_M, expressed in g/Kg, numerically equal with its molecular mass m_m, $[G_M]$ = g mol^{-1}/Kg kmol^{-1} (e.g. $G_{M,H_2O} = 17.03$ Kg kmol^{-1}, $G_{M,NH_3} = 18.02$ Kg kmol^{-1}). According to Avogadro law, one kmol of every ideal gas contains the same number of molecules, $N_A = 6.022045 \times 10^{26}$ kmol^{-1}, determined experimentally and named the Avogadro-Loschmidt number. The gas mass G is proportional to N_m and with the mass of a molecule, m_m, $G = N_m \times m_m$. From here, it results that the ratio of two different masses of gas, G_1 and G_2, contained by two equal volumes of gas and found to the same pressure and same temperature, equals the ratio of theirs molar masses, $G_1/G_2 = m_{m,1}/m_{m,2} = G_{M,1}/G_{M,2}$. Reciprocally, if the ratio of two different masses of gas, G_1 and G_2, having the same pressure and same temperature, equaled the ratio of theirs molar masses, then the masses of gas have the same volume, $G_M v = const.$ This statement can be proven in case of a kmol of ideal gas found in normal physical conditions (p = 1 atm = 760 mm Hg = 101.325 kPa and t = 0 °C). Indeed, using the Clapeyron equation (see Eqs. (1.8) and (1.9)) for every gas, it results $G_M v = 22.4$ m^3 kmol^{-1}.

- *Dalton law*. The total pressure p_{mix} of a mixture of i ideal gases, equals the sum of the partial pressures p_i, $i = 1, \ldots, n$, of the component gases:

$$p_{mix} = \sum_{i=1}^{n} p_i \tag{1.7}$$

If a gas component had the temperature and pressure equal to those of the mixture, $T_i = T_{mix}$ and $p_i = p_{mix}$, respectively, then its volume is the partial volume V_i of the i gas component, $i = 1, \ldots, n$. Defining r_i as the volumetric fraction of the i-th component, $i = 1, \ldots, n$, by:

$$r_i = \frac{V_i}{V_{mix}} \tag{1.8}$$

it is easy to show with the help of the Boyle-Mariotte law that $V_{mix} = \sum_{i=1}^{n} V_i$ and $\sum_{i=1}^{n} r_i = 1$. Further on, considering a mixture of M_i numbers of moles of i different gases, $i = 1, \ldots, n$, then the i-th component volumetric fraction is:

$$r_i = \frac{V_i}{V_{mix}} = \frac{V_i}{\sum\limits_{i=1}^{n} V_i} = \frac{M_i G_{M,i} v_i}{\sum\limits_{i=1}^{n} M_i G_{M,i} v_i} = \frac{M_i}{\sum\limits_{i=1}^{n} M_i} = N_i \tag{1.9}$$

or, expressed in words, the volumetric fraction of a mixture component equals its molar fraction, Eq. (1.14) (in Eq. (1.9) the result obtained previously $G_{M,i} v_i = const.$, $i = 1, \ldots, n$, was used).

1.1.2 Ideal Gas State Equation

Applying the Boyle-Mariotte and Gay-Lussac transformations successively, the ideal (perfect) gas equation of state, or the Clapeyron equation, is obtained:

$$pV = GRT \tag{1.10}$$

In Eq. (1.10), R, $[R] = \text{J Kg}^{-1}\ \text{K}^{-1}$, is the gas constant, characteristic for each gas and G is the gas mass, $[G] = \text{Kg}$. In order to express the Clapeyron equation in molar units we remark that the gas mass is equal to $G = M \times G_M$, where M and G_M are the number of kmol contained by the gas volume V and the gas molar mass, $[G_M] = \text{Kg kmol}^{-1}$, respectively. Further on, forming the product $G_M \times R$ it results in the universal gas constant, valid for all gases, $G_M \times R = R_M$, $R_M = 8314.41\ \text{J kmol}^{-1}\ \text{K}^{-1}$, while dividing $V = M\ G_M v$ by M results in the molar volume V_M, $V_M = V/M = G_M v$, $[V_M] = \text{m}^3\ \text{kmol}^{-1}$, so Eq. (1.10) becomes:

$$pV_M = R_M T \tag{1.11}$$

1.1.3 Mixtures

A pure substance has all molecules of the same kind. In a mixture of pure substances these become components. The composition of a mixture is expressed usually by the mass or molar fractions of the components.

1.1.3.1 Mass Fraction

$$c_i = \frac{G_i}{\sum_{i=1}^{n} G_i} = \frac{G_i}{G} \quad (1.12)$$

and

$$\sum_{i=1}^{n} c_i = 1 \quad (1.13)$$

In Eq. (1.12), G_i and c_i are the mass and mass fraction of the i-th component, respectively and n is the mixture components number. For binary mixtures, $n = 2$, the mass fractions are c_1 and $c_2 = 1 - c_1$, Eq. (1.13).

1.1.3.2 Molar Fraction

$$N_i = \frac{M_i}{\sum_{i=1}^{n} M_i} = \frac{M_i}{M} \quad (1.14)$$

and

$$\sum_{i=1}^{n} N_i = 1 \quad (1.15)$$

In Eq. (1.14), M_i and N_i are the number of moles and molar fraction of the i-th component, respectively. Similarly, the molar fractions of binary mixtures are N_1 and $N_2 = 1 - N_1$, Eq. (1.15).

1.1.4 Specific Heat

- *Real specific heat*. It is defined as the derivative of the temperature dependent heat accumulated by a body during its heating, $q(t)$, with respect to the temperature t variable:

$$c(t) = \frac{dq(t)}{dt} \quad (1.16)$$

where from:

$$q_{1-2} = \int_{t_1}^{t_2} c(t)dt \quad (1.17)$$

- *Mean specific heat.* It is calculated by the mean integral of $c(t)$ function on the $t_2 - t_1$ interval:

$$c_m = \frac{\int_{t_1}^{t_2} c(t)dt}{t_2 - t_1} \tag{1.18}$$

The specific heat can be expressed with respect to a mass of 1 kg, a mass of a mole (kmol), a volume of 1 m^3, etc. The mass and molar specific heat are used in this book.

- *Specific heat dependence on the system transformation during heating process.* The specific heat, similar to heat, is not a function of state, it depends on the transformation the system covers. There are practically two main system transformations during the heating process, isobar ($dp = 0$), and isocor ($dv = 0$), for which the specific heat is expressed:

$$c_p = \left(\frac{dq(t)}{dt}\right)_p \tag{1.19}$$

and

$$c_v = \left(\frac{dq(t)}{dt}\right)_v \tag{1.20}$$

respectively. In this book, the isobar specific heat, Eq. (1.19), is used mainly.

1.1.5 First Principle of Thermodynamics (Robert Mayer 1842)

1.1.5.1 Heat-Mechanical Work Quantitative Equivalence

The heat (S. Carnot, work published later, in 1878) and the mechanical work are two different forms of energy transmission. The heat is transferred through direct contact between different bodies (conduction, convection) and transport of radiation energy. The mechanical work transfers energy differently, implying necessarily a volume variation of the body. According to Joule experiences (1847–1850), it has been demonstrated that heat, Q, and mechanical work, L, are two equivalent forms of energy from the quantitative point of view. The proportionality coefficient is $A = 0.00234228$ kcal/kgf m, the calorific equivalent of mechanical work:

$$Q = AL \tag{1.21}$$

The reverse of A is $J = 426.935$ kgf m/kcal, the mechanical equivalent of heat, which has been calculated first by Mayer in 1842, with the help of Eqs. (1.10) and (1.45):

$$J = \frac{1}{A} \tag{1.22}$$

The coefficients A and J have been used in the past as long as the heat and work were measured in different units. Today, when the use of the S.I. system of units is compulsory, the coefficients A and J are omitted because heat and work are expressed in the same units of energy (J or kJ).

1.1.5.2 Equation of the First Principle of Thermodynamics

The heat-work quantitative equivalence was a valuable lesson for the mankind. It has been extended to other forms of energy, i.e. energy is not lost or created, but only transformed from a form of energy into another (e.g. potential gravitational, electrical, magnetic, nuclear, chemical) in a physical or chemical process, becoming the universal law of energy conservation and transformation, or the first principle of thermodynamics. The first principle resulted from experience. Its general equation, written in its differential form, is given by:

$$dQ = dU + dL \tag{1.23}$$

For a constant mass of substance equal to unit and detailing dL, Eq. (1.23) becomes:

$$dq = du + pdv + dl^* \tag{1.24}$$

Equation (1.24) shows that the elementary heat dq transferred to a body is transformed in order to increase its internal energy with the elementary amount du, in order to produce the elementary work of dilation pdv (due to the elementary volume increase dv) and for producing other forms of work (or suffering other forms of external work acting upon it, e.g. surface tension forces, see Chap. 10), except that of dilation dl^*. In Eq. (1.24), pdv and dl^* depend on transformation, so dq depends as well on the particular transformation. The internal energy is an extensive state function, $U = uG$. The function $u = \frac{U}{G}$, $[u] = \text{kJ Kg}^{-1}$, is the specific internal energy. Its increase du during a heating process results from an energy transfer which is the only independent of the particular covering way (transformation) in Eq. (1.24), i.e. it depends solely on the initial and final transformation states. Expressing this mathematically, if the curvilinear integral of the integrand du depends solely on the initial and final integration limits then the integrand is an exact total differential, while pdv and dl^* are not, that is, integrating Eq. (1.24), it results:

$$q_{1-2} = (u_2 - u_1) + \int_{v_1}^{v_2} pdv + l_{1-2}^* \tag{1.25}$$

It is a common practice to express the internal energy as a function of the specific volume and temperature, $u = u(v, T)$. Differentiating $u = u(v, T)$, one obtains:

$$du = \left(\frac{\partial u}{\partial T}\right)_v dT + \left(\frac{\partial u}{\partial v}\right)_T dv \tag{1.26}$$

Consider Eq. (1.24) for isocor transformations, $dv = 0$, and for $dl^* = 0$, wherefrom it results:

$$(dq)_v = du \tag{1.27}$$

We remember Eq. (1.20), hence with Eq. (1.26) the specific heat of an isocor transformation is:

$$c_v = \left(\frac{\partial u}{\partial T}\right)_v \tag{1.28}$$

so, Eq. (1.26) becomes:

$$du = c_v dT + \left(\frac{\partial u}{\partial v}\right)_T dv \tag{1.29}$$

For ideal gases, according to the Joule law:

$$\left(\frac{\partial u}{\partial v}\right)_T = 0 \tag{1.30}$$

(and further $\left(\frac{\partial u}{\partial p}\right)_T = 0$ as well) therefore Eq. (1.29) rewrites as:

$$du = c_v dT \tag{1.31}$$

that is the internal energy of an ideal gas depends only on temperature. The gases which cannot be considered ideal depend also on volume, because $\left(\frac{\partial u}{\partial v}\right)_T \neq 0$. For these and for a certain body, the 1st principle Eq. (1.24), considered for $dl^* = 0$, rewrites with Eq. (1.29), as:

$$dq = c_v dT + \left[\left(\frac{\partial u}{\partial v}\right)_T + p\right] dv \tag{1.32}$$

Generally, for a function of state, like the internal energy, it is important to know its absolute value in certain applications (e.g. in the chemical reactions assessment, or principally, in the phenomenologic approach of mass and heat transfer, analyzed in Chaps. 9 and 11). For the rest, in the book next heat pumping thermodynamic applications it is not needed to know the absolute value of a function of state, but its variation which occurs during the different thermodynamic transformations. In this case, the origin of reference of the state function can be chosen arbitrarily.

1.1.5.3 Enthalpy

In our further applications, the (quasi)-isobar transformations are very often met. In this case, another thermodynamic function, the enthalpy (Kamerlingh-Onnes 1909), is of much more help than the internal energy, because it simplifies the computation. The total enthalpy, noted here by $H = hG$ as an extensive parameter, equals the sum of the internal energy and of the product of pressure and volume:

$$H = U + pV \tag{1.33}$$

Similarly to internal energy, the enthalpy is obviously a function of state. The function $h = \frac{H}{G}$, $[h] = \text{kJ Kg}^{-1}$, is the specific enthalpy, equal from Eq. (1.33) to:

$$h = u + pv \tag{1.34}$$

The molar enthalpy can be derived from $h_M = \frac{H}{M}$, $[h_M] = \text{kJ kmol}^{-1}$, where M is the mols number of the system. Further, it is convenient to express the enthalpy as a function of pressure and temperature state parameters, $h = h(p, T)$. Its exact total differential as state function writes:

$$dh = \left(\frac{\partial h}{\partial T}\right)_p dT + \left(\frac{\partial h}{\partial p}\right)_T dp \tag{1.35}$$

The first principle Eq. (1.24), combined with Eq. (1.33) and written for $dl^* = 0$, has for a constant mass of substance equal to unit the differential form:

$$dq = dh - vdp \tag{1.36}$$

Considering an isobar transformation, $dp = 0$, in Eq. (1.36), it results:

$$(dq)_p = dh \tag{1.37}$$

Combining Eqs. (1.37) and (1.19), the specific heat of an isobar transformation writes:

$$c_p = \left(\frac{\partial h}{\partial T}\right)_p \tag{1.38}$$

Introducing Eq. (1.38) in Eq. (1.35), the total differential of the enthalpy is:

$$dh = c_p dT + \left(\frac{\partial h}{\partial p}\right)_T dp \tag{1.39}$$

The Joule law applied to enthalpy, shows that in case of a perfect gas this time the enthalpy doesn't depend on pressure [see Eq. (1.30)]:

$$\left(\frac{\partial h}{\partial p}\right)_T = 0 \tag{1.40}$$

Introducing Eq. (1.40) in Eq. (1.39), it results that for a perfect gas:

$$dh = c_p dT \tag{1.41}$$

and because additionally we have $\left(\frac{\partial h}{\partial p}\right)_T = 0$, it is concluded that for a perfect gas, similar to internal energy, the enthalpy depends solely on temperature. The gases which cannot be considered ideal depend also on pressure, because $\left(\frac{\partial h}{\partial p}\right)_T \neq 0$. For these and for a certain body, the 1st principle Eq. (1.36), rewrites with Eq. (1.39), as:

$$dq = c_p dT + \left[\left(\frac{\partial h}{\partial p}\right)_T - v\right] dp \tag{1.42}$$

The functions u, h and c_p are caloric properties, while p, v and T are thermal properties of a substance. Moreover, bearing in mind the definition of the equation of state, equations of type (1.31), (1.32), (1.41) and (1.42), which link thermal properties with caloric properties, are named caloric equations of state.

Let us consider the differential of Eq. (1.34) and divide it by dT in order to form the following equation with substantial derivatives:

$$\frac{dh}{dT} = \frac{du}{dT} + \frac{d(pv)}{dT} \tag{1.43}$$

From the Clapeyron equation:

$$pv = RT \tag{1.44}$$

it results that:

$$\frac{d(pv)}{dT} = R \tag{1.45}$$

Taking into account Eq. (1.45), and remarking that Eqs. (1.31) and (1.41) can be rewritten as:

$$c_v = \left(\frac{\partial u}{\partial T}\right)_v \tag{1.46}$$

and

$$c_p = \left(\frac{\partial h}{\partial T}\right)_p \tag{1.47}$$

respectively, we pass from the substantial to partial derivatives in Eq. (1.43) and obtain:

$$c_p - c_v = R \tag{1.48}$$

Equation (1.48) is named the formula of Mayer.

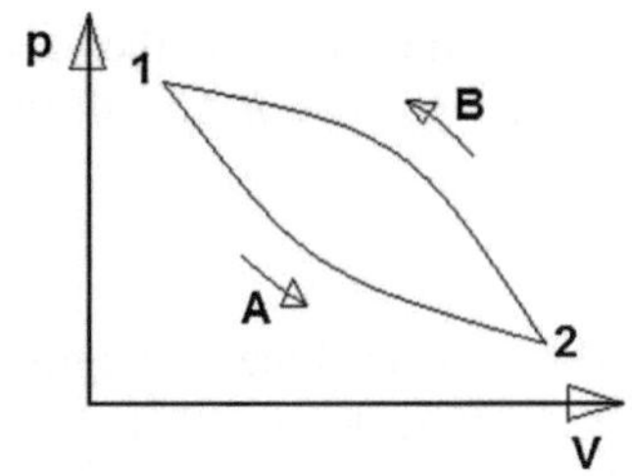

Fig. 1.1 The plot of a certain thermodynamic cycle in the *p*–*V* diagram

1.1.5.4 Cycles. First Principle Application

In this book we shall analyze mainly systems having closed transformations, that is a cyclic thermodynamic evolution. These lay down the basis of the majority of heat pumping applications. Although the application of the 1st principle has important consequences, which are general, here these will be emphasized with respect to our particular topic. Let us consider the schematic of a certain cycle, plotted in the *p*–*V* diagram, Fig. 1.1. As mentioned in Sect. 1.2.2, the dilation mechanical work l_{1-2} and therefore the heat q_{1-2} producing it, are not functions of state, but functions of transformations (covering ways) 1–*A*–2 or 1–*B*–2, etc., Fig. 1.1, that is:

$$\oint dl = \oint pdv = \int_{1A2} pdv - \int_{1B2} pdv \neq 0, \text{ therefore } (l_{1-2})_{1A2} \neq (l_{1-2})_{1B2} \quad (1.49)$$

and

$$\oint dq = \int_{1A2} dq - \int_{1B2} dq \neq 0, \text{ therefore } (q_{1-2})_{1A2} \neq (q_{1-2})_{1B2} \quad (1.50)$$

respectively. On the other side, the internal energy and enthalpy are functions of state. Considering just the total enthalpy, $H = Gh$, it can be written:

$$\oint dH = \oint d(Gh) = \int_{1A2} d(Gh) - \int_{1B2} d(Gh) = 0, \quad (1.51)$$

therefore

$$(H_2 - H_1)_{1A2} = (H_2 - H_1)_{1B2} = H_2 - H_1 \quad (1.52)$$

From the 1st principle, of energy conservation and transformation, a second important law can be derived, that of the mass conservation. The law applies to all thermodynamic systems, but here it will be derived for the cyclic transformations. For the type of applications we are considering in this book, the quantum effects, which lead to the transformation of mass in energy in certain circumstances (e.g. atomic and sub-atomic interactions), are completely negligible here, therefore are excluded from consideration. This means that the functions *G* and *h* are independent one another and performing the differential, the integrand becomes $d(Gh) = Gdh + hdG$ and Eq. (1.51) rewrites:

$$\oint d(Gh) = \oint Gdh + \oint hdG = 0 \tag{1.53}$$

The two curvilinear integrals in the right member of Eq. (1.53) are Stieltjes integrals (Nicolescu 1977–1980) and using a mean integral formula, Eq. (1.51) becomes:

$$\oint d(Gh) = G^*(\xi) \oint dh + h^*(\varsigma) \oint dG = 0 \tag{1.54}$$

In Eq. (1.54) $G^*(\xi)$ and $h^*(\varsigma)$ are mean values of functions G and h on the considered intervals of definition and

$$\oint dh = 0 \tag{1.55}$$

because h is a state function, wherefrom, Eq. (1.54) becomes:

$$\oint dG = 0 \tag{1.56}$$

Equation (1.56) proves that after one cycle the mass of the system regained its initial value, therefore the mass has been conserved in the cyclic transformation. Considering now a mixture of components having each the mass fraction c_i given by Eq. (1.12) and repeating Eqs. (1.53)–(1.56) for $G_i = c_iG$, it results:

$$\oint dG_i = \oint d(c_iG) = 0, \quad i = 1, 2, \ldots, n \tag{1.57}$$

which means that in a cyclic transformation, the mass of a species is conserved as well.

The application of the 1st principle consists in making up energy, mass or species mass balances of systems and components at hand. A balance is achieved considering that the absolute value of inputs and outputs of energy, mass or species mass are equal. This practical and precious tool for solving the design problems in various energy fields, was used in this book as well.

1.1.6 Second Principle of Thermodynamics (Sadi Carnot 1824)

According to the general belief, 1st principle goal of each heat pumping application is to maximize the effectiveness from a quantitative point of view, that is to obtain the highest useful effect with the smallest primary energy consumption, especially when the input energy in not free. However, applying solely the 1st principle recommendations will not conduct to this goal fulfilling. In order to fulfil it, Chap. 7 shows that it is necessary and enough to satisfy the qualitative recommendations of the 2nd principle as well. The 2nd principle is presented shortly in this section.

1.1.6.1 Reversible and Irreversible Transformations. Heat and Sink Sources. Thermal Efficiency and Effectiveness. Wordings of the Second Principle of Thermodynamics

Let us consider a system suffering a direct transformation. During this transformation, the system receives heat Q from a heat source and delivers the work L to the ambient. Let us consider as well the system reverse transformation, brought very near to the direct transformation. It is possible to establish an imaginary biunivoque correspondence, "in mirror", between the points of the two transformations. The pairs of points formed in such a way have the state parameters which differ one another by a small infinite, i.e. (p, v, T, x) and $(p + \delta p, v + \delta v, T + \delta T, x + \delta x)$, enabling the reverse transformation to pass through the same states of the direct transformation, but in reverse direction. This means that the system will receive from ambient the work $(-L)$ and will deliver to a sink source the heat $(-Q)$. In this way, after the system should cover the direct and the reverse transformations, it wouldn't exchange any energy (heat and work) with the ambient and heat source. In this case, the direct transformation is called a reversible transformation. An additional basic requirement for a direct transformation be reversible is that during the heat exchange the difference dT between source and working fluid temperature covering the transformations be infinite small, $T_h = T_{wf} + dT$. The need for such requirement is simple to explain, dT could be a finite temperature difference, the higher the better, during the direct transformation, only, e.g. $dT = (3\text{–}5)\ ^\circ\text{C}$, because heat Q passes naturally from the higher temperature source to the lower temperature working fluid. On the contrary, during the reverse transformation the heat $(-Q)$ cannot pass naturally from the lower temperature working fluid to the higher temperature source, only if dT had been infinite small, to the limit, zero. Suppose the system at hand is a single transformation one, and the transformation is reversible. Then, the system is in equilibrium with the single source it is supplied with and it doesn't exchange mass and energy with the ambient, therefore, according to the definition, it is an isolated system. A first conclusion regarding the work availability can be formed, according to the 1st principle of thermodynamics. Indeed, the work can be produced by a single transformation system only until the transformation becomes reversible, that is it reaches equilibrium. Further, a question is raised, what is happening with respect to the cyclic work production if a system had more reversible transformations. When the working fluid is obliged to perform a cyclic evolution and a non-zero useful effect is desired, then the cycle must be connected to at least two sources (springs), heat and sink, with a temperature gap between them, in order to remove from and bring back to the initial state the reversible transformations, on at least two different ways, respectively.

The combination between this "macroscale irreversibility" and the "microscale" reversible transformations leads in case of the cyclic evolution to continuous work production (and anticipating, to continuous heating and cooling production as well) which reaches the thermodynamic perfection. All aspects concerning the cycle possible transformations and ideal efficiency and effectiveness are the concern of the 2nd principle of thermodynamics. Continuing, an irreversible transformation is, on the contrary, that direct transformation which reverse of cannot reproduce in reverse direction the direct transformation, as described above, therefore the system is in nonequilibrium (see Chaps. 9 and 11). From experience, the irreversibility means efficiency loss. The main cause of external irreversibility of a system transformation is in practice the finite temperature difference between source and the working fluid, which cannot enable the reverse transformation to take place as described to the beginning of this section. However, our finite time life obliges us to design machines with finite time and finite surface heat exchange therefore with finite time and acceptable cost of the useful effect yield. Actually, it is therefore necessary to admit a temperature pinch in heat exchange between source and cycle working fluid, that is a compromise between task cost and efficiency (effectiveness). The remarks made here lead to the so called "source-task match" rule of thomb, which a designer must apply when choosing the best cycle for the available supplying sources and a given task. The best source-task match is achieved when the source and the working fluid have close values of theirs heat capacity (equal to the product of heat exchanging fluid mass flow rate and its specific heat). In Chap. 8 an example thereof is shown. Such favorable conditions are met for example to the sensible-sensible and latent-latent type heat exchange. In case of the coabsorbent technology, there are four absorption processes of latent type heat exchange, generation, absorption, resorption and desorption, operating normally with gliding temperature, and two heat-sink supplying sources, characterized more often than not by sensible type heat exchange. In this case, the sources and the tasks have not the best match, and the cycle efficiency is not favored. It is possible to compensate partially this drawback, practicing an internal heat recovery, of sensible–sensible and latent–latent (gax) type, according to the 1st principle recommendations. Besides this, a second partial compensation consists in using free low grade heat sources practically for all heat pumping applications. Beside the external irreversibility, the internal irreversibility is another source of loosing effectiveness. Indeed, taking into account the experience, all real systems have natural spontaneous internal transformations, which are irreversible, e.g. heat transfer from a warmer body to a colder body, work losses transformed through friction in heat, electrical conduction, gas adiabatical (polytropical) compression, gas diffusion, expansion, or lamination, mixing processes. Moreover, when a cyclic evolution is performed by several irreversible transformations, the effect of irreversibility cumulates. In this case, the higher the irreversibility of transformations, the poorer the efficiency (effectiveness) of work (heating and cooling) production. A cycle (process, transformation) perfection degree can be evaluated using again the 2nd principle of thermodynamics, through its tool named availability, or exergy (see Chap. 7). Concluding, the reversible transformation, although ideal, represents

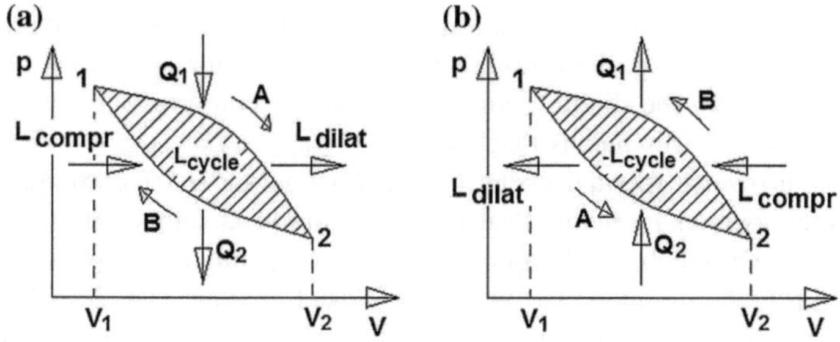

Fig. 1.2 The plot of direct (**a**) and reverse (**b**) thermodynamic cycles in the p–V diagram

a fundamental notion of thermodynamics. At the same time, it is a very useful tool in solving real systems problems, first because in certain circumstances these can approach real systems and second, a reversible transformation can be the comparison term for the degree of irreversibility.

Let us consider the schematic of two certain cycles covered clockwise and counter-clockwise, plotted in the p–V diagram, Fig. 1.2a, b, respectively. As motivated priorly, the cycles need to be connected to external sink/heat sources. For a better understanding of the cycle-source heat exchange, in many cases, not all, the sources can be considered external infinite reservoirs of heat, which temperatures of are not changing during the cycles operation. The cycle of Fig. 1.2a receives heat Q_1 from an external (high temperature) heat source and rejects heat Q_2 to an external (low temperature) sink source. In Fig. 1.2b, the cycle heat Q_1 is delivered to the heat source, while Q_2 is extracted from the sink source. Besides heat, both cycles involve a dilation and compression (pumping) work exchange with the ambient, as well. Let us calculate the mechanical work of dilation and of compression (the computation is valid for both cycles, Fig. 1.2a, b):

$$L_{12,dilat} = \int_{V_1}^{V_2} pdv \tag{1.58}$$

and

$$L_{12,compr} = -\int_{V_1}^{V_2} pdv \tag{1.59}$$

respectively. The integration is done along the transformations of dilation and of compression 1–A–2 and 2–B–1, respectively. The work is proportional to the area subtended by a curve, in the p–V diagram, therefore the relative above/below curves represent from this point of view higher/lower work, comparatively. In case of the cycle plotted in Fig. 1.2a, the dilation curve is above the compression curve, therefore $L_{12,dilat} > L_{12,compr}$. This means that $L_{cycle} = L_{12,dilat} - L_{12,compr} > 0$ and therefore the cycle is a direct thermal motor cycle, which receives heat and delivers external useful work (e.g. the Rankine cycle). On the contrary, in case of the cycle plotted in Fig. 1.2b, the dilation curve is below the compression curve, therefore $L_{12,dilat} < L_{12,compr}$. This means that $L_{cycle} = L_{12,dilat} - L_{12,compr} < 0$ and therefore the cycle is a reverse thermal cycle, typical for cooling (refrigeration)/heat pump cycles, which extracts / delivers heat from/to a sink/heat source, on the account of a work input (e.g. mechanical vapor compression cooling/heating cycles). The 1st principle Eq. (1.23) is integrated along the closed contour 1–*A*–2–*B*–1, Fig. 1.2a. The curvilinear integral of internal energy, as function of state, vanishes, and obtain:

$$\oint dQ = \oint dL \tag{1.60}$$

wherefrom:

$$Q_{cycle} = Q_1 - Q_2 = L_{cycle} > 0 \tag{1.61}$$

In case of the direct cycle, the work production from heat is the only useful effect. From Eq. (1.61) it results that $Q_1 > Q_2$, and further, $\frac{Q_2}{Q_1} < 1$. Taking advantage of this result, with the help of Eq. (1.61), the direct thermal (motor) cycle efficiency is calculated by:

$$\eta_{work} = \frac{L_{cycle}}{Q_1} = \frac{Q_1 - Q_2}{Q_1} = 1 - \frac{Q_2}{Q_1} < 1 \tag{1.62}$$

In case of the reverse cycle, Fig. 1.2b, Eq. (1.60) holds true as well, but Eq. (1.61) writes this time:

$$Q_{cycle} = Q_2 - Q_1 = -L_{cycle} < 0 \tag{1.63}$$

In case of the reverse cycle, two useful effects could be considered separately, or simultaneously, of cooling and of heating, respectively. From Eq. (1.63), it results again that $Q_1 > Q_2$, and further $\frac{Q_1}{Q_2} > 1$. With this result, the cooling and heating effects effectivenesses are given by:

$$COP_c = \frac{Q_2}{|L_{cycle}|} = \frac{Q_2}{|Q_1| - Q_2} = \left(\frac{|Q_1|}{Q_2} - 1\right)^{-1} > 1 \tag{1.64}$$

and

$$\begin{aligned} COP_h &= \frac{|Q_1|}{|L_{cycle}|} = \frac{|Q_1|}{|Q_1| - Q_2} = \left(1 - \frac{Q_2}{|Q_1|}\right)^{-1} \\ &= 1 + \frac{Q_2}{|Q_1| - Q_2} = 1 + COP_c > 1 \end{aligned} \tag{1.65}$$

respectively. If both cooling and heating effects are simultaneously useful, i.e. a cogeneration takes place, the effectiveness results in:

$$COP_{c,h} = \frac{|Q_1| + Q_2}{|L_{cycle}|} = \frac{|Q_1| + Q_2}{|Q_1| - Q_2} = COP_c + COP_h > 1 \tag{1.66}$$

The thermal performances of the direct and reverse cycles are less and higher than unit, which reason for they have been named efficiency and effectiveness (or "coefficient of performance" (COP)), respectively, in order to differentiate them one another.

As it could be remarked, the 1st principle of thermodynamics emphasizes the quantitative rules governing the energy transformations and shows how to establish an energy balance of a transformation. At the same time, it is not providing any indication about the possibility of a certain transformation to take place or not, or about the direction of a transformation unfolding. The second principle of thermodynamics completes the first principle, establishing qualitative rules for these transformations unfolding and indicates which transformations are actually possible. Similar to the 1st principle, the 2nd principle is formulated as well according to the experience. The most general wording is the following: any actual spontaneous transformation is an irreversible transformation. All other 2nd principle wordings are particular cases of this general wording. In 1850, R. Clausius formulated the 2nd principle as: it is impossible to pass heat from a colder body to a warmer body, without consuming mechanical work. Also, in 1851, W. Thomson expressed the 2nd principle as: in the nature, it is impossible to consume cyclically heat and produce mechanical work with a machine connected to a single source. As a corollary of the thermal processes, it results that using cycles the heat cannot be transformed entirely in work [see Eq. (1.62)] or equivalent forms (electrical work, magnetic forces work etc.), a certain amount of it must be rejected to a sink source. Similarly, the work or its equivalent forms cannot be transformed entirely in heat.

1.1.6.2 The Carnot Cycle

The cycle was named Carnot after its inventor, the French engineer Sadi Carnot who elaborated it 1824. It lays down the basis of the thermal motors theory and is of great importance in thermodynamics. Next, the cycle will be analyzed using intensive properties. The direct Carnot cycle is plotted in the *p–V* diagram in Fig. 1.3. Suppose the Carnot cycle is operated by an ideal gas. It consists of two isothermal transformations, *A–B*, with $T_1 = const.$, and *C–D*, with $T_2 = const.$, and two adiabatic transformations, *B–C* and *D–A*, respectively. During the *A–B* and *C–D* isothermal transformations, the cycle receives heat q_1, supplied by a heat source of constant temperature $T_h = const.$, and rejects heat q_2 to a sink source of constant temperature $T_s = const.$, respectively. Suppose, first the cycle operates in a reversible way, that is:

$$T_1 = const. = T_h \tag{1.67}$$

and

$$T_2 = const. = T_s \tag{1.68}$$

The efficiency of the Carnot cycle is calculated by (see Eq. (1.62)):

$$\eta_{work,rev.} = \frac{q_1 - |q_2|}{q_1} \tag{1.69}$$

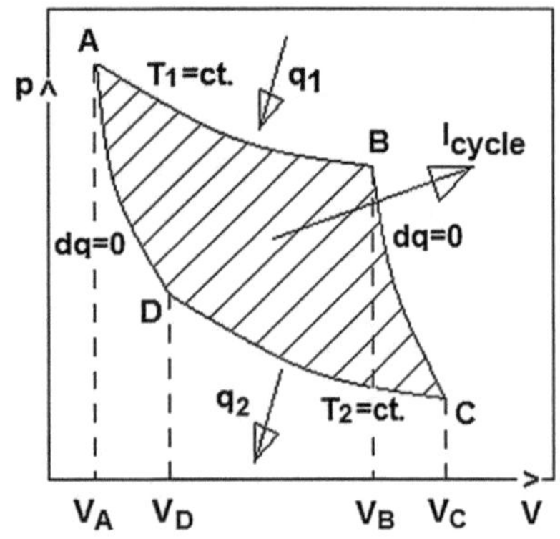

Fig. 1.3 The plot of the direct Carnot cycle in the *p–V* diagram

The heats q_1 and q_2 can be found applying Eq. (1.32) in case of an isothermal transformation, $dT = 0$, and of an ideal gas operation, resulting in:

$$dq = pdv \tag{1.70}$$

The pressure in Eq. (1.70) is expressed from the Clapeyron equation (1.10), for $G = 1$, as:

$$p = \frac{RT}{v} \tag{1.71}$$

wherefrom, Eq. (1.70) becomes:

$$dq = RT\frac{dv}{v} \tag{1.72}$$

Integrating Eq. (1.72) between v_A and v_B on one side, and between v_C and v_D, on the other side, the heats q_1 and q_2 result:

$$q_1 = RT_1 \ln \frac{v_B}{v_A} \tag{1.73}$$

and

$$-|q_2| = RT_2 \ln \frac{v_D}{v_C} \tag{1.74}$$

respectively. In Eqs. (1.73) and (1.74), $v_B > v_A$ and $v_D < v_C$, hence $q_1 > 0$ and $q_2 < 0$, respectively. Next, an adiabatic transformation will be characterized in case of an ideal gas operation. For $dq = 0$, Eq. (1.32) becomes:

$$c_v dT = -pdv \tag{1.75}$$

Dividing it by Clapeyron equation of state (1.10), where $G = 1$ is considered, it results:

$$\frac{c_v}{R}\frac{dT}{T} = -\frac{dv}{v} \tag{1.76}$$

In Eq. (1.76), the ratio $\frac{c_v}{R}$ is expressed with the help of Mayer formula (1.48),

$$\frac{c_v}{R} = \frac{1}{k - 1} \tag{1.77}$$

where $k = \frac{c_p}{c_v}$. Introducing Eq. (1.77) in Eq. (1.76), the latter results in:

$$\frac{dT}{T} = -(k - 1)\frac{dv}{v} \tag{1.78}$$

Integrating Eq. (1.78) between states B and C, on one side, and between states D and A, on the other side, it is obtained:

$$\frac{T_C}{T_B} = \left(\frac{v_B}{v_C}\right)^{k-1} \tag{1.79}$$

and

$$\frac{T_A}{T_D} = \left(\frac{v_D}{v_A}\right)^{k-1} \tag{1.80}$$

Remarking that $T_A = T_B = T_1$ and $T_C = T_D = T_2$, from Eqs. (1.79) and (1.80) it results that:

$$\frac{v_C}{v_B} = \frac{v_D}{v_A} \tag{1.81}$$

Introducing Eqs. (1.73) and (1.74) in Eq. (1.69) and taking into account Eq. (1.81), the direct reversible Carnot cycle efficiency results:

$$\eta_{work,rev.} = \frac{T_1 - T_2}{T_1} \tag{1.82}$$

The function $\eta_{work,rev} = f(T_1, T_2)$ of Eq. (1.82) was found by W. Thomson. The reversible direct Carnot cycle has the highest efficiency out of all reversible direct cycles. This means that beside reversibility, a cycle configuration matters as well. In Chap. 8, an example of two reversible direct cycles is given, which have different efficiencies because they have different configurations, Rankine and Trilateral Flash. If the direct Carnot cycle were not reversible, that is finite temperature differences $\Delta T_{1(2)} > 0$ existed between working fluid and sources temperatures, i.e.:

$$T_1 = T_h - \Delta T_1 \tag{1.83}$$

and

$$T_2 = T_s + \Delta T_2 \tag{1.84}$$

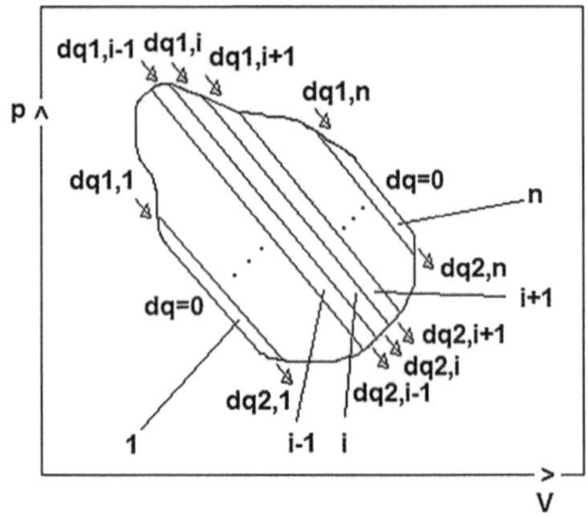

Fig. 1.4 A cycle with some configuration represented as a sum of reversible Carnot cycles

then, replacing in Eq. (1.82) temperatures T_1 and T_2 with the values given by Eqs. (1.83) and (1.84), respectively, it would results that the direct irreversible Carnot cycle efficiency:

$$\eta_{work,irrev.} = \frac{T_h - T_s - (\Delta T_1 + \Delta T_2)}{T_h - \Delta T_1} < \eta_{work,rev.} \tag{1.85}$$

is smaller than the efficiency of the reversible Carnot cycle, given by Eq. (1.82).

All cycles can be reversible, not only the Carnot cycle, provided that theirs heat input and output processes be isothermal transformations. A cycle with some configuration can be "reversible-ized", representing it as a sum of i elementary reversible Carnot cycles, infinitesimally outdistanced one another, $i = 1, \ldots, n$, $n \to \infty$, as shown in Fig. 1.4. The reception and rejection of elementary heats, $dq_{1,i} > 0$ and $dq_{2,i} < 0$, are considered isothermal processes, which take place to constant temperatures $T_{1,i}$ and $T_{2,i}$ respectively, $i = 1, \ldots, n$, while the work inputs and outputs cancel each other, except in adiabatic transformations of the first and last cycles, $i = 1$ and $i = n$.

1.1.6.3 Clausius Integral. Entropy. Entropy Variation of Irreversible Transformations. General Equation of Thermodynamics

From Eqs. (1.73), (1.74) and (1.81) it results for a reversible Carnot cycle [see also Eqs. (1.67) and (1.68)]:

$$\frac{q_1}{T_1} - \frac{|q_2|}{T_2} = 0 \tag{1.86}$$

The ratios $\frac{Q}{T}$ or $\frac{q}{T}$ are named after Lorenz, reduced heats. A similar equation can be written for each elementary Carnot cycle belonging to the cycle with some configuration, represented in Fig. 1.4. Totalizing these equations member by member, it results:

$$\sum_{i=1}^{n} \left(\frac{dq_{1,i}}{T_{1,i}} - \frac{|dq_{2,i}|}{T_{2,i}} \right) = 0 \tag{1.87}$$

If in Eq. (1.87) $n \to \infty$, the sum becomes a curvilinear integral with closed contour, and passing to extensive properties it is obtained:

$$\oint \frac{dQ}{T} = 0 \tag{1.88}$$

The integral in Eq. (1.88) is the Clausius integral. It equals zero for all reversible cycles. The integrand is noted by:

$$dS = \frac{dQ}{T} \tag{1.89}$$

The left member of Eq. (1.89) is the total (exact) differential of a new function of state, S, $s = \frac{S}{G}$, the entropy, introduced by Clausius. Although the entropy was defined and is used in conjuncture to the reversible cycles, it can be a very useful thermodynamic function in the irreversible processes analysis, as well (see Chap. 9). Being a function of state, the entropy variation depends solely on the initial and final states of a transformation. If the system covered completely a reversible Carnot cycle, or other similar cycles with some configuration, then the entropy variation must equal zero. This result is not surprising, if we bore in mind that all functions of state have this property (see the internal energy and enthalpy). Indeed, integrating Eq. (1.89) and taking into account Eq. (1.86), it results that:

$$\Delta S_{rev.} = \frac{q_1}{T_1} - \frac{|q_2|}{T_2} = \frac{q_1}{T_h} - \frac{|q_2|}{T_s} = 0 \tag{1.90}$$

However, what is different from the behavior of other functions of state and very important, is that in case of the irreversible Carnot cycle or other similar cycles with some configuration, the closed contour curvilinear integral of entropy, that is the Clausius integral, is negative this time, as it is shown next. Indeed, let us consider an irreversible Carnot cycle, including the sources as well. Than, in Eqs. (1.73) and (1.74) the temperatures T_1 and T_2 are being replaced by T_1 and T_2, given by Eqs. (1.83) and (1.84), respectively. Hence, Eqs. (1.73) and (1.74) can be rewritten as:

$$\frac{q_1}{T_h - \Delta T_1} = R \ln \frac{v_B}{v_A} \tag{1.91}$$

and

$$-\frac{|q_2|}{T_s + \Delta T_2} = R \ln \frac{v_D}{v_C} \tag{1.92}$$

respectively. Further, using infinite series expansions in order to approximate the temperature factors of heats, Eqs. (1.91) and (1.92) can be expressed as:

$$\frac{q_1}{T_h - \Delta T_1} \cong \frac{q_1}{T_h}\left(1 + \frac{\Delta T_1}{T_h}\right) = R \ln \frac{v_B}{v_A} > \frac{q_1}{T_h} \tag{1.93}$$

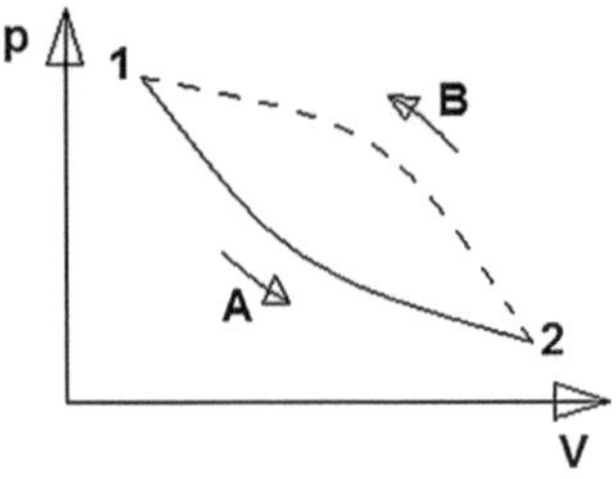

Fig. 1.5 Irreversible cycle with some configuration

and

$$-\frac{|q_2|}{T_s + \Delta T_2} \cong -\frac{|q_2|}{T_s}\left(1 - \frac{\Delta T_2}{T_s}\right) = R \ln \frac{v_D}{v_C} > -\frac{|q_2|}{T_s} \quad (1.94)$$

respectively. Adding member by member ineqns. (1.93) and (1.94), it is obtained:

$$0 = R\left(\ln \frac{v_B}{v_A} + \ln \frac{v_D}{v_C}\right) > \frac{q_1}{T_h} - \frac{|q_2|}{T_s} = \Delta S_{irev.} \quad (1.95)$$

or,

$$\Delta S_{irev.} = \frac{q_1}{T_h} - \frac{|q_2|}{T_s} < 0 \quad (1.96)$$

Referring to the cycle of Fig. 1.4 again, which operates irreversibly and totalizing in Eq. (1.87) entropy differential forms based on Eq. (1.96), the Clausius integral for irreversible processes is obtained, which is negative, as mentioned:

$$\oint \frac{dQ}{T} < 0 \quad (1.97)$$

Let us consider now an irreversible cycle with some configuration, Fig. 1.5, consisting of two transformations, a first irreversible, 1–*A*–2 and a second reversible, 2–*B*–1, respectively. We apply the Clausius integral for irreversible cycles, Eq. (1.97), to this cycle, and obtain:

$$\int_{1-A-2} \frac{dQ}{T} + \int_{2-B-1} \frac{dQ}{T} < 0 \quad (1.98)$$

In in eqn. (1.98), we take into account that the integrand of integral performed along the reversible way 2–*B*–1 is a total differential. This will lead to:

$$\int_{2-B-1} \frac{dQ}{T} = \int_2^1 \frac{dQ}{T} = S_1 - S_2 \quad (1.99)$$

wherefrom, in eqn. (1.98) becomes:

$$S_2 - S_1 > \int_{1-A-2} \frac{dQ}{T} \quad (1.100)$$

Further on, considering the differential form of in eqn. (1.100), the important result is obtained:

$$dS > \frac{dQ}{T} \tag{1.101}$$

valid for all irreversible transformations of some system. Ones of the most important systems in thermodynamics are the isolated systems, because such systems can be met in practice very often, if it were considered together with theirs supplying sources. If a system were isolated ($dQ_{syst.} = 0$), and it performed an irreversible transformation, then, from in eqn. (1.101) it results

$$dS > 0 \tag{1.102}$$

In eqn. (1.102) shows that the entropy of an isolated system increases when covering an irreversible transformation. Equations (1.101) and (1.102) can be written with two signs, equal and inequal, in order to be valid for both reversible and irreversible system transformations, respectively:

$$dS \geq \frac{dQ}{T} \tag{1.103}$$

and

$$dS \geq 0 \tag{1.104}$$

respectively. In eqn. (1.103) is the analytical expression of the second principle of thermodynamics (Kirilin et al. 1985). It can be written also as:

$$TdS \geq dQ \tag{1.105}$$

In in eqn. (1.105), the heat differential can be expressed from the first principle of thermodynamics, given by Eq. (1.23), wherefrom the general equation of thermodynamics results:

$$TdS \geq dU + dL \tag{1.106}$$

or detailing the work differential:

$$TdS \geq dU + pdV + dL^* \tag{1.107}$$

For a constant mass of substance equal to unit and making $dL^* = 0$, Eq. (1.107) becomes:

$$Tds \geq du + pdv \tag{1.108}$$

Alternatively, the heat differential in ineqn. (1.105) can be expressed using enthalpy [see Eq. (1.36)], wherefrom, in eqn. (1.108) becomes:

$$Tds \geq dh - vdp \tag{1.109}$$

1.2 Exergy and Anergy. Heat Exergy. Exergy of Closed Systems. Exergy of Open Systems. Relationship Between Exergy Dissipation and Entropy Creation. Non-equilibrium Linear Phenomenological Connection Between Generalized Forces and Currents

The exergy method is an analysis tool of thermal processes, based on the first and second principles of thermodynamics. It enables to drow up correctly balances and to ascertain losses and efficiencies of the thermal processes. In this book, it will be applied to Chaps. 7 and 9. The exergy is found in some books also under the name of availability. The maximum useful mechanical work which can be produced by a system found in a certain state, during a transformation, until the system reaches the equilibrium with an ambient reservoir, characterized by the temperature and pressure T_0 and p_0, respectively, is called exergy, or availability. A system reaching the equilibrium with an ambient reservoir becomes "relaxed" with respect to it. The availability calculus lays down the basis of the exergy method. For this reason, in this section the exergy concept and computation are introduced. First, the heat exergy calculus is given, which is of great help to exergy method application in case of the coabsorbent heat pumping cycles we are dealing with in this book (see Chap. 7). Second, the exergy of the closed and open systems is analyzed and further the connection between exergy dissipation and entropy creation is emphasized. The entropy creation or entropy source is further used to describe with non-equilibrium thermodynamics tools the irreversible phenomena which take place in binary two-phase non-ideal mixtures, met in the absorption technology, and to create a phenomenological (two-point) theory of mass and heat transfer, presented to Chap. 9.

A certain amount of Energy (e.g. heat) consists in two parts, *Energy* = *Exergy* + *Anergy* = *const.* In this expression, the *Exergy* is the energy part (e.g. electrical and mechanical) which ideally could be transformed in any energy form, and the *Anergy* is the energy part, which cannot be transformed further in any other energy form (e.g. ambient energy). From exergetical point of view, the first principle of thermodynamics could be formulated as: the sum of exergy and anergy of a system is constant (Popa and Vintila 1977).

1.2.1 Heat Exergy

The source of heat exergy is a reversible cyclic thermal motor, Fig. 1.6. The motor is supplied with heat Q_{12} from a heat source of temperature T, $T_1 \leq T \leq T_2$, is producing the work L_{cycle}, equal to the exergy E_{12} of the heat Q_{12},

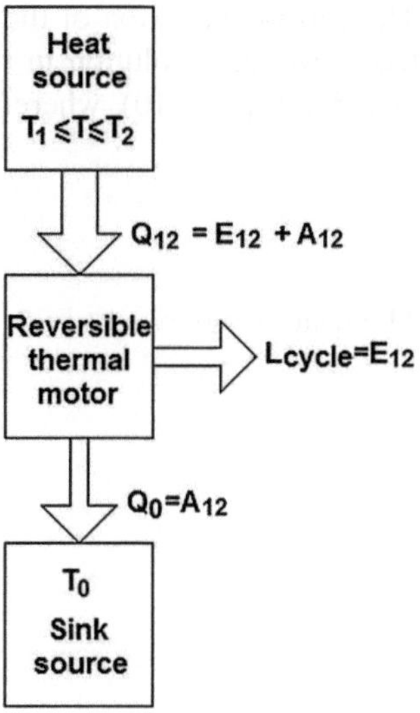

Fig. 1.6 Schematic of heat exergy calculation (Popa and Vintila 1977)

and is rejecting an amount of heat Q_0, equal to the anergy A_{12} of the heat Q_{12}, to the sink source of temperature $T_0 < T_1$. The ensemble consisting of the heat source, motor and sink source, is an isolated adiabatical system, for which the entropy variation is null:

$$\Delta S_h + \Delta S_{motor} + \Delta S_0 = 0 \tag{1.110}$$

The heat yielded by the heat source:

$$Q_{12} = -\int_1^2 TdS \tag{1.111}$$

is received by the motor working fluid, which produces the mechanical work:

$$L_{cycle} = E_{12} = Q_{12} - Q_0 = Q_{12} - A_{12} \tag{1.112}$$

The entropy variation of the heat source is obtained from Eq. (1.111):

$$\Delta S_h = -\int_1^2 \frac{dQ}{T} \tag{1.113}$$

while that of the sink source is given by:

$$\Delta S_0 = \frac{|Q_0|}{T_0} = \frac{A_{12}}{T_0} \tag{1.114}$$

The entropy variation of the motor is null, $\Delta S_{motor} = 0$, because its working fluid has a cyclical evolution in the machine. Equations (1.113) and (1.114) are introduced in Eq. (1.110), wherefrom the heat anergy becomes:

$$A_{12} = T_0 \int_1^2 \frac{dQ}{T} \tag{1.115}$$

The heat exergy results with the help of Eqs. (1.115) and (1.112) as follows:

$$E_{12} = Q_{12} - T_0 \int_1^2 \frac{dQ}{T} \tag{1.116}$$

or

$$E_{12} = \int_1^2 \left(1 - \frac{T_0}{T}\right) dQ \tag{1.117}$$

(Popa and Vintila 1977). Equation (1.117) shows that the heat exergy depends only on the heat and sink sources temperatures, through the function:

$$\theta_w(T_0, T) = 1 - \frac{T_0}{T} \tag{1.118}$$

which is the exergetical temperature factor or the heat-to-work Carnot efficiency factor (see Chap. 7). Equation (1.117) is integrable Stieltjes. Performing the integral (1.117), it results:

$$E_{12} = \left(1 - \frac{T_0}{T}\right)(\xi) \int_1^2 dQ = \left(1 - \frac{T_0}{T}\right)(\xi) Q_{12} = \theta(\xi) Q_{12} \tag{1.119}$$

The mean value of factor $\theta(\xi) = \left(1 - \frac{T_0}{T}\right)(\xi)$, $\xi \in (T_1, T_2)$, in Eq. (1.119), equals the mean Riemann integral value of function $\theta_w(T_0, T)$, Eq. (1.118), on the interval $T \in [T_1, T_2]$, given by:

$$\theta(\xi) = \frac{1}{T_2 - T_1} \int_1^2 \left(1 - \frac{T_0}{T}\right) dT = 1 - \frac{T_0}{T_2 - T_1} \ln \frac{T_2}{T_1} \tag{1.120}$$

The use of Eq. (1.120) in Eq. (1.119) results in the following expression for the heat exergy:

$$E_{12} = \left(1 - \frac{T_0}{T_2 - T_1} \ln \frac{T_2}{T_1}\right) Q_{12} \tag{1.121}$$

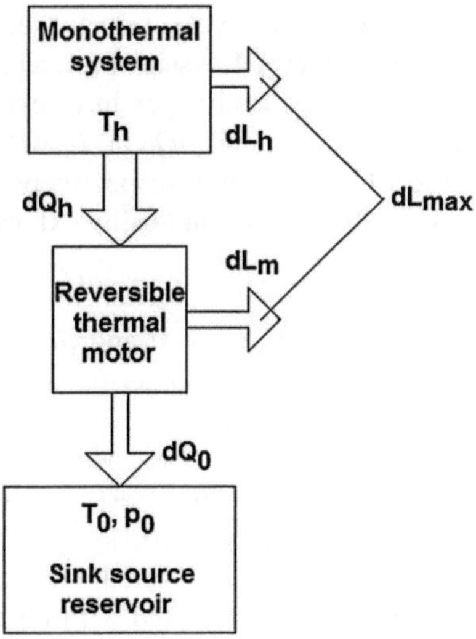

Fig. 1.7 Schematic of the monothermal system producing maximum work (Chartier et al. 1975)

Besides Eq. (1.120), found here for $\theta(\xi)$, at the end of this chapter it is shown that $\theta(\xi)$ has an equivalent and simpler mathematical expression, calculated with the help the arithmetical mean of the heat source temperatures T_1 and T_2 (see Appendix 1):

$$\theta(\xi) = 1 - \frac{T_0}{T_2 - T_1} \ln \frac{T_2}{T_1} \equiv 1 - \frac{T_0}{\frac{T_1+T_2}{2}} \quad (1.122)$$

Introducing it in Eq. (1.119), the heat exergy results to be expressed in a second equivalent way:

$$E_{12} = \left(1 - \frac{T_0}{\frac{T_1+T_2}{2}}\right) Q_{12} \quad (1.123)$$

1.2.2 Exergy of Closed Systems

The maximum useful work which could be extracted from closed systems having an isothermal and isocore reversible transformation, on one side, or an isothermal and isobar reversible transformation, on the other side, is found calculating the functions of free energy of Helmholz, $F \equiv U - TS$, or of free enthalpy of Gibbs, $\Phi \equiv H - TS$, respectively (see next section, Kirilin et al. 1985). When we deal with systems suffering some reversible transformations, the function of exergy is helpful. This is derived next, first for systems which are connected to a single source, called monothermal systems. Such a closed reversible monothermal ensemble, depicted in Fig. 1.7, is capable to produce maximum useful work. It consists of a top monothermal system, of a bottom sink source reservoir and of a Carnot reversible direct cycle, inserted between the top and the bottom sub-ensemblies. The monothermal system produces the work dL_h. The Carnot cycle produces the work dL_m and exchanges in a reversible way the heats dQ_h at T_h with the monothermal system and dQ_0 at T_0 with the sink source reservoir at (T_0, p_0), $T_h > T_0$ (direct Carnot cycle), respectively. We apply the 1st principle to the monothermal system, Eq. (1.24), and using extensive parameters, it is obtained:

$$dQ_h = dU_h + p_0 dV_h + dL_h \tag{1.124}$$

In Eq. (1.124) $dQ_h < 0$ and $dL_h > 0$. Also, $dU_h < 0$ and $p_0 dV_h > 0$ are the differentials of the monothermal system internal energy and of the work performed by the system against the ambient pressure forces, respectively. Alternatively, in Eq. (1.124), if the case, instead of the internal energy, U_h, corresponding to the monothermal system at rest, the system total energy could be used, which means the exergy coming of the system internal energy + other forms of energy which the system is possibly acted with (e.g. kinetic, potential, electrical, surface tension, chemical), and calculated with respect to the sink source reservoir at 0 K. Further, the 1st principle is utilized too for making up the energy balance of the direct Carnot cycle, obtaining (see Eq. (1.61) as well):

$$dQ_h + dQ_0 = dL_m \tag{1.125}$$

In Eq. (1.125), $dQ_h > 0$ this time, while $dQ_0 < 0$ and $dL_m > 0$. We add member by member the two Eqs. (1.124) and (1.125), resulting in:

$$dL_{u,\max} = dL_m + dL_h = -dU_h + dQ_0 - p_0 dV_h \tag{1.126}$$

The entropy balance of the reversible Carnot cycle is written, Eq. (1.86):

$$\frac{dQ_h}{T_h} = \frac{dQ_0}{T_0} \tag{1.127}$$

In Eq. (1.127) $\frac{dQ_h}{T_h} \equiv dS_h$, therefore the heat delivered to the sink source results:

$$dQ_0 = T_0 dS_h \tag{1.128}$$

wherefrom, introducing it in Eq. (1.126), the differential of the maximum useful work is expressed by:

$$dL_{u,\max} = -dU_h + T_0 dS_h - p_0 dV_h = -d(U_h - T_0 S_h + p_0 V_h) = -dE \tag{1.129}$$

In Eq. (1.129) we noted:

$$E(U_h, S_h, V_h) = U_h - T_0 S_h + p_0 V_h + Const. \tag{1.130}$$

The maximum work of the system covering a reversible transformation between two states 1 and 2 is found integrating Eq. (1.130):

$$L_{u,\max} = -\int_1^2 dE = E_1 - E_2 \tag{1.131}$$

The *Const.* in Eq. (1.130) is determined relaxing the monothermal system with respect to the sink source reservoir, when $E(U_h, S_h, V_h) = E(U_0, S_0, V_0) = 0$. Considering this boundary condition, the *Const.* results:

$$Const. = -E(U_0, S_0, V_0) \tag{1.132}$$

and Eq. (1.130), written without the index "h", becomes:

$$E(U, S, V) = (U - U_0) - T_0(S - S_0) + p_0(V - V_0) \tag{1.133}$$

The function E, $[E] = [kJ]$, of Eq. (1.133), is the exergy (availability) of the monothermal system (Chartier et al. 1975; Kirilin et al. 1985). Equation (1.131) shows that E is a function of state, therefore dE is a total differential, so $L_{u,\max}$ does not depend on the integration way, for a chosen reservoir. The exergy, Eq. (1.133), holds true for the assessment of maximum useful work of non-isobar and non-isothermal transformations, either physical or chemical, therefore the function E is more general than the functions $F \equiv U - TS$ and $\Phi \equiv H - TS$, but the value of E depends on the chosen reservoir.

The entropy and exergy are conserved in reversible transformations, only. In the irreversible transformations, $dS > 0$ holds true (see ineqn. (1.102)), and Eq. (1.129) becomes $-dE > dL_{u,\max}$, or

$$-dE = dL_{u,\max} - \frac{\partial E}{\partial t} dt = -(dE)_{irrev} \tag{1.134}$$

where $-(dE)_{irrev}$ is the exergy dissipation (Chartier et al. 1975). In the next section it will be shown that in irreversible transformations the exergy dissipation is connected with the entropy creation. This coupling is emphasized in the next section, in conjuncture with the exergy calculation of stationary open systems.

1.2.3 Exergy of Open Systems. Relationship Between Exergy Dissipation and Entropy Creation

Let us consider the open system depicted in Fig. 1.8. The exergy calculation is confined to stationary open systems. All currents are noted by the same symbol j, but are indexed differently, with i, q and l, for mass, heat and work, respectively. The currents which get in and out of the system, are marked as prime ($'$) and secondary ($''$), respectively. The currents which get in and out of the system, are considered positive and negative, respectively, either are of mass or of heat, but are considered negative and positive, respectively, if they were of work. In this way, the heat and work currents directions respect the thermodynamics convention of sign, mentioned in introduction, Sect. 1.1. In physical interactions, the nature of a current is not changing during the transformation, that is $j_i' = -j_i''$ (e.g. the water in a water heater). In a chemical interaction, in most cases the nature of a reactant is different from that of all other reactants and of produces of reaction, that is $j_i' \neq j_j'$, $i, j = 1, \ldots, n'$, $i \neq j$

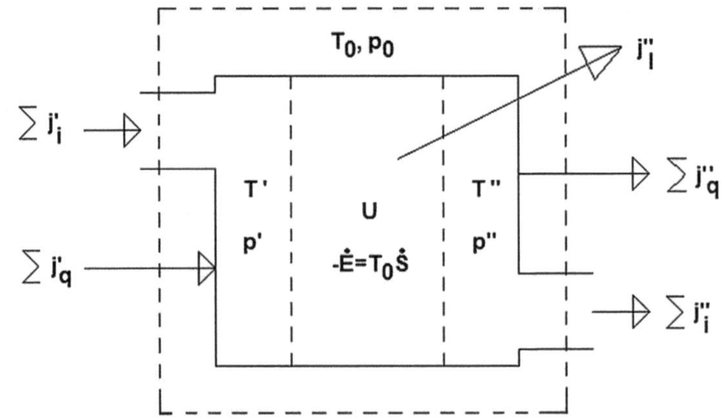

Fig. 1.8 Schematic of a thermodynamic open system (Chartier et al. 1975)

and $j''_i \neq j''_j$, $i,j = 1,\ldots,n''$, $i \neq j$ and $j'_i \neq j''_i$. The temperatures of the mass and heat currents getting in and out from the system are T' and T'', respectively, while that of the sink source reservoir is T_0.

1.2.3.1 Energy Balance

The 1st principle is applied to the schematic of Fig. 1.8. The energy balance requires that the algebraic sum of energy, included by the input and output currents, annulates. The energy of a mass current j'_i equals its (total) internal energy, $j'_i u'_i$, plus the energy gained when is being introduced in the system by an ideal pump, $p' j'_i v'_i$. It results that the energy of the j'_i mass current equals its enthalpy, that is:

$$j'_i \left(u'_i + p' v'_i\right) = j'_i h'_i, \quad i = 1,\ldots,n, \tag{1.135}$$

A similar result is obtained for the output mass currents, that is theirs energy is $i''_i h''_i$, $i = 1, \ldots, n$. The energy balance can be written now, as:

$$\sum j'_i h'_i + \sum j''_i h''_i + \sum j'_q + \sum j''_q - j''_l = 0 \tag{1.136}$$

Equation (1.136), derived from the application of the 1st principle, is very useful in the analysis of the open stationary systems met in energetics.

1.2.3.2 Exergy Computation

Unlike energy, in exergy computation the inlet currents exergy equals that of the outlet currents if the transformations were reversible, only. When the transformations are irreversible, there will be exergy differences between system inlet and outlet, for both mass and heat currents, but not for the work. The differences come of the entropy term T_0S in Eq. (1.133), which is not conserved but in reversible transformations, only. Dimensionally, the mass current exergy is specific energy, [kJ kg^{-1}(kmol^{-1})], multiplied by the dimension of a mass current, [kg(kmol) s^{-1}]. It results that the dimension of a mass current exergy is [kJ s^{-1}] = [kW]. The heat and work currents exergy has the same dimension. Bearing in mind Eq. (1.134), the expression $\frac{\Delta E}{\Delta t}$ has as well the dimension of power, $\left[\frac{\Delta E}{\Delta t}\right] = \left[kJ\ s^{-1}\right]$ = [kW] and is related intrinsically to the exergy dissipation in irreversible transformations, therefore, the function:

$$\dot{E} = \left(\frac{\Delta E}{\Delta t}\right)_{irrev}, \tag{1.137}$$

called exergy dissipation speed, is the function we are looking for in order to calculate the inlet-outlet exergy differences per second of our open stationary irreversible system. The exergy dissipation speed will be computed accounting for the cumulated exergy dissipation of the three mass, heat and work currents, given next.

Exergy Dissipation Speed of Mass Currents

The way the exergy j'_{E_i} of the i-th mass current is calculated is similar to that used for energy currents. Indeed, to the exergy of mass at rest, Eq. (1.133), it is added the work $(p' - p_0)v'_i$ of an ideal pump producing the current, which afterwards could be recovered by an ideal turbine. The exergy brought to system by the j'_{E_i} current is therefore (see Eq. (1.133)):

$$j'_{E_i} = j'_i\left[\left(u'_i - u'_{i,0}\right) + p_0\left(v'_i - v'_{i,0}\right) - T_0\left(s'_i - s'_{i,0}\right) + \left(p' - p_0\right)v'_i\right] \tag{1.138}$$

In Eq. (1.138) the terms can be conveniently grouped in order to emphasize the enthalpy, according to Eq. (1.135), resulting:

$$j'_{E_i} = j'_i\left[\left(h'_i - T_0s'_i\right) - \left(h'_{i,0} - T_0s'_{i,0}\right)\right] \tag{1.139}$$

An expression similar to Eq. (1.139) can be obtained for the exergy j''_{E_i} removed from the system. The total exergy of the mass currents brought into and removed from the system results by totalizing the currents j'_{E_i} and j''_{E_i}:

$$-\dot{E}_{mass} = \sum j'_i\left(h'_i - T_0s'_i\right) + \sum j''_i\left(h''_i - T_0s''_i\right) > 0 \tag{1.140}$$

Exergy Dissipation Speed of Heat Currents

The exergy dissipation speed of the heat currents is calculated using Eq. (1.123) and considering $T'_1 = T'_2 = T'$ and $T''_1 = T''_2 = T''$. It is obtained:

$$-\dot{E}_q = \sum j'_q\left(1 - \frac{T_0}{T'}\right) + \sum j''_q\left(1 - \frac{T_0}{T''}\right) \tag{1.141}$$

Work Produced Current as Exergy Current

The work is pure exergy, therefore:

$$\dot{E}_l = -j_l'' \tag{1.142}$$

1.2.3.3 Total Exergy Dissipation Speed. Relationship Between Exergy Dissipation and Entropy Creation. Exergy Efficiency

The system total inlet-outlet difference per second of exergy is found totalizing the contributions of the mass, heat and work currents, expressed by Eqs. (1.140)–(1.142), respectively:

$$-\dot{E} = \sum j_i'(h_i' - T_0 s_i') + \sum j_i''(h_i'' - T_0 s_i'') + \sum j_q'\left(1 - \frac{T_0}{T'}\right) + \sum j_q''\left(1 - \frac{T_0}{T''}\right) - j_l'' \tag{1.143}$$

Equation (1.143) can be rearranged as:

$$\begin{aligned}-\dot{E} = &\left(\sum j_i'h_i' + \sum j_i''h_i'' + \sum j_q' + \sum j_q'' - j_l''\right)\\ &- T_0\left[\left(\sum j_i's_i' + \sum j_i''s_i''\right) + \sum \frac{j_q'}{T'} + \sum \frac{j_q''}{T''}\right]\end{aligned} \tag{1.144}$$

The first bracket in the right member of Eq. (1.144) is null, according to the 1st principle balance, Eq. (1.136), therefore:

$$-\dot{E} = -T_0\left(\sum j_i's_i' + \sum j_i''s_i''\right) - T_0\left(\sum \frac{j_q'}{T'} + \sum \frac{j_q''}{T''}\right) \tag{1.145}$$

Noting by $\dot{S} \equiv \left(\frac{dS}{dt}\right)_{irrev}$ the creation of entropy per second of the open system (known also as entropy source) and by $\dot{E} \equiv \left(\frac{dE}{dt}\right)_{irrev}$ the dissipation (destruction) of entropy per second of the system, from Eq. (1.145) it is obtained (Chartier et al. 1975; Kirilin et al. 1985):

$$-\dot{E} = T_0\left(\dot{S}_{mass} + \dot{S}_{heat}\right) = T_0\dot{S}_{total} > 0 \tag{1.146}$$

where:

$$T_0\dot{S}_{mass} = -T_0\left(\sum j_i's_i' + \sum j_i''s_i''\right) \tag{1.147}$$

and

$$T_0\dot{S}_{heat} = -T_0\left(\sum \frac{j_q'}{T'} + \sum \frac{j_q''}{T''}\right) \tag{1.148}$$

Equation (1.146) shows the connection existing between the creation of entropy $\left(\dot{S} > 0\right)$ and exergy dissipation (destruction) $\left(\dot{E} < 0\right)$ in an open stationary system. In the literature it is known as the Gouy-Stodola equation, after the French physicist M. Gouy, who proposed it in 1889 (see also the theorem of Maxwell-Gouy, Chartier et al. 1975) and the Slovakian engineer A. Stodola, who used it first in order to solve some technical problems (Kirilin et al. 1985). Equation (1.146) has large applications in analyzing the efficiency of the thermal installations. In this book in will be used in Chap. 9.

The exergy efficiency is defined as (Bosnjakovic 1965):

$$\eta_E = \frac{Retrievable\_exergy}{Input\_exergy} < 1 \tag{1.149}$$

The more a transformation takes place in the vicinity of reversibility, the more η_E approaches unit. Matching the sources with the tasks are the 2nd principle recommendation and goal for increasing η_E. Regarding this goal from the exergy destruction point of view, it means that the sources and the tasks should be chosen in such a way that theirs exergy should be as close as possible. When this is actually possible, primary energy and investment costs can be saved.

1.2.4 Non-equilibrium Linear Phenomenological Connection Between Generalized Forces and Currents

From the last section it resulted that exergy and entropy are not conserved but in the reversible transformations, only. The rest of actual transformations (processes) are more or less far from equilibrium. The distance to equilibrium is expressed qualitatively by Eq. (1.146). Quantitatively, it is calculated by the product of the current which is carrying the exergy and entropy and of an associated generalized force, governing the irreversible transformation, expressed by:

$$X_k \equiv \frac{\dot{S}_k}{j_k} = -\frac{\dot{E}_k}{T_0 j_k} \tag{1.150}$$

The generalized force X_k and the current (flux) j_k are mutually united. The definition given here is valid for one-dimensional irreversible transfer phenomena. However, in practice, it is possible that, on one side, the transfer takes place owing to the simultaneous influence of different natural forces, or, on the other side, the system is hosting not only one, but more irreversible transformations. The experience shows that in this case a certain flux j_k is related not only to its force X_k which this is mutually united with, but is related with all the other forces participating to the total entropy increase of the system, as well. It is said that the simultaneous irreversible processes are coupled. Suppose i, $i = 1, \ldots, n$, simultaneous irreversible processes are coupled, taking into account Eq. (1.150) the entropy source is written as:

$$\dot{S} = \sum_{i=1}^{n} j_i X_i > 0 \tag{1.151}$$

According to the non-equilibrium linear phenomenological theory (Chartier et al. 1975), a current j_k of a one-dimensional irreversible transfer phenomenon is mutually united with its force X_k and with a phenomenological coefficient of proportionality L_{kk}, in such a way that:

$$j_k = L_{kk} X_k \tag{1.152}$$

The phenomenological coefficient L_{kk} is supposed to be independent of the force X_k. The experience shows that when i, $i = 1, \ldots, n$, simultaneous irreversible processes are coupled, the flux j_k, $k \in [1, \ldots, n]$, is influenced not only by the force X_k which it is mutually united with, but by the other non-mutually united forces X_i, $X_i \neq X_k$ and fluxes, as well, which the flux j_k coexists with. Mathematically, this influence is expressed by:

$$j_k = \sum_{i=1}^{n} L_{ki} X_i, \; k = 1, \ldots, n \tag{1.153}$$

Introducing Eq. (1.153) in Eq. (1.151), the entropy source results as a positively defined function with respect to forces X_i, $i = 1, \ldots, n$:

$$\dot{S} = \sum_{i=1}^{n} \sum_{k=1}^{n} L_{ik} X_i X_k > 0 \tag{1.154}$$

The goal of the non-equilibrium linear phenomenological theory is to determine the currents j_k, $k = 1, \ldots, n$, of a system hosting simultaneous irreversible coupled processes. The forces X_k are known from the physics laws, but the phenomenological coefficients are more difficult to be found, needing semi-empirical methods for their approach. There are three laws the phenomenological coefficients must respect and which are of help in verifying the correctness of the values obtained: (a) $L_{ii} > 0$; (b) $\sqrt{L_{ii} L_{jj}} > |L_{ij}|$ (Van Rysselberghe 1963); (c) $L_{ij} = L_{ji}$ (reciprocity law of Onsager 1931), $i, j = 1, \ldots, n$. The non-equilibrium linear phenomenological theory has been employed as computation tool in many physical phenomena, for which the restriction to linear form is satisfactory, e.g. mass transfer in ionized membranes, independent and coupled chemical reactions, gas interdiffusion in porous media, osmose and reverse osmose, electrokinetical phenomena, thermomechanical effect, thermoelectrical effects (Peltier, Seebeck, Thomson) (Chartier et al. 1975).

1.3 Equilibrium of Thermodynamic Systems and Phase Transformations

A homogeneous system is the system which chemical composition and physical properties of are the same in all directions, or are varying continuously from a system point to another (e.g. drink water, which is a mixture of water and dissolved mineral salts, or an air column). A heterogeneous system is composed by two or more homogeneous domains, named phases. The phases are separated one another by interfaces, where the chemical composition and/or physical properties are changing in jumps (e.g. water and ice, ammonia–water liquid and vapor mixtures). A homogeneous system and each phase of a heterogeneous system can be composed by one or more pure substances, named also components. A homogeneous system or a phase of a heterogeneous system containing more pure substances is named solution or mixture. All pure substances and all solutions can have three physical states: gaseous, liquid and solid.

1.3.1 Thermodynamic Stability and Equilibrium

A working fluid ensures the thermodynamic stability of a system or transformation if it fulfilled unconditionally and simultaneously the thermal and the mechanical stability conditions:

$$c_v = \left(\frac{dq(t)}{dt}\right)_v > 0 \tag{1.155}$$

and

$$\left(\frac{\partial p}{\partial v}\right)_T < 0 \tag{1.156}$$

respectively. The two conditions can be explained applying the Le Chatelier-Braun equilibrium shift principle: if a system were removed from the equilibrium state, the corresponding parameters are changing in order to bring the system back to equilibrium. This principle will be used in the book Chap. 11 to explain the Marangoni effect.

A relaxed system with respect to the external conditions (sources) is in equilibrium state and keeps it as long as external conditions remain the same. Suppose the external conditions changed for a while and then came back as at the beginning. In this case, the equilibrium may follow the external conditions variation and come back to the initial state, or, on the contrary, it can suffer a perpetual change, passing into a state different from the initial one. The systems proving the first or the latter behavior are called to have stable or unstable equilibrium states, respectively. A third category of equilibria are mentioned in the literature as being metastable, occupying a position situated between the two mentioned above. The systems we are dealing with in this book are supposed to have stable equilibrium states, therefore respect the conditions expressed by Eqs. (1.155) and (1.156).

The interactions type of a source-system (conjugation conditions) determine precisely a system reaction and implicitely its equilibrium state. The 2nd principle will be used next to establish qualitative rules for a system possible transformations unfolding, leading to different equilibrium states when different source-system conjugation conditions are imposed.

Indeed, in Sect. 1.1.6.3. we have seen that for the isolation conjugation conditions, $U = const.$ and $V = const.$, a system reaches equilibrium when its entropy $dS \geq \frac{dU}{T} + \frac{p}{T}dV$ becomes maximum:

$$S = S_{\max};\ dS = 0;\ d^2S < 0 \tag{1.157}$$

In this case, the entropy $S = S(U, V)$ is an equilibrium criterion for the isolated system. Other four most important conjugation conditions are given in the 1st column of Table 1.1. Each equilibrium state, 3rd column, is indicated this time by the minimum value of other equilibrium criteria named characteristic functions, given in the 2nd column: the internal energy $U = U(S, V)$, enthalpy $H(S, p)$, free energy $F(T, V)$ and the free enthalpy $\Phi(T, p)$. The knowledge of a characteristic function, expressed by its defining state parameters, enables one to calculate all other thermodynamic functions. In the 4th column there are given just two of them for each characteristic function. Also, by cross differentiation of the derived thermodynamic state parameters, the Maxwell relationships (1.160), (1.163), (1.167) and (1.171) result, 4th column (Kirilin et al. 1985). Although the entropy S is an equilibrium criterion, similar to U, H, F and Φ, this state function is not a characteristic function. However, similar to entropy, all characteristic functions mentioned above are state functions as well, as being combinations of the state functions U, H and S. The characteristic functions are called also thermodynamic potentials. The "potential" term comes of mechanics, because at equilibrium the decrease of the characteristic functions calculates in each conjunction conditions the maximum work the system could deliver (exergy). It is easy to verify for instance that in case of an isothermal transformation at equilibrium the availability is given by the free energy potential decrease $dF = -pdV - dL^* = -dL$, where $-dL$ holds for the total work available (dilation work + other work types except that of dilation, see Eqs. (1.23 and (1.24)), while in case of an isobar and isothermal transformation this is calculated by the free enthalpy potential decrease $d\Phi = -dL^*$. The term of "free energy" is also explained, next. The explanation holds true for the "free enthalpy" as well. Indeed, let us consider an isocore and isothermal transformation. The work is produced in this case by the decrease of function F, only, and arranging the free energy relationship as $U = F + TS$, Eq. (1.155), it results that the "free" part of the internal energy, equal to F, can be converted in work, only, the complementary part TS is bounded energy, which cannot be transformed in work. Otherwise expressed, the F part is the exergy, while the TS part is the anergy of a transformation at equilibrium (see Sect 1.2. introductory part).

Out of the thermodynamic potentials, the specific free enthalpy potential, $\varphi = \frac{\Phi}{G}$, occupies an apart position. Indeed, a thermodynamic system analyzed so far at equilibrium was considered to have a constant amount of substance, G. However, in order to solve physical and chemical problems involving phase equilibria, it is useful to consider G as a third variable of the thermodynamic potentials describing the equilibrium of given systems, i.e. $U = U(S, V, G)$, $H(S, p, G)$, $F(T, V.G)$ and $\Phi(T, p, G)$. Further, performing the potentials partial derivation

Table 1.1 Most important source-system equilibrium interactions and characteristic functions

Interaction conditions	Characteristic function	Equilibrium conditions	Derived thermodynamic state parameters
$V = const.$ $S = const.$	$dU \leq TdS - pdV$ Internal energy: $U = U(S, V)$	$dU \leq 0$ $U = U_{\min}$ $dU = 0$ $d^2U > 0$	$p = -\left(\frac{\partial U}{\partial V}\right)_s$ (1.158); $T = \left(\frac{\partial U}{\partial S}\right)_V$ (1.159); $\left(\frac{\partial p}{\partial S}\right)_V = -\left(\frac{\partial T}{\partial V}\right)_S$ (1.160)
$P = const.$ $S = const.$	$dH \leq TdS + Vdp$ Enthalpy : $H(S, p) = U + pV$	$dH \leq 0$ $H = H_{\min}$ $dH = 0$ $d^2H > 0$	$T = \left(\frac{\partial H}{\partial S}\right)_p$ (1.161); $V = \left(\frac{\partial H}{\partial p}\right)_S$ (1.162); $\left(\frac{\partial T}{\partial p}\right)_S = \left(\frac{\partial V}{\partial S}\right)_p$ (1.163)
$V = const.$ $T = const.$	$dF \leq -SdT - pdV$ Helmholtz free energy: $F(T, V) = U - TS$ (1.164)	$dF \leq 0$ $F = F_{\min}$ $dF = 0$ $d^2F > 0$	$p = -\left(\frac{\partial F}{\partial V}\right)_T$ (1.165); $S = -\left(\frac{\partial F}{\partial T}\right)_V$ (1.166); $\left(\frac{\partial p}{\partial T}\right)_V = \left(\frac{\partial S}{\partial V}\right)_T$ (1.167)
$P = const.$ $T = const.$	$d\Phi \leq -SdT + Vdp$ Gibbs free enthalpy: $\Phi(T, p) = H - TS$ (1.168)	$d\Phi \leq 0$ $\Phi = \Phi_{\min}$ $d\Phi = 0$ $d^2\Phi > 0$	$V = \left(\frac{\partial \Phi}{\partial p}\right)_T$ (1.169); $S = -\left(\frac{\partial \Phi}{\partial T}\right)_p$ (1.170); $\left(\frac{\partial V}{\partial T}\right)_p = -\left(\frac{\partial S}{\partial p}\right)_T$ (1.171)

with respect to G, with the help of the 1st and 2nd principles equations the remarkable result holds true:

$$\left(\frac{\partial U}{\partial G}\right)_{S,V} = \left(\frac{\partial H}{\partial G}\right)_{S,p} = \left(\frac{\partial F}{\partial G}\right)_{T,V} = \left(\frac{\partial \Phi}{\partial G}\right)_{T,p} = \varphi \tag{1.172}$$

The value of the specific free enthalpy potential at constant temperature and pressure, φ, enables to calculate for every system the variation of the characteristic function in function of the system substance amount. For this reason the function φ was named chemical potential. The chemical potential plays a great role in analyzing the phase changing processes with mass and heat transfer and particularly in the chemical reactions. Bearing in mind that all absorption processes are considered isobar processes, the chemical potential is much used to model phase equilibria of the working fluids met in the coabsorbent technology. Such models, of ammonia–water and water–lithiumbromide working combinations, have been extensively used throughout the book in order to present results of the various analyzed coabsorbent heat pumping cycles.

Continuing the equilibrium criteria analysis, next the entropy $S = S(U, V, G)$ variation with respect to the substance amount G variation is emphasized in case of an isolated system having the conjunction conditions already specified, $U = const.$ and $V = const.$. Although the entropy is neither a characteristic function, nor a thermodynamic potential, its partial derivative with respect to G is proportional to the chemical potential φ, by:

$$\left(\frac{\partial S}{\partial G}\right)_{U,V} = -\frac{\varphi}{T} \tag{1.173}$$

Equation (1.173) is derived in Appendix 1 and will be useful in sections to follow.

1.3.2 Equilibrium Conditions of a Homogeneous Isolated System. TPT Equilibrium Point and Static Equilibrium

As already mentioned, out of all thermodynamic systems having various conjugation conditions, the highest interest is presenting the isolated system, schematically shown in Fig. 1.9. Its equilibrium conditions are studied considering that it is divided in two homogeneous parts, 1 and 2, each of them including the substance amounts G_1 and G_2, respectively. The system entropy has additive properties, hence $S_{syst} = S_1 + S_2$. We bear in mind that the system entropy annulates at equilibrium, as being the isolated system criterion of equilibrium, that is:

$$dS_{syst} = d(S_1 + S_2) = dS_1 + dS_2 = 0. \tag{1.174}$$

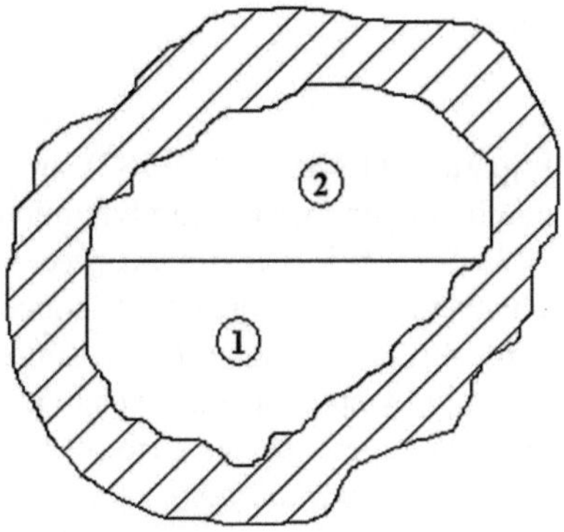

Fig. 1.9 Schematic of a thermodynamic isolated system

The entropy is function of each part internal energy and volume, $S = S(U, V)$. Differentiating it totally, results in:

$$dS = \left(\frac{\partial S}{\partial U}\right)_V dU + \left(\frac{\partial S}{\partial V}\right)_U dV \tag{1.175}$$

and remembering the 2nd principle equation at equilibrium:

$$TdS = dU + pdV \tag{1.176}$$

Equation (1.175) becomes:

$$dS = \frac{1}{T}dU + \frac{p}{T}dV \tag{1.177}$$

The isolated system is characterized by $V_{syst} = V_1 + V_2 = const.$ and $U_{sist} = U_1 + U_2 = const.$. Differentiating these relations for $G_1 = const.$ and $G_2 = const.$, it results $dV_1 = -dV_2$ and $dU_1 = -dU_2$. Further, we write Eq. (1.177) for each part of the system at hand and taking into account the last expressions, we obtain:

$$dS_1 = \frac{1}{T_1}dU_1 + \frac{p_1}{T_1}dV_1 \tag{1.178}$$

$$dS_2 = \frac{1}{T_2}dU_2 + \frac{p_2}{T_2}dV_2 = -\frac{1}{T_2}dU_1 - \frac{p_2}{T_2}dV_1, \tag{1.179}$$

respectively. Introducing Eqs. (1.178) and (1.179), in Eq. (1.174), it results:

$$\left(\frac{1}{T_1} - \frac{1}{T_2}\right)dU_1 + \left(\frac{p_1}{T_1} - \frac{p_2}{T_2}\right)dV_1 = 0 \tag{1.180}$$

In Eq. (1.180) dU_1 and dV_1 are independent one another and different from zero, therefore the expression (1.180) left member cannot be identically null but only if the brackets were null, wherefrom it is obtained:

$$T_1 = T_2 \tag{1.181}$$

and

$$p_1 = p_2 \tag{1.182}$$

Equations (1.181) and (1.182) show that the temperature and pressure are the same in a homogeneous thermodynamic isolated system at equilibrium.

In Sect. 9.5, a *two-point theory (TPT) of mass and heat transfer* is presented. The theory gives more details concerning the source-cycle working fluid interaction. It results, for instance, that a source-working fluid equilibrium cannot be reached but only by spontaneous exchange of heat currents, not coupled with the mass currents. We speak in this case of a *static equilibrium*, which defines the existence of the true *equilibrium point* of a cycle-source interaction. The equilibrium point is the first type of points which the TPT is based on. It is possible to find the equilibrium conditions of a cycle-source interaction type, using the isolated system of Fig. 1.9 and considering for example that its parts 1 and 2 are the system and the source at hand, respectively. The thermodynamic approach covered above can be applied in this new case as well and Eqs. (1.181) and (1.182) are written this time:

$$T_s = T_{wf} \tag{1.183}$$

and

$$p_s = p_{wf} \tag{1.184}$$

Equations (1.183) and (1.184) characterize an equilibrium point. These equations show that the temperature and pressure of the system working fluid found in equilibrium with a source are the same. According to TPT, a generalized force X_k and the current (flux) j_k, which this is mutually united with, Eq. (1.150), are defined in an equilibrium point. Further comments about TPT point of view concerning phase equilibria will be made in the next section.

1.3.3 Phase Equilibrium Conditions of Monocomponent and Binary Systems. TPT Ideal Point and Dynamic Equilibrium

Now, an isolated system with two phases is analyzed, in order to determine the equilibrium conditions between these phases. Suppose first the thermodynamic system consists in two homogeneous monocomponent subsystems 1 and 2, depicted in Fig. 1.9. The method used in the preceding section is applied here as well, provided that this time the system substance G_{syst} is in the two phases in different quantities and the phases exchange between them mass and heat currents. In order to be more specific, suppose the subsystems 1 and 2 are water and its vapor, respectively. The system is characterized by $V_{syst} = const.$, $G_{syst} = const.$ and $U_{sist} = const.$. Hence it can be written (indices 1 and 2 hold for the 1st and 2nd phases):

$$V_{syst} = V_1 + V_2 \tag{1.185}$$

$$G_{syst} = G_1 + G_2 \tag{1.186}$$

$$U_{syst} = U_1 + U_2 \tag{1.187}$$

From Eqs. (1.185)–(1.187) it results:

$$dV_1 = -dV_2 \tag{1.188}$$

$$dG_1 = -dG_2 \tag{1.189}$$

$$dU_1 = -dU_2 \tag{1.190}$$

Also, similar to the precedent section, we have $dS_{syst} = 0$, and $S_{syst} = S_1 + S_2$, therefore:

$$dS_{syst} = dS_1 + dS_2 = 0 \tag{1.191}$$

This time, the entropy of each subsystem is a function of the substance quantity G, as well, not only of its internal energy and volume, $S = S(U, V, G)$. Its total differentiation is expressed by

$$dS = \left(\frac{\partial S}{\partial U}\right)_{V,G} dU + \left(\frac{\partial S}{\partial V}\right)_{U,G} dV + \left(\frac{\partial S}{\partial G}\right)_{U,V} dG \tag{1.192}$$

In Eq. (1.192), the partial derivatives have the following expressions [see Eqs. (1.175)–(1.177) and (1.173)]:

$$\left(\frac{\partial S}{\partial U}\right)_{V,G} = \frac{1}{T} \tag{1.193}$$

$$\left(\frac{\partial S}{\partial V}\right)_{U,G} = \frac{p}{T} \tag{1.194}$$

$$\left(\frac{\partial S}{\partial G}\right)_{U,V} = -\frac{\varphi}{T} \tag{1.195}$$

Equations (1.193)–(1.195) are introduced in Eq. (1.192). Further, writing Eq. (1.192) for the two phases, theirs entropy results:

$$dS_1 = \frac{1}{T_1} dU_1 + \frac{p_1}{T_1} dV_1 - \frac{\varphi_1}{T_1} dG_1 \tag{1.196}$$

$$dS_2 = \frac{1}{T_2} dU_2 + \frac{p_2}{T_2} dV_2 - \frac{\varphi_2}{T_2} dG_2 \tag{1.197}$$

The entropy expressions given by Eqs. (1.196) and (1.197) are introduced in Eqs. (1.191) and taking into account Eqs. (1.188)–(1.190), it is obtained:

$$\left(\frac{1}{T_1}-\frac{1}{T_2}\right)dU_1+\left(\frac{p_1}{T_1}-\frac{p_2}{T_2}\right)dV_1-\left(\frac{\varphi_1}{T_1}-\frac{\varphi_2}{T_2}\right)dG_1=0 \quad (1.198)$$

The differentials dU_1, dV_1 and dG_1 of Eq. (1.198) are independent one another and different from zero, therefore the expression (1.198) left member cannot be identically null but only if the brackets were null, wherefrom it is obtained:

$$T_1 = T_2 \quad (1.199)$$

$$p_1 = p_2 \quad (1.200)$$

$$\varphi_1 = \varphi_2 \quad (1.201)$$

From Eqs. (1.199)–(1.201) it can be remarked that when the isolated system replaces its homogeneous two-part structure by one heterogeneous with two monocomponent homogeneous phases, the equilibrium conditions (1.181) and (1.182) are completed with the condition (1.201), requiring that phases chemical potentials be the same at equilibrium.

The thermodynamical method used so far can be extended in order to determine the equilibrium conditions of binary or polycomponent and two (poly)-phase working combinations, as well. In the coabsorbent technology, the working combinations are especially binary ($c1$, $c2$) and two-phase ($p1$, $p2$), like ammonia–water, water–lithiumbromide etc. It is not difficult to show that for such systems, the equilibrium is determined by the following conditions:

$$T_1 = T_2 \quad (1.202)$$

$$p_1 = p_2 \quad (1.203)$$

$$\varphi_{p1,c1} = \varphi_{p2,c1} \quad (1.204)$$

$$\varphi_{p1,c2} = \varphi_{p2,c2} \quad (1.205)$$

The first two equations are the same as to the precedent systems analyzed. However, the last two equations, show that for equilibrium, a component chemical potential is the same to each phase, and this is valid for every component.

Next, we shall continue our comments about TPT point of view, concerning the equilibrium in isolated systems (see precedent section). Here, the classic and TPT points of view are compared this time in conjuncture with the equilibrium conditions of the two phase isolated systems. In this case, according to TPT, equations like (1.201), referring to monocomponent systems, and like (1.204) and (1.205), regarding the binary systems, hold true, as it express statistically an equal and mutual mass and heat exchange between phases around an equilibrium state. However, concerning conditions (1.199), (1.200), or (1.202) and (1.203), these define the well known equilibrium points of the equilibrium thermodynamics, but TPT point of view is different. Indeed, according to TPT, a point defined by the conditions mentioned above is named *ideal point* and is the second type of points which the (*TPT*) bases on. A generalized force X_k and the current (flux) j_k, which this is mutually united with, Eq. (1.150), are not defined in an ideal point, because, ones this reached, a *dynamic equilibrium* transfer of coupled mass and heat currents occurs between the two phases. Therefore, the classic equilibrium point has in fact a statistical dimension, not a punctual one, as it results from Eqs. (1.199), (1.200), or (1.202) and (1.203), wherefrom the TPT *ideal point* denomination comes of. Concluding, the same term of equilibrium point used by equilibrium thermodynamics for describing both equilibrium with a source and between two phases, is replaced in the TPT non-equilibrium thermodynamics by two separate terms, equilibrium and ideal point, respectively. Section 9.5 is justifying in detail why this replacement is more appropriate.

1.3.4 Phase Transformations. Gibbs Rule of Phases

The passing of a substance from a phase to another is called phase (state) transformation. Keeping the pressure constant and varying the temperature, a substance can cover all phase transformations, solid, liquid and gas. Also, the change of pressure alters the temperature of a phase transformation. The passing from a phase to another is accompanied by heat absorption or heat rejection. This property lays down the basis of heat and/or cooling production utilizing working fluids covering thermodynamic absorption and/or mechanical vapor compression cycles. The values of the state parameters characterizing a certain phase transformation define a point of phase transformation. It has been adopted the following classification of the points of phase transformation: the point of "liquid–vapor" passing is called point of *boiling* or point of *condensation*, the point of "solid–liquid" passing is called point of *melting* or point of *solidification*, and the point of "solid–vapor" passing is called point of *sublimation* or point of *desublimation*. The absorption technology processes have in case of binary mixtures their own classification: the point of "liquid–vapor" passing is called point of *generation* or *desorption* or point of *absorption* or *resorption*, while the point of "liquid–vapor" passing in case of the monocomponent working fluids is called point of *evaporation*.

The Gibbs rule of phases is important in the analysis of phase equilibria and phase transformations. This rule establishes a relationship between the number of intensive independent variables which determine the state of a thermodynamic system in equilibrium (often these independent variables are named *freedom degrees* of the system; also, a system is *mono-*, *bi-* or *multi-variant* if it had one, two, or more independent variables), v, the number of phases, f, the number of components of the system, c, and the number of independent chemical reactions, r. The rule of phases is expressed at equilibrium by:

$$v = (c - r) + 2 - f \tag{1.206}$$

In the relationship (1.206), $(c - r)$ holds for the number of independent components when chemical reactions occur. Obviously, rel. (1.206) holds true in both equilibrium and non-equilibrium thermodynamics. A system in a non-equilibrium state has more independent variables than a system in equilibrium, comparatively (see Chap. 9).

The coabsorbent technology operates usually with binary working fluid-absorbent combinations (mixtures) (e.g. NH_3–H_2O, H_2O–LiBr, NH_3–NaSCN, NH_3–$LiNO_3$), without chemical reactions, therefore $c = 2$ and $r = 0$. Also, in this work, the two-phase (liquid + vapor) equilibria are considered exclusively, therefore $f = 2$. Hence, from rel. (1.206) it results that the absorption processes we shall meet in this book have two independent variables, $v = 2$ (the systems are bivariant). This means that, out of a total of four possible state parameters describing the equilibrium, p, v, T, x, or y, only two are independent, the rest are functions of the independent variables. The absorption processes are considered quasi-isobar, more often than not, $p = const.$, therefore one independent variable is a priori established. The second independent variable is usually chosen the temperature, T(K), or t(°C). It results that the rest of variables are dependent of p and T (volume, concentration, enthalpy, entropy, etc.). In this respect, the log $(p) - 1/T$ charts are mostly used to plot absorption cycles (see next section). In such a chart, the concentration x, Eq. (1.12), becomes a function of the pressure and temperature, $x = x(p, T)$. The log $(p) - 1/T$ chart proves to be very useful in the first phase of an absorption cycle design and outlook. The American newer versions of this chart are more elaborated, enabling one for example to do also the graphical thermal computation of a cycle. However, starting from the remark that absorption processes are bivariant, Merkel and Bosnjakovic have been chosen the enthalpy and the concentration as independent variables and created the $h - x$ chart. This chart considered the rest of variables i.e. temperature, pressure, as dependent variables. This chart completes the log $(p) - 1/T$ chart, enabling one to do the cycle graphical thermal computation (see further in this section). Besides binary, the operation with ternary (three-component) mixtures has been also proposed (e.g. NH_3–H_2O–LiBr or CH_3NH_4–H_2O–LiBr), $c = 3$. From rel. (1.206) it results that such systems are three-variant, $v = 3$. Also, monovariant systems are operating in the coabsorbent technology. These are met for instance in the air conditioning and refrigeration applications, where practically monocomponent working fluids are suffering two-phase condensation and evaporation processes. It is the case for example of the working fluids coming of rectification processes (e.g. ammonia, which enters the evaporator with a high concentration, of $x \geq 0.998$), or from the last generation process of a truncated coabsorbent plant, or from the generation processes of non-volatile absorbent mixtures (e.g. NH_3 and H_2O coming of NH_3–NaSCN, NH_3–$LiNO_3$ and H_2O–LiBr generation processes, respectively). The monovariant systems are met obviously as well to the heat pumping cycles operating with monocomponent working fluids (e.g. ammonia) and mechanical vapor compression (see Chap. 8).

1.3.5 Clapeyron–Clausius Equation

Let us consider two phases, 1 and 2, in equilibrium, Fig. 1.9. In this case, Eqs. (1.199)–(1.201) hold true, $p_1 = p_2 = p$, $T_1 = T_2 = T$, and:

$$\varphi_1(p, T) = \varphi_2(p, T), \tag{1.207}$$

respectively. Equation (1.207) is differentiated, and with Eqs. (1.199) and (1.200), it is obtained:

$$\left(\frac{\partial \varphi_1}{\partial p}\right)_T dp + \left(\frac{\partial \varphi_1}{\partial T}\right)_p dT = \left(\frac{\partial \varphi_2}{\partial p}\right)_T dp + \left(\frac{\partial \varphi_2}{\partial T}\right)_p dT \tag{1.208}$$

Equation (1.208) partial derivatives are given by Eqs. (1.160) and (1.161) of Table 1.1, resulting:

$$v_1 dp - s_1 dT = v_2 dp - s_2 dT \tag{1.209}$$

Equation (1.209), arranged as Eq. (1.210) below, is known as the Clapeyron–Clausius equation:

$$\frac{dp}{dT} = \frac{s_2 - s_1}{v_2 - v_1} \tag{1.210}$$

The important Eq. (1.210) establishes in the (*p*, *T*) diagram a relationship between the *p*(*T*) derivative and the entropy and volume differences, corresponding to a certain phase transformation. The monocomponent systems change the phase at constant pressure and temperature. The 2nd principle, Eq. (1.109), writes in this case at equilibrium:

$$Tds = dh \tag{1.211}$$

Integrating Eq. (1.211) between states 1 and 2, it results:

$$T(s_2 - s_1) = h_2 - h_1 \tag{1.212}$$

The right member ($h_2 - h_1$) of Eq. (1.212) represents the heat released or absorbed by the system during the phase transformation. It is noted by r, $[r] = \text{kJ kg}^{-1}$ for a mass equal to unit of substance changing the phase,

$$h_2 - h_1 = r \tag{1.213}$$

With this notation, Eq. (1.212) rewrites:

$$s_2 - s_1 = \frac{r}{T} \tag{1.214}$$

Introducing Eq. (1.214) in Eq. (1.210), this becomes:

$$\frac{dp}{dT} = \frac{r}{T(v_2 - v_1)} \tag{1.215}$$

Equation (1.215) is further used to liquid–vapor phase transformations in the next section.

1.4 Absorption Heat Pumping Selected Topic

1.4.1 Absorption Cycle Introduction

As already shown, the reverse thermodynamic cycle of the p–V diagram, Fig. 1.2 and Fig. 1.10, extracts the heat $Q_2 < 0$ from a lower temperature, T_s, sink source and "pumps" it to a higher temperature T_h, $T_h > T_s$, heat source, as heat $Q_1 < 0$, consuming the work $L_{cycle} = L_{12,dilat} - L_{12,compr} < 0$, in such a way that $|Q_1| = |L_{cycle}| + |Q_2| > |Q_2|$. The operation of transporting an amount of heat from a lower temperature level to a higher temperature level is known in thermodynamics as "heat pumping", and refers to two useful effects obtained simultaneously, a first of cooling the sink source and a second, of heating the heat source [see also Eqs. (1.64)–(1.66)]. If only the heating effect were had in view, then, this operation is called "the sink heat ($Q_2 < 0$) upgrade". It is very important to point out that the role played by the work input is to enable the energy be transported from the sink to the heat source. This kind of heating/cooling is an issue from the 2nd principle (see Clausius formulation of the 2nd principle, Sect. 1.1.6.1) and is more effective than the similar effects obtainable as an issue from the 1st principle (see next in this section). However, as a general remark, this kind of heat pumping task handles usually heats of low temperature levels (e.g. −10 to +60 °C), while the source is work input, situated on the highest level on a scale of energy quality. In this case, the source and the task are mismatched, which is against the 2nd principle recommendations. In order to increase the source-task match, it is possible to replace the high quality (low entropy) work by other types of sources as energy input. One of the most utilized types is the heat. The heat-driven technologies include solid absorption (adsorption), ejector, magnetic cooling and the Stirling cycle, but absorption is by far the most applied worldwide. The heat-driven absorption heat pumping has a completely changed operation principle. Indeed, instead of the mechanical vapor compression (mvc) cycle, Fig. 1.10, in this case the thermal compression cycle is utilized. In this case, the high pressure working fluid is obtained by increasing the temperature of an absorbent (mixture) which contains the absorbed working fluid, Fig. 1.11a, b. The source temperature and mixture pressure and temperature are chosen in such a way that the working fluid released (generation) through heat input, is correlated with the condenser temperature and pressure (the same, principally, as that of the mvc cycle). After the working fluid has performed the useful effect, the reverse process, of evaporated working fluid absorption and mixture initial parameters regeneration, takes place, in order to prepare the cycle for a new run. The same principle of absorption thermal compression can be utilized to produce either cooling or heating. Unlike the mvc cycle, these cycles consume work only for mixture pumping from the cycle low pressure side to its high pressure side. Another advantage of the absorption cycle is that it is capable to recover the low grade heat sources, which result from the industrial processes, for cooling and/or heat production. Nowadays,

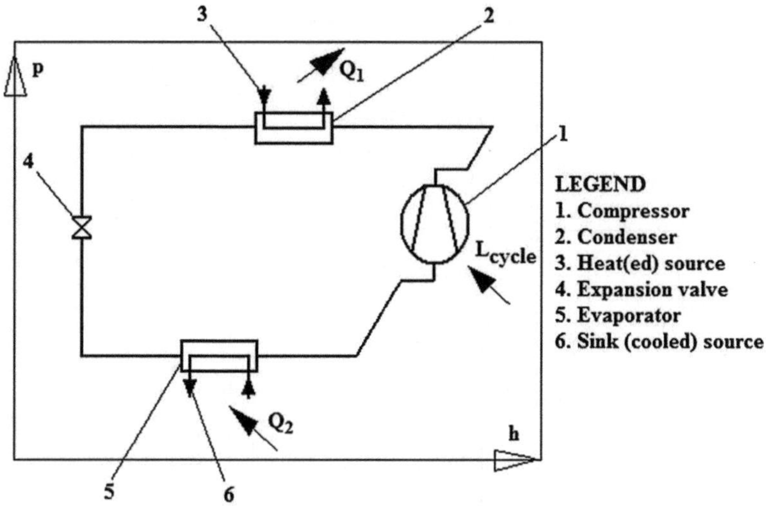

Fig. 1.10 Schematic of mechanical vapor compression cooling cycle

the absorption cycles extended theirs applications to the work production from depleted sources, as well. Historically, the absorption cycles begun with intermittent working (early of the 19th century, see also Ferdinand Carre plant) and it carried on later with the continuous operation. The absorbent (mixture) in most applications (high, industrial and domestic capacity) is liquid (e.g. NH_3–H_2O and H_2O–LiBr). For house applications (small capacity) the solid adsorbents can be used as well. In this book, we are analyzing solely the coabsorbent heat pumping cycles operated by NH_3–H_2O and H_2O–LiBr working fluid-absorbent working combinations (pairs).

Except the low temperature domain of cryogenics, with special applications like gases separation processes, supraconductibility, etc., the most cooling applications are in the industrial (−70 to −35 °C), medium (−35 to −10 °C) and normal, domestic (−10 to +5 °C) range. Concerning the heating applications, the domestic, social (50–60 °C) and district, industrial (60–95 °C) sectors are the most important. Today, these cooling and heating needs are satisfied mostly by the mvc-driven heat pumping technology, based mainly on ammonia as working fluid. It is estimated that about (15–40) % out of the electrical energy produced worldwide is consumed for cooling and heating production. The absorption challenges the mvc-driven heat pumping technology in applications providing especially free heat sources. The absorption machines used are mainly of NH_3–H_2O type for cooling and heating and of H_2O–LiBr type for air conditioning. Their main quality is first the reliability.

An NH_3–H_2O absorption machine uses ammonia as refrigerant (working fluid), while water is the absorbent. Although ammonia is toxic, its thermal and transport properties prevail upon considering it as the best refrigerant. Because water has its own vapor pressure, the generated vapor needs usually rectification,

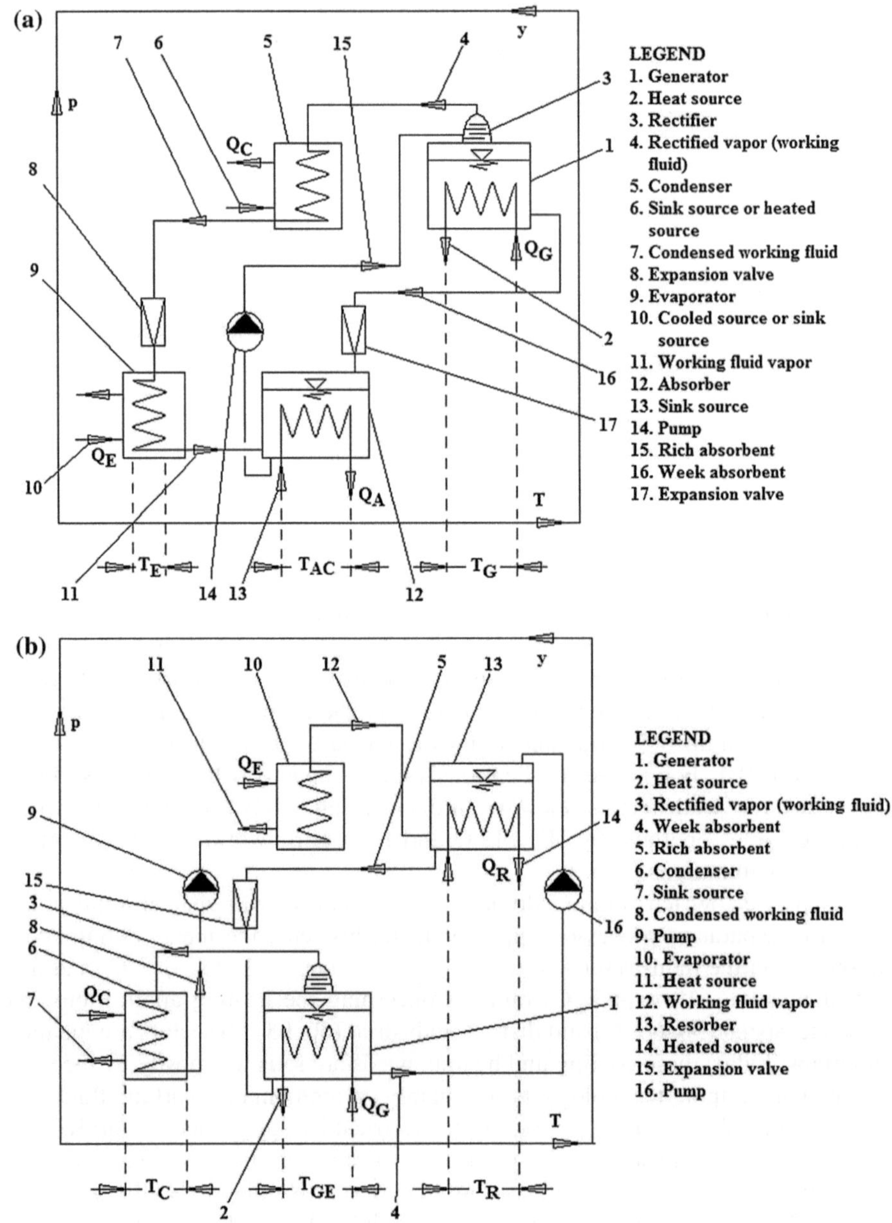

Fig. 1.11 Cooling (**a**) and heating (**b**) cycles, working with absorption thermal compression

which diminishes the coefficient of performance (COP). It can provide cooling up to −75 °C (the freezing temperature of ammonia is −77.7 °C). Usually it is supplied by steam (4–6 bar, 140–160 °C) in single-stage configuration

(COP = ~0.25) or by waste heat (up to 100 °C) in two-stage configuration, for medium and industrial cooling. It is produced and sold as well as gas-fired units with air-cooling (COP = ~0.5 and capacity up to 90 kW).

The H_2O–LiBr absorption machine works with water as refrigerant and lithium bromide as absorbent. Because the refrigerant is water, all cooling applications are above 0 °C. This working combination doesn't need vapor rectification. For this reason, the H_2O–LiBr absorption machine is more effective, COP = 0.7–1.2, in the single-stage and double effect configurations, respectively. This machine is produced for large capacities (up to 427 kW) for water chilling (5–7 °C). The heat sources powering such single-stage configuration air conditioning units are mainly the waste heat (80–95 °C), or the solar recently, or the heat coming of gas burning. For the double effect configuration, the steam, the gas-burning and the concentrated solar are an usual supply.

1.4.2 Basic Absorption Cycles

The cycles depicted in Fig. 1.11a, b describe principally, the two main types of heat pumping cycles met in absorption. These cycles are the simplest, with three temperature levels shown in the abscissa. The cycle of Fig. 1.11a has a single heat input, Q_G, to the highest temperature level, T_G, rejects heat twice, $Q_C + Q_A$, to a middle level, T_{AC}, and extracts heat only once, Q_E, to the lowest level, T_E, $T_E < T_{AC} < T_G$. The cycle is of *cooling type* (for example $T_E < 0$ °C) and additionally, if T_{AC} were enough high (say $T_{AC} = (50–60)$ °C), an useful heating effect could be obtained, as well. The cooling type cycle is very familiar in the cooling applications performed with the help of absorption technology, because in most cases is feasible. In the literature, occasionally, it is referred to also as "type I absorption heat pump". The cycle of Fig. 1.11b receives heat twice, $Q_G + Q_E$, to the middle level, T_{GE}, and rejects heat twice, but at two different levels, Q_C at T_C, and Q_R at T_R, respectively, $T_C < T_{GE} < T_R$. This last cycle is of *heating* or *heat transformer* type. In the literature it is referred to at times also as "type II absorption heat pump". The heating cycle having the configuration depicted in Fig. 1.11b, is less used in the heating applications because it did not prove so far its feasibility in most cases. However, great hopes are put into the coabsorbent version of this cycle, i.e. the truncated heating cycle (see Sect. 4.1.2.)

1.4.3 Ideal Cycles

This section provides the necessary knowledge concerning the ideal limits of the heat pumping cycles introduced so far. The cycles analyzed are plotted in the *temperature-entropy, T–s,* diagram, the most indicated in this case.

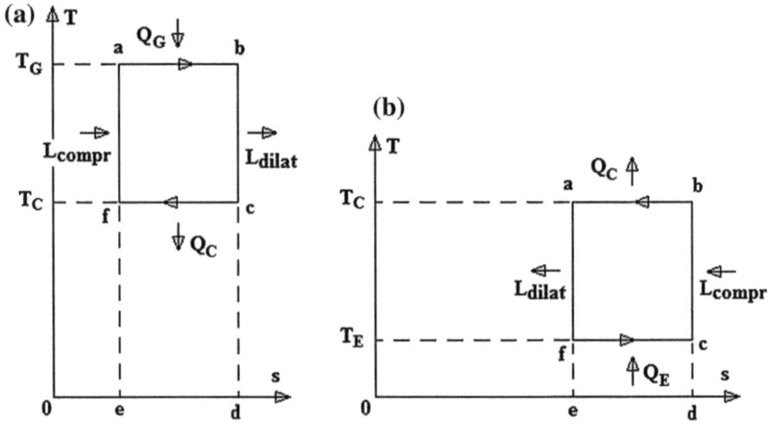

Fig. 1.12 Reversible direct (**a**) and reverse (**b**) Carnot cycles

1.4.3.1 Effectiveness of Direct and Reverse Carnot Cycles

In Sect. 1.1.6.1, the direct and reverse cycles with some configuration have been analyzed, Fig. 1.2. The upper (ideal) thermodynamic limits of these cycles are given by the reversible direct and reverse Carnot cycles, plotted in 1.12a and b, respectively.

Referring to the direct cycle, Fig. 1.12a, it is supplied with the heat $Q_G > 0$ delivered by a constant temperature T_G heat source, it rejects the heat $Q_C < 0$ to a constant temperature T_C sink source and produces the work $L_{cycle} = L_{dilat} - L_{compr} > 0$. In a *T–s* diagram the area subtended by a curve representing a reversible transformation is proportional to the amount of heat exchanged by the system during this transformation with the sources, provided the temperature origin be 0 K. In case of the direct cycle, the areas *abdea* and *fcdef* are proportional to the cycle heat input $Q_G > 0$ and output $Q_C < 0$, respectively. Bearing in mind the 1st principle energy balance, Eq. (1.61), it results that the area *abcfa* is proportional to the cycle work output, $L_{cycle} = L_{dilat} - L_{compr} = Q_G - Q_C > 0$. The efficiency of the reversible direct Carnot cycle is obtained rewriting Eq. (1.62) for reversible transformations (see Eq. (1.82) as well):

$$\eta_{work} = \frac{L_{cycle}}{Q_G} = \frac{Q_G - |Q_C|}{Q_G} = \frac{T_G - T_C}{T_G} < 1 \qquad (1.216)$$

The reverse cycle, Fig. 1.12b, is supplied with the heat $Q_E > 0$ delivered by a constant temperature T_E sink source, it rejects the heat $Q_C < 0$ to a constant temperature T_C heat source and consumes the work $|L_{cycle}| = L_{dilat} - |L_{compr}|$, $L_{cycle} < 0$. Applying the same reasoning to the reverse cycle, it is found that the area *abcfa* is proportional to the cycle work input, $|L_{cycle}| = L_{dilat} - |L_{compr}| = |Q_C| - |Q_E|$. Suppose a particular application needs that the temperatures T_E and T_C be separately or simultaneously lower than the lowest available sink source and higher than the highest available heat source, respectively. In this case, regarding things from a source viewpoint, the supply with heat $Q_E > 0$ at T_E and the rejection of heat $Q_C < 0$ at T_C, become cooling and heating processes, respectively. The COP's of the cooling and heating processes have been calculated by Eqs. (1.64) and (1.65). Writing them in conditions of reversibility, the COP's of the reverse Carnot cycle are obtained as:

$$COP_{c,C} = \frac{Q_E}{|L_{cycle}|} = \frac{Q_E}{|Q_C| - Q_E} = \frac{T_E}{T_C - T_E} > 1 \qquad (1.217)$$

and

$$COP_{h,C} = \frac{|Q_C|}{|L_{cycle}|} = \frac{|Q_C|}{|Q_C| - Q_E} = \frac{T_C}{T_C - T_E} > 1 \qquad (1.218)$$

for cooling and for heating purposes, respectively. From the two equations above the following result holds true: $COP_{c,C} + 1 = COP_{h,C}$.

As mentioned earlier, the heating/cooling processes issuing from the 2nd principle are more effective than the similar processes issuing from the 1st principle. Here, it is easy to prove this assertion in case of the heating processes. Indeed, suppose, according to the 1st principle, the ideal work-to-heat conversion effectiveness equals unit (see Sect. 1.1.5.1) (actually, the effectiveness of this conversion is well below unit). Let the work involved in conversion be $|L_{cycle}|$ and the heat produced $|Q_C|_I$, wherefrom, the conversion effectiveness is:

$$\eta_{h,I} = \frac{|Q_C|_I}{|L_{cycle}|} = 1 \qquad (1.219)$$

In our case, Eq. (1.218) writes:

$$COP_{h,C} = \frac{|Q_C|_{II}}{|L_{cycle}|} = \frac{T_C}{T_C - T_E} > 1 \qquad (1.220)$$

From the comparison of Eqs. (1.219) and (1.220) it results clearly that for same work use, $|L_{cycle}|$, the heating effect $|Q_C|_{II}$, issued from the 2nd principle is $COP_{h,C} > 1$ times higher than the heating effect $|Q_C|_I$, issued from the 1st principle, for the same working parameters. The explanation of this result is simple, according to the method issued from the 2nd principle, the heat is transported from the sink to the heat source, rather than produced (converted), like it is happening with the heating issued from the 1st principle.

1.4.3.2 Effectiveness of Carnot Absorption Cooling Cycle

Combining the two direct and reverse Carnot cycles, Fig. 1.12a, b, the ideal cycle, Fig. 1.13, of the absorption cooling cycle depicted in Fig. 1.11a is obtained. The absorption cooling cycle of Fig. 1.11a becomes reversible, if its transformations,

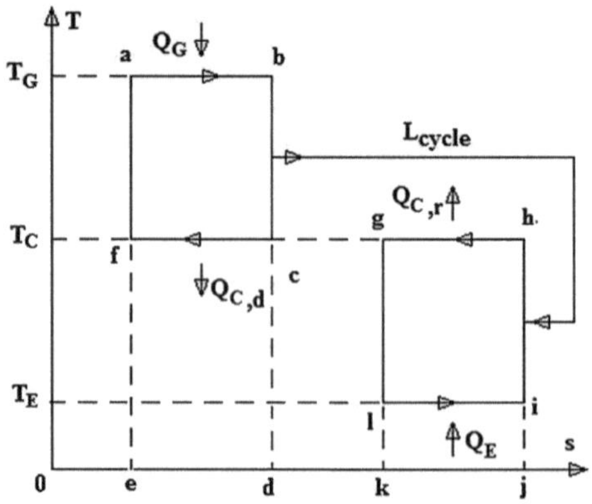

Fig. 1.13 Reversible Carnot absorption cooling cycle

noted in Fig. 1.13 as vapor generation (*ab*), vapor condensation (*cf*) and (*hg*) and evaporation (*li*), had taken place on infinite small degasing intervals of temperature and concentration and exchanged heat to constant temperatures T_G, T_C and T_E, differing only by infinite small values from the heat, sink and sink heat sources temperatures, respectively, which the cycle is connected to. In order to find the ideal cooling COP, it is supposed that the work L_{cycle}, produced by the direct Carnot cycle *abcfa* through heat input Q_G at T_G and heat rejection $Q_{C,d}$ at T_C, is utilized by the reverse Carnot cycle *hglih* to extract heat Q_E at T_E and reject heat $Q_{C,r}$ at T_C. The cooling COP is calculated by the ratio of the useful heat Q_E extracted through the heat Q_G input.

$$COP_{cc,C,a} = \frac{Q_E}{Q_G} \tag{1.221}$$

The heating COP is given by the ratio of the useful heat $Q_{C,d} + Q_{C,r}$ output through heat Q_G input.

$$COP_{ch,C,a} = \frac{Q_{C,d} + Q_{C,r}}{Q_G} \tag{1.222}$$

The ideal cooling and heating COP's of the absorption cooling cycle are obtained using Eqs. (1.217) and (1.218):

$$COP_{cc,C,a} = \frac{T_G - T_C}{T_G} \frac{T_E}{T_C - T_E} \tag{1.223}$$

and

$$COP_{ch,C,a} = \frac{T_G - T_E}{T_G} \frac{T_C}{T_C - T_E} \tag{1.224}$$

respectively. From the two equations above the following result holds true: $COP_{cc,C,a} + 1 = COP_{ch,C,a}$.

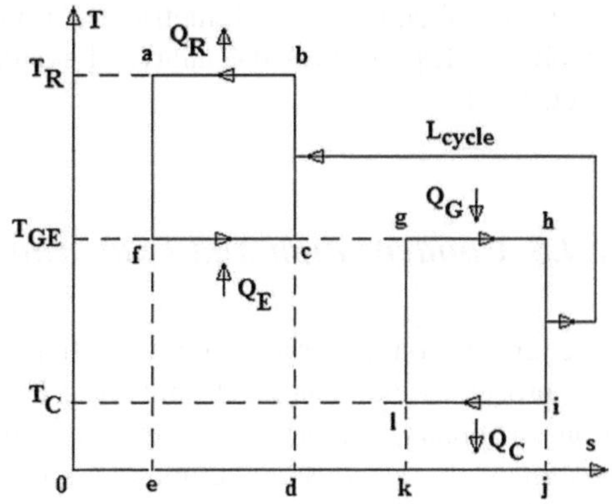

Fig. 1.14 Reversible Carnot heating (heat transformer) absorption cycle

1.4.3.3 Effectiveness of Carnot Absorption Heating (Heat Transformer) Cycle

Reverting the covering direction in the ideal cycle plotted in Fig. 1.13, it is obtained the ideal cycle, Fig. 1.14, of the absorption heating (heat transformer) cycle, depicted in Fig. 1.11b. The reversibility conditions described to the ideal cooling cycle remain valid in case of the heating cycle as well. The COP of the heating absorption cycle is defined by the ratio of the useful heat output Q_R through the heat input, $Q_E + Q_G$:

$$COP_{h,C,a} = \frac{Q_R}{Q_E + Q_G} \tag{1.225}$$

Using Eqs. (1.217) and (1.218), the ideal COP of the heating (heat transformer) absorption cycle is given by:

$$COP_{h,C,a} = \frac{T_{GE} - T_C}{T_{GE}} \frac{T_R}{T_R - T_C} \tag{1.226}$$

1.4.4 Selected Topic of Solutions Thermodynamics

As already mentioned, the absorption cycles we are analyzing in this book operate with liquid binary working combinations (called also "mixtures" further on), of working fluid (refrigerant)-absorbent type. The absorption systems at hand are heterogeneous, with two homogeneous phases, vapor in equilibrium with its liquid it comes of, respecting the phase equilibrium Eqs. (1.202)–(1.205). The mixtures are characterized by a certain liquid and vapor mass, Eq. (1.12), or molar, Eq. (1.14), fractions or concentrations, noted by y and Y for the refrigerant and by $(1 - y)$, $(1 - Y)$ for the absorbent, respectively. The gas phase is considered a mixture of ideal gases respecting Dalton law, Eq. (1.7), and which gas molar fractions of equal the volumetric fractions, Eq. (1.8). According to Gibbs rule of phases, Eq. (1.206), the analyzed systems are bivariant. For other details, see Sect. 1.3.4.

1.4.5 Condensation and Evaporation of Binary Mixtures

The absorption processes are essentially isobar and for this reason this type of processes is analyzed mostly. The isobar condensation and evaporation of binary (binary) mixtures are taking place with variable temperature. Alternatively, at constant temperature these processes occur with variable pressure. To detail the physical behavior asserted above, let us consider for example the NH_3–H_2O working pair suffering an evaporation process at the atmospheric pressure, $p_{atm} = 1.01325$ bar. Out of the two components, ammonia is more volatile because it has a lower normal boiling point, $\left[T_{b,NH_3}\right]_{p=1\text{bar}} = -33.4\ °\text{C}$ and $\left[T_{b,H_2O}\right]_{p=1\text{bar}} = 100\ °\text{C}$. In a continuous evaporation process, the ammonia is therefore the component which gets out with a higher rate from the mixture as compared to water, approaching zero concentration in the liquid phase when the temperature approaches the pure water component normal boiling temperature. The process taking place with a temperature increase within an isobar degasing interval, that is with $[T(y_1)]_p > [T(y_2)]_p$, when $y_1 < y_2$, is called to unfold with *gliding temperature*. The gliding temperature operation is typical for absorption processes and particularly for the coabsorbent technology and will be much analyzed throughout this book from the heat exchange (recovery) point of view. The unfolding of evaporation-condensation processes is more suggestively presented in a *T*–*y* diagram, Fig. 1.15. In this figure it is considered an isocore, $y_1 = const.$, and isobar, $p = p_{atm} = const.$, transformation, 1–2–3–4–5, characterized by the temperature parameter increase, from T_1 to T_5. The equilibrium state 1 is in the single-phase subcooled liquid domain, with three independent parameters, $1(p_{atm}, T_1, y_1)$, according to the rule of phases. Increasing the temperature up to T_2, the liquid phase reaches the equilibrium state 2*l*, the system becomes bivariant, $2(p_{atm}, T_2)$, and the evaporation (boiling) starts. The total of the evaporation points marking the beginning of evaporation process is called evaporation or bubble curve, Fig. 1.15, and has been found experimentally, covering the whole liquid phase concentration range, $y \in [0, 1]$. In this state 2*l*, the liquid phase is in equilibrium with the vapor phase, marked in Fig. 1.15 by point 2*v*, of the highest concentration Y_1, $Y_1 < y_1$. The point 2*v* belongs to a second significant curve in the binary mixture thermodynamics, marking the beginning of condensation or the end of evaporation, that is the dew curve. This curve is determined experimentally, as well, simultaneously with the evaporation curve, from the vapor temperature and concentration data. The domain comprised between the two curves is called evaporation lens. The point 2*l* is marking also the entering in the two-phase domain, which is called also the domain with humid vapor. The evaporation curve has as extreme states, those corresponding to the pure components, defined by $y = 0$ and $y = 1$, with

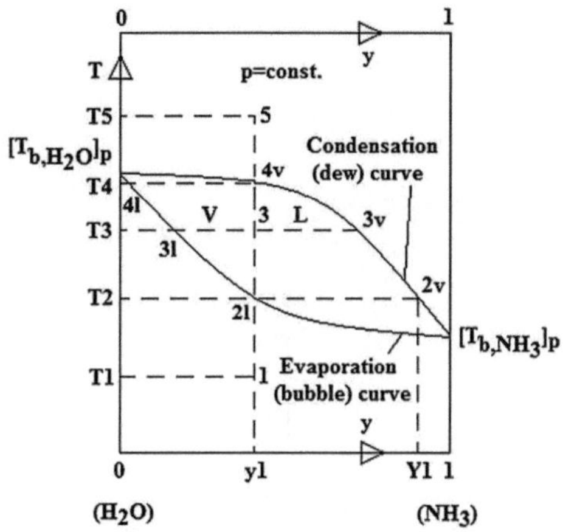

Fig. 1.15 Representation of binary mixture evaporation-condensation processes, *T*–*y* diagram

$T = [T_{b,H_2O}]_{p=1\text{ bar}}$ and $T = [T_{b,NH_3}]_{p=1\text{ bar}}$, respectively, Fig. 1.15. Increasing further the temperature to T_3, the system reaches state 3, consisting in a liquid phase, marked by $3l$ on the evaporation curve. The state $3l$ is in equilibrium with its vapor phase, marked by $3v$ on the condensation curve and situated on the same isotherm T_3. The system in state 3 is bivariant, $3(p_{atm}, T_3)$. The position of the point 3 on the isotherm $3l - 3v$ is determined next. Considering a unit mass of mixture with state 3, consisting of m_l kg of liquid phase and m_v kg of vapor phase, the mass and species balances can be written as follows:

$$m_l + m_v = 1 \tag{1.227}$$

and

$$m_l y_l + m_v y_v = y_1 \tag{1.228}$$

respectively. Solving Eqs. (1.227) and (1.228) for m_l and m_v, it results:

$$m_v = \frac{y_1 - y_{3l}}{y_{3v} - y_{3l}} \tag{1.229}$$

and

$$m_l = \frac{y_{3v} - y_1}{y_{3v} - y_{3l}} \tag{1.230}$$

The position of point 3 on isotherm $3l - 3v$ is determined by the ratio:

$$\frac{m_l}{m_v} = \frac{y_{3v} - y_1}{y_1 - y_{3l}} = \frac{L}{V} \tag{1.231}$$

The increase of temperature up to T_4, brings the system in state $4v$, situated on the condensation curve and in equilibrium with the liquid phase, marked by $4l$ on the evaporation curve. The state $4v$ represents the end of the evaporation process. The system in this state is still bivariant, $4(p_{atm}, T_4)$. The last temperature

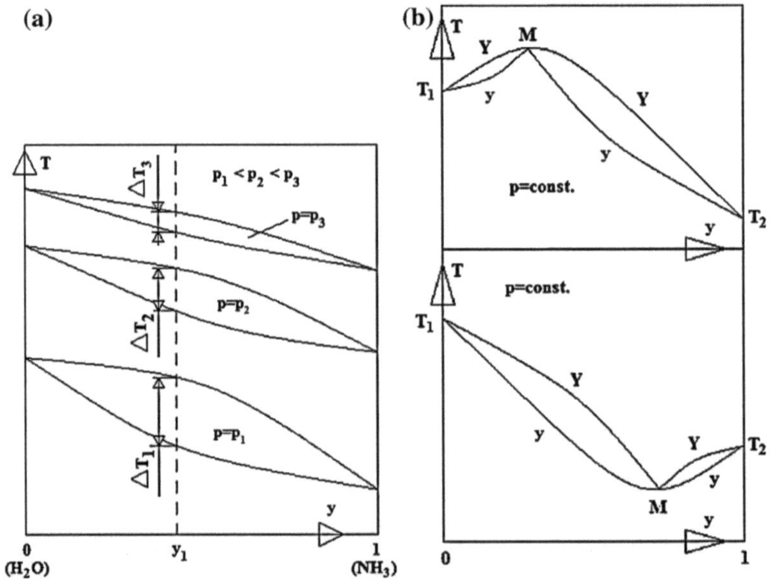

Fig. 1.16 Schematic of evaporation lenses at different pressures (**a**). Azeotropic binary mixtures (**b**)

increase, up to T_5, brings the system in equilibrium state 5, which belongs to the single-phase superheated vapor domain and similarly to state 1, is three-variant, $5(p_{atm}, T_5, y_1)$. Summarizing, in the domains situated inside and outside of the evaporation lens, the binary mixture is bivariant and three-variant, respectively. The temperature increase from T_2 to T_4, obliges the liquid and vapor phases of the system to cover the portions found in mutual equilibrium, $2l$–$4l$ of the evaporation curve with concentrations $y_{2l} > y_{3l} > y_{4l}$, and $2v$–$4v$ of the condensation curve with concentrations $Y_{2v} > Y_{3v} > Y_{4v} = y_{2l}$, respectively.

The condensation process takes place by cooling the vapor, humid vapor and liquid phases within the interval T_5–T_1 and can be followed covering the evaporation process in the reverse direction.

Repeating the experiences to higher pressures, it is found that the evaporation lens becomes smaller, Fig. 1.16a. Indeed, it can be remarked that the higher the pressure, $p_1 < p_2 < p_3$, the smaller the temperature interval of the mixture vaporization and condensation, $\Delta T_1 > \Delta T_2 > \Delta T_3$.

The diagram of Fig. 1.15 is typical for a binary working combination capable to form stable mixtures to any concentration y and have uninterrupted evaporation lenses comprising the whole concentration range, $y \in [0, 1]$. This type of mixture is called zeotropic. The mixture at hand, NH_3–H_2O, belongs to this category. Besides the zeotropic, there are also other categories of mixtures, e.g. azeotropic, or those presenting mixture fissure, etc. (Stamatescu 1972). Figure 1.16b, shows in the T–y diagram two plots of azeotropic binary mixtures where these behave in point M like a monocomponent system which evaporation and condensation of are isobar and isotherm, therefore take place without a temperature glide (e.g. ethyl spirits–chlorophorm, upper plot, and water–ethyl spirits or water–sulphuric acid, bottom plot).

1.4.6 Dissolution (Mixing) Heat of Binary Mixtures

Unlike ideal solutions, during the real solutions mixing irreversible or quasi-irreversible physical chemical processes are produced, accompanied by heat exchange with the ambient and by other volume (density) and state parameters changes. These phenomena are so much more important for the coabsorbent technology, where the mixing processes are used more than any other absorption plant did it so far.

Suppose the two components of a real solution had the same temperature prior mixing. After their dissolution it is found that their temperature is changing significantly more often than not. This change depends on the components nature, their initial temperature and on the final concentration of the solution. The gases which are not reacting chemically and have the same temperature prior mixing, it keep it after mixing as well. The dissolution of gases in liquid solutions are accompanied with small thermal adjacent phenomena, too. It results that only the liquid–liquid mixing is accompanied by a significant temperature change. It is possible that the mixture temperature increases, or on the contrary, to decrease as compared to the initial temperature of the components. In order to bring the mixture temperature to its initial value of the components, it is therefore necessary to introduce or to extract heat from the system. This heat is named total (integral) heat of dissolution (mixing). Bearing in mind that the absorption processes are isobar, the dissolution heat is established by its dissolution enthalpy.

It is considered a mixture consisting of two components of quantities m_1 and m_2, found to the same temperature T and having the specific enthalpies h_1 and h_2, Fig. 1.17. The enthalpy prior mixing is:

$$H_i = m_1 h_1 + m_2 h_2 \tag{1.232}$$

The enthalpy after mixing, found to the same temperature T, is:

$$H = (m_1 + m_2)h \tag{1.233}$$

It is noted:

$$\Delta H = H - H_i \tag{1.234}$$

The dissolution can occur with heat rejection (exothermal dissolution), or with heat absorption (endothermal dissolution). In case of the ammonia dissolution in the ammonia–water mixtures, or of water dissolution in the water–lithium bromide mixtures, the dissolution is exothermal, and a quantity of heat equal to Q_{mix} must be evacuated from the system, which according to the sign convention is negative.

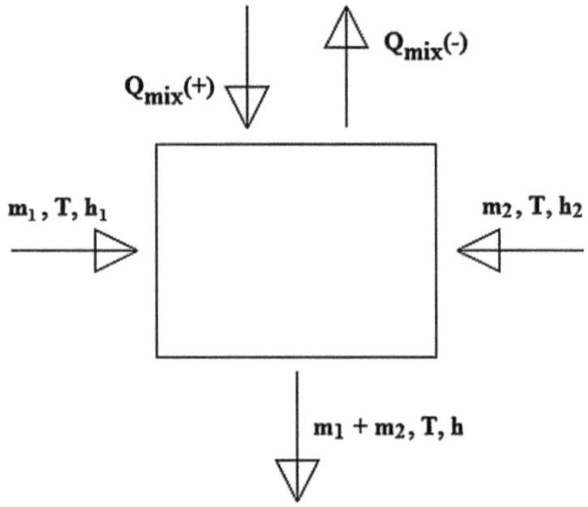

Fig. 1.17 Mixing heat

On the contrary, in case of the endothermal dissolution it is needed to introduce in the system the heat quantity Q_{mix}, which is positive. The energy balance, Fig. 1.17, is written:

$$m_1h_1 + m_2h_2 + Q_{mix} = (m_1 + m_2)h \qquad (1.235)$$

Writing Eq. (1.235) for the ammonia–water mixtures, where m_1 is the solute quantity (ammonia) and m_2 is the solvate quantity (water or ammonia–water mixtures), and dividing it by $m_1 + m_2$, it results:

$$y_1m_1 + (1 - y_1)m_2 + q_{mix} = h \qquad (1.236)$$

where

$$q_{mix} = \frac{Q_{mix}}{m_1 + m_2} \qquad (1.237)$$

Equation (1.236) is plotted in Fig. 1.18. The mixing heat of ammonia in water is negative for any concentration $y \in (0, 1)$, Fig 1.19a. However, there are solutions with both negative and positive mixing heat as well, as it is the case of that formed by the dissolution of the ethylic alcohol in water, Fig. 1.19b.

Equation (1.236) is valid for all binary mixing processes, either the components are both liquid prior to be mixed, or just one component is liquid, while the other is a gas which is absorbed/desorbed in/from the first component. The two cases mentioned above are named shorter *liquid–liquid* and *liquid–vapor*, respectively. In both cases, when Eq. (1.236) applies, it is considered that the mixing heat q_{mix} is evacuated or introduced in the system during components interaction, depending whether this is exothermal, or endothermal, respectively. According to Lorenz (Niebergall 1959), q_{mix} can be approximated by $q_{mix} \cong r + l$, in case of the liquid–vapor interactions, and by $q_{mix} \cong l$, in case of the liquid–liquid interactions. In these equations, r holds for the liquid–gas phase change heat (see Eq. (1.213)) and l is the liquid–liquid mixing heat. Let us apply next the dissolution heat topic outlined so far to a coabsorbent cycle. In a coabsorbent cycle both types of processes mentioned

Fig. 1.18 Concentration mixing heat dependence (Radcenco et al. 1983)

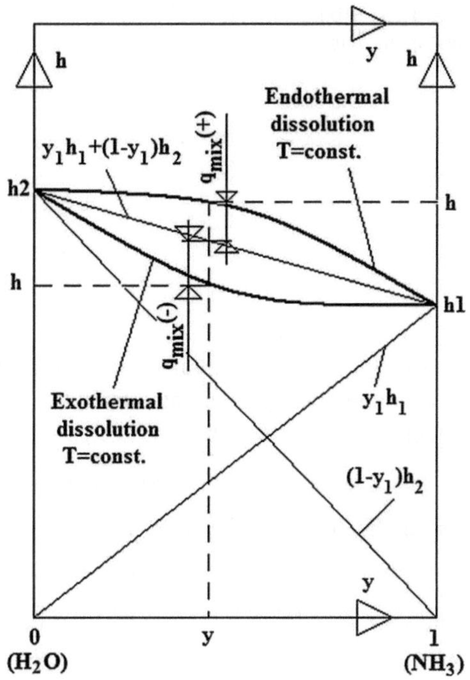

Fig. 1.19 Mixing heat of ammonia in water (**a**) and of water in ethylic spirit (**b**)

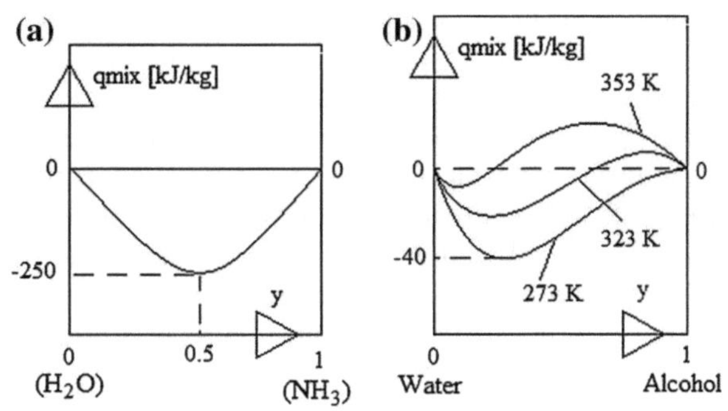

above are taking place simultaneously. Concerning the liquid–vapor interactions, Eq. (1.236) applies as it is, because generation (desorption) processes have generation heat supply from the heat source, while absorption (resorption) processes evacuate heat by the sink source. However, in case of the second type of interactions, the coabsorbent operation uses liquid–liquid mixing processes which take place in special devices, called mixers, and where the mixing heat is no more evacuated or supplied. The schematic of such a mixing process is depicted in Fig. 1.20.

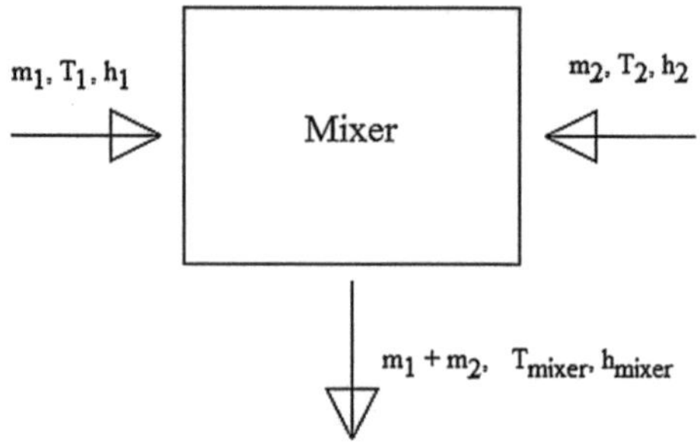

Fig. 1.20 Mixing process taking place in a mixer

The liquid mixing components shown in Fig. 1.20 have the inlet parameters, mass, temperature and specific enthalpy noted by (m_1, T_1, h_1) and (m_2, T_2, h_2). The mixer energy balance results in:

$$m_1 h_1 + m_2 h_2 = (m_1 + m_2) h_{mixer} \tag{1.238}$$

Dividing Eq. (1.238) by $m_1 + m_2$, it is obtained:

$$y_1 h_1 + (1 - y_1) h_2 = h_{mixer} \tag{1.239}$$

The enthalpy h_{mixer} in Eq. (1.239) includes this time the mixing heat as well, which is as mentioned $q_{mix} \cong l < r + l$. The values of this heat are plotted in Fig. 1.18 by the straight line connecting the enthalpies h_1 and h_2. Concerning the T_{mixer} temperature, in case of NH_3–H_2O and H_2O–LiBr working combinations this is higher than the temperatures the components had prior mixing, $T_{mixer} > T_1$ and $T_{mixer} > T_2$. This relative temperature increase is beneficial in a coabsorbent cycle operation, as in fact is a preheating of the absorbent supplying the cycle generation processes.

Introducing a very small quantity of solute in a very high quantity of solvate, an elementary (differential) mixing heat is generated. Adding a very small mass quantity to the total mass of components is practically considered a mixing process unfolding with constant mass. Let us note the solute and solvate quantities by dm_1 and m_2, respectively and the differential heat by q_{d1}. The differential heat q_{d1} is in fact the dissolution heat partial derivative with respect to the first component mass, when the second component mass is constant, that is:

$$q_{d1} = \left[\frac{\partial(\Delta H)}{\partial m_1}\right]_{m_2} \tag{1.240}$$

In Eq. (1.240), $\Delta H = H - H_i$, Eq. (1.234), or:

$$\Delta H = (m_1 + m_2) q_{mix} \tag{1.241}$$

With Eq. (1.241), Eq. (1.240) becomes:

$$q_{d1} = q_{mix} + (m_1 + m_2)\left[\frac{\partial q_{mix}}{\partial m_1}\right]_{m_2} \tag{1.242}$$

Applying the chain rule to the partial derivative of Eq. (1.242), it results:

$$\left[\frac{\partial q_{mix}}{\partial m_1}\right]_{m_2} = \left[\frac{\partial q_{mix}}{\partial y}\right]_{m_2}\left[\frac{\partial y}{\partial m_1}\right]_{m_2} \tag{1.243}$$

The mass fraction y is expressed as $y = \frac{m_1}{m_1+m_2}$ and performing its derivative with respect to m_1, it results:

$$\left[\frac{\partial y}{\partial m_1}\right]_{m_2} = \frac{m_2}{(m_1+m_2)^2} \tag{1.244}$$

Introducing Eq. (1.244) in Eq. (1.243), it is obtained:

$$q_{d1} = q_{mix} + (1-y)\left[\frac{\partial q_{mix}}{\partial y}\right]_{m_2} \tag{1.245}$$

Similarly, adding the quantity dm_2, it is obtained the differential heat q_{d2}:

$$q_{d2} = q_{mix} + (m_1+m_2)\left[\frac{\partial q_{mix}}{\partial m_2}\right]_{m_1} \tag{1.246}$$

or,

$$q_{d2} = q_{mix} + (m_1+m_2)\left[\frac{\partial q_{mix}}{\partial(1-y)}\right]_{m_1}\left[\frac{\partial(1-y)}{\partial m_2}\right]_{m_1} \tag{1.247}$$

Bearing in mind that $y = \frac{m_1}{m_1+m_2}$, the last product of brackets in Eq. (1.247) equals:

$$\left[\frac{\partial q_{mix}}{\partial(1-y)}\right]_{m_1}\left[\frac{\partial(1-y)}{\partial m_2}\right]_{m_1} = -y\left[\frac{\partial q_{mix}}{\partial y}\right]_{m_2} \tag{1.248}$$

and with the last result, Eq. (1.246) becomes:

$$q_{d2} = q_{mix} - y\left[\frac{\partial q_{mix}}{\partial y}\right]_{m_2} \tag{1.249}$$

The differential heats of mixing q_{d1} and q_{d2}, given by Eqs. (1.245) and (1.249), can be determined graphically, provided that the function $q_{mix} = q_{mix}(T, y)$ be known, as Fig. 1.21 shows. Further, the functions q_{d1} and q_{d2}, given by Eqs. (1.243) and (1.246), are derived partially with respect to m_2 for $m_1 = const.$ and with respect to m_1 for $m_2 = const.$, respectively. Remarking that $\left[\frac{\partial^2 q_{mix}}{\partial m_1 \partial m_2}\right]_{m_1,m_2} = 0$ and subtracting $\left[\frac{\partial q_{d2}}{\partial m_1}\right]_{m_2}$ from $\left[\frac{\partial q_{d1}}{\partial m_2}\right]_{m_1}$, the following result holds true:

$$\left[\frac{\partial q_{d1}}{\partial m_2}\right]_{m_1} - \left[\frac{\partial q_{d2}}{\partial m_1}\right]_{m_2} = 0 \tag{1.250}$$

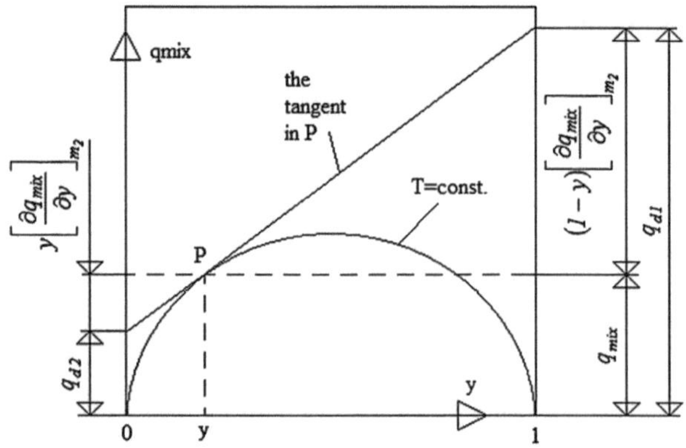

Fig. 1.21 Differential heat of mixing (Radcenco et al. 1983)

Equation (1.250) can be rewritten using the chain rule of derivation. Proceeding similarly as in Eqs. (1.243) and (1.248), it results:

$$\left[\frac{\partial q_{d1}}{\partial(1-y)}\right]_{m_1}\left[\frac{\partial(1-y)}{\partial m_2}\right]_{m_1}-\left[\frac{\partial q_{d_2}}{\partial y}\right]_{m_2}\left[\frac{\partial y}{\partial m_1}\right]_{m_2}=0 \tag{1.251}$$

and further:

$$(1-y)\left[\frac{\partial q_{d_1}}{\partial y}\right]_{m_2}+y\left[\frac{\partial q_{d_2}}{\partial y}\right]_{m_1}=0 \tag{1.252}$$

Equations (1.245) and (1.249) of q_{d1} and q_{d2} are further partially derived with respect to y for $m_2 = const.$ and with respect to $(1-y)$ for $m_1 = const.$, respectively. Taking into account that $\left[\frac{\partial^2 q_{mix}}{\partial y^2}\right]_{m_1,m_2}=\left[\frac{\partial^2 q_{mix}}{\partial y\partial(1-y)}\right]_{m_1,m_2}=0$, from the two derivated equations it results:

$$\left[\frac{\partial q_{d_1}}{\partial y}\right]_{m_2}+\left[\frac{\partial q_{d_2}}{\partial y}\right]_{m_1}=0 \tag{1.253}$$

and

$$\left[\frac{\partial q_{d_1}}{\partial y}\right]_{m_2}-\left[\frac{\partial q_{d_2}}{\partial y}\right]_{m_1}=-2\left[\left[\frac{\partial q_{mix}}{\partial y}\right]_{m_1}-\left[\frac{\partial q_{mix}}{\partial y}\right]_{m_2}\right] \tag{1.254}$$

Equations (1.252) and (1.253) are solved together for $\left[\frac{\partial q_{d_1}}{\partial y}\right]_{m_2}$ and $\left[\frac{\partial q_{d_2}}{\partial y}\right]_{m_1}$, obtaining:

$$\left[\frac{\partial q_{d_1}}{\partial y}\right]_{m_2}=\left[\frac{\partial q_{d_2}}{\partial y}\right]_{m_1}=0 \tag{1.255}$$

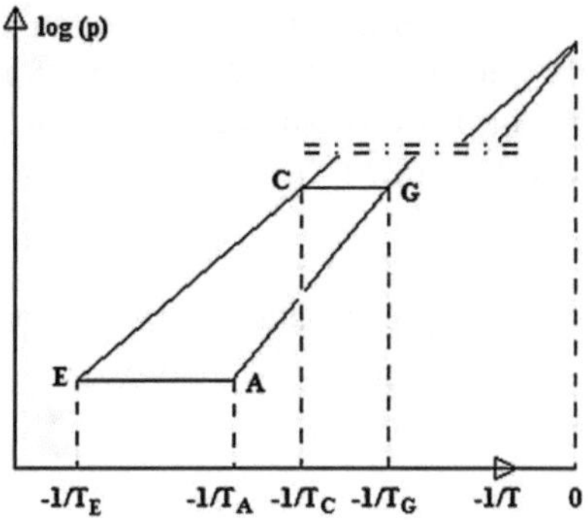

Fig. 1.22 Absorption cooling cycle plotted in the log (p) − 1/T chart (adapted from Ziegler 2005), showing the Trouton's rule application (G, C, A and E stand for cycle generator, condenser, absorber and evaporator, respectively, Fig. 1.11a)

wherefrom, introducing Eq. (1.255), in Eq. (1.254), the following result holds true:

$$\left[\frac{\partial q_{mix}}{\partial y}\right]_{m_1} = \left[\frac{\partial q_{mix}}{\partial y}\right]_{m_2} \tag{1.256}$$

The introduction of Eq. (1.256) in Eqs. (1.245) and (1.249) leads to the following final relationship, written for isothermal and isobar conditions and at equilibrium:

$$q_{mix} = yq_{d1} + (1 - y)q_{d2} \tag{1.257}$$

1.4.7 Absorption Cycle Charts

In Sect. 1.3.4, two useful charts in the absorption cycle design and calculus, namely log (p) − 1/T and enthalpy-concentration h–y, have been introduced. Here, it will be gone deeper in these charts elaboration and use.

1.4.7.1 log (p) − 1/T Chart

The choice of pressure p and absolute temperature T as thermodynamic parameters of an absorption chart is explained by the facility which this chart is used with in absorption cycle plot. Indeed, in a log (p) − 1/T chart of a refrigerant-absorbent working combination, a curve plotted for a constant concentration (isostere), $y = const.$, can be considered with enough accuracy a straight line. This property is described schematically in Fig. 1.22 for an absorption cooling cycle, by the empirical rule of Trouton (Ziegler 2005): "all vapor pressure lines are linear in the Raoult's plot (log (p) − 1/T), and converge for 1/T approaching zero".

A thermodynamic confirmation of Trouton's rule is commonly obtained integrating the Clapeyron–Clausius equation (1.215) for the liquid–vapor phase transformations taking place in refrigerant-absorbent working combinations use. In Eq. (1.215), the liquid specific volume v_l is 2–3 orders of magnitude smaller than the vapor specific volume, $v_2 = v$, $v_1 << v_2 = v$, so it can be neglected comparatively in calculus, $v_1 \cong 0$. Further on, for pressures limited to values of tens of bars, the vapor phase is satisfactorily considered ideal gas, so the Clapeyron equation (1.10) can be applied, $G = 1$ kg:

$$pv = RT \tag{1.258}$$

Taking into account the above remarks and notings and Eq. (1.258), after arranging the terms Eq. (1.215) writes:

$$\frac{dp}{p} = \frac{L}{R}\frac{dT}{T^2} \tag{1.259}$$

In Eq. (1.259), the pure component phase change heat r was replaced by the sum $r + l = L$, which is accounting for the phase change and mixing heats involved simultaneously in the binary refrigerant-liquid physical interactions of the absorption processes. Equation (1.259) is integrated obtaining:

$$\ln p = a - b\frac{1}{T} \tag{1.260}$$

In Eq. (1.260), a is an integration constant and $b = \frac{L}{R}$. Both a and b constants depend on the modeled particular working pair. Equation (1.260) shows that the pressure curve of a working pair plotted in a chart of log (p) and $-1/T$ coordinates is linear (see Trouton's rule). The constants a or/and b depend on the mixture concentration y parameter, which defines a family of curves in the chart at hand.

Ammonia–Water Working Pair

The first who applied Eq. (1.260) to the NH_3–H_2O working combination was H. Mollier. He plotted the curves family within a limited domain, $0.12 \leq y \leq 0.5$, $1 \leq p \leq 10$ ata and $-1/423.15 \leq -1/T \leq -1/293.15$ K, and worked with the relationship (1.261), (Planck 1953):

$$\log p = 7.973 - \frac{1220}{0.00466 * y + 0.656}\frac{1}{T} \tag{1.261}$$

where $[y] = \frac{\text{kg\_NH}_3}{\text{kg\_(NH}_3\text{+H}_2\text{O)}}$. Today, the limits of the y, p and T have been extended. A typical log (p) – $1/T$ chart of NH_3–H_2O is shown in Fig. 1.23. In order to facilitate its use the temperature in abscissa in currently expressed not in $-1/T$, with $[T^{-1}] = \text{K}^{-1}$, but in t, $[t] = °\text{C}$.

1.4.7.2 Dühring Chart

The behavior of certain working pairs can be described by a second empirical rule, commonly referred to in the absorption literature as the Dühring rule (McNeely 1979; Ziegler 2005): "if the boiling (bubble) point temperature of a solution is plotted against the boiling (dew) point temperature of its solvent at the

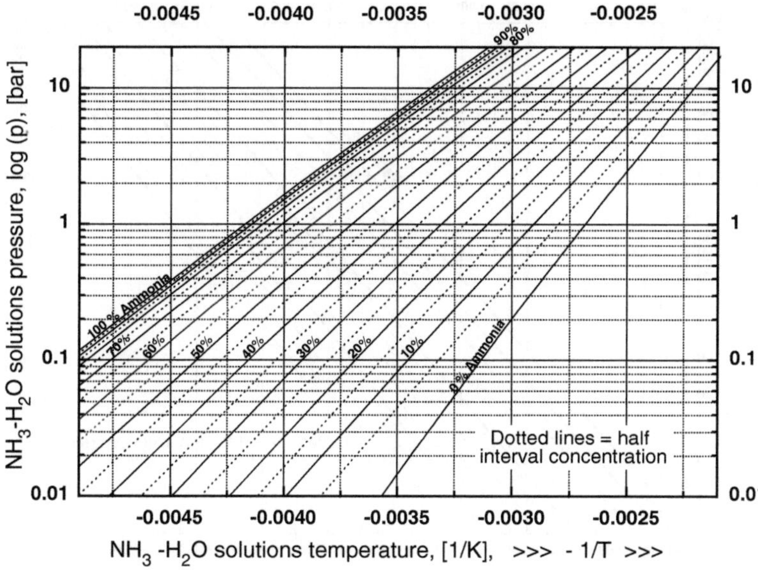

Fig. 1.23 A typical NH_3–H_2O log $(p) - 1/T$ chart

same pressure (see Fig. 1.15), all the resulting isosteres are straight lines and converge for T approaching zero". There is no thermodynamic relation that specifies the straight line, but if it is assumed, or if the data justifies the straight line from the range of pressures observed, the Dühring equation can be derived from the Clapeyron equation, as shown below (McNeely 1979). Indeed, Eq. (1.215) is written for the solvent and solution phase change in the dew points and bubble points temperatures, respectively, at any given concentration and pressure, as:

$$dT_{dew} = \left[\frac{dpT\left(v' - v\right)}{r}\right]_{dew} \tag{1.262}$$

and

$$dT_{bubble} = \left[\frac{dpT\left(v' - v\right)}{r + l}\right]_{bubble} \tag{1.263}$$

respectively. Equations (1.262) and (1.263) are written for the same pressure, $p_{dew} = p_{bubble}$, therefore $dp_{dew} = dp_{bubble}$. Moreover, $dT_{dew} = dt_{dew}$ and $dT_{bubble} = dt_{bubble}$. Taking into account the last results and dividing Eqs. (1.262) and (1.263) term by term, the linear form of the isosteres allows one to write the following result:

$$\frac{dt_{bubble}}{dt_{dew}} = \frac{\left[T\left(v' - v\right)\right]_{bubble} r_{dew}}{\left[T\left(v' - v\right)\right]_{dew} (r + l)_{bubble}} = A(y) = const. \tag{1.264}$$

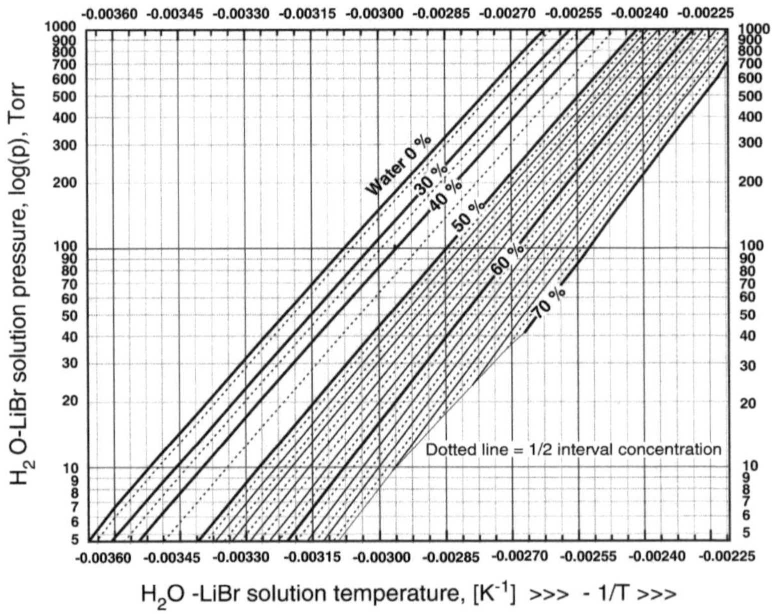

Fig. 1.24 H_2O–LiBr chart in the log (*p*), $-$ 1/*T* coordinates (Patek 2006)

The integration of Eq. (1.264) along a certain isostere ($y = const.$), leads to the mathematical expression of the Dühring rule (McNeely 1979):

$$t_{bubble} = A(y)t_{dew} + B(y) \tag{1.265}$$

In Eq. (1.265), $A(y)$ and $B(y)$ are constants for a given isostere and $B(y)$ is an integration constant. The Dühring rule is graphically applied to an absorption cooling cycle in Fig. 1.25. The vapor–liquid interaction of H_2O–LiBr working pair obeys besides the Trouton's rule (see Fig. 1.24), the Dühring rule as well. The chart of this working pair, plotted in the T_{bubble}, T_{dew} coordinates, was constructed using the equilibrium data calculated according to Patek and Klomfar (2006a) model. The Dühring H_2O–LiBr chart given in Fig. 1.26 has the largest use in the air conditioning applications. It is best known with two ordinates, showing besides T_{dew}, the log (*p*) coordinate, as well.

1.4.7.3 Enthalpy-Concentration Chart

Until recently, the graphical-analytical charts were the main tools used in the absorption cooling and heating applications calculus. The enthalpy-concentration *h*–*y* chart belongs to this category. However, the computer science and technology progress, recorded lately, enabled the old charts be replaced by computer codes and models, capable to simulate the working pairs physical and thermal data and perform an absorption cycle calculus automatically. For this reason, in this book

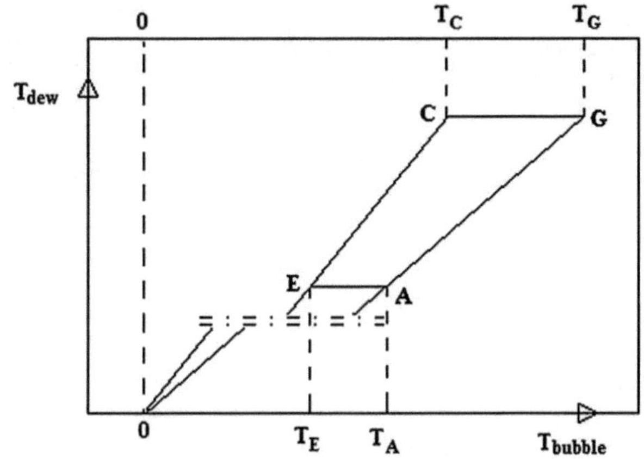

Fig. 1.25 Absorption cooling cycle plotted in the T_{dew}, T_{bubble} coordinates (after Ziegler 2005), showing the Dühring's rule application (for G, C, A and E notations see Fig. 1.22)

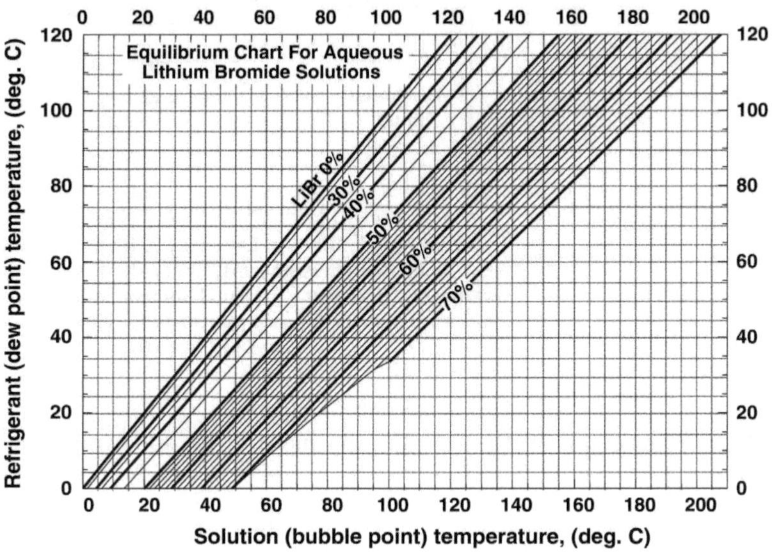

Fig. 1.26 H_2O–LiBr chart in the T_{dew}, T_{bubble} coordinates (McNeely 1979; Patek and Klomfar 2006a, b)

the *h–y* chart is not used extensively, but only once, in Chap. 2, and not for quantitative computation purposes, but for qualitative assessments of the absorption cycle design solely. Instead, the models of Ziegler and Trepp (1984) and Patek and Klomfar (2006a, b) for the ammonia–water and water–lithium bromide, respectively, have been used in order to calculate all cooling and heating cycles of this book. Consequently, the *h–y* chart is presented next schematically, only, for the ammonia–water combination. This chart was set up experimentally, by Merkel and

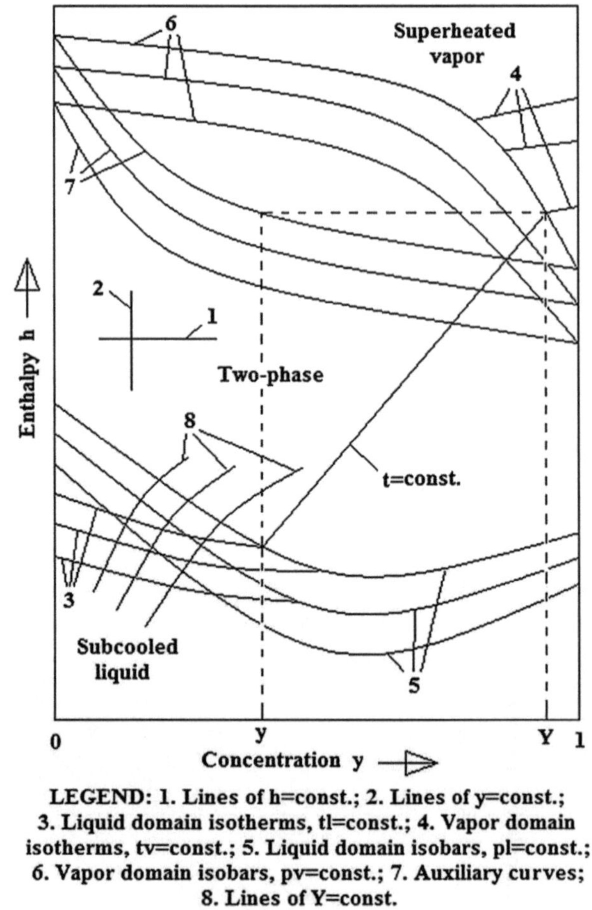

Fig. 1.27 Shematic of *h–y* enthalpy-concentration chart of the ammonia–water working pair

Bosnjakovič (Niebergall 1959). The diagram includes the following families of curves, Fig. 1.27:

- *Curves of constant enthalpy, 1, h = const.*, which are the horizontal straight lines;
- *Curves of constant concentration (isosters), 2, y = const.*, which are the vertical straight lines;
- *Liquid domain isothermals, 3,* $t_l = const.$, for boiling temperatures between −70 and 200 °C. The shape of these curves is typical for the ammonia–water which desorption and absorption of are endothermal and exothermal processes, respectively;
- *Vapor domain isothermals, 4,* $t_v = const.$; these are straight lines within the same boiling temperature limits, mentioned above; the mixture heat between dry ammonia and water vapor is negligible, therefore their mixing proces is practically ideal; the isotherms 4 are not shown in the diagram;

- *Liquid domain isobars, 5, $p_l = const.$* are represented within the pressure limits of 0.002 and 2 MPa, which the cooling and heating absorption plants are working usually with; these lines represent the bubble point temperatures for each isobar shown (see Fig. 1.15);
- *Vapor domain isobars, 6, $p_v = const.$*, represented within the same pressure limits; the isobars 5 and 6 of an *h–y* diagram can be drawn with the help of a *t–y* diagram, Fig. 1.15, as shown in Radcenco et al. (1983);
- *Auxiliary curves, 7,* are drawn for each isobar and serve for determining the vapor state in equilibrium with the respective liquid, in order to help representing the isothermals in the two-phase (wet vapor) domain;
- *Curves of constant vapor concentration, 8, $Y = const.$*, are drawn in the liquid domain and enable to find rapidly the vapor concentration in equilibrium with the liquid of origin.

Some completing remarks follow in case of the ammonia–water: (a) the liquid isothermal 3 is modifying to a very small extent with pressure variation, therefore practically a single isothermals net results for the diagram, whatever the pressure is; (b) in superheated vapor domain, the specific heat depends on both temperature and pressure, therefore each isobar normally generates an isothermal family of curves; this would complicate the diagram and for this reason the superheated vapor lines have not been represented, as mentioned, but, if intended to be done, they can be easily drawn, because are linear curves; (c) the diagram origin was settled as follows: $h = 0$ for water ($y = 0$; $t = 0$ °C); in this case, the pure liquid ammonia enthalpy is $h = 349.5$ kJ/kg at $t = 0$ °C. Concluding, computationally speaking, the use of the *h–y* diagram in absorption technology will decrease probably much in the future. However, the interest for this diagram will be kept, because of the rich qualitative information this single chart gathers and offers to people involved in absorption cycle study, research and design (e.g. in graphical representation of cycle main devices processes, heats and vapor rectification).

1.4.8 Working Fluid-Absorbent Mixtures Model

The chemical and physical interactions and the thermodynamical relationships which govern them have been a permanent concern of physicists and chemists since long time ago. A major development of the chemical thermodynamics dates back in the 19th century, due to Gibbs (1876) and Helmholz (1902), for to mention just ones of the most remarkable scientists involved in this domain. More recent works of chemical thermodynamics and liquids mixtures have been elaborated by Lewis and Randall (1961) and Rowlinson and Swinton (1982). Today, the global warming and the limited availability of the fossil fuel and other primary energy sources force the mankind to find rapidly, more effective solutions in clean energy production, generally, and in the cooling and heating production, particularly. New energy sources (e.g. renewables, low grade recovery) are put to work, and many technologies, matching best with these sources, are investigated and applied. To this extent, the absorption technology is credited with considerable hope. Consequently, in the recent past years and today, we assist to an increased experimental and thermodynamical modelling activity of the existing and new absorption working pairs. Amongst the authors of models and the modeled working pairs, the followings can be mentioned: Ziegler and Trepp (1984), Bourseau and Bugarel (1986), Xu and Goswami (1997), Patek and Klomfar (1995), for NH_3–H_2O, Infante Ferreira (1984), Bourseau and Bugarel (1986) for NH_3–NaSCN, Radermacher and Alefeld (1982), Balamuru et al. (2000) for ternary NH_3–H_2O–salt, McNeely (1979), Patek and Klomfar (2006a) for LiBr–H_2O, Patek and Klomfar (2006a, b) for solid phase equilibrium of LiBr–H_2O, and LiCl–H_2O, Patek and Klomfar (2007) for LiCl–H_2O, Infante Ferreira (1984) for NH_3–$LiNO_3$.

In this section, the computational model of the ammonia–water working combination, due to Schultz (1971) and Ziegler and Trepp (1984), is analyzed. We have chosen this working combination given its importance in refrigeration and heat production. Also, the Schultz-Ziegler-Trepp (SZT) model was selected because it is well known and representative for the way it combines the experimental data and the thermodynamical relationships in order to create an useful and satisfactory accurate tool of investigation of the NH_3–H_2O applications.

1.4.8.1 Equation of State

Ziegler extended the Schultz model, limited to 25 bar, to a broader range, $230\ \text{K} < T < 500\ \text{K}$ and $0.2\ \text{bar} < p < 50\ \text{bar}$, in order to meet not only the refrigeration needs with this mixture, but those of heating as well. The NH_3–H_2O system has two components, NH_3 and H_2O, therefore, at equilibrium, according to Gibbs rule of phases, Sect. 1.3.4, is bivariant. Its equation of state bases on the specific Gibbs free enthalpy assessment, φ, $\varphi = \frac{\Phi}{G}$ [$\Phi(T, p) = H - TS$, Table 1.1, Eq. (1.159)], therefore the system independent state parameters are T and p. The specific Gibbs free enthalpy is at the same time the chemical potential of the system, $\varphi = \left(\frac{\partial \Phi}{\partial G}\right)_{T,p}$, according to Eq. (1.163). Schultz considered two separate equations for the liquid and the gaseous phases, linked by the conditions of phase equilibrium [see Eqs. (1.202)–(1.205)]:

$$\begin{aligned} T^l &= T^g \\ p^l &= p^g \end{aligned} \tag{1.266}$$

$$\begin{aligned} \varphi^l_{NH_3} &= \varphi^g_{NH_3} \\ \varphi^l_{H_2O} &= \varphi^g_{H_2O} \end{aligned} \tag{1.267}$$

These have the following form:
Liquid phase:

$$\varphi^l(T,p,y) = (1-y)\varphi^l_{H_2O}(T,p) + y\varphi^l_{NH_3}(T,p) + RT\left[(1-y)\ln(1-y) + y\ln y\right] + \varphi_E(T,p,y) \tag{1.268}$$

Gas phase:

$$\varphi^g(T,p,Y) = (1-Y)\varphi^g_{H_2O}(T,p) + Y\varphi^g_{NH_3}(T,p) + RT[(1-Y)\ln(1-Y) + Y\ln Y] \tag{1.269}$$

The Gibbs free enthalpy of a phase is the sum of the contributions of the components free enthalpy, represented by the first two terms of each equation, the ideal mixing enthalpy, given by the third term of each equation, and the excess enthalpy. The ideal mixing enthalpy is bound, because is negative, that is it diminishes the free enthalpy of the mixture, as compared to the sum of the free enthalpy of the separate components (see Gibbs paradox, Murgulescu and Vîlcu 1982). The gas phase is assumed to be an ideal mixture of real components. Equations (1.268) and (1.269) characterize the free enthalpy of a system in nonequilibrium. At equilibrium, these equations must be completed with Eqs. (1.266) and (1.267).

As stated in Sect. 1.3.1, the knowledge of a characteristic function, here Φ and φ, enables to derive other thermodynamic functions, such as: *Molar volume*, Eq. (1.169):

$$v^{l(g)} = \left(\frac{\partial\varphi^{l(g)}(T,p,y)}{\partial p}\right)_{T,y} \tag{1.270}$$

Molar entropy, Eq. (1.170):

$$s^{l(g)} = -\left(\frac{\partial\varphi^{l(g)}(T,p,y)}{\partial T}\right)_{p,y} \tag{1.271}$$

Molar enthalpy:

$$h^{l(g)} = -T^2\left(\frac{\partial\left(\frac{\varphi^{l(g)}(T,p,y)}{T}\right)}{\partial T}\right)_{p,y} \tag{1.272}$$

Chemical potential

A key of solving this problem is found for example in Sect. 1.4.6. Indeed, the dissolution heat of the binary mixture analyzed there, Eq. (1.234), is exchanged at constant temperature and pressure, therefore is the Gibbs free enthalpy of our system at hand, because only the free enthalpy can be exchanged. We consider first an exothermal vapor absorption process in the NH_3–H_2O liquid mixture. In this case, the final result will be to obtain the chemical potentials, $\varphi^l_{NH_3}$ and $\varphi^l_{H_2O}$ of the liquid phase. Taking the above remarks into account, from Eqs. (1.234) and (1.241) we can write for the liquid phase:

$$\Delta H = H - H_i = \left(m_{NH_3} + m_{H_2O}\right)q_{mix} = \Phi = \left(m_{NH_3} + m_{H_2O}\right)\varphi \tag{1.273}$$

From Eq. (1.273) it results:

$$\varphi = q_{mix} \tag{1.274}$$

that is the liquid phase specific Gibbs free enthalpy equals the specific mixing (dissolution) heat. Further on, according to our approach and bearing in mind Eqs. (1.172) and (1.274), the differential heat, given by Eq. (1.240), leads to the liquid phase chemical potentials:

$$q_{d,NH_3(H_2O)} = \left[\frac{\partial(\Delta H)}{\partial m_{NH_3(H_2O)}}\right]_{m_{H_2O(NH_3)}} = \left[\frac{\partial \Phi}{\partial m_{NH_3(H_2O)}}\right]_{m_{H_2O(NH_3)}} = \varphi_{NH_3(H_2O)} \tag{1.275}$$

Performing the differential calculus in Eq. (1.275), the final resulting Eqs. (1.245) and (1.249) will become our final issues for the liquid chemical potentials we are looking for:

$$\varphi^l_{NH_3} = \varphi^l(T,p,y) + (1-y)\left[\frac{\partial \varphi^l(T,p,y)}{\partial y}\right]_{T,p,(1-y)} \tag{1.276}$$

and

$$\varphi^l_{H_2O} = \varphi^l(T,p,y) - y\left[\frac{\partial \varphi^l(T,p,y)}{\partial y}\right]_{T,p,(1-y)} \tag{1.277}$$

In case of the gas phase chemical potentials, a desorption process is considered, which generates vapor. Proceeding similarly as for the absorption process, the chemical potentials of the gas phase, $\varphi^g_{NH_3}$ and $\varphi^g_{H_2O}$ will result as well.

Equations (1.270)–(1.272) give the volume, entropy and enthalpy of the mixtures as explicit functions of temperature, pressure and composition. The phase equilibrium is analyzed next.

Phase equilibrium data calculation

According to our approach, prior to calculate the phase equilibrium we need to prove first the following (Staicovici 2013):

Lemma 1 *In a physical binary bivariant (T, p) refrigerant-absorbent gas–liquid system at equilibrium, the excess enthalpy equals the ideal mixing enthalpy.*

Proof The equalities (1.267), characterizing the conditions of phase equilibrium, can be written in an explicit form as (see Eqs. (1.276) and (1.277) for liquid and similar for gas phase):

$$\varphi^l(T,p,y) + (1-y)\left[\frac{\partial \varphi^l(T,p,y)}{\partial y}\right]_{T,p,(1-y)}$$
$$= \varphi^g(T,p,y) + (1-Y)\left[\frac{\partial \varphi^g(T,p,y)}{\partial Y}\right]_{T,p,(1-Y)} \tag{1.278}$$

and

$$\varphi^l(T,p,y) - y\left[\frac{\partial \varphi^l(T,p,y)}{\partial y}\right]_{T,p,(1-y)} = \varphi^g(T,p,y) - Y\left[\frac{\partial \varphi^g(T,p,y)}{\partial Y}\right]_{T,p,(1-Y)} \tag{1.279}$$

respectively. Subtracting term by term the last two equations, the following result holds true:

$$\left[\frac{\partial \varphi^l(T,p,y)}{\partial y}\right]_{T,p,(1-y)} = \left[\frac{\partial \varphi^g(T,p,y)}{\partial Y}\right]_{T,p,(1-Y)} \tag{1.280}$$

Further on Eq. (1.257) is written for the liquid and gas phases. With the help of Eqs. (1.274) and (1.275) this becomes:

$$\varphi^l(T,p,y) = (1-y)\varphi^l_{H_2O}(T,p) + y\varphi^l_{NH_3}(T,p) \tag{1.281}$$

and

$$\varphi^g(T,p,Y) = (1-Y)\varphi^g_{H_2O}(T,p) + Y\varphi^g_{NH_3}(T,p) \tag{1.282}$$

respectively. Subtracting the last two equations term by term and taking into account Eq. (1.267), it is obtained:

$$\varphi^g(T,p,Y) - \varphi^l(T,p,y) = (Y-y)\left[\varphi^g_{NH_3}(T,p) - \varphi^g_{H_2O}(T,p)\right] \tag{1.283}$$

In Eq. (1.283), the last bracket of the right member can be calculated from Eqs. (1.276) and (1.277) and the left hand member can be expressed with the help of Eqs. (1.268) and (1.269). Further on, in the resulting Eq. (1.283), Eqs. (1.267) and (1.280) are taken into account, wherefrom, the Lemma 1 is proved:

$$RT\ln\frac{(1-Y)^{1-Y}Y^Y}{(1-y)^{1-y}y^y} = \varphi_E(T,p,y) \quad \text{q.e.d.} \tag{1.284}$$

The Lemma 1 holds true also for systems having more than two components and for a gas phase with non-ideal mixture. The phase equilibrium can be calculated proceeding according to the next (Staicovici 2013):

Lemma 2 *The phase equilibrium data* (*T*, *p*, *y*, *Y*) *of a physical binary bivariant* (*T*, *p*) *refrigerant-absorbent gas-liquid system can be found solving a system consisting of equation resulted from Lemma 1 and of this equation derivation with respect to the refrigerant concentrations* (*y*, *Y*) *of each phase.*

Proof Equation (1.280) is applied to Eqs. (1.268) and (1.269). Taking into account Eq. (1.267), it results:

$$RT\frac{d}{dy}\left[(1-y)\ln(1-y)+y\ln y\right]+\left[\frac{\partial\varphi_E(T,p,y)}{\partial y}\right]_{T,p,(1-y)}$$
$$=RT\frac{d}{dY}[(1-Y)\ln(1-Y)+Y\ln Y] \quad (1.285)$$

or

$$\ln\frac{y}{1-y}+\frac{\left[\frac{\partial\varphi_E(T,p,y)}{\partial y}\right]_{T,p,(1-y)}}{RT}=\ln\frac{Y}{1-Y} \quad (1.286)$$

Equations (1.285) and (1.286) are nothing else but Eq. (1.284) coming of the Lemma 1, which have been derived with respect to refrigerant concentrations (y, Y) of each phase. In this way, the first part of the Lemma 2 has been proved. In order to prove the whole Lemma 2, Eqs. (1.284)–(1.286) are rewritten in a form which emphasizes a system of two equations with two unknowns (y, Y), provided the independent state parameters (T, p) are known:

$$(1-Y)^{1-Y}Y^Y=(1-y)^{1-y}y^y\exp\left(\frac{\varphi_E(T,p,y)}{RT}\right) \quad (1.287)$$

and

$$Y(T,p,y)=\frac{y\exp\left(\frac{1}{RT}\left[\frac{\partial\varphi_E(T,p,y)}{\partial y}\right]_{T,p,(1-y)}\right)}{(1-y)+y\exp\left(\frac{1}{RT}\left[\frac{\partial\varphi_E(T,p,y)}{\partial y}\right]_{T,p,(1-y)}\right)} \quad (1.288)$$

respectively. The system of Eqs. (1.287) and (1.288) is transcendental with respect to (y, Y) and can be solved by the trial-and-error method (or using a simple iterative method, Dodescu 1979), in such a way that for an initial guess value y Eq. (1.288) gives an Y value which must be confirmed by Eq. (1.286) with a required accuracy. If this is not fulfilled, a new value for y is used and the process is continued until the required accuracy in determining the Y value is reached. In this way the second Lemma 2 part is proved and the whole Lemma 2 as well, q.e.d. We remember that the phase equilibrium points (T, p, y, Y) found with the help of system of Eqs. (1.287) and (1.288) are the *ideal points* of the physical binary bivariant (T, p) refrigerant-absorbent gas-liquid system at hand, according to the TPT, Sect. 1.3.3.

The free excess enthalpy of the liquid phase, $\varphi_E(T, p, y)$, is expressed by a relation proposed by Redlich and Kister, and has the following form, (Ziegler and Trepp 1984):

$$\varphi_E(T,p,y)=y(1-y)\left[f_1(T,p)+f_2(T,p)(2y-1)+f_3(T,p)(2y-1)^2+\ldots\right] \quad (1.289)$$

In Eq. (1.289), the progression is restricted to three terms. The functions f_1, …, f_3, are given by Ziegler and Trepp (1984):

$$f_1(T,p)=e_1+e_2p+(e_3+e_4p)T+e_5T^{-1}+e_6T^{-2} \quad (1.290)$$

$$f_2(T,p) = e_7 + e_8 p + (e_9 + e_{10} p)T + e_{11} T^{-1} + e_{12} T^{-2} \tag{1.291}$$

$$f_3(T,p) = e_{13} + e_{14} p + e_{15} T^{-1} + e_{16} T^{-2} \tag{1.292}$$

In Eqs. (1.290)–(1.292), e_1, …, e_{16} are coefficients given in Ziegler and Trepp (1984). The system of Eqs. (1.287) and (1.288) is verified by the pure components concentrations, ($y = 0$; $Y = 0$) and ($y = 1$; $Y = 1$), as well. And last, but not the least, the Lemmas 1 and 2 can be used to find the bubble and the dew point curves for an absorption/generation process, taking place with constant pressure p. Indeed, let us consider the evaporation lens depicted in Fig. 1.15. The problem consists in determining the values of the dew point parameters (T_4, y_{4l}, Y_{4v}) and of the bubble point parameters (T_2, y_{2l}, Y_{2v}), in such a way that $Y_{4v} = y_{2l}$. In this case, in a first step the pair of unknowns (T_2, Y_{2v}) is determined with Eqs. (1.287) and (1.288), while (p, y_{2l}) are known. In a second step, the pair of unknowns (T_{4l}, y_{4l}) are found, while p and $Y_{4v} = y_{2l}$ are known values. The problem has a unique solution.

1.4.8.2 Equation of State of Components

In this section, the free enthalpy of components for liquid and gas phases, $\varphi^l_{NH_3}$, $\varphi^l_{H_2O}$, $\varphi^g_{NH_3}$ and $\varphi^g_{H_2O}$, will be given, in order to calculate the phases free enthalpy, Eqs. (1.268) and (1.269), and further to evaluate the mixture volume, entropy and enthalpy, Eqs. (1.270)–(1.272). The common way to proceed in an experimental computational modelling, is to determine first, experimentally, caloric and thermal properties of the mixture. In this respect, the caloric and thermal properties are selected in order to allow experimental data acquisition with the required accuracy and be processable further with existing thermodynamics tools. The experimental data are then processed with mathematical tools (e.g. selection, regression, inter- and/or extrapolation) and the caloric and thermal properties are expressed analytically, as functions of state parameters (here T and p). Finally, the analytical functions obtained are linked to the task (e.g. here the analytical expression of phases free enthalpy) with the help of thermodynamics relationships. In our case, as thermal and caloric parameters the volume and the specific heat at constant pressure, respectively, have been selected, for both phases. The caloric equations of state are due to Funk (1948) and Mollier (1926):

Liquid phase:

$$v^l(T,p) = a_1 + a_2 p + a_3 T + a_4 T^2 \tag{1.293}$$

Gas phase:

$$c^l_p(T,p_0) = b_1 + b_2 T + b_3 T^2 \tag{1.294}$$

$$v^g(T,p) = \frac{RT}{p} + c_1 + \frac{c_2}{T^3} + \frac{c_3}{T^{11}} + \frac{c_4 p^2}{T^{11}} \tag{1.295}$$

$$c_p^{g,0}(T) = d_1 + d_2 T + d_3 T^2 \quad (1.296)$$

In Eqs. (1.293)–(1.296), a_1, …, a_4, b_1, …, b_3, c_1, …, c_4 are coefficients which have different values for each component (Ziegler and Trepp 1984). The d_1, …, d_3 coefficients of regression are coming of values for the ideal gas specific heat $c_p^{g,0}(T)$, published by Haar (1968) for ammonia and by Keenan et al. (1978) for water. The lower subscript "0" indicates a reference state (T_0, p_0), characterized by the same free enthalpy in both phases.

The liquid phase free enthalpy is calculated with Eqs. (1.293) and (1.294) by:

$$\varphi^l(T,p) = \left(h^l(T_0,p_0) + \int c_p^l(T,p_0)dT\right) - T\left(\int_0^{T_0} ds + \int_{T_0}^{T} ds\right) \quad (1.297)$$

As already mentioned in Sect. 1.4.7.3, completing remark (a), the liquid isothermal of the mixture at hand is modifying to a very small extent with pressure variation, that is $\left(\frac{\partial h}{\partial p}\right)_T \cong 0$. Bearing this in mind, and considering at the same time Eqs. (1.42) and (1.109), the last integral in Eq. (1.297) is written as:

$$\int_{T_0}^{T} ds = \int_{T_0}^{T} \frac{c_p^l(T,p_0)}{T} dT - \int_{p_0}^{p} \frac{v^l(T,p)}{T} dp \quad (1.298)$$

Introducing Eq. (1.298) in Eq. (1.297), it results:

$$\varphi^l(T,p) = h^l(T_0,p_0) + \int_{T_0}^{T} c_p^l(T,p_0)dT - Ts^l(T_0,p_0)$$
$$- T\int_{T_0}^{T} \frac{c_p^l(T,p_0)}{T} dT + \int_{p_0}^{p} v^l(T,p)dp \quad (1.299)$$

Further on, with Eqs. (1.293) and (1.294), the liquid phase free enthalpy, Eq. (1.299), becomes:

$$\begin{aligned}\varphi^l(T,p) =& h^l(T_0,p_0) - Ts^l(T_0,p_0) + b_1(T - T_0) + \frac{b_2}{2}\left(T^2 - T_0^2\right) + \frac{b_3}{3}\left(T^3 - T_0^3\right) \\ & - T\left[b_1 \ln\frac{T}{T_0} + b_2(T - T_0) + \frac{b_3}{2}\left(T^2 - T_0^2\right)\right] \\ & + \left(a_1 + a_3 T + a_4 T^2\right)(p - p_0) + \frac{a_2}{2}\left(p^2 - p_0^2\right) \quad (1.300)\end{aligned}$$

Concerning the analytical expression of gas phase free enthalpy, the model must take into account that in our case the specific heat depends on both temperature and pressure in the subcooled domain (between 0 K and T_0). A similar assertion has been made in Sect. 1.4.7.3, completing remark (b), for the superheated vapor domain. In the equilibrium domain (between T_0 and T) the ideal gas specific heat depends only on temperature, as Eq. (1.296) shows. It is, therefore, necessary to find the function $\left(\frac{\partial c_p^g(T,p)}{\partial p}\right)_T$ for the subcooled domain. From Eq. (1.109) it results:

$$\left(\frac{\partial h}{\partial p}\right)_T = v + T\left(\frac{\partial s}{\partial p}\right)_T \tag{1.301}$$

The use of Maxwell equation (1.171) in Eq. (1.301), yields:

$$\left(\frac{\partial h}{\partial p}\right)_T = v - T\left(\frac{\partial v}{\partial T}\right)_p \tag{1.302}$$

Derivating partially with respect to temperature in (1.302) and reverting to the old notation, it results the function we look for:

$$\left(\frac{\partial c_p^g(T,p)}{\partial p}\right)_T = -T\left(\frac{\partial^2 v^g(T,p)}{\partial T^2}\right)_p \tag{1.303}$$

Integrating Eq. (1.303) with respect to pressure it is obtained:

$$c_p^g(T,p_0) - c_p^g(T,p=0) = -T\int_0^{p_0}\left(\frac{\partial^2 v^g(T,p)}{\partial T^2}\right)_p dp \tag{1.304}$$

The second member of the left side of Eq. (1.304) is obviously zero, therefore Eq. (1.304) becomes:

$$c_p^g(T,p_0) = -T\int_0^{p_0}\left(\frac{\partial^2 v^g(T,p)}{\partial T^2}\right)_p dp \tag{1.305}$$

Derivating partially in Eq. (1.295) with respect to temperature and performing the integral, it results:

$$c_p^g(T,p_0) = -\left(3*4p_0\frac{c_2}{T^4} + 11*12p_0\frac{c_3}{T^{12}} + \frac{11*12}{3}p_0^3\frac{c_4}{T^{12}}\right) \tag{1.306}$$

The free enthalpy of gas phase is given by:

$$\begin{aligned}\varphi^g(T,p) = &\int_0^{T_0} c_p^g(T,p_0)dT - T\int_0^{T_0}\frac{c_p^g(T,p_0)}{T}dT + h^g(T_0,p_0)\\ &- Ts^g(T_0,p_0) + \int_{T_0}^{T} c_p^{g,0}(T)dT - T\int_{T_0}^{T}\frac{c_p^{g,0}(T)}{T}dT + \int_0^p v^g(T,p)dp\end{aligned} \tag{1.307}$$

Introducing Eqs. (1.295), (1.296) and (1.306) in Eq. (1.304), the analytical expression of the gas phase free enthalpy results:

$$\begin{aligned}\varphi^g(T,p) =& h^g(T_0,p_0) - Ts^g(T_0,p_0) + d_1(T-T_0) + \frac{d_2}{2}\left(T^2-T_0^2\right) + \frac{d_3}{3}\left(T^3-T_0^3\right)\\ &- T\left[d_1 \ln\frac{T}{T_0} + d_2(T-T_0) + \frac{d_3}{2}\left(T^2-T_0^2\right)\right] + RT\ln\frac{p}{p_0} + c_1(p-p_0)\\ &+ c_2\left(\frac{p}{T^3} - 4\frac{p_0}{T_0^3} + 3p_0\frac{T}{T_0^4}\right) + c_3\left(\frac{p}{T^{11}} - 12\frac{p_0}{T_0^{11}} + 11p_0\frac{T}{T_0^{12}}\right)\\ &+ \frac{c_4}{3}\left(\frac{p^3}{T^{11}} - 12\frac{p_0^3}{T_0^{11}} + 11p_0^3\frac{T}{T_0^{12}}\right)\end{aligned} \tag{1.308}$$

The model analyzed above may be expressed in a dimensionless form, and in molar units (Ziegler and Trepp 1984), as well. It has been used in the book in order to calculate the coabsorbent heat pumping cycles run by ammonia–water. The author extended this model in order to be used in the phenomenological theory of mass and heat transfer (see Chap. 9).

Appendix 1

(i) In Eq. (1.120), $\ln\frac{T_2}{T_1}$ is written as:

$$\ln\frac{T_2}{T_1} = \ln\frac{1+x}{1-x} = f(x) \tag{A1.1}$$

where

$$x = \frac{T_2-T_1}{T_2+T_1},\ x \in [0,1),\ x << 1. \tag{A1.2}$$

The function $f(x)$ of Eq. (A1.1), arranged in a different form:

$$f(x) = \ln(1+x) - \ln(1-x) \tag{A1.3}$$

is derivated, in order to obtain:

$$f'(x) = \frac{1}{1+x} + \frac{1}{1-x} \tag{A1.4}$$

The ratios in Eq. (A1.4) are expressed using the infinite series of power:

$$\frac{1}{1\pm x} = 1 \pm \sum_{i=1}^{\infty}(-x)^i \tag{A1.5}$$

We introduce Eq. (A1.5) in Eq. (A1.4) and totalize. The same rank terms of Eq. (A1.5) are vanishing through summation, except the first ones, hence Eq. (A1.4) results as:

$$f'(x) = 2 \tag{A1.6}$$

Integrating Eq. (A1.6) between T_1 and T_2, the following result holds true (see Eq. A1.2):

$$f(x) = 2x = \ln\frac{1+x}{1-x} \tag{A1.7}$$

Introducing Eqs. (A1.7) and (A1.2) in Eq. (A1.1), it results:

$$\ln\frac{T_2}{T_1} = 2\frac{T_2 - T_1}{T_2 + T_1} \tag{A1.8}$$

and further, with it, Eq. (1.122) holds true:

$$\theta(\xi) = 1 - \frac{T_0}{\frac{T_1+T_2}{2}} \tag{A1.9}$$

(ii) The 2nd principle equation is written with intensive parameters and at equilibrium, as follows:

$$Tds = du + pdv \tag{A1.10}$$

In Eq. (A1.10) the specific entropy, internal energy and volume are written in function of the extensive values as $s = \frac{S}{G}$, $u = \frac{U}{G}$ and $v = \frac{V}{G}$, respectively. Differentiating these expressions, it is obtained:

$$\begin{aligned} ds &= \frac{GdS - SdG}{G^2} \\ du &= \frac{GdU - UdG}{G^2} \\ dv &= \frac{GdV - VdG}{G^2} \end{aligned} \tag{A1.11}$$

Introducing Eq. (A1.11) in Eq. (A1.10), after arranging some terms, it results:

$$TGdS = GdU - (U - TS)dG + p(GdV - VdG) \tag{A1.12}$$

In Eq. (A1.12):

$$(U - TS) = F = Gf \tag{A1.13}$$

Introducing Eq. (A1.13) in Eq. (A1.12) and dividing the entire Eq. (A1.12) by G, it is obtained:

$$TdS = dU - fdG + pdV - p\frac{V}{G}dG \tag{A1.14}$$

In Eq. (A1.14) we take into account that $\frac{V}{G} = v$. Further Eq. (A1.14) is partially derivated with respect to G for $U = const.$ and $V = const.$, wherefrom the following result holds true:

$$T\left(\frac{\partial S}{\partial G}\right)_{U,V} = \left(\frac{\partial U}{\partial G}\right)_{U,V} - f + p\left(\frac{\partial V}{\partial G}\right)_{U,V} - pv \qquad \text{(A1.15)}$$

In Eq. (A1.15) right side the members in the brackets are vanishing and bearing in mind that $f + pv = \varphi$, the needed entropy partial derivative is obtained:

$$\left(\frac{\partial S}{\partial G}\right)_{U,V} = -\frac{\varphi}{T} \qquad \text{(A1.17)}$$

References

Balamuru, V. G., Ibrahim, O. M., & Barnett, S. M. (2000). Simulation of ternary ammonia–water–salt absorption refrigeration cycles. *International Journal of Refrigeration, 23*(2000), 31–42.

Bosnjakovic, F. (1965). *Technical Thermodynamics*. New York: Holt, Rinehart et Wiston. (translated from German).

Bourseau, P., & Bugarel, R. (1986). Réfrigération par cycle à absorption–diffusion: Comparaison des performances des systèmes NH_3–H_2O et NH_3–NaSCN. *International Journal of Refrigeration, 9*(1986), 206–214.

Chartier, P., Gross, M., & Spiegler, K. S. (1975). *Applications de la thermodinamique du non-équilibre*. Paris: Hermann.

Dodescu, G. (1979). Numerical methods in Algebra (in Romanian). Technical Publishing House, Bucharest.

Patek, J., & Klomfar, J. (1995). Simple functions for fast calculations of selected thermodynamic properties of the ammonia-water system. *International Journal of Refrigeration, 18*(4), 228–234.

Patek, J., & Klomfar, J. (2006a). A computationally effective formulation of the thermodynamic properties of LiBr–H_2O solutions from 273 to 500 K over full composition range. *International Journal of Refrigeration, 29*(2006), 566–578.

Patek, J., & Klomfar, J. (2006b). Solid-liquid phase equilibrium in the systems of LiBr–H_2O and LiCl–H_2O. *Fluid Phase Equilibria, 250*(2006), 138–149.

Patek, J., & Klomfar, J. (2007). Thermodynamic properties of LiCl–H_2O system at vapor-liquid equilibrium from 273 to 400 K. *International Journal of Refrigeration, 31*, 1–17.

Plank, R. (1953). *Thermodynamische grundlagen*. Berlin: Springer.

Funk, H. (1948). *Mitt Kältetechn Inst Techn Hochschule Karlsruhe* **3**, 33.

Gibbs, J. W. (1876). On the equilibrium of heterogeneous substances. *Transaction on Connectcut Academy of Arts and Sciences, 2*, 309, 382(1873), *3*, 108, 343, 520 (1875–1878).

Haar, L. (1968). Properties of Ammonia as an ideal gas. *Journal of Research of the National Bureau of Standards Section A Physics and Chemistry, 72A*(2), 207–216.

Helmholz, v. H. (1902). Abhandlungen zur Thermodynamik chemischer Vorgänge, *Ostwalds Klassiker, nr.* **124**, Leipzig.

Infante Ferreira, C. A. (1984). Thermodynamic and physical property data equations for ammonia–lithium nitrate and ammonia–sodium thiocyanate solutions. *Solar Energy, 32*(2), 231–236.

Keenan, J. H., Keyes, F. G., Hill, P. G., & Moore, J. G. (1978). *Steam tables*. New York: Wiley.

Kirilin, V. A., Sicev, V. V., & Seindlin, A. E. (1985). *Thermodynamics*. Scientific and Technical Publishing House, Bucharest, 540p (in Romanian).

Lewis, G. H., & Randall, M. (1961). *Thermodynamics*, revised by K.S. Pitzner and L. Brewer, New York, Toronto, London, p. 445.

McNeely, A. L. (1979). Thermodynamic properties of aqueous solutions of lithium bromide, *ASHRAE Transactions*, **85**(T1)(3), 413, PH-79-3.

Mollier, R. (1926). Neue Tabellen und Diagrame für Wasserdampf, 4 Auflage, Springer, Berlin.

Murgulescu, I. G., & Vîlcu, R. (1982). *Introduction in the physical chemistry* (Vol. 3). Bucharest: Publishing House of the Romanian Socialist Republic Academy. (in Romanian).

Nicolescu, M. (1977–1980) *Mathematical Analysis, Vol. I, II*. Bucharest: The Didactic and Pedagogic Publishing House.

Niebergall, W. (1959). *Handbuch der Kältetechnik, Sorptions-Kältemaschinen* (Vol. 7). Berlin: Springer.

Popa, B., & Vintila, C. (1977). *Termotechnics and thermal machines*. Bucharest: The Didactic and Pedagogic Publishing House. (in Romanian).

Radermacher, R., & Alefeld, G. (1982). Lithiumbromid-Wasser—Lösungen als Absorber für Ammoniak oder Methylamin, *Brennstoff-Wärme-Kraft, 34*(1), 31–38.

Radcenco, V., Porneala, S., & Dobrovicescu, A. (1983). *Processes in refrigeration installations*. Bucharest: Didactic and Pedagogic Publishing House. (in Romanian).

Rowlinson, J. S., & Swinton, F. L. (1982). *Liquids and liquid mixtures, 3rd En*. London: Butteworths.

Schultz, S. (1971). Eine Fundamentalgleichung für das Gemisch aus Ammoniak und Wasser und die Berechnung von Asorptionskältemascinen-Prozessen. Habilitationsschrift Abt. Maschienenbau und konstructiven Ingenieurbau. Ruhr Univesität Bochum.

Staicovici, M. D. (2013). On the phase equilibrium computation in binary mixtures involved in absorption technology, Sent for publication to Revue Ruomaine des Sciences Techniques, série Électrotechnique et Énergetique, Romanian Academy.

Stamatescu, C. (1972). *Cooling technology, part I*. Bucharest: Technical Publishing House. (in Romanian).

Rysselberghe, V. (1963). *Thermodynamics of irreversible processes*. Paris: Herman.

Ziegler, B. & Trepp, Ch. (1984). Equation of state for ammonia–water mixtures. *International Journal of Refrigeration,* **7**(2), 101.

Ziegler, F. (2005). Absorption cycle basics revisited. In *International Sorption Heat Pump Conference*, Denver, ISHPC-082-2005, June 22–24, 2005.

Xu, F., & Goswami, D. Y. (1997). Thermodynamic properties of ammonia–water mixtures for power-cycle applications. *Energy, 24*(1999), 525–536.

Chapter 2
Mass and Heat Exchange Analysis of the Absorption Processes: The Divided Device Method

2.1 Heat Exchange Analysis of Isobar Absorption Processes with Gliding Temperature

The coabsorbent cycles, described in the next chapters, have typical absorption technology processes, of desorption, generation, resorption and absorption. Generally, they will be named absorption processes. These take place usually on large temperature intervals, or simpler, operate with gliding temperatures. In this paragraph we shall present a method to be used in the heat exchange analysis of an absorption process with a source which this is thermally interacting with. Prior to do it we shall consider a single process from the viewpoint of its constitution mass and heat transfer. This aspect will be detailed in Chap. 9. At this moment we only anticipate that a current of mass can couple with a heat current, in order to generate a coupled process of mass and heat transfer, with phase change. In this case, the two currents are influencing reciprocally. On the contrary, in a noncoupled process of mass and heat transfer, the two currents do not influence reciprocally and the resulting process is not changing the phase and is characterized by a heat transfer only. The absorption processes belong to the coupled transfer category, with a complex heat exchange named next also "latent", while the absorbent or source fluid cooling/heating phenomena, taking place without phase change, belong to the "sensible" heat exchange category.

There are three types of thermal interactions occuring in a coabsorbent cycle:

(a) latent-sensible, or coupled-noncoupled mass and heat transfer, like, for example, that occuring between an NH_3-H_2O absorbent absorbing NH_3 vapor and the cooling water which takes over the heat released by the former fluid;
(b) latent-latent, or coupled-coupled mass and heat transfer, like, for example, that occuring between an NH_3-H_2O absorbent absorbing NH_3 vapor and the NH_3–H_2O absorbent generating NH_3 vapor; in this particular case, the former fluid releases heat which is taken over by the latter fluid in a recovering way, in order to generate useful NH_3 vapor within an overlapped temperature glide;

M.-D. Staicovici, *Coabsorbent and Thermal Recovery Compression Heat Pumping Technologies*, Heat and Mass Transfer, DOI: 10.1007/978-3-642-54684-6_2,

(c) sensible-sensible, or noncoupled-noncoupled mass and heat transfer, like, for example, that occuring in a low concentration NH_3–H_2O absorbent subcooling process, when the fluid releases its sensible heat to a high concentration NH_3–H_2O absorbent which receives this heat in a recovering preheating way, but without changing its phase.

A spontaneous (natural) heat exchange is conditional upon the simultaneous fulfilment of the followings events:

i. there is a thermal capacity match between the fluids exchanging heat, or, in other words, the capacity of the former fluid to release (receive) heat equals that of the latter fluid to receive (release) heat;
ii. the temperature of the fluid releasing heat is always higher than that of the fluid receiving heat with a certain temperature pinch.

To monitor and design a heat exchange means to check the consistency of the conditions (i) and (ii) fulfillment in each reasonable large interval of temperature throughout the thermal process at hand. The most working combinations in the absorption technology, not all, are binary (e.g. NH_3-H_2O, NH_3-LiNO_3, $NH_3-NaSCN$, $H_2O-LiBr$). A (c) heat transfer of subcooling-superheating type occurs usually in non-equilibrium conditions, at constant pressure and concentration ($p = const.$, $y = const.$), and the process suffers a temperature glide, only. In this case, the check of the (i) and (ii) conditions fulfillment is simpler. An experienced designer doesn't need even to perform this check, it calculates directly the preheated absorbent outlet temperature based on the knowledge of the input parameters of the subcooled and preheated absorbents. However, this is not the case of the heat exchange of the (a) type and so much more of the (b) type, when the systematic check of the (i) and (ii) conditions fulfillment is mandatory. In this case, even in the simplest situation, of the isobar processes, the heat exchange becomes complex. Indeed, this time the binary fluids are two-variant, so, if the temperature changes, the concentration will change as well, $y = \lfloor y(T) \rfloor_p$, in a quantitatively way difficult to know a priori and the (i) and (ii) conditions can be verified but only using a mathematical tool which uses the working combination thermodynamic equilibrium data. At this point, from the considerations emphasized so far, it becomes clear that the heat exchange analysis must start with the local (elementary) mass and heat transfer of an individual absorption process. In Fig. 2.1a, b, the two main opposite absorption processes, of absorption (resorption) and of generation (desorption), are schematically represented, respectively. Next, the word 'absorbent' is referring to a mixture of absorbent and refrigerant. In Fig. 2.1a, a mass flow, m (kg mixture), of concentration y (kg refrigerant/kg mixture) and enthalpy h (kJ/kg mixture), combines with an infinite small quantity of vapor mixture dm of concentration Y and enthalpy H, in order to result in a mixture (m, h) of higher concentration, $y + dy$. The process depicted in Fig. 2.1b is the reverse of that shown in Fig. 2.1a. The mass and heat balances of both isobar processes are written by Eqs. (2.1) and (2.2), respectively:

$$ym(y) + Y_{G(D)}dm = (y + dy)m(y + dy) \tag{2.1}$$

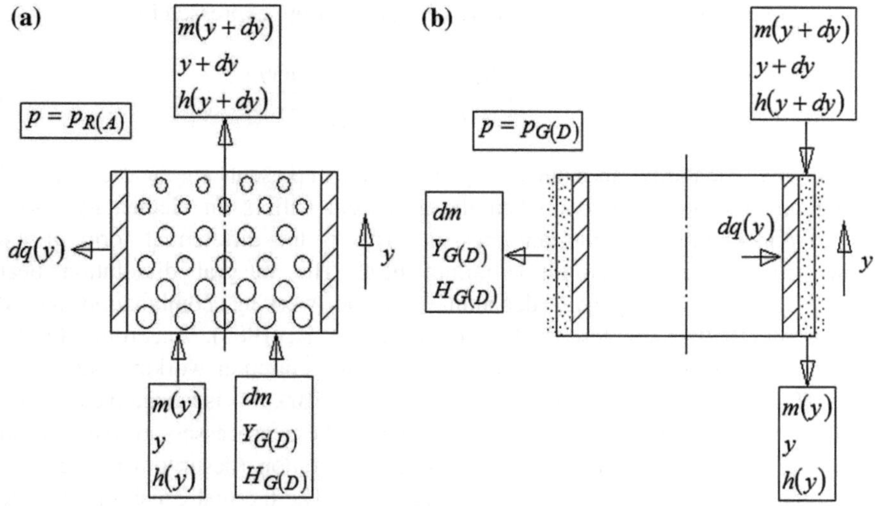

Fig. 2.1 The two opposite isobar absorption (**a**) and generation (**b**) processes, schematically representation serving to the local (elementary) mass and heat transfer analysis

$$m(y)h(y) + H_{G(D)}dm = dq + m(y + dy)h(y + dy) \tag{2.2}$$

In Eqs. (2.1) and (2.2), the lowercase and uppercase letters hold for the liquid and vapor phases, respectively, and m is the liquid phase mass flow rate. In order to avoid confusion, q will be used throughout this chapter, instead of Q, although q is an extensive property. Performing the calculation in Eq. (2.1) and neglecting the second order terms, it results:

$$\frac{dm}{m} = \frac{dy}{Y_{G(D)} - y} \tag{2.3}$$

Proceeding similarly with Eq. (2.2), we obtain:

$$dq = m\left[\left(H_{G(D)} - h(y)\right)\frac{dm}{m} - dh(y)\right] \tag{2.4}$$

Equation (2.4) can be rewritten as Eq. (2.5), taking into account Eq. (2.3) in its expression:

$$dq = m\left[\frac{H_{G(D)} - h(y)}{Y_{G(D)} - y} - \frac{\partial h(y)}{\partial y}\right]dy \tag{2.5}$$

Equation (2.5) can be put under a different form:

$$\left[\frac{\partial q}{\partial y}\right]_p = [X(y)]_p \tag{2.6}$$

which emphasizes the isobar thermal capability function, expressed by:

$$\left[X(y)\right]_p = m\left[\frac{H_{G(D)} - h(y)}{Y_{G(D)} - y} - \frac{\partial h(y)}{\partial y}\right]_p \tag{2.7}$$

The thermal capability, an extensive quantity (kJ), represents the heat exchanged during an absorption process when the absorbent suffers an elementary variation of concentration at constant pressure. This is the differential form of the known finite integral dissolution/boiling heat. The integral dissolution heat can be positive or negative, depending on the working combination nature (Niebergall 1959; Stamatescu 1972; Radcenco et al. 1983). According to the thermodynamics rule of a heat sign, in case of the common working pairs (e.g. $NH_3{-}H_2O$, $NH_3{-}LiNO_3$, $NH_3{-}NaSCN$, $H_2O{-}LiBr$), this is a negative quantity for the absorption (resorption) processes, that is these processes are exothermal (Fig. 2.1a), and is a positive quantity for the generation (desorption) processes, that is these processes are endothermal (Fig. 2.1b). The thermal capability has the same sign as that of the dissolution/boiling heat, but here we considered it a positive quantity, preferred in the practical calculus. The two terms of Eq. (2.7) bracket, given by Eqs. (2.8) and (2.9),

$$\tan(a) = \left[\frac{\partial h(y)}{\partial y}\right]_p \tag{2.8}$$

and

$$\tan(b) = \left[\frac{H_{G(D)} - h(y)}{Y_{G(D)} - y}\right]_p \tag{2.9}$$

can be derived graphically in an enthalpy-concentration diagram. These are shown for the $NH_3{-}H_2O$ in Fig. 2.2 for a certain operating point characterized by the state parameters y, Y, H, h. The simple approach presented helps finding at a glance the most recommended operating region in the solubility field of a working combination for a particular application be more effective. For instance, if the mass flow m were kept constant, a heat pump should increase the capability of its resorber in order to increase its COP. This can happen for $NH_3{-}H_2O$ and similar working combinations in its low concentration region, where in Eq. (2.8) the derivative is negative and both terms in Eq. (2.7) bracket are positive. However, because a generation process is the opposite of the resorption one, we expect to have in this region a high generation capability, as well. This further recommends to avoid generation processes be designed in a low concentration region of a working pair diagram. In the high concentration region, the liquid phase isobar has a positive derivative. The second term of Eq. (2.7) is therefore negative, diminishing the capability, which reason for this region is recommended for the generation processes. Similarly, extending the generation notion to the desorption ones, we expect that a high concentration region will enable to obtain lower temperatures

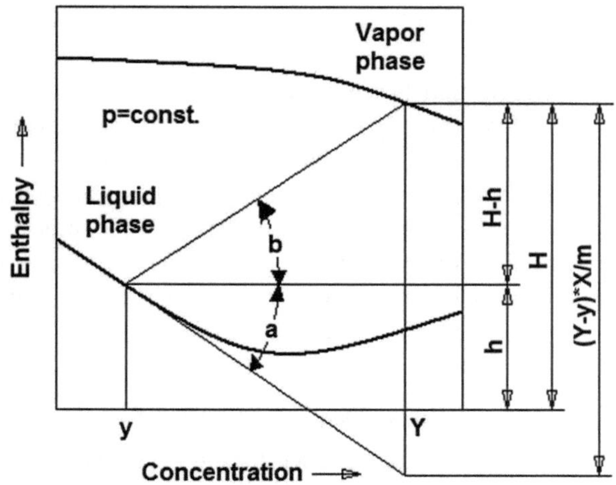

Fig. 2.2 Thermal capability main parameters represented in an enthalpy-concentration diagram for NH_3–H_2O

(e.g. negative), but with a smaller capability. Additionally, bearing in mind that a resorption process is the opposite of that of generation, we should not design a resorption based useful heating processes in a high concentration region, because here the capability and therefore the heated fluid temperature increase are reduced.

The above remarks entitle us to make a few useful higher order comments, next. In practice, not all processes of a cycle can take place simultaneously in the best region of a working pair, therefore a compromise should be made and priorities should be established: (1) The main process (that we are looking for) should be designed first in the most recommended region; (2) The rest of processes should be placed in favored regions, as much as possible, having in view, however, the second principle respect, of minimizing the exergy loss. To this extent, the cogeneration, and so much more the trigeneration, if produced by processes opposite to the main process, are increasing the exergy loss and therefore are diminishing the effectiveness wich the main useful effect is produced with. This is valid for all thermodynamic cycles (e.g. a Rankine power cycle working in cogeneration of power and heat), and particularly for our coabsorbent cycles. As an example it could be mentioned the nontruncated coabsorbent cooling cycle working in cogeneration of cooling and heating, analyzed in the next chapter. Indeed, if a cogenerated heating were performed besides cooling, the cycle exergy loss increases through the useful heating output, and consequently, there will be a cooling effect reduction, as results will show. However, the cycle effectiveness of a combined energy production could be regarded from a different point of view, too. Indeed, suppose the application (society) had a synchronized need of combined energy production. In this case, the simultaneous production-consumption of two or more different forms of energy produced by a single primary heat source means a higher order application, characterized by a lower entropy value, defined as a measure of an energetic system order degree. Consequently, a cycle operating in cogeneration, or trigeneration, etc., and fulfiling the required level of output parameters, becomes of course of higher effectiveness than a higher entropy single-effect cycle. This might be the case of some coabsorbent cycles, as it will be shown later in this book.

An algorithm based on Eq. (2.7) has been introduced in a past work by the author (Staicovici 1995). With it we evaluated a heat exchange of type (b) in the ternary working combination medium ($NH_3-H_2O-LiBr$, Radermacher and Alefeld 1982). However, this older method is less friendly in applications. For this reason, in the next section a powerful, simple, practical and easy to use method is given for the first time, perfectly integrated in a cycle thermodynamic computation algorithm, in order to assess an isobar absorption heat exchange.

2.2 The Divided Device Method for Isobar Absorption Processes Heat Exchange Assessment

The isobar thermal capability of Eq. (2.7) is a function of absorbent concentration, usually considered in the dissolution heat aspects of an individual absorption process. In a heat exchange analysis, however, it is more indicated to work with the temperature as variable, for obvious reasons. The translation from concentration to temperature results in:

$$dq(T) = \left[X(y(T))\right]_p dy(T) = \left[X(y(T))\frac{\partial y(T)}{\partial T}\right]_p dT = [X(T)]_p dT \tag{2.10}$$

wherefrom, the new thermal capability (kJ/K), as well an extensive quantity as it depends on the involved mass, writes:

$$\left[\frac{\partial q(T)}{\partial T}\right]_p = [X(T)]_p \tag{2.11}$$

The left member first order partial derivative of Eq. (2.11) can be approximated by one of the finite difference formula, known from the numerical solving. Here it can be used for instance the forward difference (Nicolescu 1977–1980; Bakhvalov 1977):

$$\left[\frac{\partial q(T)}{\partial T}\right]_p \cong \left[\frac{q(T+\Delta T)-q(T-\Delta T)}{2\Delta T}\right]_p \tag{2.12}$$

In Eq. (2.12), the heat $q(T)$ is supposed to be a real, continuous function, defined on real intervals of temperature, $T \in \left[T_{DI(GI)}, T_{DO(GO)}\right]$ and $T \in \left[T_{RO(AO)}, T_{RI(AI)}\right]$, which the desorbtion (generation) and resorbtion (absorbtion) processes at hand operate on, and ΔT is the temperature increment, chosen at convenience. It is enough to find $q(T)$ for the two opposite processes, desorbtion and resorbtion, schematically presented in Fig. 2.3a, b, respectively, in order to cover all

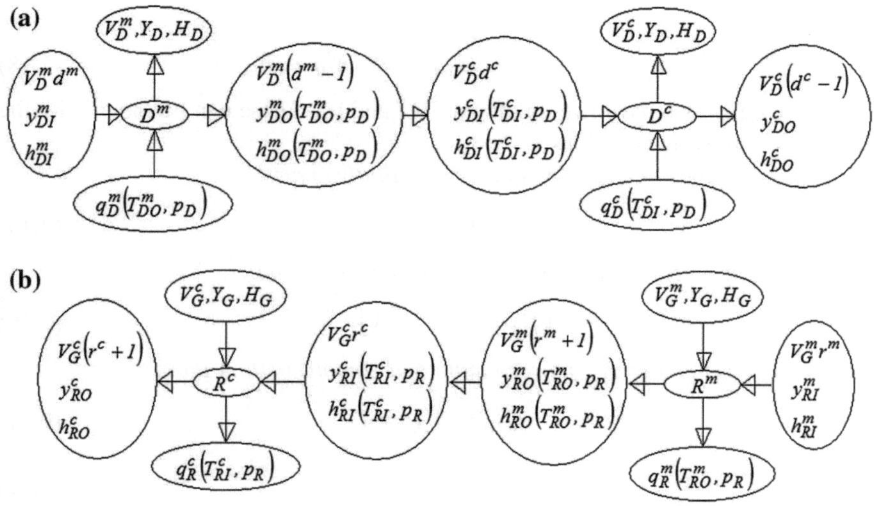

Fig. 2.3 A desorption (**a**) and resorption (**b**) processes schematic, divided in two parts, serially connected, serving for the divided device method presentation

absorption processes analysis. Each device housing an absorption process is divided in two parts, serially connected and labeled with the superscripts "*m*" from main and "*c*" from complementary, where the "Divided Device" (DD) method name comes from (Staicovici 2011). We consider first Fig. 2.3a and write the mass and heat balance for both parts. For the main part, D^m, this results in:
Mass balance:

$$V_D^m d^m y_{DI}^m = V_D^m Y_D + V_D^m (d^m - 1) y_{DO}^m \tag{2.13}$$

wherefrom, the main desorber mass flow factor is expressed by:

$$d^m = \frac{Y_D - y_{DO}^m}{y_{DI}^m - y_{DO}^m} \tag{2.14}$$

Heat balance:

$$q_D^m = V_D^m H_D + V_D^m (d^m - 1) h_{DO}^m - V_D^m d^m h_{DI}^m \tag{2.15}$$

wherefrom, the main desorber desorption heat is expressed by:

$$q_D^m = V_D^m \left[(H_D - h_{DO}^m) + d^m (h_{DO}^m - h_{DI}^m) \right] \tag{2.16}$$

Similarly, for the complementary part, D^c, this results in:
Mass balance:

$$d^c = \frac{Y_D - y_{DO}^c}{y_{DI}^c - y_{DO}^c} \tag{2.17}$$

Heat balance:

$$q_D^c = V_D^c\left[\left(H_D - h_{DO}^c\right) + d^c\left(h_{DO}^c - h_{DI}^c\right)\right] \tag{2.18}$$

The above equations have two unknowns, V_D^m and V_D^c. These can be found with the help of the following equations:
Vapor mass balance on desorber and its constitutive parts:

$$V_D^m + V_D^c = V_D \tag{2.19}$$

Mass flow continuity between main and complementary parts:

$$V_D^m\left(d^m - 1\right) = V_D^c d^c \tag{2.20}$$

State parameters continuity between main and complementary parts:

$$T_{DO}^m = T_{DI}^c = T \tag{2.21}$$

$$p_{DO}^m = p_{DI}^c = p_D \tag{2.22}$$

The heat $q(T)$ is calculated in a point of temperature T always situated to the boundary between the two m and c parts of a device, as Eqs. (2.21) and (2.31) show. From Eqs. (2.19) and (2.20), it results:

$$V_D^m = V_D\frac{d^c}{d^c + d^m - 1} \tag{2.23}$$

and

$$V_D^c = V_D\frac{d^m - 1}{d^c + d^m - 1} \tag{2.24}$$

Equations (2.13)–(2.24) hold true for a generation device also, provided that the index D be changed by G and the mass flow factor d be changed by g. We consider now Fig. 2.3b. Proceeding similarly, the items given below hold true in case of a resorption process. For the main part, R^m, we obtain:
Mass balance:

$$r^m = \frac{Y_G - y_{RO}^m}{y_{RO}^m - y_{RI}^m} \tag{2.25}$$

Heat balance:

$$q_R^m = V_G^m\left[\left(H_G - h_{RO}^m\right) + r^m\left(h_{RI}^m - h_{RO}^m\right)\right] \tag{2.26}$$

For the complementary part, R^c, it results:
Mass balance:

$$r^c = \frac{Y_G - y_{RO}^c}{y_{RO}^c - y_{RI}^c} \tag{2.27}$$

Heat balance:

$$q_R^c = V_G^c\left[\left(H_G - h_{RO}^c\right) + r^c\left(h_{RI}^c - h_{RO}^c\right)\right] \tag{2.28}$$

The two unknowns, V_G^m and V_G^c can be found with the help of the following equations:

$$V_G^m + V_G^c = V_G \tag{2.29}$$

$$V_G^m\left(r^m + 1\right) = V_G^c r^c \tag{2.30}$$

$$T_{RO}^m = T_{RI}^c = T \tag{2.31}$$

$$p_{RO}^m = p_{RI}^c = p_R \tag{2.32}$$

From Eqs. (2.29) and (2.30), the complementary vapor results this time:

$$V_G^m = V_G\frac{r^c}{r^c + r^m + 1} \tag{2.33}$$

and

$$V_G^c = V_G\frac{r^m + 1}{r^c + r^m + 1} \tag{2.34}$$

Equations (2.25)–(2.34) hold true for an absorption device also, provided that the index G and R be changed by D and A, respectively, and the mass flow factor r be changed by a. The thermal capability of an absorption processes is found within the limits of the DD method considering Eq. (2.12) for the appropriate heat $q(T)$ expressed by one of Eqs. (2.15, 2.18, 2.26, 2.28), etc. in a consistent way for all temperature T values which cover the entire process operation. Additionally, we stress that it is not important which part of a device heat expression is used in Eq. (2.12), the result is the same whether for instance we use the q_D^m or q_D^cheat for the desorber. This equal choice results from equaling Eq. (2.12) numerator of both parts:

$$q_i^m(T + \Delta T) - q_i^m(T - \Delta T) = q_i^c(T - \Delta T) - q_i^c(T + \Delta T), \quad i = D, R, G, A \tag{2.35}$$

Rearranging the terms, it results the true result for all T, if we beared in mind equations of type (2.21) or (2.31):

$$q_i^m(T + \Delta T) + q_i^c(T + \Delta T) = q_i^m(T - \Delta T) + q_i^c(T - \Delta T) = q_i = const. \tag{2.36}$$

Further on, the heat expression utilized in Eq. (2.12) was that corresponding to the main part. The main part of a device was considered computationally that part which allows the absorbent access into the device. Proceeding in this way it enables the thermodynamic state of absorbent going into device be taken into account, which obviously is crucial for a correct absorbtion process heat exchange analysis.

The method, completed by details of application, is covered once for the analysis of the type (a) heat exchange, with a single absorption process, and twice in a correlated way for the type (b) heat exchange. The DD method analysis results will be shown in the next chapters wherever it will be necessary.

References

Bakhvalov, N. S. (1977). *Numerical methods*. Moscow: Publishing House "Nauka", MIR Publishers.

Nicolescu, M. (1977–1980). *Mathematical analysis* (Vol. I, II). Bucharest: The Didactic and Pedagogic Publishing House. (in Romanian).

Niebergall, W. (1959). *Handbuch der Kältetechnik* (Vol. 7). Berlin: Sorptions-Kältemaschinen, Springer.

Radcenco, V., Porneala, S., & Dobrovicescu, A. (1983). *Processes in refrigeration installations*. Bucharest: Didactic and Pedagogic Publishing House. (in Romanian).

Radermacher, R., & Alefeld, G. (1982). Lithiumbromid-Wasser Lösungen als Absorber für Ammoniac oder Methylamin. *Brennstoff-Wärme-Kraft, 34*(1), 31–38.

Staicovici, M. D. (1995). Poybranched regenerative GAX cooling cycles. *International Journal of Refrigeration, 18*, 318.

Staicovici, M. D. (2011). Heat exchange analysis of coabsorbent cycle absorption processes. In: *Proceedings of ISHPC 2011 Conference*, Padua, Italy.

Stamatescu, C. (1972). *Cooling technology, Part I.* Bucharest: Technical Publishing House. (in Romanian).

Chapter 3
Coabsorbent Cycles

3.1 Introduction

The coabsorbent cycles have been disclosed recently by their author in published works on heat pumping and power applications (Staicovici 2006a, b, c, d, e, 2007a, b, c, d, e, f, g, h, 2008a, b, c, 2009, 2011; Staicovici and Staicovici 2013). In the following chapters, the theoretical research laying down the basis of coabsorbent technology is presented, including the nontruncated and truncated coabsorbent cycles, the theory of truncation, the cooling and heating fractals, the truncation columns and column cycles, the coabsorbent cycles ideal COP estimate, the new cycles with pressure and concentration stages and the multi-effect cycles transposition into the coabsorbent technology.

3.2 Nontruncated Heating and Cooling Coabsorbent Cycles

So far, the absorbent administration of an absorption cycle, including two or more subcycles (interconnected by mass and/or heat transfer), bases on a common, known practice that, in short, could be expressed by "a separate absorbent flow in each individual subcycle". In a previous work (Staicovici 2006a), the author is proposing a new type of absorption cycles, with a different absorbent administration and named "with co-absorbent", or simpler, *coabsorbent cycles*. In this way, using the coabsorbent cycles all the condensation and resorption cycle problems (e.g. reduced solubility field, rectification need, absorbent migration, or cycle complexity increase with COP improvement) could be avoided to a great extent. The key of a coabsorbent cycle is the resorption one (Niebergall 1959). Some relative recent works concerning the resorption cycles have been accomplished for cooling with the water/lithium bromide (Perez 1993), or for the evaluation of the absorption and resorption heat pumps COP with the ammonia-water (Baehr 1981). The coabsorbent cycle is built up by joining the resorption cycle subcycles (see Figs. 3.1, 3.28) along a common isostere ($y = y_M = const.$), so that the opposed

M.-D. Staicovici, *Coabsorbent and Thermal Recovery Compression Heat Pumping Technologies*, Heat and Mass Transfer, DOI: 10.1007/978-3-642-54684-6_3,

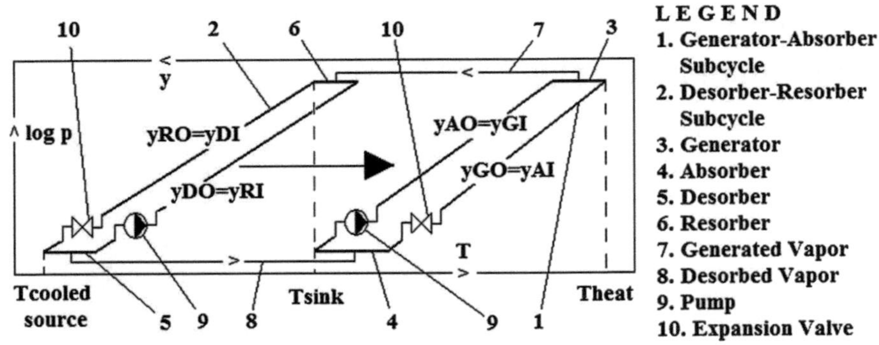

Fig. 3.1 The resorption cooling cycle serving as a formal explanation of the nontruncated coabsorbent cooling cycle and coabsorbent technology genesis

pair isobar or non-isobar processes, of generation and resorption, and of absorption and desorption, are in mass (vapor) and heat exchange, and the subcycles separate absorbents have a common point "*M*" (mixing point), where they are mixing up and cyclic generate the absorbent of mean concentration, y_M. By joining in accordance with the temperature decrease two cooling + heating type subcycles, it is obtained a *nontruncated cooling coabsorbent cycle*, and by joining in accordance with the temperature decrease two heating + cooling type subcycles, it is obtained a *nontruncated heating (heat transformer) coabsorbent cycle* (Staicovici 2006a). The coabsorbent cycles are new, individual ones and generalize the thermodynamic absorption cycles. They benefit of a different flow and heat properties as compared to the already known condensing cycles with separate absorbent flow, which become a particular case of the coabsorbent cycle big comprehensive class. First, we shall deal with the nontruncated cycles. Second, we shall introduce the truncated coabsorbent cycles. Both have theoretical and practical importance.

3.2.1 Nontruncated Cooling Coabsorbent Cycle

3.2.1.1 Cycle General Presentation

The nontruncated coabsorbent cooling cycle formally comes of the resorption cooling cycle, depicted in the Fig. 3.1. The left side subcycle starts to work on a broader desorber-resorber concentration interval (see arrow) until its $y_{DO} = y_{RI}$ isostere overlaps the $y_{AO} = y_{GI}$ isostere of the right side subcycle. Thus, the two subcycles of resorption cycle will have a single internal isostere $y_{RI} = y_{GI} = y_M$, resulted through joining (mixing up) of absorbents coming of the desorber and the absorber (*mixing absorbents*). The benefic consequences of this joining are the followings:

(a) The resorption process temperature prolongs until $T_{RI} = T_{GI}$, so the nontruncated coabsorbent cooling cycle is capable to cogenerate heat in certain circumstances, besides cooling;

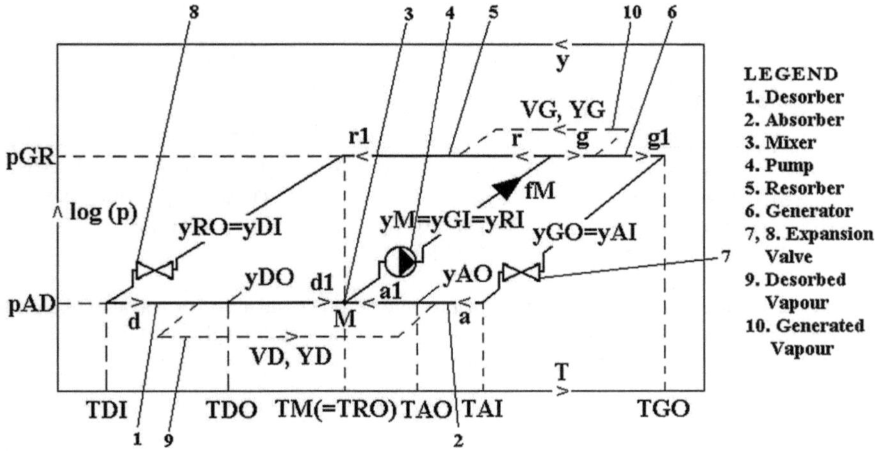

Fig. 3.2 The *3-temperature* nontruncated cooling coabsorbent cycle

(b) The desorption process takes place on a broader temperature interval, conveniently chosen by a designer, as we shall see, increasing more often than not the COP_c;
(c) The known solution inventory problem, which the resorption cycle is confronted with, is solved for good and all in a general and elegant way;
(d) It is enabling the low pressure desorption and absorption processes to overlap theirs operation temperatures as a unique and very effective method to increase the COP_c;
(e) The two pumps of the resorption cycle are replaced by a single one.

The nontruncated coabsorbent cooling cycle is plotted in a $\log(p) - 1/T$ diagram in Fig. 3.2. Its operation is given next (see arrows in Fig. 3.2). The absorption and desorption on one side, and the generation and resorption on the other side, are isobar and coupled processes. They take place on large concentration intervals with gliding temperatures. The equilibrium evolution of the absorbents is represented by solid lines.

The generated and desorbed vapor are represented by dotted lines. The cooling effect takes place in desorber, externally heated by the source which must be cooled. Here, the low pressure (p_{AD}) absorbent covers a continuous desorption process, characterized by different equilibrium temperatures and corresponding concentrations, noted by pairs of state parameters (T, y), from the desorber inlet, where $(T, y) = (T_{DI}, y_{DI})$, till the desorber outlet, where $(T, y) = (T_{DO}, y_{DO})$, with $T_{DI} < T_{DO}$ and $y_{DI} > y_{DO}$. The desorption interval $(T, y) \in [(T_{DI}, y_{DI}), (T_{DO}, y_{DO})]$ can be chosen at convenience, depending on each particular application. The desorbed vapor, noted by V_D, Y_D, is absorbed in the coupled absorber, externally cooled by the cycle sink source. In absorber, the low pressure (p_{AD}) absorbent covers similarly a continuous absorption interval characterized by different equilibrium temperatures and corresponding concentrations, from the absorber inlet, where $(T, y) = (T_{AI}, y_{AI})$, till the absorber outlet, where $(T, y) = (T_{AO}, y_{AO})$,

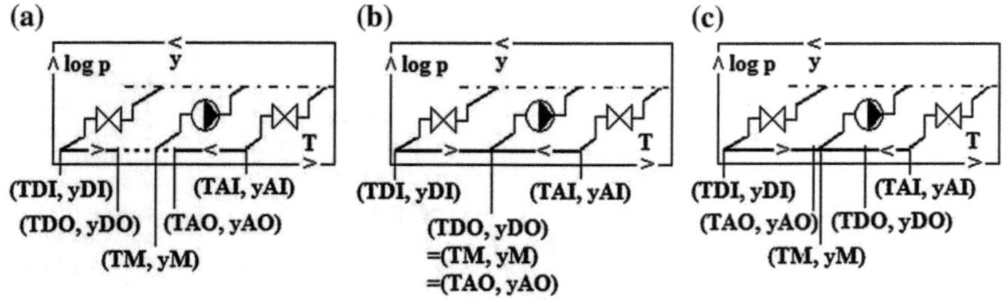

Fig. 3.3 **a–c** The relative position of the low pressure desorber and absorber operation temperature and concentration intervals: **a** nonoverlapping (T, y) interval; **b** single point (balance) overlapping (T, y) interval; **c** overlapping (T, y) interval

with $T_{AI} > T_{AO}$ and $y_{AI} < y_{AO}$. The absorption interval $(T, y) \in [(T_{AI}, y_{AI})$ $(T_{AO}, y_{AO})]$, is this time no more independent, the absorption outlet parameters $(T, y) = (T_{AO}, y_{AO})$, depend on all cycle parameters and especially on those of the desorption process. Further, the absorbents exiting the low pressure desorber and absorber are being mixed up in the mixer, noted by M. The mixing process is considered adiabatic and perfect. It can be performed naturally, for example during the gravity falling of mixing absorbents, or using very low power consuming mixing pumps. The resulting cumulated flow f_M leaves the mixer M with the temperature and the one mean concentration $(T, y) = (T_M, y_M)$. The T_M is with a good approximation the internal sink value corresponding to cycle external sink source temperature, otherwise the precise internal sink value is T_{AO}. However, the pair $(T, y) = (T_M, y_M)$ has a larger meaning in the coabsorbent technology, that of a design mark which remains unchanged during a coabsorbent cycle operation. Other cycle parameters can refer theirs value to this mark. To this extent, it is meaningful to analyse the relative position of the T_{DO} and T_{AO} *mixing temperatures* and y_{DO} and y_{AO} *mixing concentrations* on one side and of the one mixer concentration y_M and temperature T_M on the other side, depicted in Fig. 3.3a–c. Three relative positions are met, described shortly, next:

(a) $T_{DO} < T_M < T_{AO}$ and $y_{AO} < y_M < y_{DO}$, when the sink source internal temperature is higher than that of the cooled source, Fig. 3.3a. This is a normal operation mode of the nontruncated cooling coabsorbent cycle. In this case, the low pressure desorber and absorber have nonoverlapping in theirs operation temperature ranges, that is $[T_{DI}, T_{DO}] \cap [T_{AI}, T_{AO}] = \Phi$;

(b) $y_{AO} = y_M = y_{DO}$, when the sink source internal temperature equals that of the cooled source, Fig. 3.3b. In this case, the desorber and absorber have a single point (balance) overlapping in theirs operation temperature ranges, so $[T_{DI}, T_{DO}] \cap [T_{AI}, T_{AO}] = T_{DO} = T_{AO} = T_M$;

(c) $T_{DO} > T_M > T_{AO}$ and $y_{AO} > y_M > y_{DO}$, when the sink source internal temperature is smaller than that of the cooled source, Fig. 3.3c. In this case, the cycle operation and COP_c are favoured and the desorber and absorber have overlapping over a finite interval of temperature, that is $[T_{DI}, T_{DO}] \cap [T_{AI}, T_{AO}] = [T_{AO}, T_{DO}]$.

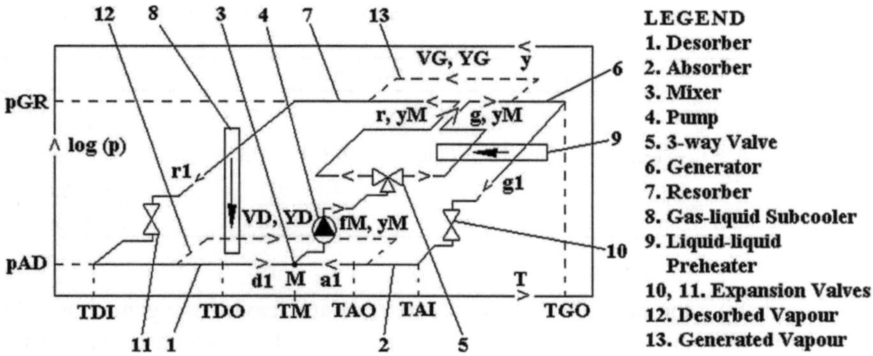

Fig. 3.4 Nontruncated cooling coabsorbent cycle optimum internal heat recovery configuration

Continuing with the cycle operation, after being generated the cumulated flow f_M is pumped and distributed in pre-established complementary quantities to the high pressure (p_{GR}) generator and resorber. The vapor generated in generator, externally heated by the cycle heat source, is resorbed in resorber. It is noted by V_G, Y_G in Fig. 3.2. The resorber is cooled externally by a source which is heated in certain conditions, as a second cycle useful effect. Finally, the absorbents exiting the resorber and generator with temperatures and concentrations $(T, y) = (T_{RO}, y_{RO})$ and $(T, y) = (T_{GO}, y_{GO})$, respectively, are expanded to the low pressure desorber and absorber, in order to close the cycle.

An application with the cycle at hand should provide the following minimum number of input data for a cooling cycle design (see Fig. 3.2):

(a) the resorber outlet temperature and concentration (T_{RO}, y_{RO}), depending on the sink source temperature;
(b) the desorber inlet/outlet temperatures, T_{DI}/T_{DO}, depending on the lowest/highest temperatures the useful cooling effect requires, or on the cooled source required minimum/maximum temperatures;
(c) the generator outlet temperature, T_{GO}, depending on that of the heat source.

Out of these four parameters it results the rest of the others parameters, p_{AD}, p_{GR}, $T_M (= T_{RO})$, y_M, and y_{GO}.

Finally, the cycle internal heat management is considered. The internal heat recovery offers in case of the nontruncated cooling coabsorbent cycle several solutions. The optimum configuration ensuring one of the highest cycle cooling COP is depicted in Fig. 3.4 (see the big arrow heat exchangers). According to it, the absorbent performing resorption process is subcooled through desorbed vapor superheating, prior to be absorbed in the absorber, as a first internal heat recovery. This solution is already known and practiced in the condensing absorption (Niebergall 1959). According to the second thermal recovery, solely the absorbent suffering the generation process is preheated prior to enter the generator. In the diagram of Fig. 3.4, it is identified as the absorbent having the mass flow rate equal to V_{Gg} and the concentration y_M. This is done through the subcooling of the

Table 3.1 The specific inlet and outlet absorber, generator, resorber and desorber mass flow factors

Device	Absorber		Generator		Resorber		Desorber	
Inlet	$a = \frac{Y_D - y_{AO}}{y_{AO} - y_{AI}}$	(3.1)	$g = \frac{Y_G - y_{GO}}{y_{GI} - y_{GO}}$	(3.2)	$r = \frac{Y_G - y_{RO}}{y_{RO} - y_{RI}}$	(3.3)	$d = \frac{Y_D - y_{DO}}{y_{DI} - y_{DO}}$	(3.4)
Outlet	$a_1 = a + 1 = \frac{Y_D - y_{AI}}{y_{AO} - y_{AI}}$	(3.5)	$g_1 = g - 1 = \frac{Y_G - y_{GI}}{y_{GI} - y_{GO}}$	(3.6)	$r_1 = r + 1 = \frac{Y_G - y_{RI}}{y_{RO} - y_{RI}}$	(3.7)	$d_1 = d - 1 = \frac{Y_D - y_{DI}}{y_{DI} - y_{DO}}$	(3.8)

absorbent having the mass flow rate equal to $V_G g_1$ and leaving the generator, till a temperature close to the T_M value, before to be expanded and to enter the absorber.

Next, we consider the following specific inlet and outlet mass flow factors (kg absorbent/kg refrigerant) for absorber, generator, resorber, and desorber. They are given in Table 3.1. These factors will be obtained writing the mass balances of each cycle device, shown to Sect. 3.2.1.2. In these equations, the mean generator vapor concentration appears. It is a cycle important parameter. Its value is conditioning a full coabsorbent operation mode of a cooling cycle. The mean generator vapor concentration value must be higher than a certain minimum value, named here the *(vapor) threshold concentration.* A cooling cycle has the y_{RO} value as threshold concentration. This means that for the resorber inlet mass flow rate r be a real, positively defined quantity, $r > 0$, the generator vapor which is being resorbed by resorber must have a mean concentration higher than the threshold value, that is $Y_G > y_{RO}$, in order that the resorber be capable to reach its y_{RO} designed outlet concentration [see Eq. (3.3)]. A similar situation is met to a resorption cycle. Having problems with the threshold concentration means that this condition can't be fulfilled. Sometimes, when a nontruncated cooling coabsorbent cycle is confronted with a threshold concentration problem, a small rectification of the generated vapor can be practised in order to eliminate it. This, however, is not recommended, but only in special cases, when the use of a nontruncated cooling coabsorbent cycle matches best to a particular application (e.g. when there is a need of a desorber temperature glide). The cycle main features are expressed by the equations given in Table 3.2. The devices partial mass balance result in Eqs. (3.9) and (3.10), where V_{GD} is the ratio of the vapor generated by generator, V_G, through that of the desorber, V_D (chosen at random, $V_D = 1$, here), Eq. (3.11). From the division term by term of Eqs. (3.9) and (3.10), we obtain the *typical nontruncated cooling coabsorbent cycle equation*, Eq. (3.12). This shows the relation between the cycle unknown (a and d) and known (g_1 and r_1) specific flow factors and the concentrations y_{RO}, y_{GO} and y_M. Also, it defines the cycle S *symmetry factor* [see the last ratio in Eq. (3.12)]. This factor reveals how much the flow factors and the specific heat of each adjacent opposite processes (generation/resorption and absorption/desorption) are different each other. The close to unit S values indicate a cycle design good symmetry with respect to the y_M isostere and relative small differences between flows and specific heats, and are preferred in operation, while those far from unit indicate the contrary. The cycle simple mass balance results also in Eq. (3.13), wherefrom, with Eq. (3.12), we obtain the Eqs. (3.14) and (3.15). The cycle a and

Table 3.2 The most important nontruncated cycle parameters

Cycle	Cooling nontruncated		Heating nontruncated	
Devices partial mass balance	$V_{GD} g_1 = a$	(3.9)	$V_{DG} a_1 = g$	(3.20)
	$V_{GD} r_1 = d$	(3.10)	$V_{DG} d_1 = r$	(3.21)
	$V_{GD} = \frac{V_G}{V_D};\ V_D = 1$	(3.11)	$V_{DG} = \frac{V_D}{V_G};\ V_G = 1$	(3.22)
Typical equation	$\frac{a}{d} = \frac{g_1}{r_1} = S = \frac{y_{RO} - y_M}{y_M - y_{GO}} = const. > 0$	(3.12)	$\frac{g}{r} = \frac{a_1}{d_1} = S = \frac{y_M - y_{DO}}{y_{AO} - y_M} = const. > 0$	(3.23)
Cycle mass balance	$f_M = V_{GD}(g_1 + r_1) = a_1 + d_1$	(3.13)	$f_M = V_{DG}(a_1 + d_1) = r_1 + g_1$	(3.24)
Derived equations	$\frac{V_{GD} g_1}{f_M} = \frac{a}{f_M} = \frac{S}{S+1}$	(3.14)	$\frac{V_{DG} a_1}{f_M} = \frac{g}{f_M} = \frac{S}{S+1}$	(3.25)
	$\frac{V_{GD} r_1}{f_M} = \frac{d}{f_M} = \frac{1}{S+1}$	(3.15)	$\frac{V_{DG} d_1}{f_M} = \frac{r}{f_M} = \frac{1}{S+1}$	(3.26)
Mixing concentrations	$y_{DO} = y_{DI} - \frac{Y_D - y_{DI}}{d-1}$	(3.16)	$y_{GO} = y_{GI} - \frac{Y_G - y_{GI}}{g-1}$	(3.27)
	$y_{AO} = y_{AI} + \frac{Y_D - y_{AI}}{a+1}$	(3.17)	$y_{RO} = y_{RI} + \frac{Y_G - y_{RI}}{r+1}$	(3.28)
Mixer mass balance	$d_1 y_{DO} + a_1 y_{AO} = (d_1 + a_1) y_M$	(3.18)	$r_1 y_{RO} + g_1 y_{GO} = (r_1 + g_1) y_M$	(3.29)
Dependent mixing concentration	$y_{AO}(y_{DO})$	(3.19)	$y_{RO}(y_{GO})$	(3.30)
	$= \frac{d_1(y_{DO} - y_M) y_{AI} + (Y_D - y_{AI}) y_M}{d_1(y_{DO} - y_M) + (Y_D - y_{AI})}$		$= \frac{g_1(y_{GO} - y_M) y_{RI} + (Y_G - y_{RI}) y_M}{g_1(y_{GO} - y_M) + (Y_G - y_{RI})}$	
	or		or	
	$y_{DO}(y_{AO})$		$y_{GO}(y_{RO})$	
	$= \frac{a_1(y_M - y_{AO}) y_{DI} + (Y_D - y_{DI}) y_M}{a_1(y_M - y_{AO}) + (Y_D - y_{DI})}$		$= \frac{r_1(y_M - y_{RO}) y_{GI} + (Y_G - y_{GI}) y_M}{r_1(y_M - y_{RO}) + (Y_G - y_{GI})}$	

d flow factors depend on the desorber and absorber end mixing concentrations, y_{DO} and y_{AO}, respectively. They can be determined from Eqs. (3.1), (3.4), (3.9) and (3.10), resulting in Eqs. (3.16) and (3.17), respectively. The computation is direct, if $Y_D = 1$ (the non-volatile absorbent case), or an iterative one, if $Y_D < 1$ (the volatile absorbent case). Equation (3.18) is giving the mixer mass balance of refrigerant species. In applications, y_{DO} is considered an independent (known) mixing concentration, y_M is known as well, and y_{AO} is a dependent (resulting) mixing concentration. Its value can be calculated with the help of Eq. (3.18), resulting in Eq. (3.19). Once y_{DO} and y_{AO} are known, the *d* and *a* flow factors can be computed and this is of help in the coabsorbent cooling cycle model, as it will result later. Alternatively, it is possible to follow an iterative computation process, choosing a value for the V_{GD} vapor and calculating the y_{DO} and y_{AO} concentrations so that they have desired values usually in the range of $y_M \leq y_{DO} \leq y_{DI}$, and $y_M \geq y_{AO} \geq y_{AI}$, respectively. Bearing this in mind, in this case the desorber and absorber processes end temperatures are smaller than the mixer temperature, $T_{DO} \leq T_M$, and higher than that, $T_{AO} \geq T_M$, respectively. As it was already written, this situation would characterize a normal cycle operation, Fig. 3.3a.

However, if this were not true, that is accidentally or systematically we have $T_{DO} \geq T_M$ and $T_{AO} \leq T_M$, respectively, when the sources favour the cycle COP increase, then the desorber outlet concentration y_{DO} decreases below the mixing point concentration, $y_M \geq y_{DO}$, and the absorber outlet concentration y_{AO} exceeds it, $y_M \leq y_{AO}$, Fig. 3.3c, but in all cases equations (3.16) and (3.17) verify unconditionally the mass balance on the mixer *M*, given by Eq. (3.18) for a species. In other words, the mixing concentration y_M is achievable with mixing absorbents having alternatively the end concentration below and above the y_M value around the *balance point*, which is characterizing the cycle operation with the desorber and absorber end concentrations equal to that of the mean concentration, $y_{DO} = y_{AO} = y_M$, Fig. 3.3b. The balance point corresponds to the V_{GD} ratio equal to unit in case of the pure refrigerant, only, that is $Y_D = 1$ (see Fig. 3.16 heat cycle case study). In Fig. 3.5 we present the results of an ammonia/water first case study (Staicovici 2006a). This shows the cycle features described so far. Here we considered the resorber and generator outlet concentrations as $y_{RO} = 0.9$ and $y_{GO} = 0.37$, respectively and the mean concentration of $y_M = 0.55$. The mean generator vapor concentration is $Y_G = 0.974$ and that of the desorber was variable, $Y_D = 0.982$–1.0. The pressure and temperature data in the cycle main points are: $p_{GR} = 11$ bar, $p_{AD} = 5$ bar, $T_{GO} = 90$ °C, $T_{GI} = T_{RI} = 55$ °C, $T_{RO} = T_M = 30$ °C, $T_{DI} = 7$ °C, $T_{AI} = 60$ °C. The concentration threshold condition is fulfilled because $Y_G > y_{RO}$. Applying Eqs. (3.2) and (3.3), the generator and resorber inlet flow factors resulted in $g = 3.356$ and $r = 0.2114$. The outlet generator and resorber flow factors resulted applying Eqs. (3.6) and (3.7), respectively, as $g_1 = 2.356$ and $r_1 = 1.2114$. The balance point $y_{DO} = y_{AO} = y_M = 0.55$ is reached when the vapor ratio is $V_{GD} \cong 1.06$, different from unit, because $Y_D \leq 1$. The mixer mean concentration can be achieved with the absorbent flows a_1 of concentrations $y_{AO} < y_M = 0.55$ and temperatures $T_{AO} > T_M = 30$ °C, mixed up with the absorbents d_1 of concentrations $y_{DO} > y_M = 0.55$ and temperatures $T_{DO} < T_M = 30$ °C, if the cycle worked in the

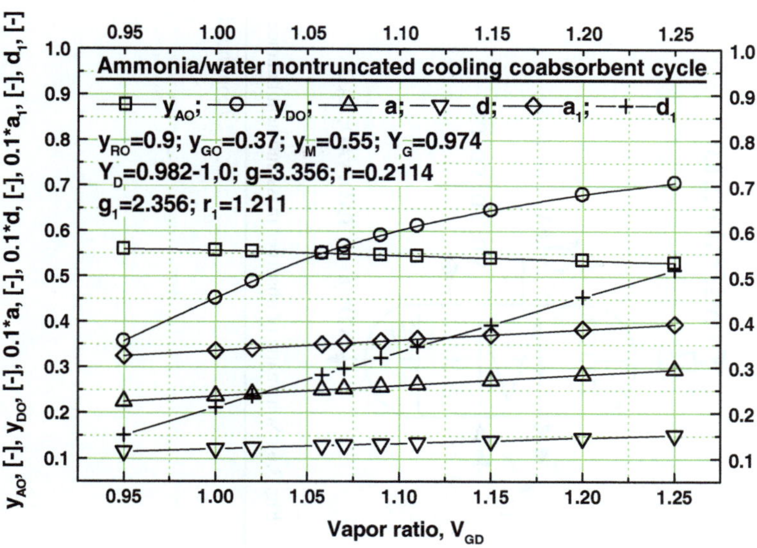

Fig. 3.5 Flow analysis case study of a nontruncated cooling coabsorbent cycle

right side of the balance point, where $V_{GD} > 1.06$ and a normal cycle operation is met. Alternatively, the mixer mean concentration can be achieved with the absorbent flows a_1 of concentrations $y_{AO} > y_M = 0.55$ and temperatures $T_{AO} < T_M = 30$ °C, mixed up with the absorbents d_1 of concentrations $y_{DO} < y_M = 0.55$ and temperatures $T_{DO} > T_M = 30$ °C, if the cycle worked in the left side of the balance point, where $V_{GD} < 1.06$ and sources temperatures favour the cycle COP increase. The absorbent exits the desorber and absorber with temperatures varying within $T_{DO} = (63–16)$ °C and $T_{AO} = (28.5–32.5)$ °C, respectively, when the vapor ratio V_{GD} covers the abscissa in Fig. 3.5. From the graph we can also remark that a high variation of the y_{DO} value is bringing about only a small variation of the y_{AO} value. The first Eq. (3.19) is showing this behavior directly. As prior mentioned, the balance point left side is a higher cooling COP region as compared to the right one. Indeed, here the cycle works on the broader desorption and absorption temperature intervals, with comparatively higher and lower mean values, respectively. This, according to the Lorentz ideal absorption cooling COP (Niebergall 1959), is favouring the efficiency increase. Figure 3.5 plots as well the desorber and absorber flow factors for this case study. They have a linear behaviour with the vapor ratio variation.

3.2.1.2 Cycle Model

The nontruncated cooling coabsorbent cycle was modeled considering the internal heat recovery configuration given in Fig. 3.4. The model is simple, stationary. It includes first Eqs. (3.1)–(3.20) of Tables 3.1 and 3.2, selectively. To these equations it is added a new set of Eqs. (3.31)–(3.42), Tables 3.3 and 3.4, resulting from the

Table 3.3 Refrigerant mass balance and heat balance of absorber and generator

Device	Balance equations		Balance figure
Absorber	Refrigerant mass balance: $V_D a y_{AI} + V_D Y_D = V_D(a+1) y_{AO}$ Resulting equation: $a = \frac{Y_D - y_{AO}}{y_{AO} - y_{AI}}$ Heat balance: $q_A + V_D(a+1)h_{AO} = V_D H_D + V_D a h_{AI}$ Resulting equation: $q_A = V_D((H_D - h_{AO}) + a(h_{AI} - h_{AO}))$	(3.31) (3.1) (3.32) (3.33)	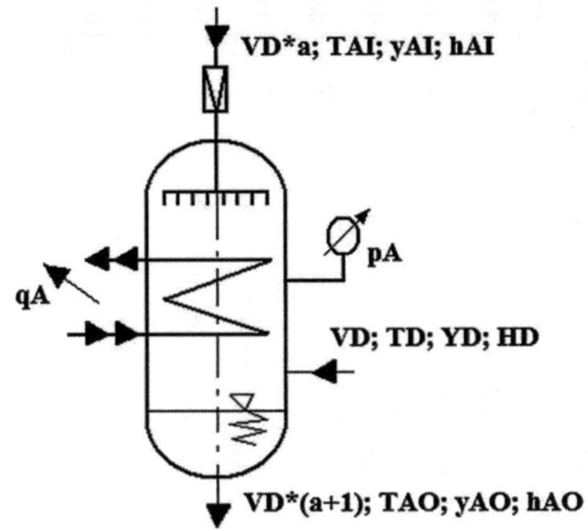 **Fig. 3.6** Schematic of absorber mass and heat balance

(continued)

Table 3.3 (continued)

Device	Balance equations		Balance figure
Generator–rectifier	Generator refrigerant mass balance: $V_G g y_{GI} = V_G Y_G + V_G (g-1) y_{GO}$	(3.34)	See Fig. 3.7
	Resulting equation: $g = \frac{Y_G - y_{GO}}{y_{GI} - y_{GO}}$	(3.2)	
	Rectifier refrigerant mass balance: $V_G(1+x)Y_G = V_G Y_{rctf} + V_G x y_{rctf}$		
	Resulting equations: $x = \frac{Y_{rctf} - Y_G}{Y_G - y_{rctf}}; \; x_{id} = \frac{Y_{rctf} - Y_G}{Y_G - y_{GI}}$ $y_{rctf} = Y_G - \left(\frac{x_{id}}{x}\right)(Y_G - y_{GI}); \; \left(\frac{x_{id}}{x}\right) = 0.7 - 0.92$ (NH_3–H_2O, Niebergall 1959)		
	Generator heat balance: $q_G + V_G g h_{GI} = V_G (g-1) h_{GO} + V_G H_G$		
	Resulting equation: $q_G = V_G((H_G - h_{GO}) + g(h_{GO} - h_{GI}))$	(3.35)	
	Rectifier heat balance: $q_{rctf} + V_G H_{rctf} + V_G x h_{rctf} = V_G (1+x) H_G$		
	Resulting equation: $q_{rctf} = V_G\left(\left(H_G - H_{rctf}\right) + x\left(H_G - h_{rctf}\right)\right)$		
	Generator–Rectifier heat: $q_{Grctf} = V_G\left(\left(H_{rctf} - h_{GO}\right) + g(h_{GO} - h_{GI})\right) + q_{rctf}$	(3.36)	

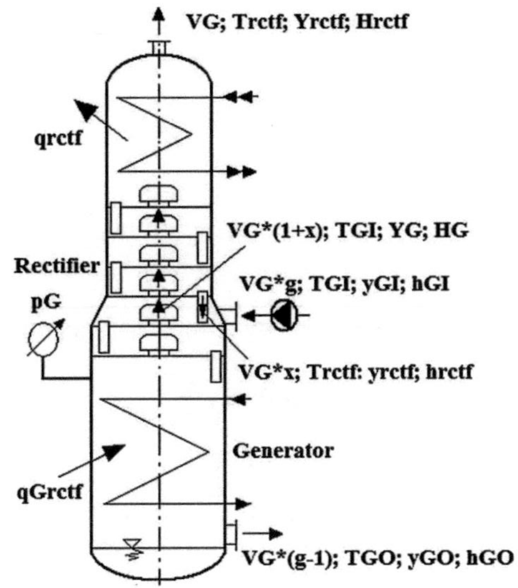

Fig. 3.7 Schematic of generator-rectifier mass and heat balance (according to Niebergall 1959)

Table 3.4 Refrigerant mass balance and heat balance of resorber and desorber

Device	Balance equations	Balance figure
Resorber	Refrigerant mass balance: $V_G r y_{RI} + V_G Y_G = V_G(r+1)y_{RO}$ (3.37) Resulting equation: $r = \frac{Y_G - y_{RO}}{y_{RO} - y_{RI}}$ (3.3) Heat balance: $q_R + V_G(r+1)h_{RO} = V_G H_G + V_G r h_{RI}$ (3.38) Resulting equation: $q_R = V_G \begin{pmatrix} (H_G - h_{RO}) + \\ r(h_{RI} - h_{RO}) \end{pmatrix}$ (3.39)	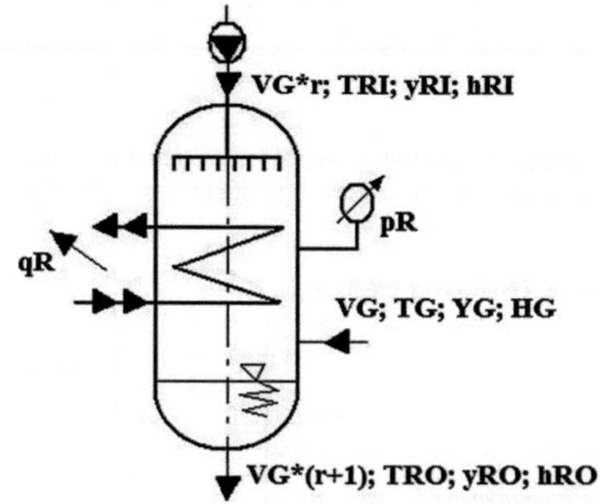 **Fig. 3.8** Schematic of resorber mass and heat balance

(continued)

Table 3.4 continued

Device	Balance equations		Balance figure
Desorber	Refrigerant mass balance:		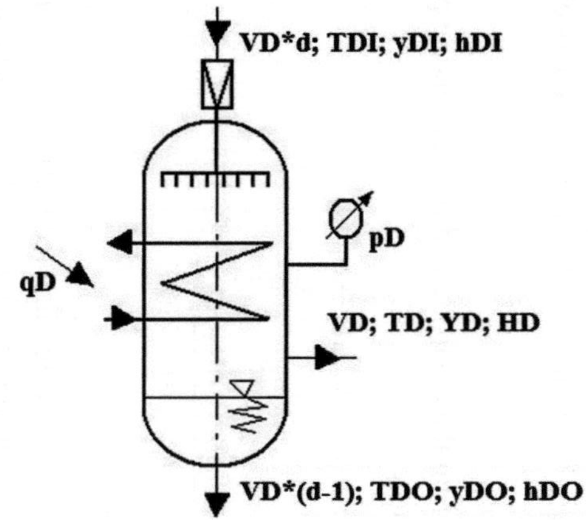 **Fig. 3.9** Schematic of desorber mass and heat balance
	$V_D d y_{DI} = V_D Y_D$	(3.40)	
	$+ V_D (d-1) y_{DO}$		
	Resulting equation:		
	$d = \frac{Y_D - y_{DO}}{y_{DI} - y_{DO}}$	(3.4)	
	Heat balance:		
	$q_D + V_D d h_{DI} =$	(3.41)	
	$V_D (d-1) h_{DO} + V_D H_D$		
	Resulting equation:		
	$q_D =$	(3.42)	
	$V_D \begin{pmatrix} (H_D - h_{DO}) + \\ d(h_{DO} - h_{DI}) \end{pmatrix}$		

refrigerant mass balance and heat balance written for absorber, generator, resorber and desorber and schematically presented in Figs. 3.6, 3.7, 3.8 and 3.9. Next we give also the calculus of other necessary functions, which complete the model.

Mixer heat balance:

The enthalpy of absorbent resulting from the adiabatic and perfect mixing process is calculated from the mixer heat balance:

$$V_D d_1 h_{DO} + V_D a_1 h_{AO} = V_D(d_1 + a_1)h_M \tag{3.43}$$

resulting in,

$$h_M = \frac{d_1 h_{DO} + a_1 h_{AO}}{d_1 + a_1} \tag{3.44}$$

Solution to solution recovery heat exchange:

The absorbent supplying the generator is preheated in a recovering way in the heat exchanger 9 of Fig. 3.4. This is done by subcooling the absorbent exiting the generator, prior to be expanded and to enter the absorber. Its enthalpy can be calculated from the heat balance to this heat exchanger:

$$V_G g_1(h_{GO} - h_{AI}) = V_G g(h_{GI} - h_M) \tag{3.45}$$

In Eq. (3.45), the state function $h_{AI} = h_{AI}(p_G, T_M + \delta T)$, where δT is a necessary heat exchange positive temperature pinch. The searched function results from Eq. (3.45) as:

$$h_{GI} = h_M + \frac{g_1}{g}(h_{GO} - h_{AI}) \tag{3.46}$$

The heat exchange is controlled by the mean logarithmic temperature difference value, calculated by Popa and Vintila (1977):

$$\Delta T_{lg,m} = \frac{(T_{GO} - T_{GI}) - \delta T}{\ln \frac{T_{GO} - T_{GI}}{\delta T}} \tag{3.47}$$

In Eq. (3.47), T_{GI} is the effective temperature of the absorbent entering the generator and having the h_{GI} enthalpy, given by Eq. (3.46).

Gas to solution recovery heat exchange:

The absorbent leaving the resorber is subcooled in a recovering way prior to be expanded and to supply the desorber. This is done in the heat exchanger 8, Fig. 3.4, by superheating the desorbed vapor prior to be absorbed in the absorber. Its enthalpy can be calculated from the heat balance of this heat exchanger:

$$V_G r_1(h_{RO} - h_{DI}) = V_D\big(H_D(p_D, T_{RO} - \delta T) - H_D\big(p_D, T_{D,m}\big)\big) \tag{3.48}$$

In Eq. (3.48), the vapor is superheated from a non-equilibrium state characterized by the parameters (p_D, $T_{D,m}$, Y_D) till a non-equilibrium state, with state parameters (p_D, $T_{RO} - \delta T$, Y_D), where $T_{D,m}$ is the initial mean vapor temperature and δT is again a necessary heat exchange positive temperature pinch. The searched function results from Eq. (3.48) as:

$$h_{DI} = h_{RO} - \frac{V_D}{V_G r_1}\left(H_D(p_D, T_{RO} - \delta T) - H_D\left(p_D, T_{D,m}\right)\right) \tag{3.49}$$

The heat exchange is controlled by the mean logarithmic temperature difference value, calculated here by Popa and Vintila (1977):

$$\Delta T_{lg,m} = \frac{\left(T_{RO,sc} - T_D\right) - \delta T}{\ln \frac{T_{RO,sc} - T_D}{\delta T}} \tag{3.50}$$

In Eq. (3.50), $T_{RO,sc}$ is the subcooling temperature.

Solution pumping work:

The cycle theoretical mechanical work consumed for pumping the absorbent is given by:

$$w_p = f_M \frac{V_D(d_1 v_{DO} + a_1 v_{AO})}{V_D(a_1 + d_1)}(p_G - p_A) \tag{3.51}$$

The middle factor in Eq. (3.51) is the value of the mixed absorbent specific volume, calculated as a weighted mean of the components specific volumes. Bearing in mind Eq. (3.13), the searched work becomes:

$$w_p = V_D(d_1 v_{DO} + a_1 v_{AO})(p_G - p_A) \tag{3.52}$$

Cycle theoretical coefficients of performance (COP):

Cooling COP:

$$COP_c = \frac{Useful\_cooling\_effect}{Heat\_input} = \frac{q_D}{q_G} \tag{3.53}$$

Heating COP:

$$COP_h = \frac{Useful\_heating\_effect}{Heat\_input} = \frac{q_R}{q_G} \tag{3.54}$$

COP relative to the cycle mechanical work consumption:

$$COP_w = \frac{Useful\_main\_cooling(heating)\_effect}{Work\_input} = \frac{q_D(q_R)}{w_c + w_p} \tag{3.55}$$

In Eq. (3.55), w_c holds for the mechanical vapor compression, which intervenes in the hybrid operation cycles (see further in the text).

Cogeneration of cooling and heat COP:

$$COP_{cg} = \frac{Useful\_cooling + heating\_effect}{Heat\_input} = \frac{q_D + q_R}{q_G} \tag{3.56}$$

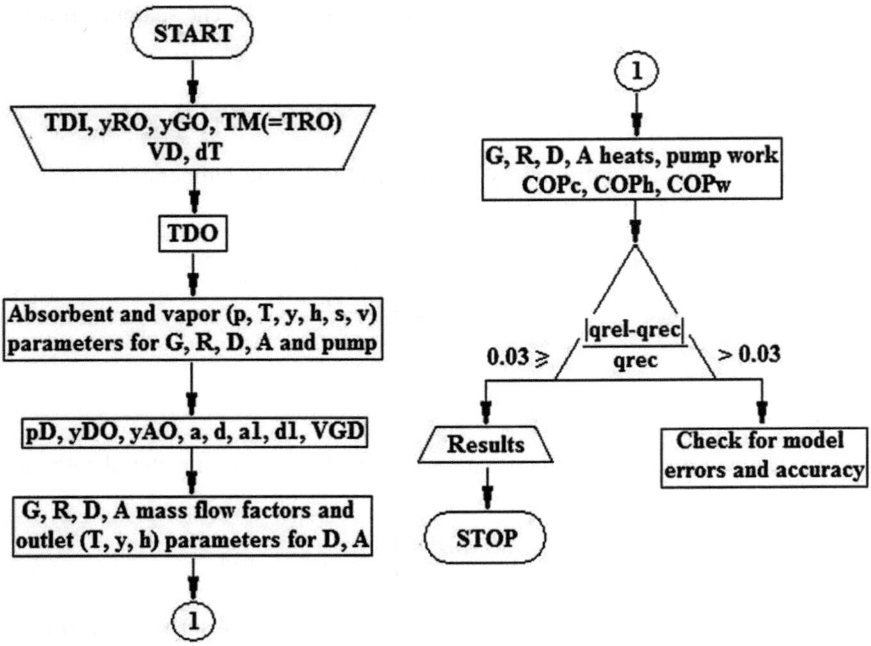

Fig. 3.10 The cooling cycle model organigrame

Cycle heat balance accuracy check:

Cycle (output) released energy (heat):

$$q_{rel} = q_A + q_R \tag{3.57}$$

Cycle (input) received energy:

$$q_{rec} = q_G + q_D + w_p(+w_c) \tag{3.58}$$

Check for the cycle heat balance accuracy and algorithm and computation correctness:

$$\Delta q_{rel,rec} = \frac{|q_{rel} - q_{rec}|}{q_{rec}} \leq 0.03 \tag{3.59}$$

The model solution consisting of Eqs. (3.1)–(3.59) recommends a computer use. The organigrame of a most probable steady-state non-iterative model computer code is presented in Fig. 3.10. The input data are those mentioned in Sect. 3.2.1.1, points (a–c), except T_{GO}, which was replaced by y_{GO}. A specific application is characterized by a required T_{DO} value, the next model input data. With the first input data, the code calculates the cycle configuration, that is the absorbent and vapor p, T, y, h, s, v state parameters of the main devices. Using the T_{DO} value, the y_{DO} concentration is found from the equilibrium diagram of the working combination, $y_{DO} = y_{DO}(T_{DO}, p_D)$ (e.g. enthalpy-concentration, $h - y$, Chap. 1). Further, the y_{AO} concentration is determined with the help of the first Eq. (3.19). These two concentrations enable the a and d cooling cycle flow factors computation, Eqs. (3.1) and (3.4), and further the V_{GD} vapor generated is found, Eq. (3.9) or Eq. (3.10).

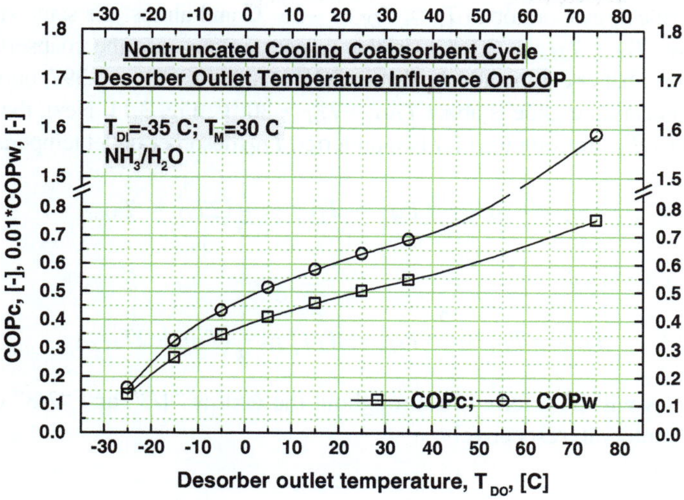

Fig. 3.11 The desorber outlet temperature (T_{DO}) influence on COP_c for $T_{RO} = T_{M,c}$, $T_{GO} = 123.1C$, $T_{RI} = 110.1C$, $r \cong 0.2$; $y_{RO} \cong 0.71–0.74$; $y_M \cong 0.226–0.238$

Finally, the main devices flow factors, energy and cycle COP's are derived. The model output values are valid when Eq. (3.59) is fulfilled.

3.2.1.3 Heat Transfer Analysis of Cycle Desorption and Resorption Processes and Model Results

Desorption Interval Importance

A first concern was to study the influence which the desorber outlet temperature, T_{DO}, is having on the COP_c and COP_w effectivenesses. Let us consider the study case shown in Fig. 3.11, done for $T_{DI} = -35$ °C and $T_{M,c} = 30$ °C constant temperatures. The $COP_c(T_{DO})$ and $COP_w(T_{DO})$ functions highly increase with the T_{DO} increase. According to the chosen example, when T_{DO} covers the range $T_{DO} = T_{DI} + 10$ °C to $T_{DO} = T_{DI} + 110$ °C, the cooling effectiveness increases by about 5 times. Consequently, a nontruncated cooling coabsorbent cycle should operate with desorption temperature intervals as large as possible, for the $COP_c(T_{DO})$ benefit. On the contrary, if the $T_{DO} - T_{DI}$ desorption temperature interval is enough small, the COP_c of the cooling condensing cycle can be particularly much better than the $COP_c(T_{DO})$ of the coabsorbent cycle, even the condensing cycle uses the rectification. An example proving this and matching closely to our data, is provided here by the work of (Niebergall 1959, pp. 217–219). Indeed, he shows that a condensing ammonia-water medium cooling cycle with vapor rectification has a cooling effectiveness of $COP_c = 0.397$ if its evaporator and condenser operated to $T_E = -34$ °C and $T_C = 30.4$ °C temperatures, respectively. In case of the coabsorbent cycle (see Fig. 3.11), its cooling effectiveness is more than 2.5 times less, comparatively, $COP_c(T_{DO}) \cong 0.15$, for a desorber temperature glide of $T_{DO} - T_{DI} = 10$ °C (acceptable cooling heat exchanger temperature pinch for both cycles), evaporator to $T_E \cong T_{DI} = -35$ °C and almost the same sink source temperature ($T_C \cong T_{M,c} = 30$ °C). In this respect, in our case the coabsorbent cycle could perform better comparatively, that is $COP_c(T_{DO}) \geq COP_c = 0.397$, only when its desorber temperature glide is about $T_{DO} - T_{DI} \geq 30$ °C, Fig. 3.11. Next, the functions COP_c and COP_w are computed for two preferred T_{DO} desorber outlet temperatures:

$$T_{DO} = T_{DI} + \frac{1}{3}\left(T_{M,c} - T_{DI}\right) = \frac{1}{3}\left(2T_{DI} + T_{M,c}\right)$$

and

$$T_{DO} = T_{DI} + \frac{2}{3}\left(T_{M,c} - T_{DI}\right) = \frac{1}{3}\left(T_{DI} + 2T_{M,c}\right)$$

These values will be referred to throughout the text as "1/3" and "2/3" out of the $T_{M,c} - T_{DI}$ interval, respectively.

Desorption Heat Transfer Analysis

To the beginning, a study case of normal cooling is considered (Staicovici 2011). The desorption process takes place with a short gliding temperature. It will be analyzed from the heat transfer type (a) point of view. To this extent, the DD method is employed, Chap. 2. The input data are as follows:

- Desorber inlet/outlet temperature, $T_{DI} = -10.0$ °C/$T_{DO} = 3.35$ °C;
- Cycle sink temperature, $T_{M,c} = 30$ °C;
- Resorber outlet concentration, $y_{RO} = 0.975$;
- Desorber vapor quantity, $V_D = 1$;
- Brine initial temperature, $T_{Db,0} = 5.85$ °C.
- Number of steps, $n_D = 10$.

The desorber device is exchanging heat in counter-current with a brine having a mass flow rate m and a specific heat c_p, which is supposed to have a constant value during its useful cooling process. Let us consider the desorption interval $I_D = [T_{DI}, T_{DO}]$, $T_D \in I_D$, is covered by $T_{D,i}$ tabular values equally spaced, $T_{D,i} = T_{DI} + (i - 0.5)\Delta T_D$, where $i = 1, 2, \ldots, n_D$ and

$$\Delta T_D = \frac{T_{DO} - T_{DI}}{n_D} \tag{3.60}$$

is the desorber step or tabulation interval. The specific thermal capability function (kJ/kg K) of a tabular value is noted by $X_i = X(T_{D,i})$. The brine temperature interval is $I_{Db} = \left\lfloor T_{Db,n_D}, T_{Db,0} \right\rfloor$, $T_{Db} \in I_{Db}$. The current brine temperature $T_{Db,i}$, $i = 1, 2, \ldots, n_D$, is obtained simply through:

$$T_{Db,i} = T_{Db,0} - \sum_{j=n_D+1-i}^{n_D} \Delta T_{Db,j} \tag{3.61}$$

where $T_{Db,0}$ is its initial value, $T_{Db,0} = T_{DO} + (1\text{–}5)\ °C$, and $\Delta T_{Db,j}$ is the brine temperature decrease per step j. The algorithm which will be presented enables to reduce the heat exchange analysis to a verification of the (i) and (ii) conditions fulfillment, Chap. 2, in the $T_{D,i}$ and $T_{Db,i}$ temperatures points. This way to proceed has been used for all absorption processes investigated from the heat exchange viewpoint throughout the book. Taking into account Eq. (3.60), the thermal capability mean integral value is approximated on the desorption interval by:

$$\bar{X} = \frac{1}{T_{DO} - T_{DI}} \sum_{i=1}^{n_D} X\left(T_{D,i}\right) \Delta T_D = \frac{1}{n_D} \sum_{i=1}^{n_D} X\left(T_{D,i}\right) \tag{3.62}$$

Proceeding similarly, the brine heat input can be written in function of its heat capacity mean value $\left(\overline{mc_p}\right)$ as:

$$\sum_{i=1}^{n_D} \left(mc_p\right) \Delta T_{Db,i} = \left(\overline{mc_p}\right) \left(T_{Db,0} - T_{Db,n_D}\right) \tag{3.63}$$

The local fulfillment of the conditions (i) and (ii) is expressed by:

$$\left(\overline{mc_p}\right) \Delta T_{Db,i} = X_i \Delta T_D, \quad i = 1, 2, \ldots, n_D \tag{3.64}$$

and

$$T_{Db,i} > T_{D,i} + \frac{1}{2} \Delta T_D \tag{3.65}$$

respectively. Equation (3.65) are added term by term for all i and considering Eqs. (3.61) and (3.62) it results:

$$\bar{X}(T_{DO} - T_{DI}) = \left(\overline{mc_p}\right) \left(T_{Db,0} - T_{Db,n_D}\right) \tag{3.66}$$

Equation (3.66) expresses the global condition (i). The brine temperature interval in Eq. (3.66) can be chosen at convenience. Here it is prefered to work with:

$$T_{Db,0} - T_{Db,n_D} = T_{DO} - T_{DI} \tag{3.67}$$

wherefrom, Eq. (3.66) becomes:

$$\bar{X} = \left(\overline{mc_p}\right) \tag{3.68}$$

Further, combining Eqs. (3.68) and (3.64), the brine temperature drop per step is expressed by:

$$\Delta T_{Db,i} = \frac{X_i}{\bar{X}} \Delta T_D \tag{3.69}$$

The application of DD method leads to results shown by Table 3.5. They have been obtained with Eqs. (2.12), (2.14), (2.15) and (2.23) for $n_D = 10$. From Eq. (3.60) we have $\Delta T_D = 1.333$ and with the X_i data, Table 3.5, Eq. (3.62) results in

Table 3.5 Ammonia-water results of a desorption heat exchange type (a) study case, obtained by applying the DD method for thermal capability assessment

i	1	2	3	4	5	6	7	8	9	10
$T_{D,i}$ (°C)	−9.35	−7.95	−6.65	−5.35	−4.05	−2.65	−1.35	−0.05	1.35	2.65
X_i (kJ/kg K)	528.9	106.8	42.26	22.85	14.67	10.44	7.944	6.337	5.232	4.434
$\Delta T_{Db,i}$ (°C)	9.404	1.899	0.751	0.406	0.261	0.186	0.141	0.113	0.093	0.079
$T_{Db,i}$ (°C)	−8.02	1.387	3.286	4.038	4.444	4.705	4.891	5.032	5.145	5.238

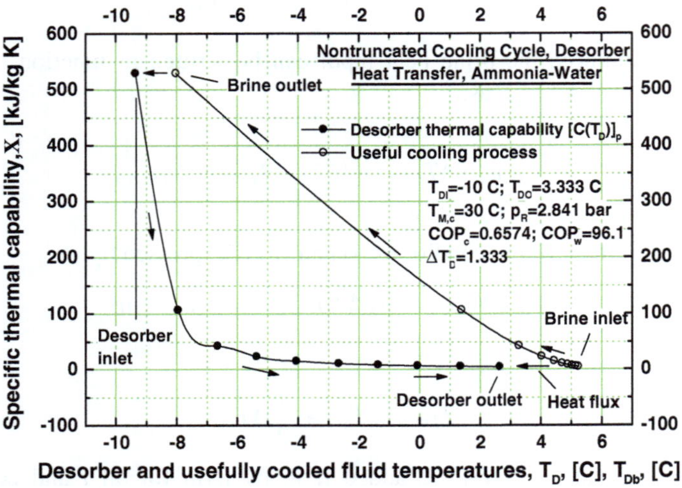

Fig. 3.12 Normal cooling heat exchange of a desorber with short gliding temperature

$\bar{X} = 74.9867$. Further on, the values of the 4th row in Table 3.5 are found with Eq. (3.69). The total brine cooling amounts to:

$$\sum_{i=1}^{10} \Delta T_{Db,i} = 13.333\,°\mathrm{C} \tag{3.70}$$

Applying Eq. 3.61, the $T_{Db,i}$ values have been calculated and are given in the last row of Table 3.5. The thermal evolution of the absorbent and brine heat exchange, expressed in terms of the specific thermal capability (kJ/kg refrigerant K), is represented versus desorber and brine temperatures in Fig. 3.12. As it is easy to remark, the (i) and (ii) conditions are respected. In practice, however, a final check must be done in conjuncture to the particular geometry of the desorber device. The current specific desorption heat (kJ/kg refrigerant) along an absorption process is generally calculated by one of Eqs. (2.16), (2.18), (2.26), (2.28). Here we used Eq. (2.16). It is completely non-linear with respect to the desorber temperature, as Fig. 3.13 shows.

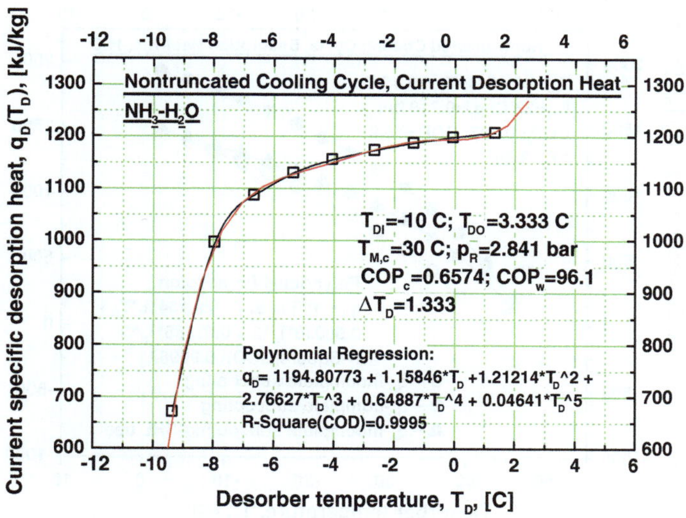

Fig. 3.13 Current specific desorption heat in case of a normal cooling

Next, it will be considered a second study case, this time for industrial cooling, with a much higher desorption temperature glide (Staicovici 2011). The input data are as follows:

- Desorber inlet/outlet temperature, $T_{DI} = -47.45$ °C/$T_{DO} = -0.45$ °C;
- Cycle sink temperature, $T_{M,c} = 23$ °C;
- Resorber outlet concentration, $y_{RO} = 0.642$;
- Desorber vapor quantity, $V_D = 1$;
- Brine initial temperature, $T_{Db,0} = 2.5275$ °C;
- Number of steps, $n_D = 14$.

The absorbent with low y_{RO} concentration is introducing a new particular aspect to be analyzed, when performing a desorption process, namely the reduction of the useful desorption temperature design interval. The current specific desorption heat and the absorbent and brine heat exchange are represented for this latter case in Figs. 3.14 and 3.15, respectively. Figure 3.14 shows that the desorption heat (lower, actual curve) has negative values to desorber inlet, $q_D < 0$, that is heat is rejected by the absorbent in this portion. Such an absorbent behavior is well known and can be met to both condensing and coabsorbent cooling cycles. It takes place whenever a nonequilibrium incomplete subcooled state fluid (mono-or poly-component substance) suffers an expansion (here from the resorber high pressure to the desorber low pressure), causing a certain amount out of its mass be evaporated in order to reach the first equilibrium state (here given by $T_D = -42.45$ °C $> T_{DI}$; $p = 0.218$ bar). Suppose the enthalpy of the absorbent which suffered an incomplete subcooling in the subcooler 8, Fig. 3.4, is noted by h_{DII}. Similarly, the enthalpy of the absorbent corresponding to design equilibrium point, characterized

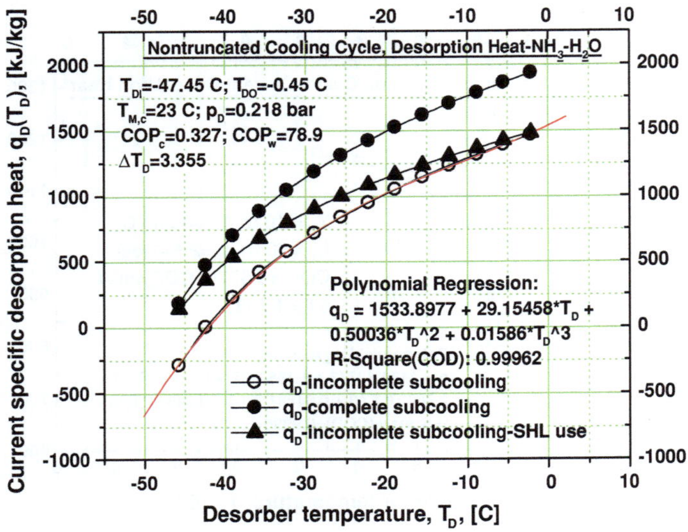

Fig. 3.14 Current specific desorption heat in case of an industrial cooling for incomplete and complete subcooling

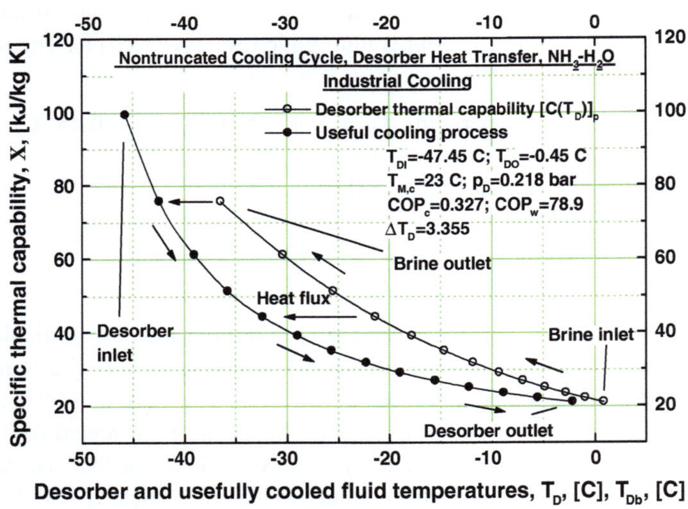

Fig. 3.15 Industrial cooling heat exchange of a desorber with large gliding temperature in case of incomplete subcooling

by the state parameters (T_{DI}, p_D) and benefiting by a complete subcooling, is $h_{DI}(T_{DI}, p_D)$. The heat of the entire desorption processes provided with complete and incomplete absorbent subcooling are given by Eqs. (3.42) and (3.71),

$$q_{D,ics} = V_D((H_D - h_{DO}) + d(h_{DO} - h_{DII})) \tag{3.71}$$

respectively. The plot of Eq. (2.16) throughout desorber, versus T_D, written for the complete and incomplete subcooling, is shown in our case in Fig. 3.14 by the upper and lower curve, respectively. Theirs maximum values are calculated with Eqs. (3.42) and (3.71), respectively. In each T_D point the two curves differ by a constant *subcooling (heat) deficiency*, computed with:

$$q_{sd} = V_D d(h_{DII} - h_{DI}) \tag{3.72}$$

(here $q_{sd} = 469$ kJ/kg). The presence of this deficiency causes two negative effects. The first one diminishes correspondingly the desorption heat. Indeed, after expansion, the absorbent becomes superheated with the subcooling heat deficiency amount. For this reason, it will extract at first the superheat out of its own mass, instead of extracting heat from the very beginning out of the brine mass, and afterwards will extract heat from the brine, according to its remaining cooling capacity, only. The second effect is met just to a coabsorbent plant (not to a condensing one) and consists in diminishing the useful desorption temperature interval by the domain corresponding to the portion where subcooling heat deficiency is extracted. Figure 3.14 shows heat is extracted/rejected until cca. −42.45 °C, so the brine cooling process can take place up to this temperature, only, instead of −47.45 °C, Fig. 3.15. In other words, a subcooling deficiency brings about an *(useful) heat and temperature double deficiency* to the desorption process in coabsorbent plants operating with far from unit concentration (e.g. $y < 0.9$) desorbing absorbents. Fortunately, the temperature deficiency can be counteracted to a major extent, providing these processes with a *subcooling hysteresis loop* (SHL), introduced in this book. Next we shall describe the SHL solution, given schematically in Fig. 3.16. From the figures already presented it results that the actual desorption process unfolding depends on the desorbing absorbent past subcooling state (where the hysteresis comes of) in such a way that the higher the subcooling, the smaller the desorption temperature deficiency. The only way to increase the subcooling is to use the own desorption cooling process. The absorbent leaving the resorber, Fig. 3.16, is first subcooled by the desorbed vapor in the subcooler 8 and then enters the SHL, position 14, where it yields the heat q_{sd} until its temperature becomes close to T_{DI}. Further on, the absorbent is expanded and goes into the desorber 1. Here it is usefully extracting an amount of heat from the source 15, which must be cooled, and the subcooling heat. The useful heat extracted amounts to $q_{D,cs} - q_{sd} = q_{D,ics}$ [see Eqs. (3.42), (3.71), (3.72)]. As it can be seen, the actual desorption heat of the desorber provided with SHL is identical to that without SHL, so the heat balance is closed. However, this time the advantage is that the temperature deficiency is almost completely eliminated, and the first desorber equilibrium state becomes close to the design one (T_{DI}, p_D). In order to study the desorber heat exchange of a nontruncated coabsorbent cooling cycle provided with SHL, let us consider that the desorber is constructively made of two sub-devices operating in parallel, each of them with a distinct role, of subcooling and of desorption, respectively (suggested by Fig. 3.16, also). Because of the SHL use, the desorber sub-devices operate with complete subcooling. The desorption heats of the subcooling and desorption sub-devices can be expressed as:

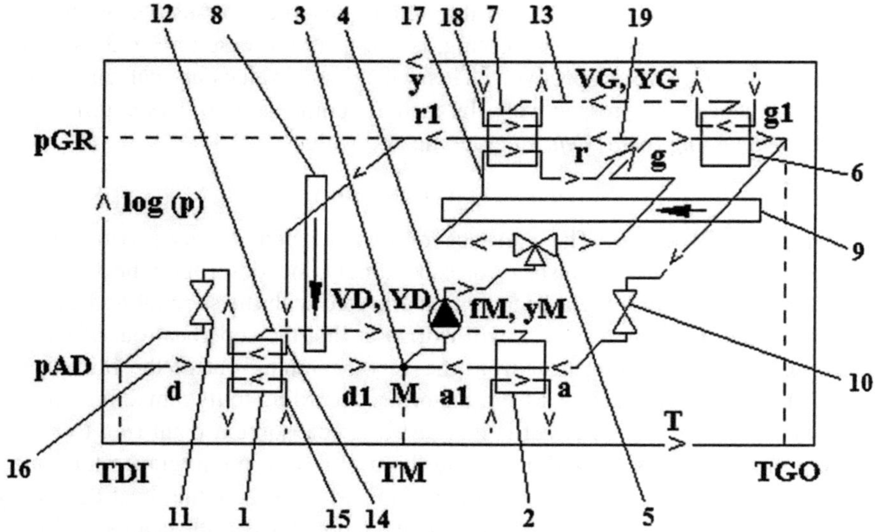

Fig. 3.16 Nontruncated cooling cycle provided with desorption subcooling and resorption preheating hysteresis loops (SHL and PHL, respectively)

$$V_{D,sd}[(H_D - h_{DO}) + d_{sd}(h_{DO} - h_{DI})] = V_D d(h_{DII} - h_{DI}) \tag{3.73}$$

$$V_{D,ics}[(H_D - h_{DO}) + d_{ics}(h_{DO} - h_{DI})] = V_D[(H_D - h_{DO}) + d(h_{DO} - h_{DII})] \tag{3.74}$$

respectively. In the above equations we have assumed that the temperature operation ranges of the subcooler 8, Fig. 3.4, and of subcooling sub-device do not overlap, which is very likely to happen in practice. Equations (3.73) and (3.74) are completed by the following two obvious equations:

$$V_{D,sd} + V_{D,ics} = V_D \tag{3.75}$$

$$d_{sd} = d_{ics} = d \tag{3.76}$$

From Eqs. (3.73)–(3.76), the unknown vapor $V_{D,sd}$ and $V_{D,ics}$ result, taking into account Eqs. (3.42), (3.71), and (3.72), as well:

$$V_{D,sd} = V_D \frac{q_{sd}}{q_{D,cs}} \tag{3.77}$$

and

$$V_{D,ics} = V_D \frac{q_{D,ics}}{q_{D,cs}} \tag{3.78}$$

respectively. The heat exchange can now be assessed, considering just the main part of the desorber (see the DD method, Chap. 2) and bearing in mind Eq. (2.16). In this equation, the factor V_D^m, given by Eq. (2.23), must be multiplied in case of the SHL use by one of the subunit vapor amounts given by Eqs. (3.77) and (3.78), depending whether the subcooling, or the desorption function of the desorber is intended to be analyzed with respect to the heat exchange, resulting in:

$$V_{D,sd}^m = V_{D,sd} V_D^m \tag{3.79}$$

and

$$V_{D,ics}^m = V_{D,ics} V_D^m \tag{3.80}$$

respectively. Results of the desorber heat exchange analysis at hand will be given here for the desorbtion sub-device, only, applying the DD method completed with the algorithm presented above. Solving Eqs. (3.77) and (3.78), resulted in $V_{D,sd} = 0.236$ and $V_{D,ics} = 0.764$, respectively. This means that, out of the total of $V_D = 1.0$ kg/kg desorbed vapor, 0.236 kg/kg is desorbed in the subcooling sub-device in order to subcool till ~T_{DI} the absorbent $V_D d$ going into desorber, and the rest of 0.764 kg/kg is desorbed in the desorption sub-device, in order to usefully extract the heat from the brine. The SHL absorbent subcooling is performed from 1.25 °C (subcooler 8 outlet temperature), till a temperature approaching $T_{DI} = -47.45$ °C. The SHL provided useful desorption heat evolution versus T_D is plotted in Fig. 3.14 (see the solid up triangle curve). As we expected, the curve is lying between the minimum value of the complete subcooling desorption heat top curve and the maximum value of the incomplete subcooling desorption heat bottom curve. The heat extracted from the brine amounts to $q_{D,ics} = 1{,}518$ kJ/kg, which is exactly the difference between the heat which would have been extracted if the absorbent had been completely subcooled, $q_{D,cs} = 1{,}987$ kJ/kg, and that lost by the desorber in the subcooling process, $q_{sd} = 469$ kJ/kg. The SHL benefit of preserving the desorption temperature design inteval is emphasized in Fig. 3.17. The brine can be cooled up to a temperature close to $T_{DI} = -47.45$ °C, which is not at all the case for the desorption process not furnished with SHL (see for comparison Fig. 3.15). The SHL method is necessary to be applied in all far from unit concentration absorbent desorption processes, suffering from a subcooling deficiency, whether the working combination has volatile refrigerant of not. The SHL manufacturing additional cost is practically null. Further on, all desorption processes needing a SHL are provided with SHL, without any mention in this respect.

The desorption heat exchange in the medium water-lithium bromide was also analyzed for the nontruncated cooling cycle (Staicovici 2011). An air conditioning study case operation has the following input data:

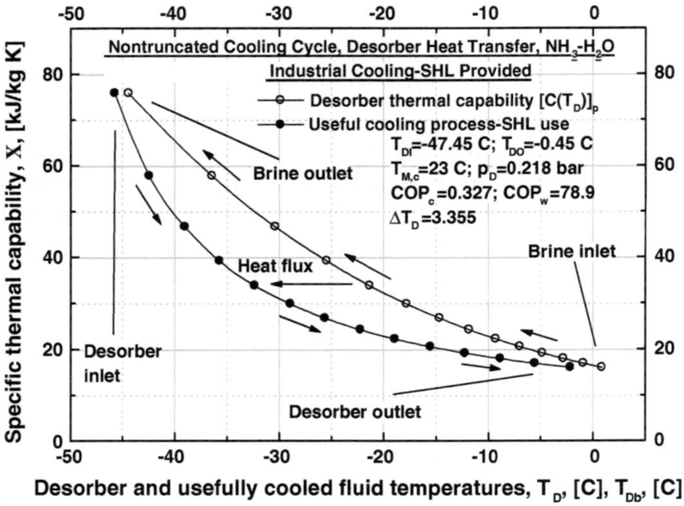

Fig. 3.17 Industrial cooling heat exchange of a desorber with large temperature glide, incomplete subcooling and low concentration absorbent desorption process provided with a subcooling hysteresis loop (*SHL*) (see for comparison Fig. 3.15 plot)

- Desorber inlet/outlet temperature, $T_{DI} = 7$ °C/$T_{DO} = 23$ °C;
- Cycle sink temperature, $T_{M,c} = 32.4$ °C;
- Resorber outlet concentration, $y_{RO} = 45$;
- Desorber vapor quantity, $V_D = 1$;
- Chilled water initial temperature, $T_{Db,0} = 26$ °C;
- Number of steps, $n_D = 20$.

The thermal evolution of the absorbent and chilled water heat exchange, expressed in terms of the specific thermal capability (kJ/kg refrigerant K), is represented versus desorber and chilled fluid temperatures in Fig. 3.18. This shows a very good source—task match.

Concluding, the desorption process with gliding temperature is physically possible from the heat exchange point of view, at least in case of the ammonia-water and water-lithium bromide working combinations. Technically, thanks to SHL use, the brine (water) temperature can be usefully decreased till values close to the T_{DI} temperature in all cooling applications (air conditioning, normal and industrial).

Resorption Heat Transfer Analysis

Further on, the heat transfer of cycle resorption process will be studied (Staicovici 2011). To this extent, we consider the process at hand of the first study case of this section. The process takes place with a large gliding temperature. The heat transfer is of type (a) as well. The additional input data are as follows:

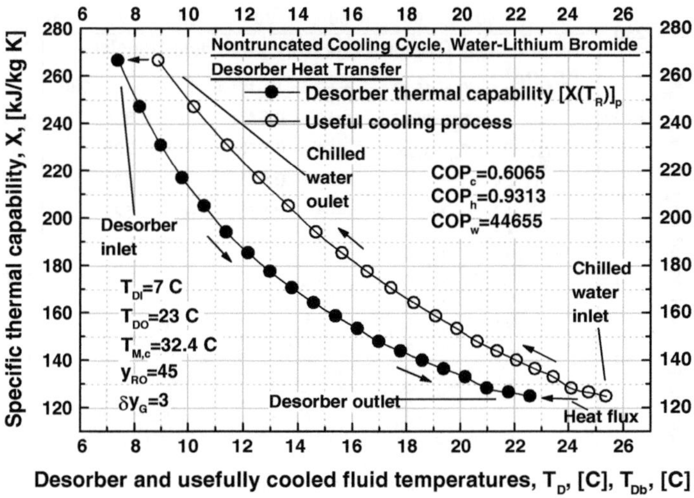

Fig. 3.18 Water chilling heat exchange process of a H_2O–LiBr desorber

- Desorber inlet/outlet temperature, $T_{RI} = 77.25$ °C/$T_{RO} = T_{M,c} = 30$ °C;
- Resorber outlet concentration, $y_{RO} = 0.975$;
- Resorber cooling (usefully heated eventually) fluid temperature, $T_{ROf,0} = 28$ °C.
- Number of steps, $n_R = 10$.

The resorber device is exchanging heat in counter-current with a fluid having again a mass flow rate m and a specific heat c_p, which is supposed to have a constant value during resorption. Let us consider the resorption interval $I_R = [T_{RO}, T_{RI}]$, $T_R \in I_R$, is covered by $T_{R,i}$ tabular values equally spaced, $T_{R,i} = T_{RI} - (i - 0.5)\Delta T_R$, where $i = 1, 2, \ldots, n_R$ and:

$$\Delta T_R = \frac{T_{RI} - T_{RO}}{n_R} \tag{3.81}$$

is the resorber step. The specific thermal capability function (kJ/kg K) of a tabular value is noted by $X_i = X(T_{R,i})$. The resorber highest thermal capability is met to the device exit, that is for the lowest $T_{R,i}$ value,

$$X_{\max} = X\left(T_{R,n_R}\right) \tag{3.82}$$

In case of the counter-current flow, taking into account Eq. (3.82), the cooling fluid must have the following heat capacity:

$$\left(mc_p\right)_{\max} = X_{\max} \tag{3.83}$$

Equation (3.83) expresses the necessary condition for the heat exchange condition (i) be fulfilled. Taking into account Eq. (3.83), the cooling fluid temperature increases each step with an amount equal to $\Delta T_{Rf,i}$, $i = 1, 2, \ldots, n_R$, in such a way that:

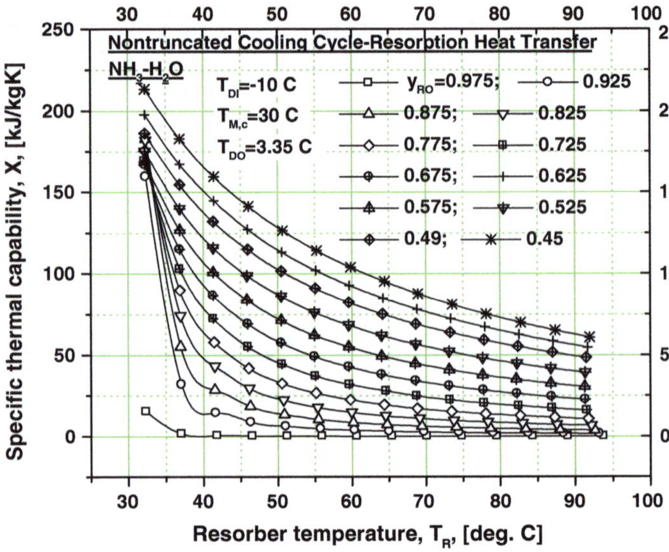

Fig. 3.19 Specific thermal capability family versus T_R, having as parameter the resorber outlet concentration of the nontruncated cooling cycle at hand

$$X_{\max} \Delta T_{Rf,i} = X_{n_R+1-i} \Delta T_R \tag{3.84}$$

wherefrom:

$$\Delta T_{Rf,i} = \frac{X_{n_R+1-i}}{X_{\max}} \Delta T_R \tag{3.85}$$

The DD method was applied using Eqs. (2.12), (2.25), (2.26) and (2.33). From Eq. (3.81) it resulted $\Delta T_R = 4.722$. The total heating effect of the cooling fluid amounts to a small value this time:

$$\sum_{i=1}^{10} \Delta T_{Rf,i} = 5.749\,^{\circ}\mathrm{C} \tag{3.86}$$

The current cooling fluid temperature $T_{Rf,i}$, $i = 1, 2, \ldots, n_R$, is obtained simply through:

$$T_{Rf,i} = T_{Rf,0} + \sum_{j=n_R+1-i}^{n_R} \Delta T_{Rf,j} \tag{3.87}$$

where $T_{Rf,0}$ is its initial value, $T_{Rf,0} = T_{RO} - (1\text{–}5)\,^{\circ}\mathrm{C}$.

Next, a first analysis is done, regarding the way the heating effect might be increased (see Eq. (3.86) poor issue). The results thereof are given in the correlated Figs. 3.19, 3.20 and 3.21. In Fig. 3.19, it is plotted a family of thermal capability curves versus T_R, keeping constant the cooling cycle temperatures T_{DI} and $T_{M,c}$ and

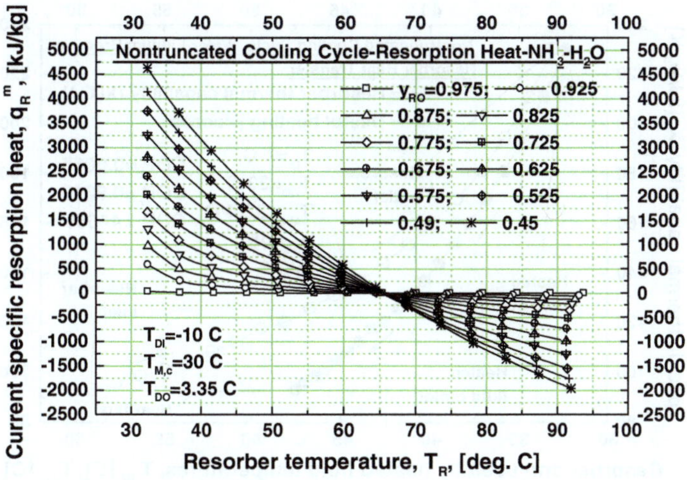

Fig. 3.20 Current specific resorption heat family, versus T_R, having y_{RO} as parameter

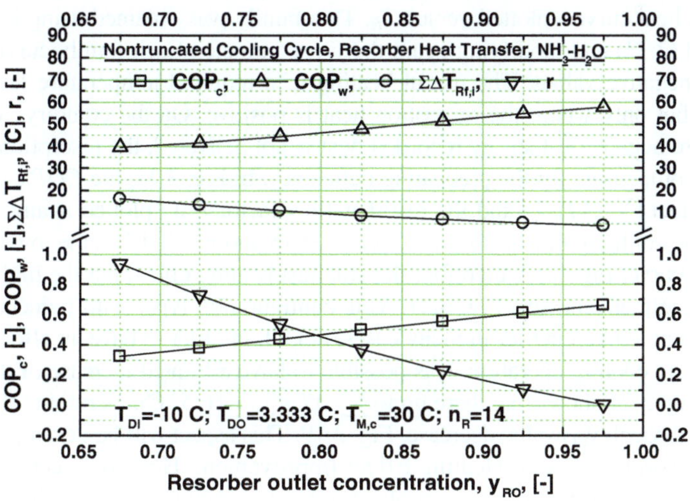

Fig. 3.21 COP_c, COP_w, heating effect $\left(\sum \Delta T_{Rf,i}\right)$ and resorber flow factor r versus resorber outlet concentration, in completion of Fig. 3.19 resorber heat exchange study

modifying the resorber outlet concentration, y_{RO}. According to this figure, the decrease of this parameter value increases the thermal capability, therefore a heating effect raise could be expected. This positive feature is counteracted, however, by a *heat and temperature preheating double deficiency*, which the cycle is confronted with. Indeed, out of the mixing flow f_M, only the absorbent going into generator is preheated, Fig. 3.4, that supplying resorber has the mixer sink temperature. The preheating deficiency is very clearly emphasized in Fig. 3.20 by a family of current specific [kJ/(kg refrigerant K)]

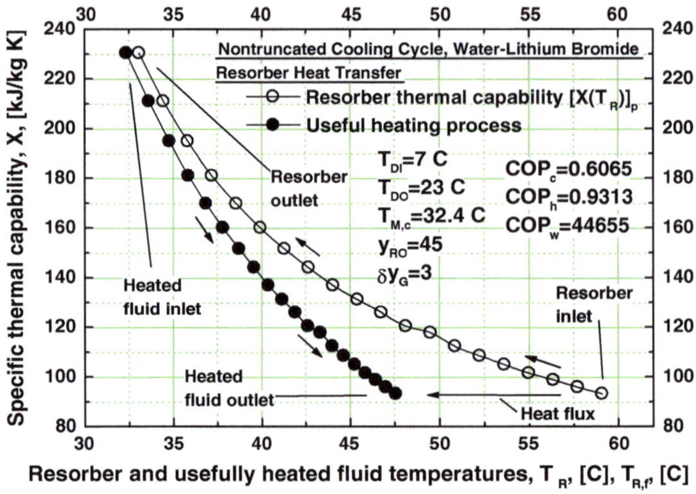

Fig. 3.22 Water heating heat exchange process of a H_2O–LiBr resorber

resorption heat curves, plotted versus T_R. This family was obtained using Eqs. (2.25), (2.26) and (2.33) and modifying again the y_{RO} parameter. Each family curve has positive and negative values. The useful heating of a required temperature level can be delivered but only where the resorption heat is positive. On the contrary, in the portions where negative values are recorded heat is not delivered, the resorption heat output must compensate for the preheating deficiency. In Fig. 3.21, the COP_c, COP_w, the heating effect $\left(\sum \Delta T_{Rf,i}\right)$ and the resorber flow factor r are plotted against y_{RO}. The initial cycle configuration, with $Y_G \cong y_{RO} = 0.975$ and $r \cong 0.0$, has the lowest thermal capability curve, Fig. 3.19, therefore the smallest heating effect, but the highest COP_c and COP_w effectivenesses, Fig. 3.21. This configuration type, with reduced r values will be promoted hereinafter in many cases, not all, in order to have an effective coabsorbent cycle cooling operation. It is possible to have a limited increase of the heating effect as Fig. 3.21 shows (e.g. here up to $\sum \Delta T_{Rf,i} = 16.5\,°\mathrm{C}$), increasing the thermal capability through y_{RO} value decrease, Fig. 3.19. This aspect will be developed more in Further Resorption Useful Heating Effect Improvement. However, such an option is probably considered but in certain applications only, when the cogeneration of cooling and heating is needed, because COP_c and COP_w decrease much (see Fig. 3.21), at least in case of the ammonia-water. The same is happening to the industrial cooling operation, when the resorption heating effect is negligible as well (not shown here).

The resorption heat exchange in the medium water-lithium bromide was analyzed for the nontruncated cooling cycle as well. Here we consider the same air conditioning study case operation as in Fig. 3.18. The resorption heat rejection process is capable to produce useful heating in case of this working combination as well. The thermal evolution of the absorbent and heated water heat exchange, expressed in terms of the specific thermal capability (kJ/kg refrigerant K), is represented versus resorber and heated fluid temperatures in Fig. 3.22. It shows a satisfactory thermal match at this low resorption pressure of a single-stage cycle. The resorption heating

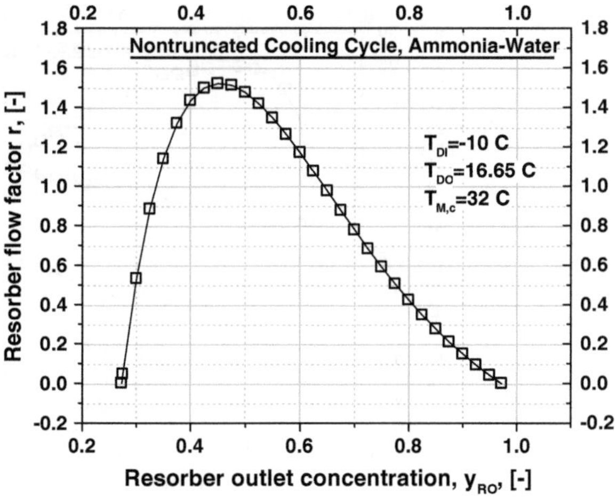

Fig. 3.23 Example of how the resorber flow factor r is altered by its outlet concentration y_{RO} variation

effect is higher to higher pressures for a same absorbent concentration range. This property is capitalized by the pressure-stage multi-effect coabsorbent cycles for the cooling (air conditioning) COP benefit, shown further on in this book.

Further Resorption Useful Heating Effect Improvement

In order to analyse the possibility to obtain a cycle heating maximum effect, we bear in mind the resorption heat expression of Eq. (3.39). We remark that its increase is favored by the raise of resorber flow factor r (e.g. see Fig. 3.20). Figure 3.23 shows r has a maximum value for a cycle given input data, when the resorber absorbent outlet concentration y_{RO} covers a certain interval of values. The maximum of r is placed between two values of zero. These result, according to Eq. (3.3), when either y_{RO} approaches Y_G, obtaining the right zero value, or when, on the contrary, Y_G approaches y_{RO}, obtaining the left zero value. This r behavior is valid for the working combinations having an own pressure absorbent (volatile absorbent), only (in our case the NH_3–H_2O) (Staicovici 2011). When the mixture has a non-volatile absorbent (e.g. H_2O–LiBr), r increases continuously from a right side zero value, till a certain maximum end value. Figure 3.24 confirms our previous assertion. Indeed, the maximum heating effect, expressed by the total raise of the resorber cooling fluid temperature, here $\sum_{i=1}^{n_R} \Delta T_{Rf,i} \cong 25\,°C$ (see Eq. (3.86) type), takes place in a zone where the resorber flow factor r has maximum values ($r = 1.4$–1.5 in Fig. 3.23). Figure 3.24 is typical for cycle operation with combined cooling and heating (Staicovici 2011). Besides heating, it plotts also the COP_c effectiveness and the maximum operating pressure. As previously remarked, the obtainment of a useful heating effect is done by sacrificing the cooling effect effectiveness (COP_c decreases from cca. 0.66, to about 0.4–0.5, with T_{DO} subtending 2/3 out of the desorption interval). This negative issue is, however, counterbalanced to a certain extent by the maximum operation pressure decrease (p_G diminishes from cca. 12 bar, to 2.5–4.0 bar, Fig. 3.24), which lightens construction and increases the safety in operation. In Fig. 3.25 it is plotted a

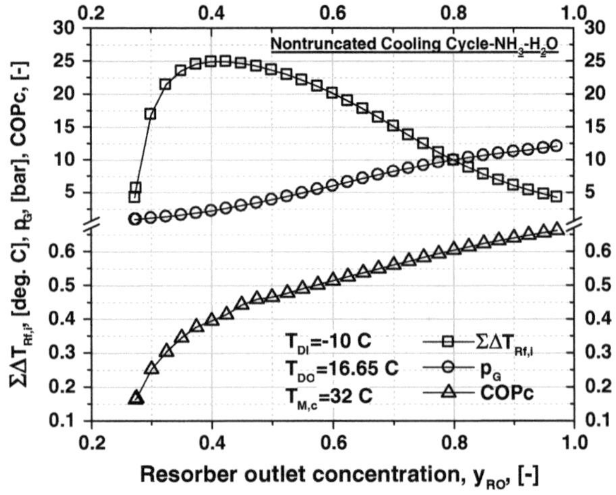

Fig. 3.24 Typical heating effect diagram: resorber cooling fluid temperature increase, cycle cooling effectiveness and the cycle maximum working pressure

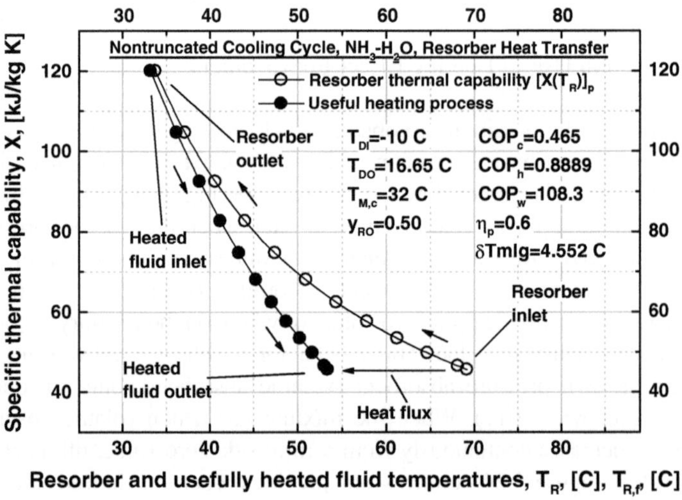

Fig. 3.25 Typical resorber heat transfer in combined cooling and heating

typical resorber heat transfer, where the temperature raise of useful heating amounts to $\sum_{i=1}^{n_R} \Delta T_{Rf,i} \cong 23.65\,°\text{C}$. Repeating the analysis begun to Fig. 3.23 for different desorber inlet temperatures, T_{DI}, it results the family of curves plotted in Fig. 3.26 (Staicovici 2011). It shows that the decrese of T_{DI} displaces the curves towards higher y_{RO} values and theirs maxima decrease gradually. The loci of maxima are almost straight lines, as it is marked in the figure by a dotted line for the study case we have chosen. The maxima locus of a family constitutes the border between the right side

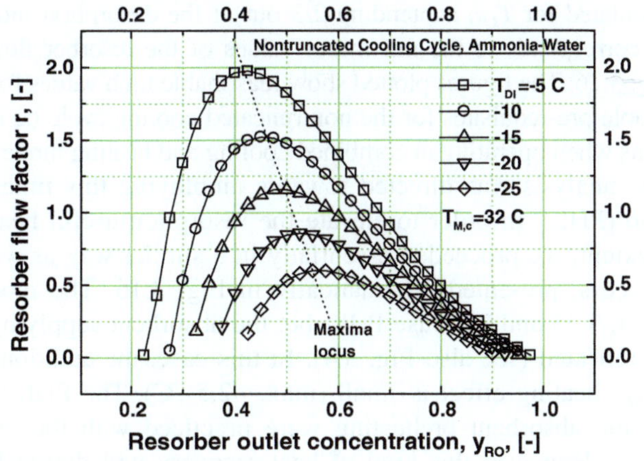

Fig. 3.26 Family of resorber flow factor curves against resorber outlet concentration, having the desorber inlet temperature as parameter

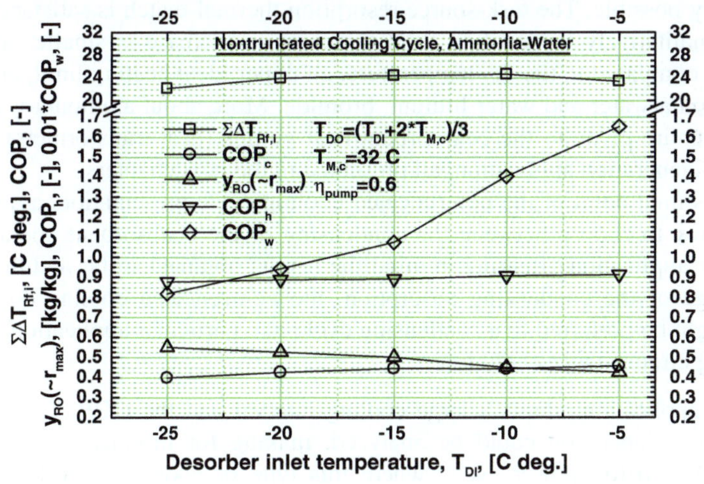

Fig. 3.27 Combined cooling and heating results: $\sum_{i=1}^{nR} \Delta T_{Rf,i}$, $r_{\max}$, COP_c, COP_h and COP_w functions versus T_{DI}. Input data: $T_{M,c} = 32$ °C, $T_{DO} = (T_{DI} + 2 \times T_{M,c})/3$, $\eta_{pump} = 0.6$, $y_{GO} = y_M - 0.05$

y_{RO} domain, recommended to cycle operation in combined cooling and heating, and the left side y_{RO} domain, which is not appropriate for any operation. If the sink temperature were covering for instance the domain $T_{M,c} = (27–37)$ °C, the curves family remains essentially unchanged with respect to the y_{RO} maxima values, but, of course, the corresponding values of r maxima will vary. Concerning the heating effect, $\sum_{i=1}^{nR} \Delta T_{Rf,i}$increases slightly towards higher $T_{M,c}$ values and decreases by 5–8 % towards lower $T_{M,c}$ values. In Fig. 3.27 we have plotted the $\sum_{i=1}^{nR} \Delta T_{Rf,i}$, COP_c, COP_h and COP_w functions versus T_{DI}, using Sect. 3.2.1.2 model (Staicovici 2011). They have been calculated for T_{DO} subtending 2/3 out of the desorption interval. The y_{RO} concentration corresponds to the maximum values of the resorber flow factor, r_{max}, taken from Fig. 3.26. The figures plotted show reasonable high values for all functions, creating favorable pre-requisites for the nontruncated cooling cycle be feasible in certain applications when operating in combined cooling and heating mode.

Finally, our analysis was directed towards employing this time a *preheating hysteresis loop* (PHL), in order to elevate the resorption useful heating temperature. To this extent, we proceeded essentially in a similar way as we did it to the desorption process, presented schematically in Fig. 3.16. The model results of Figs. 3.21, 3.24, 3.25 and 3.27 use PHL, but the absorbent supplying the resorber in not at all preheated (see also Fig. 3.4). In this case, the additional increase of the $\sum_{i=1}^{n_R} \Delta T_{Rf,i}$ heating effect is small (max. ~2.5 °C). The PHL benefit can be augmented if the absorbent preheating were practiced with the preheater 9, as Fig. 3.16 shows. However, this kind of heat recovery will diminish correspondingly the COP_c, so we avoided to study it here. A resorption process provided with PHL will be completely analyzed further to the hybrid heat pump of Sect. 3.2.5.3.

Concluding, the resorber heat exchange of the nontruncated cooling cycle is physically possible. The task-source resorption thermal match is satisfactory only, that is qualitatively the cooling fluid temperature increase is smaller to a good extent as compared to the T_{RI} temperature, an upper ideal heating limit, at least for the ammonia-water and water-lithium bromide. More about ammonia-water heating effects are given to the coabsorbent heat pumps (heat transformers) and to a special trigeneration application, all presented in the next chapters.

At this final point of the nontruncated cooling cycle presentation, a question is raised: how to choose between the condensing or the coabsorbent operation in a particular application? Bearing in mind the results obtained so far and the general knowledge of the condensing absorption, the nontruncated cooling coabsorbent cycle might be preferred to a condensing cycle in a particular application when the following circumstances hold true:

(i) The application admits a large cooling interval and the $COP_c(T_{DO}) > COP_c$. Such an operation could be analyzed, making for instance the case of the COOLENERG power cycle, where this type of cooling is cyclic necessary (Staicovici 2002, 2004), or of the district cooling. As shown, the desorbing absorbent and the cooled source fluids can exchange heat in the desorber with a small quasi-constant temperature pinch throughout the desorber. From the second principle of thermodynamics this good temperature match of the two fluids avoids an important cycle exergy loss, favoring the $COP_c(T_{DO})$ increase. On the contrary, in this kind of applications, the evaporator of the condensing cycle can be the device with a large exergy loss out of the whole cycle. A designer can shun partially this situation using several evaporators, with different evaporation temperatures (Niebergall 1959, see next multistage absorption). However, besides the constructive complexity increase, the evaporation stages introduce the absorbent inventory problem as well (Niebergall 1959), so the coabsorbent cycle might be preferred so much more in this case.

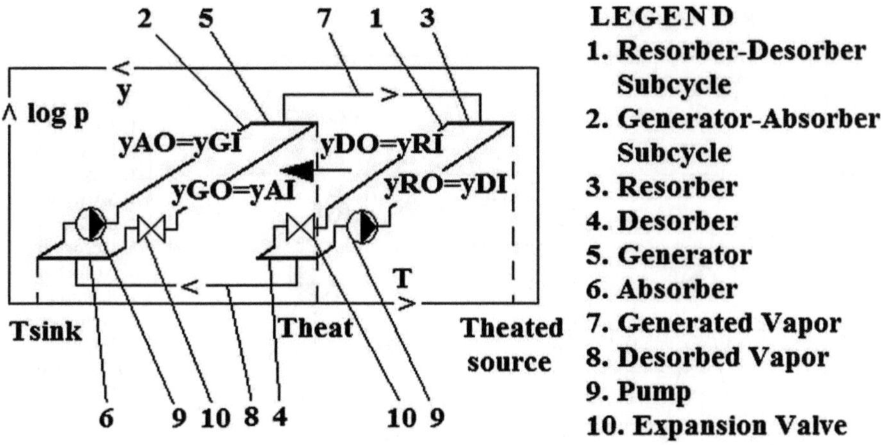

Fig. 3.28 The resorption heating cycle serving as a formal explanation of the nontruncated coabsorbent heating cycle genesis

(ii) The cogeneration of heating is an important issue as well, besides the cooling production.
(iii) The safer, lower pressure operation, is an important criterion.

3.2.2 *Nontruncated Heating (Heat Transformer) Coabsorbent Cycle*

The nontruncated coabsorbent heating (heat transformer) cycle formally comes from the resorption heating cycle, depicted in Fig. 3.28. The right side subcycle increases its resorber-desorber concentration interval of operation (see the arrow) until the $y_{DO} = y_{RI}$ isostere overlaps the $y_{GO} = y_{AI}$ isostere of the left side subcycle. Thus, the two subcycles of resorption cycle will have a single internal isostere $y_{AI} = y_{DI} = y_M$, resulted through the joining (mixing up) of the absorbents coming of generator and resorber. The consequences of this joining are the followings:

(a) The desorber absorbent concentration increases. Consequently, the desorption load will decrease and the desorbed vapor mean concentration will increase, favoring the COP_h increase;
(b) The resorption process takes place on a broader temperature interval, conveniently chosen by a designer, increasing the match with the heated sources;
(c) The solution inventory problem of the resorption heating cycle is solved;
(d) The number of the expansion valves is reduced to a single 3-way valve.

The nontruncated heating coabsorbent cycle is plotted in the $\log(p) - 1/T$ diagram in Fig. 3.29. It is contrary covered by the absorbent flow as compared to

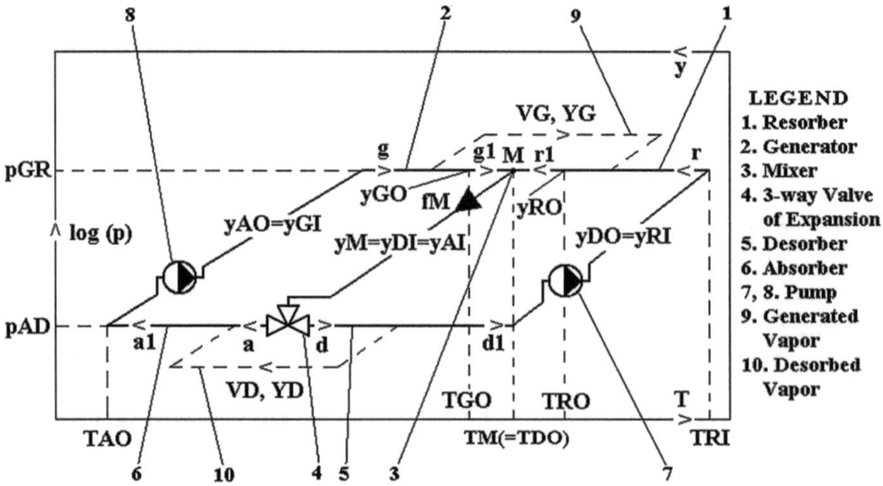

Fig. 3.29 The nontruncated heating (heat transformer) coabsorbent cycle

the cooling cycle in Fig. 3.2. Formally, it can be obtained also rotating the cooling cycle by 180° in the plot diagram plane. The cycle internal heat exchange can be improved, providing it with a structure similar to that of Fig. 3.4. A flow chart including some features of the nontruncated heating coabsorbent cycle was proposed when the "Self Regenerated Absorption Temperature Amplifier" (SRATA) was conceived (Kashiwagi 1980–1990). However, its author neither considered SRATA as a coabsorbent cycle, nor has foreseen the advantages of this more general class of absorption cycles, having condensing absorption as a particular case, as previously mentioned. Its operation it follows (see arrows). The absorbents exiting the high pressure resorber and generator are being mixed up in the *M* mixer. Similarly to the cooling case, the mixing process can be performed naturally, for example during the gravity falling of the mixing absorbents, or using very reduced power mixing pumps. Again, the mixing process is considered adiabatic and perfect, as to the cooling cycle. The relative positions of the resorber and T_M generator operation temperature intervals can be similar to those depicted in Fig. 3.3a–c and will not be repeated here. This figure applies here as well, provided that the desorber and absorber be replaced by the generator and resorber, respectively. This time, temperature is the internal heating temperature, corresponding to the external heat source. Continuing, the resulting cumulated flow f_M leaves the mixer with the concentration y_M and then is expanded and distributed in pre-established complementary quantities to the low pressure absorber and desorber. Finally, the absorbents exiting the desorber and absorber are pumped to the high pressure resorber and generator, in order to close the cycle. The heating coabsorbent cycle has as input data the desorber outlet temperature (related to the heat source) and the concentration, T_{DO} and y_{DO}, respectively, the resorber inlet/outlet temperature

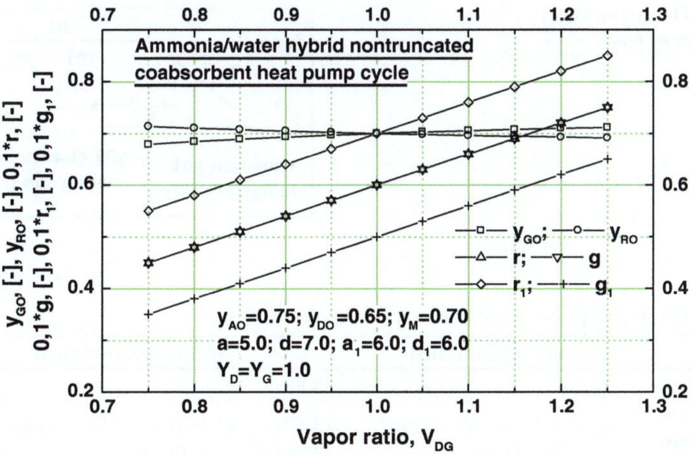

Fig. 3.30 Flow analysis case study of a nontruncated heating coabsorbent cycle

(related to the useful heating effect maximum/minimum temperature), T_{RI}/T_{RO}, and the absorber outlet temperature (related to the sink source), T_{AO}. From these five parameters it results the rest of the useful parameters: p_{AD}, p_{GR}, $T_M(= T_{DO})$, y_M, and y_{AO}. The flow factors, defined by Eqs. (3.1)–(3.8), hold here true also. The threshold concentration is in the heating coabsorbent cycle case the y_{AO} value, that is for this cycle operates as a full coabsorbent cycle we must have the $Y_D > y_{AO}$ fulfilment. The cycle most important parameters are given in Table 3.2. As compared to the classic condensing heat transformer, where $y_{AO} = y_{GI} \cong 1.0$, the coabsorbent cycle operates with a similar temperature lift, equal to $(T_{RI} - T_{DO})$, and a similar absorber end temperature T_{AO}, for a same generation end temperature $T_{M,c}(= T_{DO})$. Similarly to the condensing cycle, the temperature lift increase is requiring the absorber cooling till temperatures T_{AO}, which the sink sources not always can supply.

A second case study results are given in Fig. 3.30, for a nontruncated heating coabsorbent cycle, namely for the non-isobar heat pump, running with NH_3–H_2O and presented further in this chapter (Staicovici 2006a). In this case, the generator operates within a small concentration interval, $0.68 \leq y_G \leq 0.72$, and the generated vapor is almost pure ammonia, that is $Y_G \cong 1(\cong Y_D)$. The mixing point concentration is chosen for this study $y_M = 0.7$. This is achieved by mixing up the absorbents coming of generator and resorber and having alternatively the concentrations y_{GO} and y_{RO}, higher and lower than the y_M, in the way shown in Fig. 3.30. As priorily mentioned, in this situation the balance point corresponds to the V_{DG} ratio equal to unit, Fig. 3.30. The rest of the comments of the first study case remain valid here as well.

Applications with the nontruncated heating coabsorbent cycle will be given briefly to Sect. 3.2.5.3.

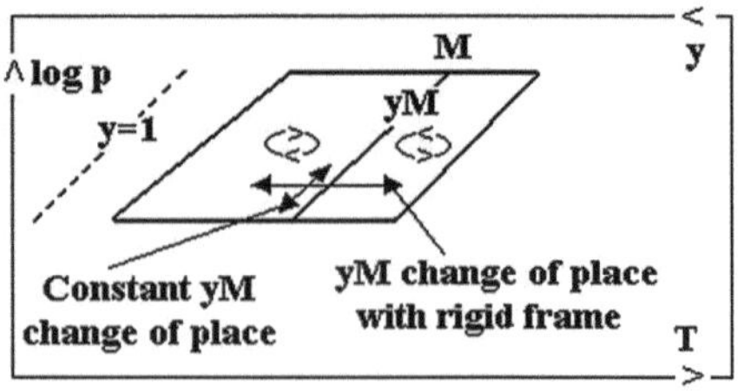

Fig. 3.31 The coabsorbent cycle change of place

Table 3.6 Change of place of the nontruncated cycles

Cycle	Cooling nontruncated		Heating nontruncated	
Change of place with $y_M = const.$	$\frac{a}{d} = S$	(3.88)	$\frac{g}{r} = S$	(3.94)
	$(a + d)V_D = f_M = const.$	(3.89)	$(r + g)V_G = f_M = const.$	(3.95)
	$a = \frac{f_M}{V_D} \frac{y_{RO} - y_M}{y_{RO} - y_{GO}}$	(3.90)	$g = \frac{f_M}{V_G} \frac{y_M - y_{DO}}{y_{AO} - y_{DO}}$	(3.96)
	$d = \frac{f_M}{V_D} \frac{y_M - y_{GO}}{y_{RO} - y_{GO}}$	(3.91)	$r = \frac{f_M}{V_G} \frac{y_{AO} - y_M}{y_{AO} - y_{DO}}$	(3.97)
Change of place with rigid frame	$y_M = y_{RO} - \frac{a}{\frac{f_M}{V_D}} (y_{RO} - y_{GO})$	(3.92)	$y_M = y_{DO} + \frac{g}{\frac{f_M}{V_G}} (y_{AO} - y_{DO})$	(3.98)
	$y_M = y_{GO} + \frac{d}{\frac{f_M}{V_D}} (y_{RO} - y_{GO})$	(3.93)	$y_M = y_{AO} - \frac{r}{\frac{f_M}{V_G}} (y_{AO} - y_{DO})$	(3.99)

3.2.3 Cycle Change of Place

During operation, the cycle-sources interaction may change. Two particular changes of place can happen when $f_M = const.$, schematically shown in Fig. 3.31. The first cycle changing of place is up and down along the mean concentration isostere, keeping it constant. During this operation, the cycle extreme isosteres values may change. The second, could happen when keeping constant the extreme isosteres and pressures (rigid frame) and changing the y_M value. Next we shall give simple equations of both cases, for the cycles analyzed, included in the Table 3.6. For the cooling cycle, from Eq. (3.12), written as Eqs. (3.88), and (3.89), we obtain the flow factors valid for the $y_M = const.$ change of place, given by Eqs. (3.90) and (3.91). In this case, the cycle frame (the concentrations y_{RO} and y_{GO} and the corresponding pressure stages) may undergo modifications, which alter the a and d values, but keep $y_M = const.$ The y_M change of place with the rigid frame operation is obtained from Eqs. (3.90) and (3.91), as Eqs. (3.92) and (3.93). As we remark, y_M depends linearly on the a or d flow factors. This behaviour could be useful in the cycle control and optimization. Concerning the heating cycle, we consider Eq. (3.23) written as Eqs. (3.94), and (3.95), wherefrom, when $y_M = const.$, the flow factors are given by Eqs. (3.96) and (3.97). Similarly, the cycle frame (the concentrations y_{AO} and y_{DO}) may undergo modifications, which alter the g and r values, but keep the $y_M = const.$ The y_M change of place with the rigid frame operation is obtained from Eqs. (3.96) and (3.97), as Eqs. (3.98) and

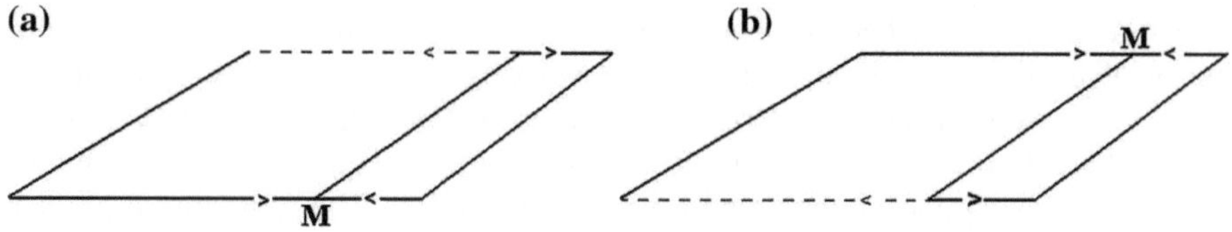

Fig. 3.32 The nontruncated cooling (**a**), and heating (heat transformer) (**b**) coabsorbent-condensing cycles, plotted in the $\log(p) - 1/T$ diagram

(3.99). They keep a same linear expression with respect to the g or r flow factors. An example of a heat pump suffering changes of place with rigid frame during its operation, is given in Staicovici 2006a.

3.2.4 Nontruncated Coabsorbent-Condensing Cycle

The nontruncated cooling and heating cycles work fully as coabsorbent cycles only if $Y_G > y_{RO}$ and $Y_D > y_{AO}$, respectively. When $Y_G = y_{RO}$ and $Y_D = y_{AO}$, the resorption and absorption processes, respectively, are replaced each by condensing processes and the cycles become nontruncated cooling and heating condensing-coabsorbent cycles, Fig. 3.32a, b, respectively. The resorption process in Fig. 3.32a and the absorption process in Fig. 3.32b, vanish, so they are represented by dotted lines. On the contrary, theirs opposed processes, namely the desorption process in Fig. 3.32a and the resorption process in Fig. 3.32b, finish by mixing theirs absorbents in the mixer, and are represented therefore by solid line. In practice, the nontruncated cooling cycle of Fig. 3.32a can be for instance the single-stage condensing cycle, operating with the NH_3/H_2O combination and with an incomplete rectification or without vapor rectification at all. Here, the cooling effect is produced on a broader evaporation interval and the resulting absorbent is mixed up with the absorbent exiting the absorber, prior to be pumped and to enter the recovering heat exchanger and the generator. For the particular cycles at hand, Eqs. (3.3) and (3.1) take this time the values $r = 0$, and $a = 0$, respectively. When $Y_G = y_{RO} \cong 1.0$, or $Y_D = y_{AO} \cong 1.0$, these cycles become simple known condensing cycles, which have not even the desorption and resorption processes, respectively. This is the case for example of the cycles operating with the $H_2O/LiBr$, $NH_3/NaSCN$ or $NH_3/LiNO_3$ working combinations, which absorbents of have not their own vapor pressure, so the generated vapor is practically the pure H_2O and NH_3, respectively. In this case, the desorption and resorption processes become simply evaporation and condensation processes, respectively. Referring further just to cooling production, as shown to Sect. 3.2.1.3, when resorber flow factor approaches the zero value ($r \cong 0.0$), the heating effect vanishes, so a nontruncated cooling coabsorbent-condensing cycle doesn't work in cogeneration of heating and cooling, but generates maximum cooling, only.

From the above it becomes obvious that a coabsorbent cycle, either of cooling or of heating type, generalizes the known single-stage condensing absorption cycle. Next we shell go a little bit deeper into this subject, finding the conditions which characterize a coabsorbent cooling cycle be less, equally, or more effective than a condensing cooling cycle with vapor rectification. The following input data are considered for the coabsorbent cycle:

(i) The desorber inlet and internal sink temperatures, T_{DI} and $T_{M,c}$, respectively, are known; also, the generator concentration operation interval, $\delta y_G = y_{GI} - y_{GO}$, is given;
(ii) The resorber outlet concentration y_{RO} is specified, wherefrom the generator vapor mean concentration Y_G and the resorber flow factor r are computed;
(iii) The effectivenesses of the coabsorbent cooling cycle and condensing cooling cycle with vapor rectification, $COP_{c,cbs-cd}$ and $COP_{c,cd}$, respectively, are in the following relationship:

$$COP_{c,cbs-cd} = k_{COP} COP_{c,cd} \tag{3.100}$$

where $k_{COP} \in R$, $k_{COP} \leq 1$ or $k_{COP} \geq 1$ is a specified real strictly positive constant. We introduce a coabsorbent cycle *relative desorption interval* function:

$$\frac{T_{DO}(r) - T_{DI}}{T_{M,c} - T_{DI}} = i_d(r) \tag{3.101}$$

as the ratio of the actual desorption interval devided by the balance desorption interval (see Fig. 3.3b). The actual desorption interval corresponds to one of the three possible desorption processes depicted in Fig. 3.3a–c, therefore, $i_d < 1$, $i_d = 1$, or $i_d > 1$, respectively, $i_d \in R$. The desorber outlet temperature is calculated solving Eq. (3.101) with respect to $T_{DO}(r)$,

$$T_{DO}(i_d(r)) = T_{DI} + i_d(r)\left(T_{M,c} - T_{DI}\right) \tag{3.102}$$

This function depends on i_d and is found through an iterative process based on the (i)–(iii) items fulfilling. We have obtained comparative results of coabsorbent versus condensing cooling cycles using Sect. 3.2.1.2 model and considering that the (i)–(iii) input data are as follows: (i) $T_{DI} = ((-5)\text{–}(-45))$ °C, $T_{M,c} = 24; 32; 40$ °C and $\delta y_G = 0.05$; (ii) $r \cong 0.0$ (coabsorbent-condensing cycle); (iii) $k_{COP} = 1.0$. The resulting $COP_{c,cbs-cd} = COP_{c,cd} = COP_c$, $COP_{w,cbs-cd}$, $COP_{w,cd}$, i_d and T_{DO} are plotted in Fig. 3.33a–c for the three $T_{M,c}$ values, respectively. In these figures, T_{DO} is a border, named the *desorber end temperature threshold*, $T_{DO}(i_d(r))$. Generally, this separates the coabsorbent cycle operation in three domains characterized by: (a) $T_{DO} < T_{DO}(i_d(r))$, $COP_{c,cbs-cd} < k_{COP}COP_{c,cd}$, (b) $T_{DO} = T_{DO}(i_d(r))$, $COP_{c,cbs-cd} = k_{COP}COP_{c,cd}$ and (c) $T_{DO} > T_{DO}(i_d(r))$, $COP_{c,cbs\text{-}cd} > k_{COP}COP_{c,cd}$. Figure 3.33a–c results show also that $i_d \leq 3/4$ is for almost all normal coabsorbent cooling applications a necessary and enough condition in order that $COP_{c,cbs-cd} \geq COP_{c,cd}$.

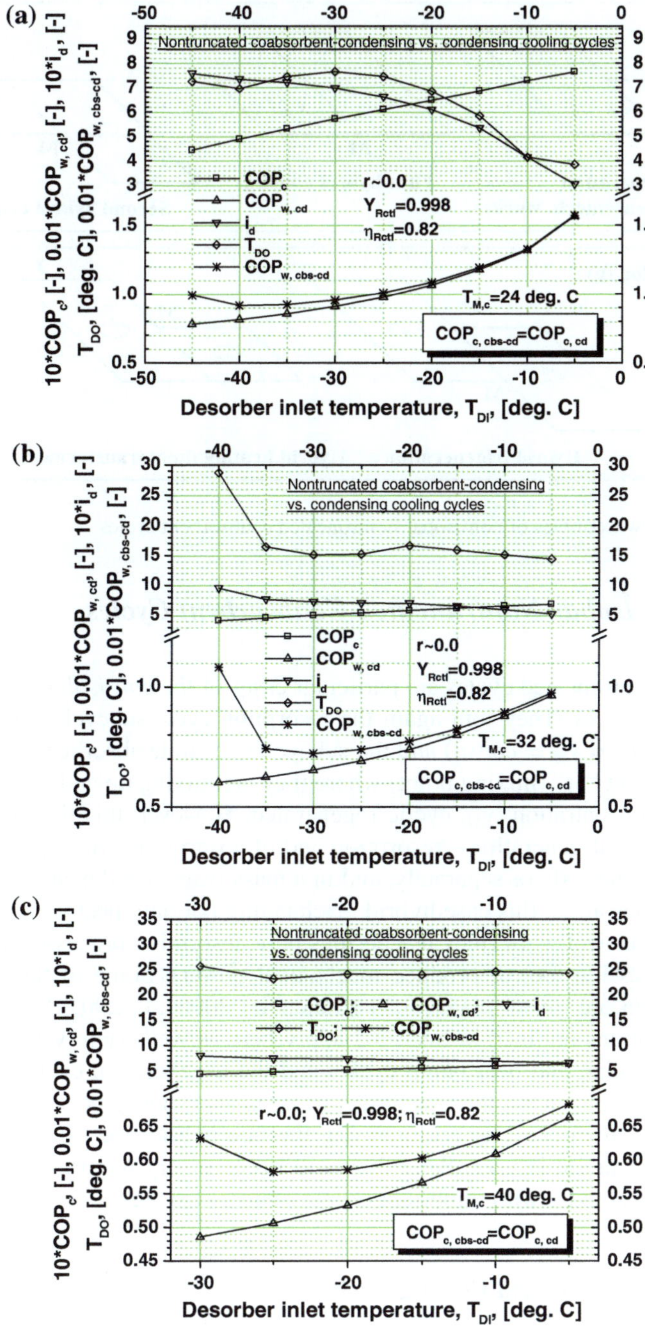

Fig. 3.33 **a–c** Comparative results of coabsorbent-condensing versus condensing cooling cycles

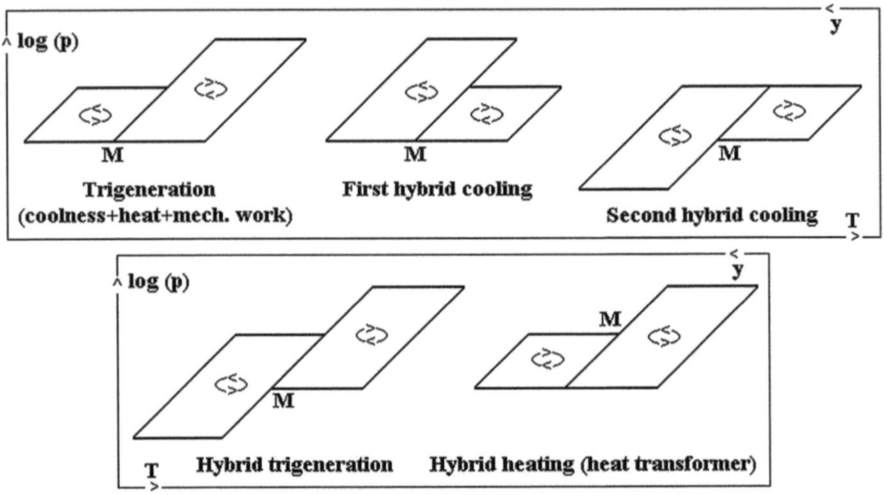

Fig. 3.34 A few examples of non-isobar nontruncated coabsorbent cycles

3.2.5 Non-isobar Nontruncated Coabsorbent Cycles

The isobar nontruncated cycles are particular cases of *the non-isobar nontruncated coabsorbent cycles*. They join again the resorption cycle subcycles along a common isostere ($y = y_M = const.$) and the subcycles separate absorbents have as well a common point "*M*" (mixing point) where they are mixing up and form the absorbent mean concentration, y_M, cyclic regenerated. However, this time the opposed pair processes, of generation—resorption and of absorption—desorption, are non-isobar, simultaneously or separately, and in a mass (vapor) and heat exchange. The resulting cycles are in this case hybrid, exchanging not only heat with the sources, but also mechanical work (they use besides pure absorption processes, mechanical vapor compression too, or produce mechanical power). Some of the most interesting of them are given in Fig. 3.34 (Staicovici 2006a, c, 2007g). The theoretical considerations outlined in Sect. 3.2 apply here as well. Here we shall analyse the hybrid cooling and heat pump cycles and the trigeneration cycle as well. For the sake of completeness, the Osenbrück cycle is also shortly presented. As it will be shown, the hybrid cycles are outdoing the non-hybrid ones, because of theirs higher COP's, and it is very likely they will become very important in the future heat pumping activity with important primary energy savings.

3.2.5.1 The First Hybrid Cooling Cycle

The hybrid nontruncated coabsorbent cycle is represented in the log(p) $-$ $1/T$ diagram of Fig. 3.35 (Staicovici 2006c, e, 2007b, d, e, f, g; Staicovici and Staicovici 2013). Next we give its short operation description. The low pressure p_l desorber

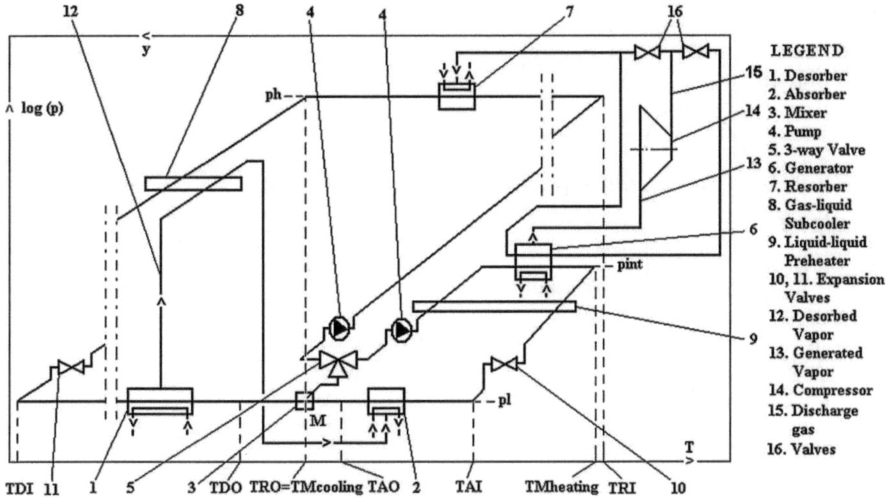

Fig. 3.35 The hybrid nontruncated coabsorbent cooling cycle log(p) − $1/T$ diagram

1 has a heat input from the source which is usefully cooled. The desorbed vapor 12 is superheated in a recovering way in the gas-liquid subcooler 8 prior to be absorbed in the low pressure p_l absorber 2. The absorbents of the desorber and absorber evolve on large temperature intervals, $T_{DI} - T_{DO}$ and $T_{AI} - T_{AO}$, respectively. They leave these devices and mix up in the M mixer 3, generating the y_M concentration absorbent. The pumps 4 supply with it the intermediate pressure p_{int} generator 6 and the high pressure p_h resorber 7 in a pre-established amount, regulated by the 3-way valve 5. The absorbent supplying generator 6 is preheated in a recovering way in the liquid-liquid preheater 9. The generated vapor 13 is compressed from p_{int} to p_h by means of the compressor 14. The compressed vapor 13 is resorbed in the resorber 7, releasing the resorption heat to the sink source. Optionally, prior the discharge gas 15 be resorbed, its superheat $|q_a|$ is recovered in the generator 6 by means of valves 16, in order to diminish the heat source input and increase the COP_c. The absorbents coming of the resorber and generator are subcooled in the heat exchangers 8 and 9, respectively, and then are expanded with the valves 10 and 11 to p_l, for supplying the desorber and absorber, respectively and closing the cycle. A nontruncated cooling hybrid cycle can benefit of SHL (see Desorption Heat Transfer Analysis), not shown here.

Concerning the compressor discharge gas superheat, in case of an adiabatic (polytropic) mvc process this equals the compressor mechanical work input ($|q_a| = |l_a|$, see Chap. 10 mvc first theorem). Bearing this in mind, the in generator recovered superheat can be evaluated by:

$$q_{G,rec} = |l_a| \frac{h_a - h_{GI}}{h_a - h_{M,c}} \cong |l_a| \frac{T_a - T_{GI}}{T_a - T_{M,c}} \tag{3.103}$$

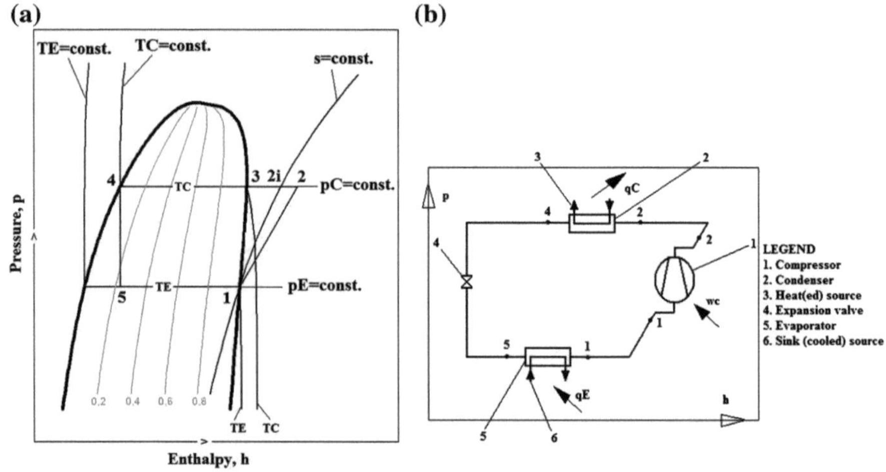

Fig. 3.36 Plot of a mvc cooling cycle in the *p-h* diagram (**a**) and its schematic (**b**)

where h_a and T_a are the discharge gas enthalpy and temperature, respectively. The actual generation heat needed from the heat source results by subtracting $q_{G,rec}$ out of the generation heat, Eq. (3.36).

A designer would be tempted to consider the hybrid cooling cycle for an application when two operation conditions are fulfilled simultaneously. First, its generation temperature $T_{M,h}$ is considerably lower than that of the isobar nontruncated cycle, Figs. 3.2 and 3.4, that is cycle heat/sink sources temperature gap is enough small. Second, its COP_w is higher to a good extent than that of a cycle producing cooling by mechanical vapor compression (mvc abbreviation), COP_{mvc}, only. We have modeled the coabsorbent operation of this cycle with the code given in Sect. 3.2.1.2, slightly adapted and completed. Concerning the theoretical mvc operation, its simple computation algorithm is given next.

Bare Mvc Versus Hybrid (Mvc + Absorption) Operation

Bare mvc operation

The theoretical operation of the known mvc cooling cycle is plotted in Fig. 3.36a in a *p-h* diagram. A schematic of this cycle is depicted in Fig. 3.36b. In this cycle, the portions (1–2) (2–4) (4–5) and (5–1) represent the state parameters of the mvc, condensing, expansion and evaporation processes. The condensing and evaporation processes are characterized by constant (p_C, T_C), and (p_E, T_E) parameters operation, respectively. The discharge gas superheat $|q_a|$ is rejected to the sink, or is partially recovered in the generator, in the portion (2–3). The reversible mvc evolution occurs along the isentropic (1–2i). The internal (isentropic) efficiency of the compression process is calculated in a covering way by Radcenco et al. (1985):

$$\eta_{c,i} \cong \frac{T_E}{T_C} \leq 1 \tag{3.104}$$

Alternatively, the internal efficiency can be expressed by (see Fig. 3.36):

$$\eta_{c,i} = \frac{h_{2i} - h_1}{h_2 - h_1} \tag{3.105}$$

In Eq. (3.105), the $h_2 = h_2(p_C, T_2)$ enthalpy is unknown. Its value can be obtained from Eqs. (3.104) and (3.105), resulting in:

$$h_2 = h_1 + \frac{h_{2i} - h_1}{\eta_{c,i}} \tag{3.106}$$

The specific compression work is obtained from Eq. (3.106), as:

$$w_c = h_2 - h_1 = \frac{h_{2i} - h_1}{\eta_{c,i}} \tag{3.107}$$

The mvc cooling cycle effectiveness is calculated by:

$$COP_{mvc} = \frac{q_E}{w_c} \tag{3.108}$$

Hybrid (mvc + absorption) operation

In this case, the compressor is connected with its gas suction and discharge lines to the generator and resorber of the hybrid coabsorbent cooling cycle, respectively. Principally, there is not any difference between the bare mvc and hybid operation, as long as the computation of the internal (isentropic) efficiency of the compression process is analysed. Indeed, Eq. (3.104) holds true here also, provided that the vapor internal temperatures of the evaporation and condensation processes be replaced by the generation and resorption vapor mean temperatures, respectively. Bearing in mind that the generation and resorption take place this time on temperature intervals, the internal efficiency of the hybrid operation compression process can be computed by:

$$\left(\eta_{c,i}\right)_{hybrid} = \frac{T_{G,m}}{\frac{T_{RI}+T_{RO}}{2}} \cong \frac{T_{GI} + T_{GO}}{T_{RI} + T_{RO}} \tag{3.109}$$

Further on, Eqs. (3.105)–(3.108) are valid as well here, and theirs enthalpies are calculated with the hybrid cooling cycle values. The specific compression work is computed by:

$$(w_c)_{hybrid} = h_2 - h_1 = \frac{h_{2i} - h_1}{\left(\eta_{c,i}\right)_{hybrid}} \tag{3.110}$$

It is worth mentioning that, given the favourable configuration of the cycle we are dealing with, in certain applications in Eq. (3.109) we can have $((T_{RI} + T_{RO})/2) < T_{G,m}$, so that $(\eta_{c,i})_{hybrid} > 1$. Indeed, unlike the bare mvc cycle, the compression process of the coabsorbent hybrid cycles can be benefited because the discharged vapor is resorbed in the resorber which operates to a good extent with temperatures close to the sink and well below those of the generator.

First hybrid cooling cycle model results

The model results for COP_c, COP_w and COP_{mvc} are given in Fig. 3.37a–f for the NH_3–H_2O working combination. All simulation results were obtained for a sink source having values within a quite large temperature interval, $T_{M,c}$ = (12–42) °C. The desorber inlet temperatures have been considered within the $T_{DI} = -5, \ldots, -50$ °C range and those of the desorber outlet are equal to $T_{DO} = T_{DI} + 3$ °C. No rectification of the generated vapor was considered in the working model of this type of hybrid cooling cycle. Also, the operation with higher desorption intervals was analyzed here as well, but the results are not much different to those plotted. The heat/sink temperature gap was considered $\Delta T_M = T_{M,h} - T_{M,c}$ = 20; 30; 40 °C. The compression work w_c intervening in the COP_w assessment, Eq. (3.55), was calculated according to the algorithm given in Bare Mvc Versus Hybrid (Mvc + Absorption) Operation. The hybrid cycle COP_w values were compared to the mvc cooling effectiveness, COP_{mvc}, in each figure mentioned above. However, the comparison is not a straight one. Indeed, the coabsorbent cycle absorbent has a bi-variant desorption, within a temperature interval, $T_D = T_D(p_D, y_D) \in [T_{DI}, T_{DO}]$, when operating isobar, as known. On the contrary, the mvc mono-component refrigerant cycle is mono-variant with a single-value temperature of isobar evaporation, $T_E = T_E(p_E)$. Hence, the two cycles produce cooling through desorption and evaporation on complete different intervals. When $[T_{DI}, T_{DO}]$ is enough large, in order to compare them, it is necessary to consider first an infinite number of mvc cycles that cover the $T_E \in [T_{DI}, T_{DO}]$ interval of evaporation. Then, their mean cooling effectiveness relative to this cooling interval, COP_{mvc}, must be found, considering each cycle effectiveness. To this purpose, the cooling effectiveness of a single mvc cycle was assessed in relation to that of the ideal cycle (Kirilin et al. 1985):

$$COP_{c,Carnot,mvc}(T_E) = \frac{T_E}{T_{M,c} - T_E}, \tag{3.111}$$

with the help of the following equation:

$$COP_{mvc}(T_E) = k_{mvc} COP_{c,Carnot,mvc}(T_E). \tag{3.112}$$

In Eq. (3.112), the constant

$$k_{mvc} = \frac{COP_{mvc}(T_{E,0})}{COP_{c,Carnot,mvc}(T_{E,0})} = const., \tag{3.113}$$

is evaluated with respect to a known practical reference value $COP_{mvc}(T_{E,0})$. The mean value we are looking for is calculated by the mean integral of the function

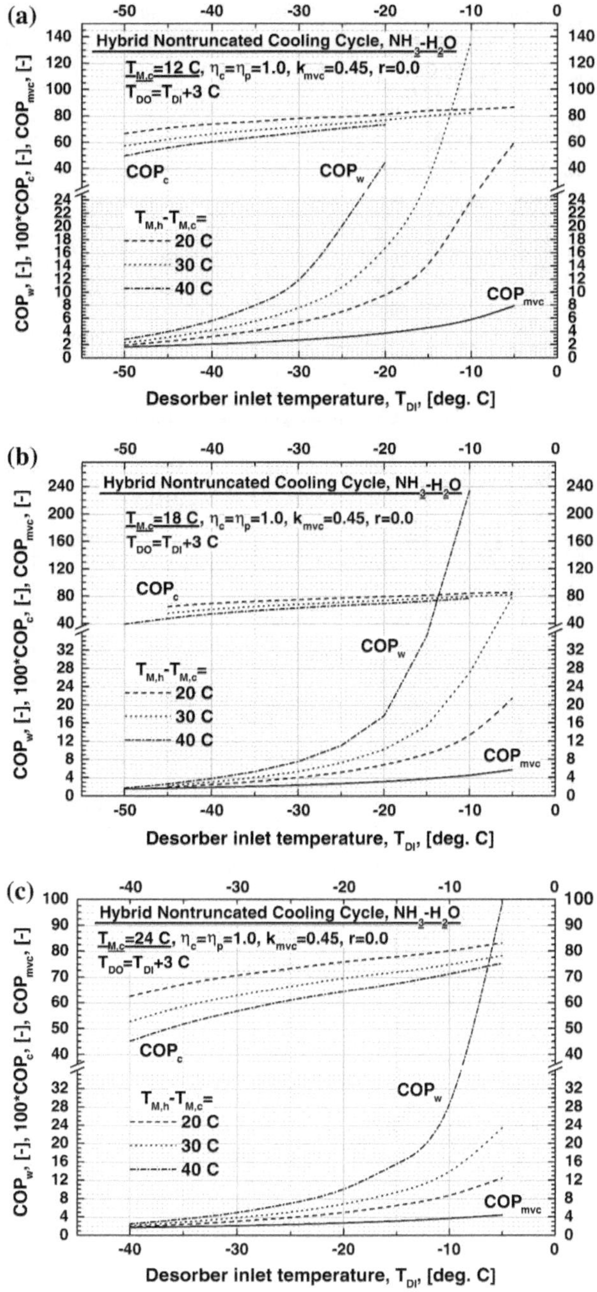

Fig. 3.37 **a** Hybrid nontruncated coabsorbent cooling cycle model results, $T_{M,c} = 12$ °C. **b** Hybrid nontruncated coabsorbent cooling cycle model results, $T_{M,c} = 18$ °C. **c** Hybrid nontruncated coabsorbent cooling cycle model results, $T_{M,c} = 24$ °C. **d** Hybrid nontruncated coabsorbent cooling cycle model results, $T_{M,c} = 30$ °C. **e** Hybrid nontruncated coabsorbent cooling cycle model results, $T_{M,c} = 36$ °C. **f** Hybrid nontruncated coabsorbent cooling cycle model results, $T_{M,c} = 42$ °C

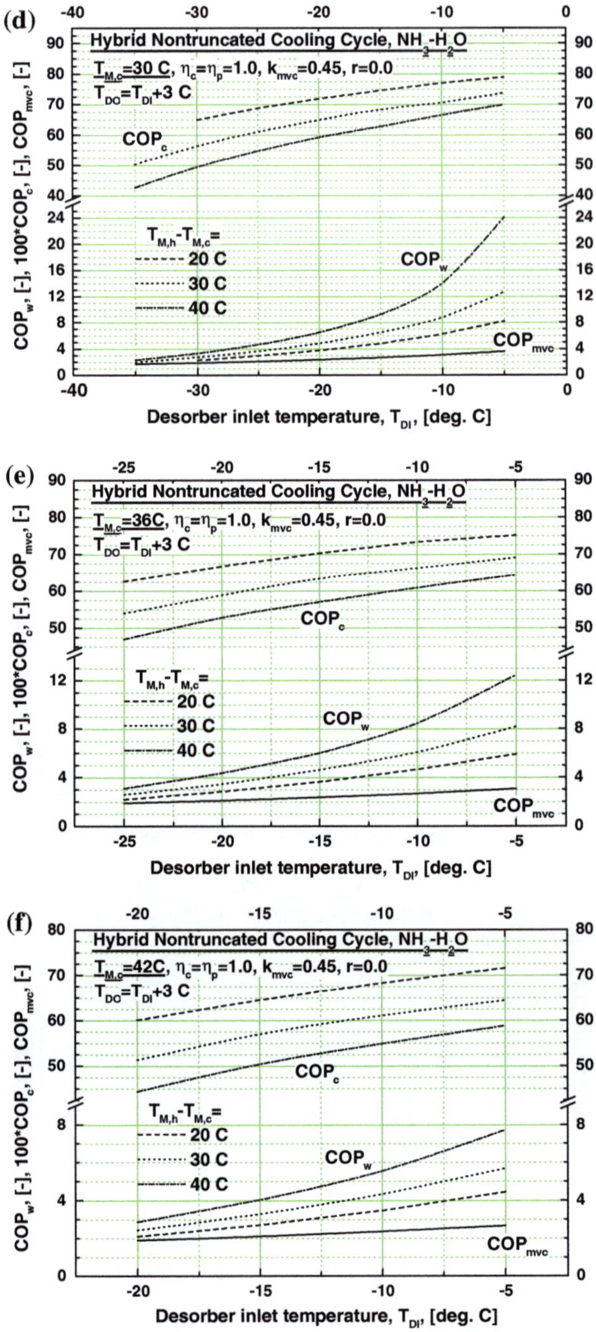

Fig. 3.37 continued

$COP_{mvc}(T)$ on the considered evaporation interval, with the help of Eqs. (3.111)–(3.113), resulting in:

$$COP_{mvc} = \frac{1}{T_{DO} - T_{DI}} \int_{T_{DI}}^{T_{DO}} COP_{mvc}(T_E)dT_E = k_{mvc}\left(-1 - \frac{T_{M,c}}{T_{DO} - T_{DI}} \ln \frac{T_{M,c} - T_{DO}}{T_{M,c} - T_{DI}}\right) \tag{3.114}$$

The COP_{mvc} mean value was computed using the industrial cooling experience results for the NH_3–CO_2 cascade. This system achieves a practical effectiveness of $COP_{mvc}(T_{E,0}) = 1.26$ for a $T_{E,0} = 223.15$ K evaporation temperature and $T_{C,0} = 305.15$ K condensation temperature (Prosper de Mulder 2004; Belozerov et al. 2007). Considering these data in Eq. (3.113), a constant value of $k_{mvc} = 0.45$ results, used hereinafter. The model output data plotted in Fig. 3.37a–f assumed a single-stage compression without mechanical loss, an ideal pumping ($\eta_{mech} = \eta_{pump} = 1$), and the resorber flow factor is null, $r \cong 0.0$. The data confirm that, for same working parameters, an appropriate synegia of components leads to better results than those of separate components (e.g. the Carnot COP of a hybrid cycle is much higher than that of each cycle ideal effectiveness, see Chap. 5). Indeed, both COP_c and COP_w are better and much better than the cooling effectiveness of the isobar nontruncated coabsorbent and bare mvc cycles, respectively. The hybrid cycle is mostly recommended to low-medium sink source temperature applications, when its external low grade heat boosting can save important electrical power in normal to industrial cooling production. In this respect, according to informal sources, the hybrid cooling cycle (Fig. 3.35) has been applied recently to a north capital district cooling (Copenhagen), given the particular favoring operating conditions this place offers to the cycle at hand. The better COP_w is explained first by an obvious reason, the compressor compression ratio reduces considerably, as compared to the bare mvc and second, the internal efficiency of the compression process is favored, as mentioned in section Bare Mvc Versus Hybrid (Mvc + Absorption) Operation. Here, however, a comment regarding the cycle total electrical power consumption in certain situations is mandatory. We are making reference mostly to applications having heat supply cogenerated by a thermal electrical station. It is a known fact that the heat cogeneration sacrifices (consumes) an amount of the station electrical output. In such case it is therefore necessary to check whether the total power consumption through cogenerated heat supply plus that of the hybrid cycle compression is still enough lower than that of the bare mvc, in order to justify a hybrid cycle choice. A report about how such an analysis could be handled is given further in this book.

3.2.5.2 The Osenbrück Cycle

The Osenbrück cycle, named like this in honour of its inventor (Osenbrück 1895), is depicted in the $\log(p) - 1/T$ diagram in Fig. 3.38. In the literature it is known also as

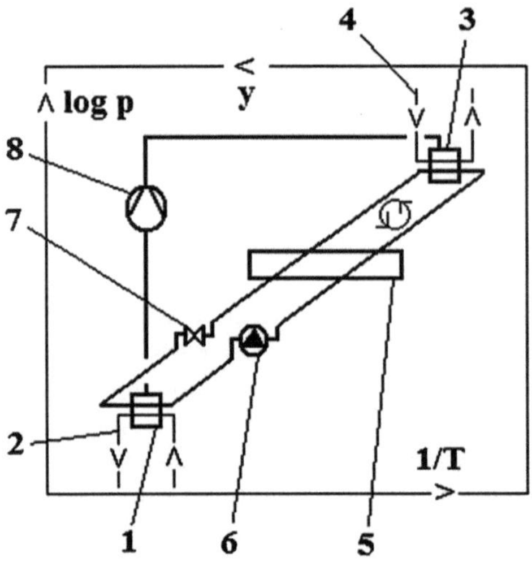

LEGEND: 1. Desorber; 2. Sink heat source; 3. Resorber; 4. Resorber cooling source; 5. Poor-rich absorbent HE; 6. Pump; 7. Expansion valve; 8. Compressor.

Fig. 3.38 The Osenbrück cycle, used in heat pumping, depicted in the log(p) − $1/T$ diagram

absorption/compression cycle, or as vapor-compression cycle with solution circuits, and is based on a reverse Clausius-Rankine process which produces heat pumping with mechanical work input. Later, in the early 20th century, the Osenbrück cycle was completed with other patents (e.g. Sloan and Roncin 1920). The first theoretical studies have been done by Altenkirch (1950, 1951) and later the cycle was included in several textbooks (e.g. Niebergall 1959). As mentioned, given its potential in energy-saving and environment protection, during the last decades the absorption/compression cycle has increasingly been the object of a research activity directed towards improving its configuration and effectiveness (Itard and Machielsen 1994; Groll 1997a). The original cycle, Fig. 3.38, includes two main isobar absorption devices, only, a low pressure desorber 1 and a high pressure resorber 3, connected through a pump 6 and an expansion valve 7, in a same single absorbent loop confined by two isosteres, $y_{RO} = y_{DI}$ and $y_{DO} = y_{RI}$. The desorber and resorber are externally heated and cooled by the sink heat 2 and sink 4, sources respectively. A low-high concentration absorbent heat exchanger 5 improves the cycle effectiveness by an internal heat recovery process. The low pressure vapor desorbed in desorber is compressed till the resorber high pressure by the compressor 8 in order to be resorbed in the resorber. Either desorber and resorber processes can have useful yield (cogeneration of cooling and heating), or just only one of them (e.g. Risberg et al. 2004, in heating). The cooling and heating take place on large gliding temperature intervals and suffer of heat and temperature double deficiency (see Sections Desorption Heat Transfer Analysis and Enhancing the Heat Pump Heating Effect by Using Preheating Hysteresis Loops (PHL)). From this point of view, improvements of both cooling and heating processes can be brought, applying the SHL and PHL techniques (see Fig. 3.16). Also, the Osenbrück cycle cannot be used to too low temperature cooling applications. Indeed, if this happened, the saturation pressure would decrease considerably as compared to that of the pure refrigerant,

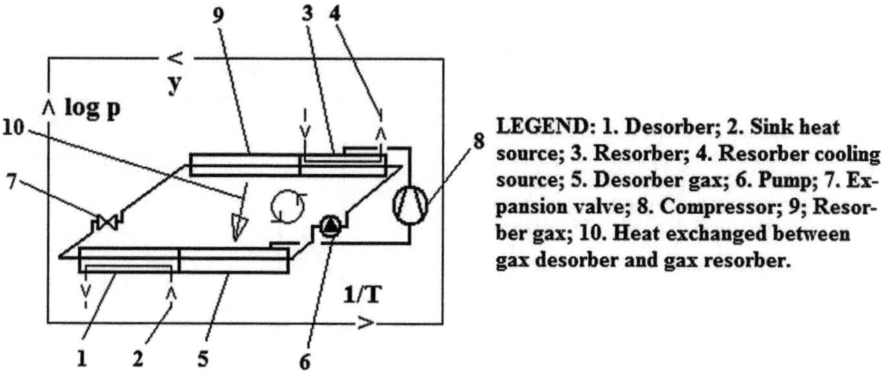

Fig. 3.39 Schematic of an absorption/compression cycle provided with desorber-resorber gax recovery

because it works with the absorbent combination of the refrigerant, so the compressor compression ratio increases much for same input data and the COP decreases much. This drawback can be partially avoided using the absorption/compression cycle with desorber-resorber gax recovery, Fig. 3.39. It comes of that of Fig. 3.38, if its concentration interval $y_{RO} - y_{DO}$ were increased and its pressure difference $p_R - p_D$ were decreased correspondingly, in such a way that the resorber and the desorber can operate with temperature overlapping upon a certain interval. The portions of these devices operating with temperature overlapping, i.e. the resorber gax 9 and desorber gax 5, could benefit of a gax recovery 10. This modified Osenbrück cycle belongs to Altenkirch and Tenckhoff (1911). The gax recovery will be analyzed in more detail to Sect. 4.2.1. The advantage of this cycle is in cooling working mode, as mentioned, when a large temperature difference can be overcome at a moderate pressure ratio, much lower as compared to that of the Osenbrück cycle, Fig. 3.38. The remaining portions of desorber and resorber connected externally, work still with large temperature glide. The type of the cycle with gax recovery has been reportedly demonstrated successfully theoretically and experimentally (Groll and Kruse 1992; Groll 1997b).

3.2.5.3 The Hybrid Heating (Heat Transformer) Cycle

The hybrid (non-isobar) nontruncated heating (heat transformer) cycle, or shorter, the hybrid heat pump (Staicovici 2006a, b, e, 2007b, c, d, e, g, 2008a; Staicovici and Staicovici 2013), is shown in the $\log(p) - 1/T$ diagram of Fig. 3.40. First, some introductory comments are necessary. It is already known that the main role of a heat pump is to extract heat from a heat source of a lower temperature level, and to "upgrade" it to a higher, useful, temperature level. An absorption heat pump uses the extracted heat to boost the useful heat delivery and releases an amount of heat to the sink (see for example the condensing or the coabsorbent heat transformer cycle). A hybrid heat pump is doing it similarly, but it uses additionally the

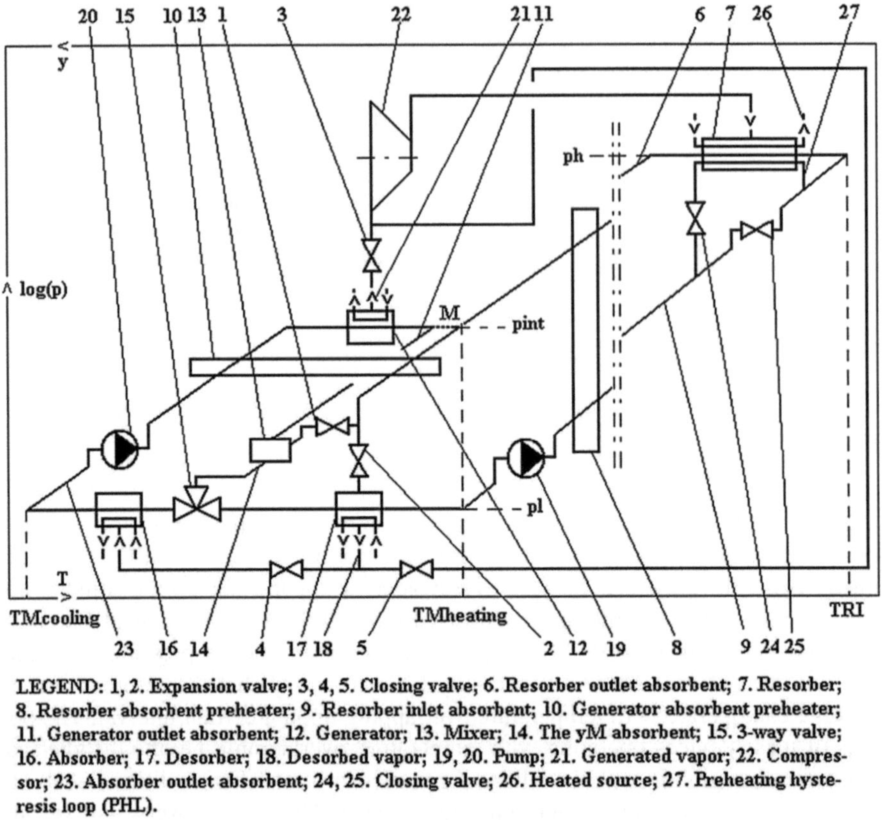

Fig. 3.40 The hybrid (non-isobar) nontruncated heating (heat transformer) cycle, shown in the $\log(p) - 1/T$ diagram

mechanical vapor compression work input. In this case, the heat input reduces the specific mechanical vapor compression consumption as compared to a bare mvc heat pump. Here are included the hybrid nontruncated and, anticipating, the hybrid truncated coabsorbent heat pump as well. Only the heat pump which needs solely the mvc work input for the heat delivery is not releasing heat directly to the sink. In this case, the heat pump sink is that of the thermopower station producing the electrical power for its supply. To this category of heat pumps not connected to a sink source belong for instance the mvc heat pump with a mono-component or with blends of refrigerants (e.g. Cox et al. 2009), and the Osenbrück cycle. The facility to be linked to a particular application and the effectiveness are amongst the first selection criteria of a heat pump. From the facility point of view, the mvc heat pump is ranked the first, because it doesn't need the sink source as mentioned (here included the Osenbrück cycle as well), but unfortunately, its exergy efficiency is not that high and depends on the thermopower station effectiveness (see Chap. 7). On the contrary, the absorption heat pump enjoys a good exergy

Table 3.7 Different pairs of heat/sink sources with theirs probable temperatures, which could supply the hybrid heat pump during the whole year

Period	Heating (heat) source	Cooling (sink) source	$T_{M,heating}$, (°C)	$T_{M,cooling}$, (°C)
Winter	Ground	Ambient air	(+5)–(+8)	(−40)–(−4)
	Underground water	Ambient air	(+8)–(+14)	(−40)–(−4)
	Thermopower station (steam or organic Rankine) condenser cooling water	Tower cooled water	(+35)–(+40) or (+65)–(+70)[a]	(+10)–(+15)
	Thermopower station (steam or organic Rankine) condenser cooling water	Ocean (sea) water	(+35)–(+40) or (+65)–(+70)[a]	(+5)–(+8)
	Thermopower station (steam or organic Rankine) condenser cooling water	Ambient air	(+35)–(+40) or (+65)–(+70)[a]	(−20)–(−4)
Winter-Summer	Thermopower station (steam or organic Rankine) condenser cooling water	Ambient air	(+35)–(+40) or (+65)–(+70)[a]	(+10)–(+20)
Summer	Ambient air, waste heat	Ground	(+30)–(+40)	(+5)–(+8)
	Ambient air, waste heat	Underground water	(+30)–(+40)	(+12)–(+14)
	Solar energy	Ground	(+25)–(+35)	(+5)–(+8)
	Solar energy	Underground water	(+25)–(+35)	(+12)–(+14)
	Thermopower station (steam or organic Rankine) condenser cooling water	Tower cooled water	(+35)–(+40) or (+65)–(+70)[a]	(+25)–(+35)
	Thermopower station (steam or organic Rankine) condenser cooling water	Ocean (sea) water	(+35)–(+40) or (+65)–(+70)[a]	(+10)–(+20)

[a]Thermopower station low temperature cogeneration or Organic Rankine Cycle (*ORC*) condensing

efficiency, but usually it experiences difficulties in application, as it is requiring a sink source, as mentioned, which not always is available or easy to apply. The hybrid heat pump is somewhere between the two extreme heat pumps, needing a sink source for the operation, but the effectiveness can be much better than that of a mvc heat pump, due to the heat-work synergia it benefits. Because of the global warming, which the earth is confronted with lately, the effectiveness became a much more important selection criterion. A heat pump highly fulfilling this criterion and enjoying a good match between its operation mode and the supplying sources has the real chances to acquire a large utilization. This might be the case of the coabsorbent hybrid heat pump at hand too, Fig. 3.40, for at least a good period of time in the year and in many parts of the globe.

The heat pump in Fig. 3.40 must be connected to the heat/sink sources, which provide an internal temperature difference $\Delta T_M = T_{M,heating} - T_{M,cooling}$ (also noted priorly simpler by $T_{M,h} - T_{M,c}$), and to the heated one, having the cycle internal heat supply within the $T_{RI} - T_{RO}$ temperature gap. A few pairs of heat/sink sources which our heat pump could be coupled with are given in Table 3.7, together with theirs probable temperatures. The cycle operation depends on the ΔT_M value. The coabsorbent operation, with all main devices, Fig. 3.40, requires that $\Delta T_M \geq$ 8–10 °C. When $\Delta T_M <$ 8–10 °C, the heat pump can be switched on the Osenbrück cycle operation. Both operation ways are described next.

The coabsorbent operation. The valves 3 and 4 are open, and the 2 and 5 are closed. The absorbent 6, exiting the high pressure p_h resorber 7, is first subcooled in the heat exchanger 8, preheating in a recovering way the absorbent 9 entering the resorber 7, externally cooled by the source which is usefully heated. Then,

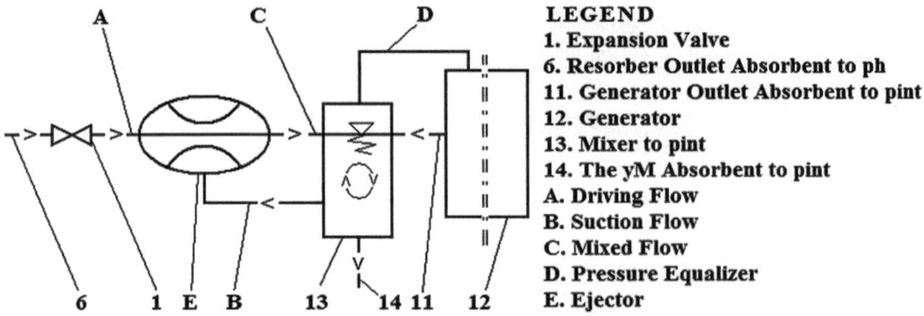

Fig. 3.41 A possible mixing solution with an ejector, feasible in case of the mixing absorbents having a significant pressure difference between them

is subcooled finally in the heat exchanger 10, preheating in a recovering way the absorbent 23 entering the intermediate pressure p_{int} generator 12, externally heated. After being twice subcooled, the absorbent 6 is expanded with the valve 1 from the high pressure p_h till the intermediate one p_{int} and then is mixed up with the absorbent 11 coming of the generator 12 in the mixer 13, in order to generate the cycle one y_M mean concentration absorbent 14. When the mixing fluids have a significant pressure difference between them, like here, it is feasible to use an ejector to help this process, as shown in Fig 3.41. Further on, the absorbent 14 is expanded from the intermediate pressure p_{int} till the low pressure p_l and is distributed with the help of the 3-way valve 15, playing also the role of an expansion device, in pre-established quantities, to the low pressure p_l desorber 17, externally heated, and to the low pressure p_l absorber 16, externally cooled, both coupled on the vapor side 18. The absorbents exiting the desorber and the absorber are pumped by the pumps 19 and 20, respectively, are preheated in the heat exchangers 8 and 10, respectively, as mentioned, and enter the resorber and the generator. The absorbent 9 can be preheated a second time, in the preheating hysteresis loop (PHL, see Enhancing the Heat Pump Heating Effect by Using Preheating Hysteresis Loops (PHL)), closing the valve 25 and opening the valve 24, and then enters the resorber 7. The generated vapor 21 is compressed by the compressor 22 from the p_{int} pressure till the p_h pressure and then is resorbed in the resorber releasing the exothermal useful effect, in order to close the cycle.

The Osenbrück (hybrid) cycle operation. The valves 3 and 4 are closed and the valves 2 and 5 are open. The absorbent 9 coming of the desorber 17, externally heated, is pumped by the pump 19 from the low pressure p_l till the high pressure p_h, is preheated in a recovering way in the heat exchanger 8 through the subcooling of the absorbent 6 exiting the high pressure p_h resorber 7 and enters the resorber. After being subcooled, the absorbent 6 is expanded from p_h till p_l, with the expansion valve 2, and enters the desorber 17. The compressor 22 compresses the desorbed vapor 18 from p_l till p_h, which subsequently is resorbed in the resorber. The resorption heat is transferred to the usefully heated source and the cycle closes.

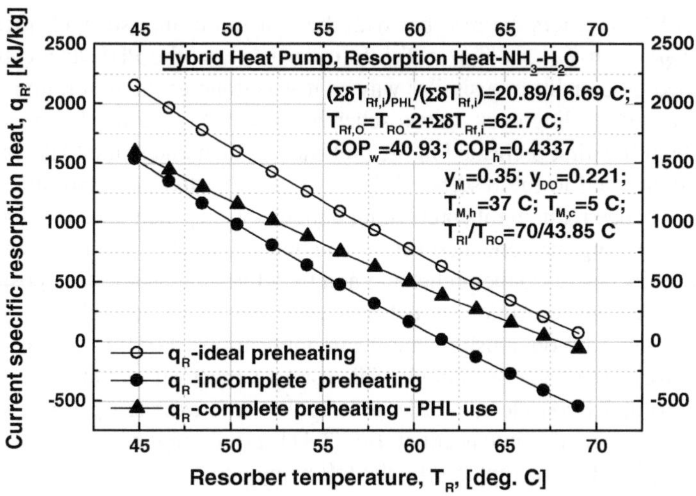

Fig. 3.42 Current specific resorption heat of a hybrid heat pump study case for incomplete, ideal and PHL complete preheating

Enhancing the Heat Pump Heating Effect by Using Preheating Hysteresis Loops (PHL)

The hybrid heat pump useful resorption process is always confronted with the *preheating double deficiency* problem. Let us study the following common heat exchange case of the hybrid heat pump cycle, characterized by the following input data:

- Heat/sink source temperatures, $T_{M,h} = 37$ °C/$T_{M,c} = 5$ °C;
- Resorber inlet/outlet temperature, $T_{RI} = 70$ °C/$T_{RO} = 43.85$ °C;
- Absorbent mean concentration, $y_M = 0.35$;
- Desorber absorbent outlet concentration, $y_{DO} = 0.221$;
- Heated source initial temperature, $T_{RO} - 2$ °C $= 41.85$ °C;
- Number of steps, $n_R = 14$.

The specific resorption heat of this case is given in Fig. 3.42, versus T_R, in three facets. The lowest curve represents the rejected resorption heat with resorber 7 absorbent supply preheated incompletely in the preheater 8, Fig 3.40, only. The absorbent enters the resorber with a nonequilibrium state, characterized by (p_R, $T_{R,icp}$, y_{RI}). The curve has negative and positive values, but only where the positive values exist the resorber can deliver useful heat. The top curve shows the resorption process rejecting heat taking place with a supplying absorbent ideally preheated (without preheating deficiency), characterized by the following equilibrium state parameters (p_R, T_{RI}, y_{RI}). It is possible to reduce the preheating deficiency, employing this time a *preheating hysteresis loop* (PHL), in order to elevate the resorption useful final heating temperature. The curve representing this heating effect enhancement is the intermediate one. The absorbent benefiting of PHL enters the resorber with the nonequilibrium state, characterized by (p_R, T_{RII}, y_{RI}), with $T_{R,ipc} < T_{RII} < T_{RI}$. In order to show the PHL application it will be proceeded essentially in a similar way as it was done to the desorption process of the nontruncated cooling cycle, given schematically in Fig. 3.40. The low concentration absorbent 9 is first preheated in the preheater 8, then enters the PHL, position 27, thermally connected to the resorber 7. Here, it is preheated a second time, receiving the heat q_{pd}, hence reducing its preheating deficiency, and then goes into the resorber, where it is resorbing the generated vapor. The heat released during the resorption process is utilized to increase the preheating and to deliver the useful heat, position 26. A complete preheating resorption (noted as ideal in Fig. 3.42), calculated according to Eq. (3.41), cannot be obviously achieved in applications. For this reason, the partial preheating we could achieve through PHL use is a practical maximum, so it will be considered formally complete, in order to keep the same expression as that it was used to desorption. Taking this into account, Eq. (3.41), expressing the ideal (complete) preheating resorption heat, is replaced by the following equation:

$$q_{R,cp} = V_G((H_G - h_{RO}) + r(h_{RII} - h_{RO})) \tag{3.115}$$

The resorption heat without PHL but with preheater absorbent preheating (pos. 8, Fig. 3.40) will be considered with incomplete preheating again, for expression consistency, and is calculated by:

$$q_{R,icp} = V_G\left((H_G - h_{RO}) + r\left(h_{R,icp} - h_{RO}\right)\right) \tag{3.116}$$

Also, the absorbent preheating deficiency heat amounts to:

$$q_{pd} = V_G r\left(h_{RII} - h_{R,icp}\right) \tag{3.117}$$

In Eqs. (3.115)–(3.117), $h_{R,icp}$ and h_{RII}, with $h_{R,icp} < h_{RII}$, hold for the enthalpies resulting from the preheater and PHL preheating, respectively. From Eqs. (3.115)–(3.117), it is easy to remark that:

$$q_{R,icp} = q_{R,cp} - q_{pd} \tag{3.118}$$

which means that the actual resorber useful resorption heat provided with PHL is identical to that without PHL, so the heat balance is closed. However, as it will be shown, the PHL use has the advantage to increase to a certain extent the usefully heated fluid output temperature. In order to study the resorber heat exchange of the hybrid heat pump cycle equiped with PHL, let us consider the resorber is constructively made of two sub-devices operating in parallel, each of them with a distinct role, of preheating and of useful heating resorption, respectively (suggested by Fig. 3.40, also). Because of the PHL use and bearing in mind the expression convention, the resorber sub-devices operate with complete preheating. The resorption heats of the preheating and resorption sub-devices can be expressed as:

$$V_{G,pd}\left[(H_G - h_{RO}) + r_{pd}(h_{RII} - h_{RO})\right] = V_G r\left(h_{RII} - h_{R,icp}\right) \tag{3.119}$$

$$V_{G,icp}\left[(H_G - h_{RO}) + r_{icp}(h_{RII} - h_{RO})\right] = V_G\left[(H_G - h_{RO}) + r\left(h_{R,icp} - h_{RO}\right)\right] \tag{3.120}$$

respectively. In the above equations, we have assumed that there is a necessary temperature pinch between the heat exchanging absorbents going into PHL and preheating sub-device, respectively. Equations (3.119) and (3.120) are completed by the following two obvious equations:

$$V_{G,pd} + V_{G,icp} = V_G \tag{3.121}$$

$$r_{pd} = r_{icp} = r \tag{3.122}$$

From Eqs. (3.119)–(3.122), the unknown vapor $V_{G,pd}$ and $V_{G,icp}$ result, taking into account Eqs. (3.115)–(3.118), as well:

$$V_{G,pd} = V_G \frac{q_{pd}}{q_{R,cp}} \tag{3.123}$$

and

$$V_{G,icp} = V_G \frac{q_{R,icp}}{q_{R,cp}} \tag{3.124}$$

respectively. The heat exchange can now be assessed, considering just the main part of the resorber (see the DD method, Chap. 2) and bearing in mind Eq. (2.26). In this equation, the factor V_G^m, given by Eq. (2.33), must be multiplied in case of PHL use by one of the subunit vapor amounts given by Eqs. (3.123) and (3.124), depending whether the preheating, or the resorption function of the resorber is intended to be analyzed with respect to the heat exchange, resulting in:

$$V_{G,pd}^m = V_{G,pd} V_G^m \tag{3.125}$$

and

$$V_{G,icp}^m = V_{G,icp} V_G^m \tag{3.126}$$

respectively.

Results of the resorber heat exchange analysis at hand have been obtained applying the DD method completed with the algorithm presented above. Solving Eqs. (3.123) and (3.124), it resulted in $V_{G,pd} = 0.2061$ and $V_{G,icp} = 0.7939$, respectively. This means that, out of the total of $V_G = 1.0$ kg/kg resorbed vapor, 0.2061 kg/kg is resorbed in the preheating sub-device in order to preheat the absorbent $V_G r$ going into resorber till ~62 °C, and the rest of 0.7939 kg/kg is resorbed in the resorption sub-device, in order to usefully heat the source to be heated. The absorbent supplying the resorber is first incompletely preheated from 37 till 42.35 °C in the preheater 8, and next is preheated in the PHL, till 63.24 °C. The heat released during the useful heating process amounts to $q_{R,icp} = 1{,}695.7$ kJ/kg. This is exactly the difference between the heat which would have been rejected if the absorbent had been completely preheated, $q_{R,cp} = 2{,}136$ kJ/kg, and that received in the PHL process in order to reduce the preheating deficiency, $q_{pd} = 440.3$ kJ/kg. The temperature raise during the useful heating provided with PHL equals $\sum_{i=1}^{n_R} \Delta T_{Rf,i} \cong 20.89\,°C$, while that without PHL is only $\sum_{i=1}^{n_R} \Delta T_{Rf,i} \cong 16.69\,°C$. The PHL method application is not confronted with an additional investment cost, there is an additional operation cost

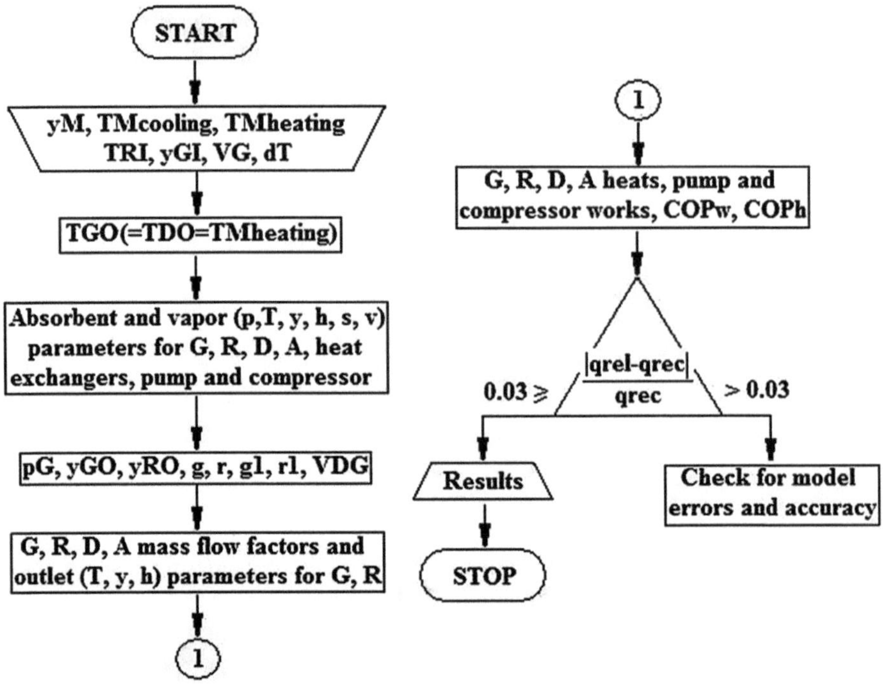

Fig. 3.43 Organigrame of the hybrid heating nontruncated cycle model

coming of the very small additional power use for absorbent pumping through the PHL sub-device, only. Further on, all resorption processes provided with PHL are mentioned in this respect.

Choosing the Best Heat Pump Operation Domain and Model Results

Both operation modes of the heat pump, coabsorbent and Osenbrück, have been modeled considering the NH_3–H_2O working combination and results thereof will follow. The computer code, non-iterative and steady-state, is similar to that shown to Sect. 3.2.1.2, and can be used here as well, provided that some changes are necessary, because we deal here with a heating cycle. The code organigrame is shown in Fig. 3.43. The code input data are those specified to Sect. 3.2.2, except the y_{DO}, which was replaced by the $y_{M,heating}$ (noted also just by y_M). In our case it is possible to ensure the generator and resorber a permanent operation of the type shown in the Fig. 3.3b, expressed here by the single point overlapping $y_{GO} = y_M = y_{RO}$. However, generally, the model introduces this condition next, as a required value for the T_{GO}. With the first input data, the code calculates the cycle configuration, that is the absorbent and vapor p, T, y, h, s, v state parameters of the main devices. Using the T_{GO} value, the $y_{GO} = y_{GO}(T_{GO}, p_G)$ concentration

Table 3.8 Checking for the best hybrid heat pump y_M concentration value

y_M[a] (–)	0.24	0.34	0.44	0.54	0.64	0.74	0.84	0.94
y_{DO} (–)	0.138	0.234	0.324	0.402	0.4645	0.508	0.534	0.551
$\sum_{i=1}^{nR} \Delta T_{Rf,i}$[b] (°C)	13.625	13.666	13.663	13.568	12.472	9.790	6.338	3.672
COP_w (–)	8.356	8.454	8.693	9.181	9.868	10.69	11.45	11.96
T_{RO} (°C)	34.05	34.25	33.75	32.45	30.75	28.75	26.95	25.85
P (bar)	0.7733	1.624	2.999	4.787	6.644	8.134	9.084	9.723
$COP_{w,Osenbrueck}$ (–)	5.509	5.522	5.541	5.556	5.554	5.534	5.496	5.459

[a]$y_{GI} - y_M = 0.05$; [b]without PHL

results from the equilibrium diagram and $y_{RO}(y_{GO})$ can be computed from the first Eq. (3.30). Further, the g and r flow factors are determined with Eqs. (3.2) and (3.3) and the generated vapor V_{DG} results from Eqs. (3.20) or (3.21). Finally, the main devices flow factors, energy and cycle COP's are derived. The model output values are valid when Eq. (3.59) is fulfilled. The compression work w_c intervening in the COP_w assessment [see Eq. (3.55)] was again calculated according to the algorithm given in Bare Mvc Versus Hybrid (Mvc + Absorption) Operation. The COP_h effectiveness was assessed with the Eq. (3.54).

The model first objective was to select the most appropriate value of the cycle absorbent y_M one mean concentration for a particular application. From Chap. 2, it resulted that a heat pump resorber should operate most effectively in the low concentration region in case of the ammonia-water for all applications. This recommendation was checked for a winter heating process. Table 3.8 gives results thereof for the $T_{M,c} = -20$ °C, $T_{M,h} = +5$ °C and $T_{RI} = 55$ °C main temperatures. Indeed, they show that the highest heating effect $\sum_{i=1}^{nR} \Delta T_{Rf,i}$ is obtained for the lowest y_M values, $y_M = 0.25$–0.55. This is done by sacrificing the COP_w, which decreases gradually from the highest to the lowest y_M values. This drawback is however, compensated to some extent by a corresponding cycle (resorber) maximum operating pressure decrease, which increases safety and lightens construction. Table 3.8 shows as well the COP_w of the Osenbrück cycle, noted as $COP_{w,Osenbrueck}$.

A second model objective was to find the hybrid heat pump results in applications which emphasize its advantages in comparison to the Osenbrück and bare mvc heat pumps operating in similar conditions. In this respect, there were considered three winter-summer heating applications with free heat sources, given in Table 3.7: (a) ground or underground water, with $T_{M,h} = +5$ °C, Fig. 3.45a–c; (b) condenser cooling water of a steam Rankine cycle thermopower station (SRC), with $T_{M,h} = +37$ °C, Fig. 3.46a–c, and (c) condenser cooling water of an organic Rankine cycle thermopower station (ORC), with $T_{M,h} = +70$ °C, Fig. 3.47a–c. Of course, the sources list having the heat temperatures mentioned or close to them is not exhaustive. In Fig. 3.44, three cycles related to our model are plotted in a $\log(p) - 1/T$ diagram: *abcdcefghbc* holds for the hybrid heat pump (hhp, Fig. 3.40), *abcdcefbc* holds for the nontruncated heat transformer (ntht, Fig. 3.29), and *cefghbc* holds for the Osenbrück cycle (Fig. 3.38). All Figs. 3.45, 3.46, 3.47 results have the ideal resorber inlet temperature T_{RI} as parameter, $T_{RI,hhp} \geq T_{RI} \geq T_{RI,ntht}$

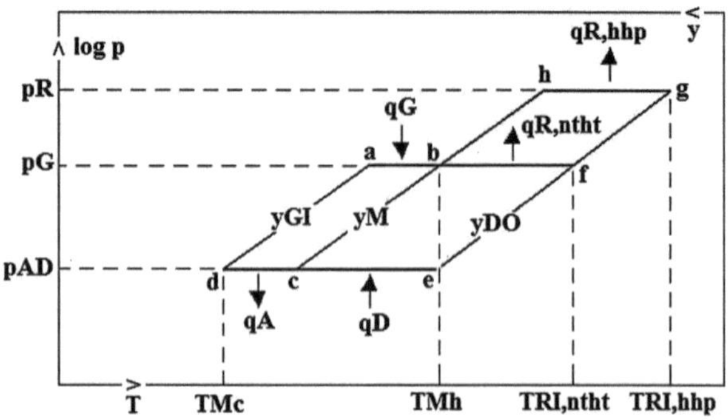

Fig. 3.44 Hybrid heat pump (*abcdcefghbc*), nontruncated heat transformer (*abcdcefbc*) and Osenbrück (*cefghbc*) cycles in the $\log(p) - 1/T$ diagram

(see Fig. 3.44), with $T_{RI,hhp}$ chosen in order that $COP_w > COP_{w,Osenbrueck}$. The functions ($T_{M,c}$, COP_w and $T_{Rf,O}$) (COP_h and $COP_{w,Osenbruck}$) and $\sum_{i=1}^{n_R} \Delta T_{Rf,i}$ are plotted versus resorber pressure, p_R, $p_R \geq p_G$ (see Fig. 3.44) in Figs. 3.45a, 3.46a, 3.47a, Figs. 3.45b, 3.46b and 3.47b and Figs. 3.45c, 3.46c and 3.47c, respectively. The model input data for the compressor mechanical efficiency and pump efficiency are equal to unit ($\eta_{mech} = \eta_{pump} = 1$). Also, the compressor internal (isentropic) efficiency was calculated using the mean values of the generator and resorber operating temperatures [see Eq. (3.109)]:

$$\eta_i = \frac{T_{GI} + T_{GO}}{T_{RI} + T_{RO}} \tag{3.127}$$

Quite remarkably, in all hybrid heat pump results plotted in Figs. 3.45, 3.46 and 3.47, the compressor internal efficiency, given by Eq. (3.127), preserved a quasi-constant value, $\eta_i \cong 0.85$. The model results in Figs. 3.45a, 3.46a and 3.47a show that those corresponding to a cycle configuration close to the nontruncated heat transformer (generator and resorber pressures have close values) exhibit the highest COP_w and the smallest usefully heated fluid (heated source) temperatures, $T_{Rf,O}$, out of all cycles analyzed and plotted in Fig. 3.44. Also, it needs a sink source supply, of temperature $T_{M,c}$, equal to that of the hybrid heat pump. On the contrary, the cycle having the configuration corresponding to or alike to that of the Osenbrück cycle (see Sect. 3.2.5.2) has the lowest COP_w, Figs. 3.45b, 3.46b and 3.47b, but it doesn't need a sink source supply, as mentioned. The COP_w of the hybrid heat pump, Figs. 3.45a, 3.46a and 3.47a, is less than that of the heat transformer, but it is much higher than that of the Osenbrück cycle, Figs. 3.45b, 3.46b and 3.47b, or similar mvc heat pumps. Besides this, its $T_{Rf,O}$ values, calculated with the following equation:

$$T_{RO} < T_{Rf,O} = T_{RO} + \sum_{i=1}^{n_R} \Delta T_{Rf,i} - \Delta T_{M,h} < T_{RI}, \quad \Delta T_{M,h} = 2, \ [^\circ C] \tag{3.128}$$

are higher as compared to those of the heat transformer. Potentially, these two advantages could offer the hybrid heat pump a larger market in the household and industrial application, as compared to the Osenbrück and heat transformer cycles. The effectiveness with respect to the heat source use, COP_h, Figs. 3.45b, 3.46b

Fig. 3.45 **a** Results of $T_{M,c}$, COP_w and $T_{Rf,O}$ versus p_R, for $T_{M,h} = 5$ °C, $y_M = 0.34$ and $T_{RI} = (55–70)$ °C. **b** COP_h and $COP_{w,Osenbruck}$ versus p_R, for $T_{M,h} = 5$ °C, $y_M = 0.34$ and $T_{RI} = (55–70)$ °C. **c** $\sum_{i=1}^{n_R} \Delta T_{Rf,i}$ versus p_R, for $T_{M,h} = 5$ °C, $y_M = 0.34$ and $T_{RI} = (55–70)$ °C

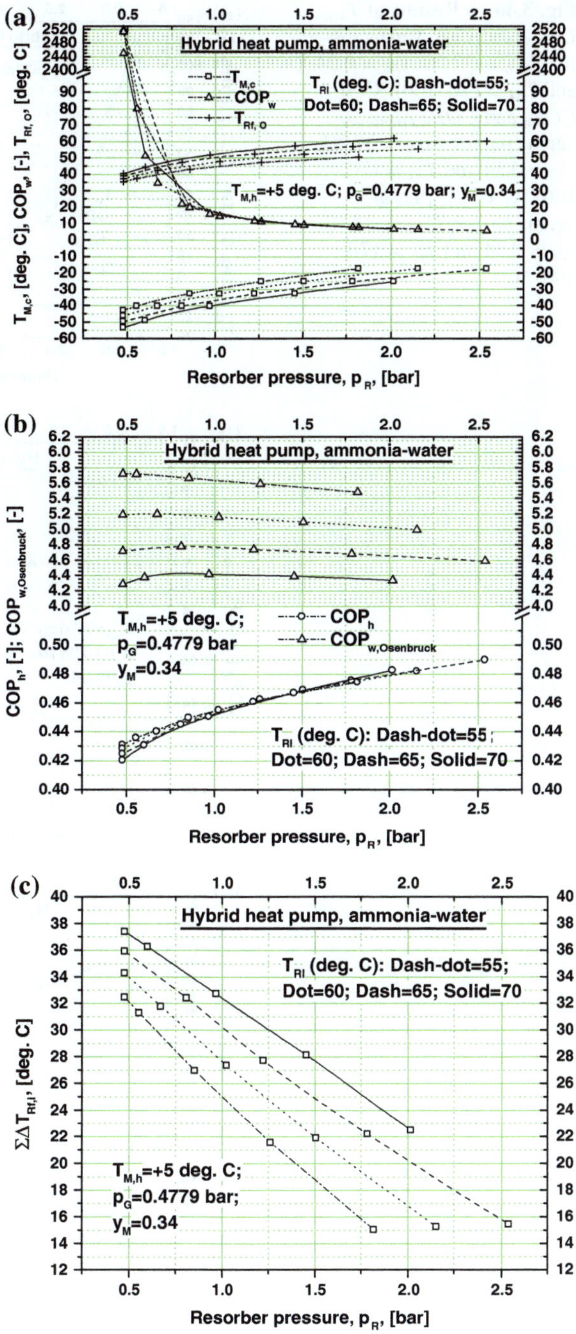

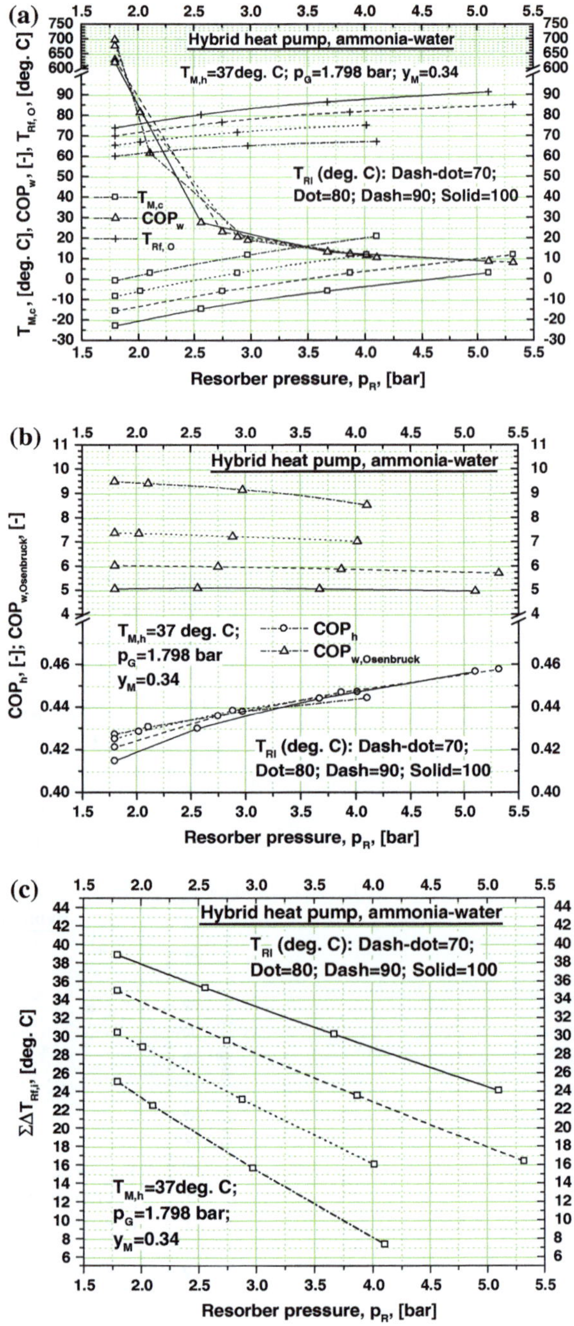

Fig. 3.46 **a** Results of $T_{M,c}$, COP_w and $T_{Rf,O}$ versus p_R, for $T_{M,h} = 37$ °C, $y_M = 0.34$ and $T_{RI} = (70–100)$ °C. **b** COP_h and $COP_{w,Osenbruck}$ versus p_R, for $T_{M,h} = 37$ °C, $y_M = 0.34$ and $T_{RI} = (70–100)$ °C. **c** $\sum_{i=1}^{n_R} \Delta T_{Rf,i}$ versus p_R, for $T_{M,h} = 37$ °C, $y_M = 0.34$ and $T_{RI} = (70–100)$ °C

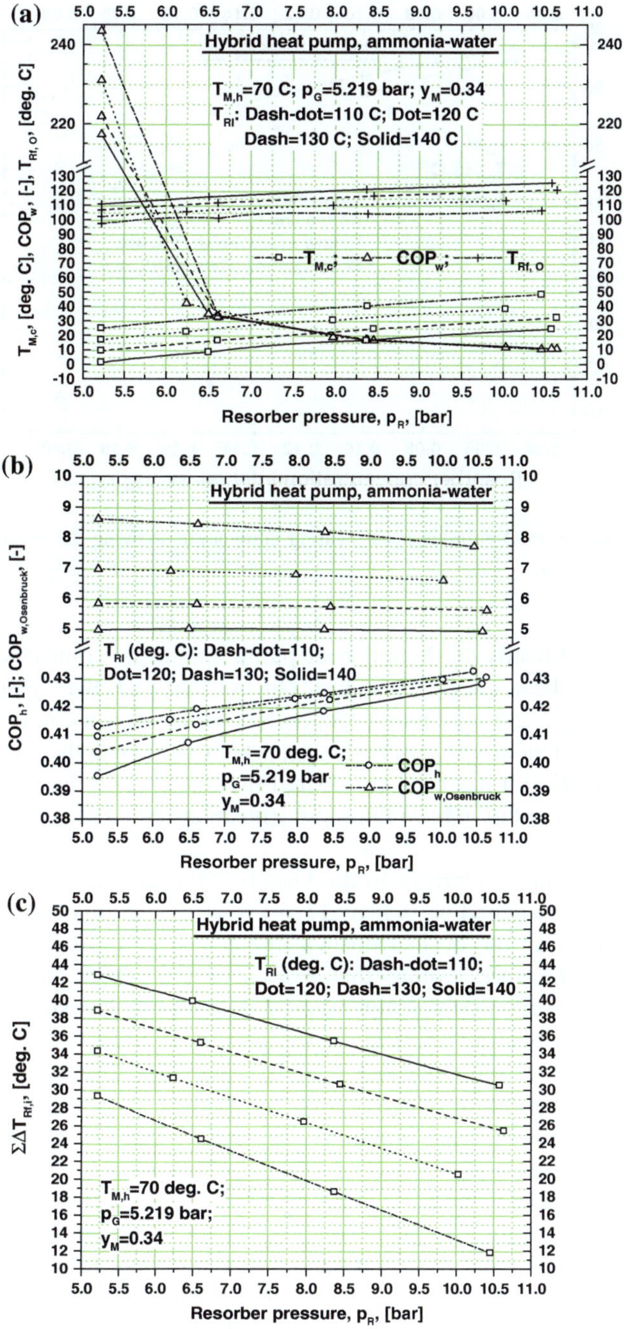

Fig. 3.47 **a** Results of $T_{M,c}$, COP_w and $T_{Rf,O}$ versus p_R, for $T_{M,h}$ = 70 °C, y_M = 0.34 and T_{RI} = (110–140) °C. **b** COP_h and $COP_{w,Osenbruck}$ versus p_R, for $T_{M,h}$ = 70 °C, y_M = 0.34 and T_{RI} = (110–140) °C. **c** $\sum_{i=1}^{n_R} \Delta T_{Rf,i}$ versus p_R, for $T_{M,h}$ = 70 °C, y_M = 0.34 and T_{RI} = (110–140) °C

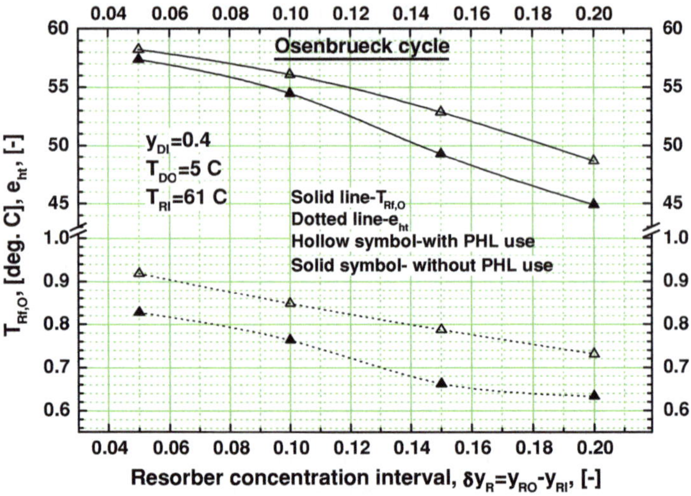

Fig. 3.48 Results of an Osenbrück cycle heating application showing the relationship of $T_{Rf,O}$ versus $\delta y_R = y_{RO} - y_{DO}$

and 3.47b, recommends the hybrid heat pump should be connected to free and enough available heat sources. And last, but not the least, the hybrid heat pump heating effect, $\sum_{i=1}^{nR} \Delta T_{Rf,i}$, Figs. 3.45c, 3.46c and 3.47c, is a continuous decreasing and linear function with respect to p_R, which facilitates the $T_{Rf,O}$ prediction. (Note: in Eq. (3.128) it has been considered that the usefully heated fluid and the resorber absorbent solution have the same [(mean specific heat capacity) × (mass flow rate)] product).

A final comment is referring to the Osenbrück cycle operating alone, which the heat pump is switched on, when $\Delta T_M < 8$–10 °C. In this case, it is strongly recommended to run the cycle with low $\delta y_R = y_{RO} - y_{DO}$ concentration intervals (devices safe work require that $\delta y_{R,\min} = 0.04$–0.05 for NH_3–H_2O), in order to maximize the useful temperature $T_{Rf,O}$ output. This statement is exemplified by a short heating application study case with $T_{RI} = 61$ °C = *const.*, $T_{DO} = 5$ °C = *const.* and $y_{DI} = 0.4 =$ *const.*, Fig. 3.48. The results of $T_{Rf,O}$ and of the heating temperature effectiveness, e_{ht}, expressed by Eq. (4.137), show a continuous decrease of these functions with $\delta y_R = y_{RO} - y_{DO}$ increase, either PHL is considered or not. Also, as expected, both heating functions have higher values comparatively, when PHL is present in the cycle structure. However, the operation of the Osenbrück cycle alone, requiring minimum δy_R values, is different as compared to that characterized by the entire hybrid cycle configuration functioning, Fig. 3.40. Indeed, in case of the full configuration operation with $T_{RI} =$ *const.*, $T_{DO} =$ *const.* and $y_M =$ *const.*, the increase of δy_R has this time a positive effect, bringing nearer one another the p_R and p_G pressures, Fig. 3.44, and hence decreasing the compressor compression ratio and increasing COP_w for same $T_{Rf,O}$.

The Sky as Cycle Sink Source

We regard now our cycle winter application for house heating and domestic hot water (dhw) preparation with emphasis solely towards an ecological prospect. In this respect, its both heat and sink sources should be natural. During the winter and depending on the geographical area, such pairs of sources, which could ensure a quasi-continuous cycle supply, are just a few (e.g. that made up by the ground (groundwater) and the ambient air, Table 3.7). For this reason, each of them is very precious. In this context, finding a new sink source, of lower temperature than the temperatures of the already known supplying sources, is a good opportunity to extend the heating or cooling coabsorbent truncated cycle applications and increase its effectiveness (here included our heat pump as well). Supposing we have n, $n \in N$, supplying sources, $ss(T_1)$, …, $ss(T_n)$, characterized by $T_1 < T_2 < \cdots < T_n$, the number of pairs of sources liable to supply a coabsorbent cycle results in C_n^k, where $k = 2$, according to the combinatorial analysis. If now a new supplying source $ss(T_0)$ is added to the already known ones, so that $T_0 < T_1 < T_2 < \cdots < T_n$, the number of pairs of sources increases to C_{n+1}^k. The supplementary potential pairs generated by adding the $ss(T_0)$ source result in $C_{n+1}^k - C_n^k = C_n^{k-1}$. The new sink source $ss(T_0)$ we are thinking of here is the sky. For example, if $n = 3$ (air, ground and ground water), then $C_n^{k-1} = C_3^1 = 3$. The sky is the sink with the lowest temperature on Earth. Indeed, this source comes of the cosmic microwave background radiation and has a thermal black body spectrum at a temperature of 2.725 K. The spectrum peaks in the microwave region with a frequency of 283 GHz, corresponding to a 1.06 mm wavelength (Penzias and Wilson 1965), if measured per unit wavelength and using Wien's law (Popa and Vintila 1977). However, the presence of the atmosphere increases this background temperature up to effective sky temperatures with much higher values. A terrestrial object can be cooled through radiative heat exchange mainly in the "window" of the atmosphere, where this is essentially transparent in the wavelength region from 8 to 14 μm, while outside the window the atmosphere has radiating bands covering much of the infrared spectrum. The black body spectrum peaking in the window has the corresponding temperatures of 362.2–206.986 K, respectively, according to Wien's law. The effective sky temperature is related to water vapor content of the air and/or air temperature (Brunt and Quart 1932; Bliss 1961). Swinbank and Quart (1963) relates the sky temperature to the local air temperature in the simple relationship:

$$T_{sky} = 0.0552 \times T_{air}^{1.5} \tag{3.129}$$

where T_{sky} and T_{air} are both in degrees Kelvin. Whillier (1967) uses an even simpler expression and substracts 6 °C from the air temperature.

$$T_{sky} = T_{air} - 6 \tag{3.130}$$

It is not clear which relationship should be used. Equations (3.129) and (3.130) give identical results when the air temperature is 315 K, Fig. (3.49). This seems to be a reasonable summer condition but Whillier's 6 °C should probably be raised

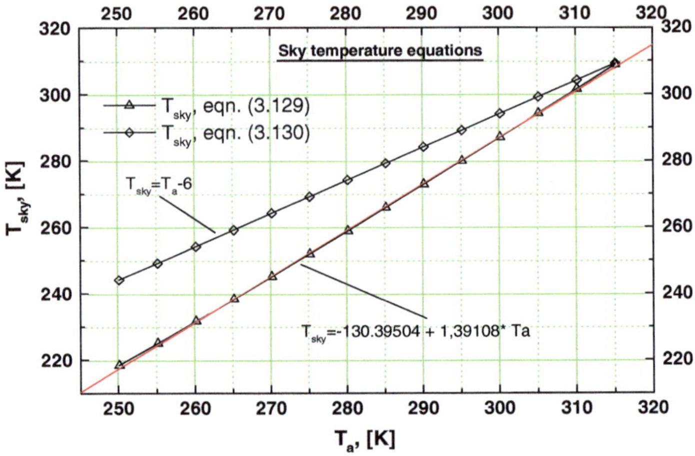

Fig. 3.49 Sky temperature versus ambient temperature, according to Eqs. (3.129) and (3.130)

to about 20 °C for winter operation (Duffie and Beckman 1974). It is possible that both the clouds and the ground, if the cooled object is tilted and can see the ground, will tend to increase the effective sky temperature over that for a clear sky. The influence of clouds and ground is not taken into account in either expression. The sky radiation heat exchange coefficient is confronted with small values. Indeed, the net radiation to a small object with emittance ε, area A and temperature T, radiating to the sky is thus found from, Duffie and Beckman (1974):

$$q = \varepsilon A\sigma\left(T_{sky}^4 - T^4\right) \tag{3.131}$$

Equation (3.131) rewrites in a form which emphasizes the radiation heat exchange coefficient h_r:

$$q = Ah_r\left(T_{sky} - T\right) \tag{3.132}$$

where

$$h_r = \varepsilon\sigma\left(T_{sky}^2 + T^2\right)\left(T_{sky} + T\right) \tag{3.133}$$

Considering in Eq. (3.133) typical winter temperature values for the temperate climate, $T \approx T_{air} = 273.15$ K, from Eq. (3.129) we obtain $T_{sky} = 249.2$ K. Further on, considering $\varepsilon = 1$ and the Stefan-Boltzmann constant value $\sigma = 5.6697 \times 10^{-8}$ Wm^{-1} K^{-4}, Eq. (3.133) gives $h_r \approx 4.05$ Wm^{-2} K^{-1}. This quite low magnitude is comparable to that of the conduction heat exchange coefficient, characterizing the applications with the ground as heat or sink source (Lund et al. 2004).

For this reason, similar to ground heat pumps, high sky heat exchange surfaces are expected in normal household applications. Moreover, because this time the sky heat exchanger is not mounted underground, but on the ground, horizontally and towards the sky, the heating applications are confined to small capacity heat pumps

Table 3.9 Hybrid heat pump results having the sky as sink source ($y_{M,h}$ = 0.34; $y_{GI} - y_M = 0.05$)

Heat source	Sink source	T_{air}/T_{sky} (°C)/(°C)	$T_{M,h}$ (°C)	$T_{M,c}$ (°C)	$T_{Rf,O}$ (°C)	T_{RO}/T_{RI} (°C)/(°C)	$COP_w/COP_{w,Ossenbruck}$ (–)	COP_h (–)
Ground water	Sky	3.0/−19.84	10.0	−10.0	50.72	40.4/54.5	8.143/6.178	0.4672
Ground		0.0/−23.95	3.0	−17.0	51.08	40.26/55.0	6.414/5.188	0.4822

use, typical for instance for passive houses (Badescu and Staicovici 2005), which require smaller occupied soil area. In Table 3.9 we have included results of two study cases of the hybrid heat pump having the sky as cycle sink source. Without this sink the results of the two cases wouldn't have been satisfactory, because in this case the air, a potential next sink, can hardly ensure a large enough ΔT_M for the cycle operation. The theoretical benefit of using our sky cooled hybrid heat pump is shown by the COP_w values of Table 3.9, which are 2–3 times higher as compared to a ground heat pump for similar working parameters, (Lund et al. 2004).

3.2.5.4 The Trigeneration (Cooling + Heating + Power) Cycle

Bearing in mind our remark of Chap. 2, motivating the employ of cycles with simultaneous production of two or more different forms of energy when supplied by a single primary heat source in effective applications, we give in Fig. 3.50 the schematic of the trigeneration cycle depicted in Fig. 3.34 (Staicovici 2007g, 2008c). This cogenerates cooling, heating and mechanical work (electricity). Next, we give a short description of its operation. The absorbent leaving the low pressure desorber 1, usefully cooling the source 2, is mixing up with the low pressure absorbent coming of the absorber 5, externally cooled by the sink source 5, in order to generate the mean concentration y_M absorbent in the mixer 6. The cold vapor 3, desorbed in the desorber 1, is superheated in a recovering way in the subcooler 22 and then is absorbed in the absorber 4. The mean concentration absorbent is divided in two complementary flows with the help of the three-way valve 7. Then, these flows are pumped with pumps 8 and 9 to the high pressure generator 10, externally heated with the heat source 11 and to the intermediate pressure resorber 18, where the external source 19 is usefully heated, respectively. In our case, the PHL (see Enhancing the Heat Pump Heating Effect by Using Preheating Hysteresis Loops (PHL)) use is not capable to boost the cycle heating effect (i.e. the $T_{Rf,O}$ increase), so it hasn't been considered in the cycle thermodynamic model, following next. Further on, prior to reach the generator, the absorbent supplying it is preheated in a recovering way in the liquid–liquid preheater 20. The vapor generated in the generator 10 is partially rectified optionally, during a process called dephlegmation. This takes place in the dephlegmator 12. The dephlegmation heat can be withdrawn, rejecting it by means of the external sink source 13, or recovering it, usefully heating the source 19. Subsequently, the generated vapor is optionally superheated in the superheater 14, recovering an amount of heat of the fuel flue gas 15, coming of the generator heat

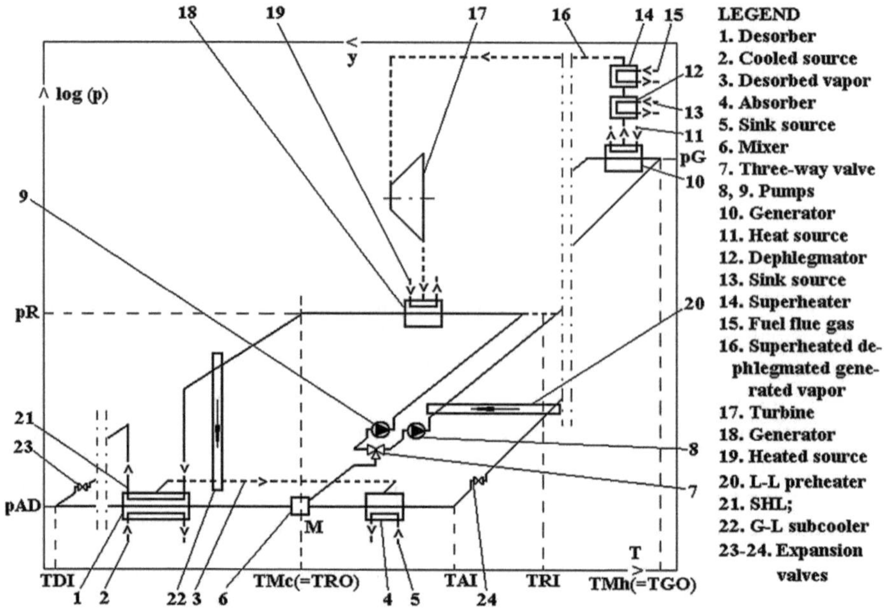

Fig. 3.50 A schematic of the trigeneration (cooling + heating + power) coabsorbent cycle plotted in the log(p) − 1/T diagram

source 11 (see also Fig. 3.51). After being superheated, the generated vapor 16 is expanded in the turbine (or turbogenerator) 17, from the high pressure to the intermediate one, producing useful mechanical work, and then is resorbed in the resorber 18. The absorbent leaving the generator is subcooled in the preheater 20, then is expanded from the high pressure to the low pressure with the help of the expansion valve 24 and finally is absorbed in the absorber 4. The absorbent leaving the resorber 18 is subcooled twice, first in the subcooler 22 and second in the SHL loop 21, and then is expanded from the intermediate pressure to the low pressure with the help of the expansion valve 23, in order to supply the desorber and finally to close the cycle. The cycle simultaneous yield of cooling, heating and power (CHP) implies a good synchronization of each energy item consumption. This could create problems in actual use. For this reason, in order to avoid them, a practical continuous operation of the trigeneration cycle requires to include in the flowchart the possibility to store partially or totally the energies produced, but this topic is not discussed here.

The Trigeneration Cycle Thermodynamic Model

The trigeneration cycle model associates the models of the cycle component devices and the effectiveness assessment equations, as it follows:

1. Desorber, absorber and resorber—see Cycle Model, Tables 3.3 and 3.4;
2. SHL loop—see Desorption Heat Transfer Analysis;

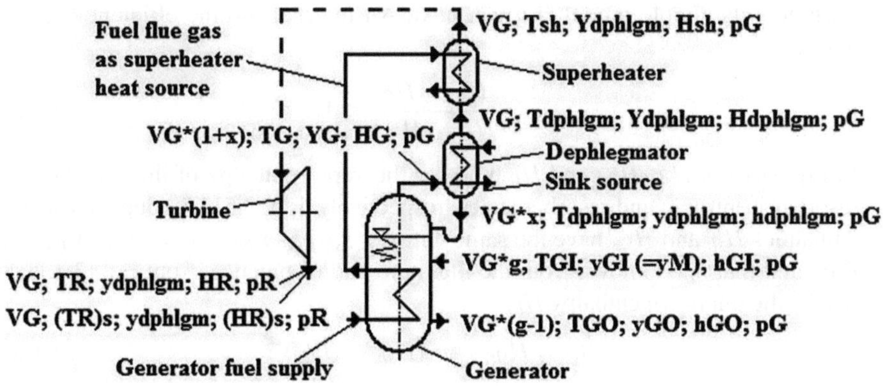

Fig. 3.51 The main thermodynamic state parameters of the generator, dephlegmator, superheater and the turbine, included in the trigeneration cycle model

3. Generator—dephlegmator—is computed with the generator-rectifier model, Eqs. (3.34)–(3.36) and Fig. 3.7 of Table 3.4, where its main thermodynamic state parameters are those given in Fig. 3.51. The dephlegmation is practiced here as a solution to augment the heating effect of the resorber, through its specific flow factor *r* increase. The dephlegmation process is known in the literature as an incomplete vapor rectification, performed with a dephlegmator which is a much less complex device than a rectifier (e.g. Niebergall 1959). Bearing in mind that we operate here with non-azeotropic refrigerant-absorbent working combinations (e.g. NH_3–H_2O, CH_3NH_2–H_2O, NH_3–H_2O–LiBr, CH_3NH_2–H_2O–LiBr, Radermacher and Alefeld 1982) which vapor concentration of is higher at equilibrium than that of the absorbent it comes of and increases with its equilibrium temperature decrease, physically, both the rectification and the dephlegmation obtain the rise of the generated vapor mean concentration, Y_G, through a number of its successive condensation and vaporization processes which take place to progressively lower temperatures. The higher this number, the higher the Y_{dphlgm}. In our case, it is enough to produce a smaller number of such condensation and vaporization processes, in order to increase Y_G with an amount δY_{dphlgm}, named here "added vapor dephlegmation", so that

$$Y_{dphlgm} = Y_G + \delta Y_{dphlgm} \tag{3.134}$$

where, for NH_3–H_2O, $Y_{dphlgm} \ll Y_{rctf} \geq 0.998$ (usual practical satisfactory rectification value, Niebergall 1959) and $Y_{dphlgm} \leq 0.85$ (our next applications maximum value). The generator-rectifier model mentioned above is applied in our case considering in equations the index "dphlgm" instead of "rctf". Also, the vapor enters the dephlegmator with the mean state parameters values, according to Fig. 3.51, instead of those corresponding to generator inlet, as Fig. 3.7 shows it is happening in a classic rectification column;
4. Generated vapor superheater—its model expresses the quasi-isobar vapor superheating, based on the partial heat recovery of the flue gas coming of the burning of the fuel supplying the generator, Fig. 3.51;

5. Turbine—its simple model is given next. An internal turbine efficiency, η_{is}, is defined by:

$$\eta_{is} = \frac{H_{sh} - H_R}{H_{sh} - H_{R,s}} \leq 1 \tag{3.135}$$

In Eq. (3.135), H_{sh}, $H_{R,s}$ and H_R hold for the vapor enthalpy of the superheated, resorber isentropic and resorber states, respectively, Fig. 3.51. Amongst these, the enthalpies H_{sh} and $H_{R,s}$ have the same entropy, $s_{sh}(p_G, Y_{dphlmg}) = s_R(p_R, Y_{dphlmg})$. Provided that η_{is} is known from the literature and/or practice, from Eq. (3.135) it results the unknown enthalpy H_R:

$$H_R = H_{sh} - \eta_{is}\left(H_{sh} - H_{R,s}\right) \tag{3.136}$$

and the turbine useful mechanical work output:

$$w_t = V_G(H_{sh} - H_R) = V_G\eta_{is}\left(H_{sh} - H_{R,s}\right) \tag{3.137}$$

The cycle work output efficiency, η_w, is defined by:

$$\eta_w = \frac{w_t}{q_{Grctf}} \tag{3.138}$$

In Eq. (3.138), q_{Grctf} is calculated by Eq. (3.36) employing the dephlegmation and superheating data;

6. Cycle global evaluation according to the first principle of thermodynamics—the trigeneration effectiveness is calculated by:

$$COP_{trgn} = \frac{q_D + q_R + w_t - w_p}{q_{Grctf}} = COP_c\left(1 - COP_w^{-1}\right) + COP_h + \eta_w \tag{3.139}$$

making use of Eqs. (3.53)–(3.55) and (3.138).

7. Cycle global evaluation according to the 2nd principle—the exergy efficiency, η_{ex}, assessment [see Eq. (7.61)].

The Trigeneration Cycle Thermodynamic Model Results

Three-Temperature Trigeneration Cycle

The first model results have been obtained considering that the cycle operation depends on three main temperatures, T_{DI} (and related T_{DO}), $T_M = T_{RO}$ (and related $T_{Rf,O}$) and T_{GO} (see Fig. 3.50 representation). The model used the following input data: $T_{GO} = 220$ °C, $T_{sh} = T_{dphlg\ m} + 140$ °C, $T_{DO} = 18$ °C, $\eta_{is} = 0.85$, $\eta_p = 0.8$. In our assessment, the y_M value was kept low, $y_M = 0.2$. This diminishes the pump work, favoring the increase of η_w (considered the most important cycle figure of merit) and increases the heating effect and $T_{Rf,O}$. On the contrary, increasing the y_M value will favor the cooling effect, but are detrimental to the work and heating production effects. The model results of the latter case are not given here. In Fig. 3.52a–f, the functions COP_c, COP_h, η_w, COP_{trgn}, η_{ex}, $T_{Rf,O}$ are plotted versus δY_{rctf}, for $T_{DI} = -10$ °C, …, $T_{DI} = -35$ °C, respectively. Each figure gives results

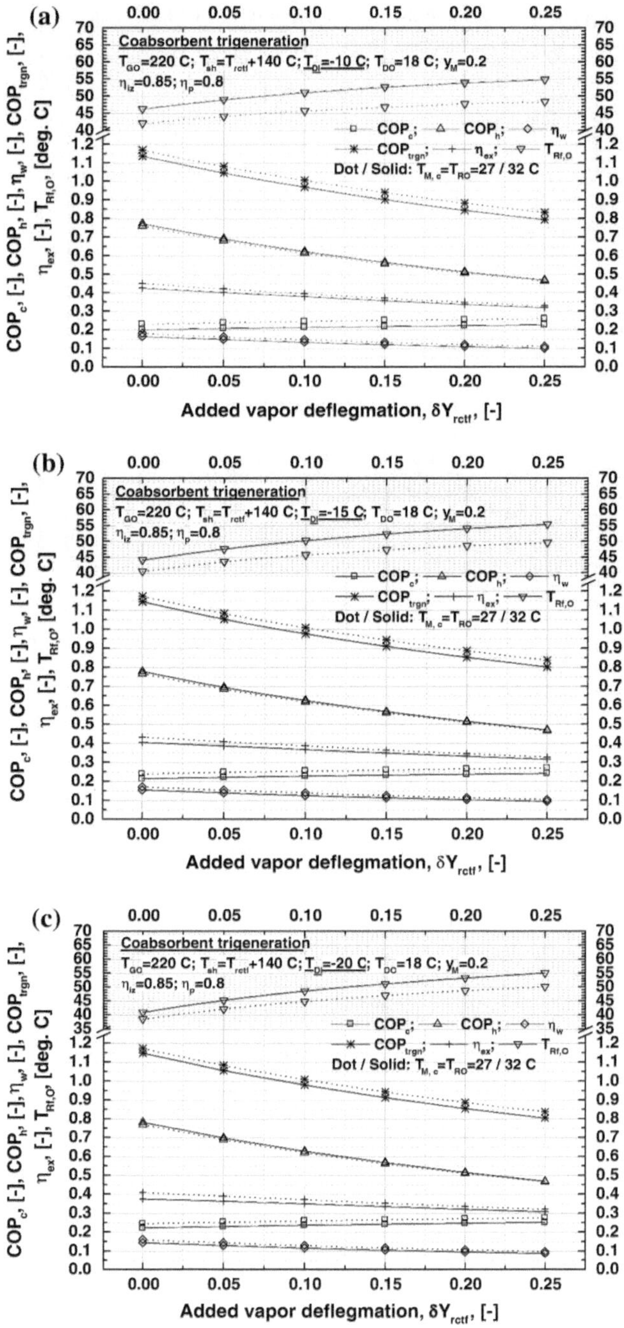

Fig. 3.52 **a** COP_c, COP_h, η_w, COP_{trgn}, η_{ex}, $T_{Rf,O}$ results versus δY_{rctf}, for $T_{DI} = -10$ °C. **b** Same as **a**, for $T_{DI} = -15$ °C. **c** Same as **a**, for $T_{DI} = -20$ °C. **d** Same as **a**, for $T_{DI} = -25$ °C. **e** Same as **a**, for $T_{DI} = -30$ °C. **f** Same as **a**, for $T_{DI} = -35$ °C

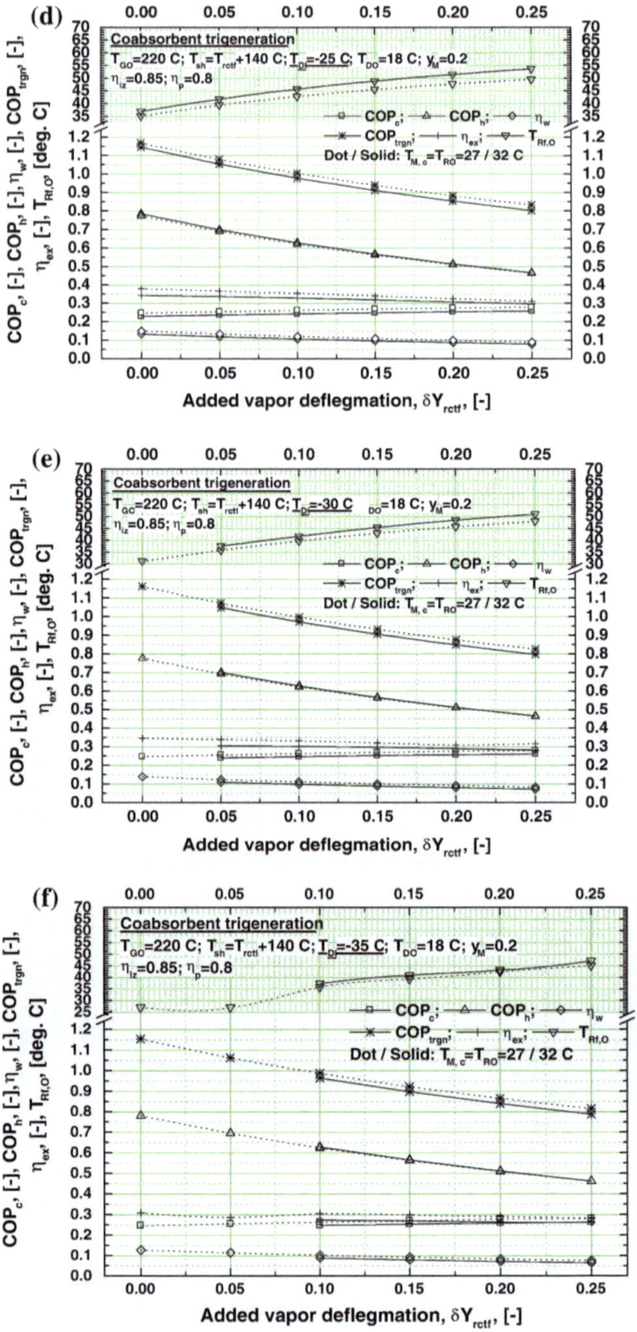

Fig. 3.52 continued

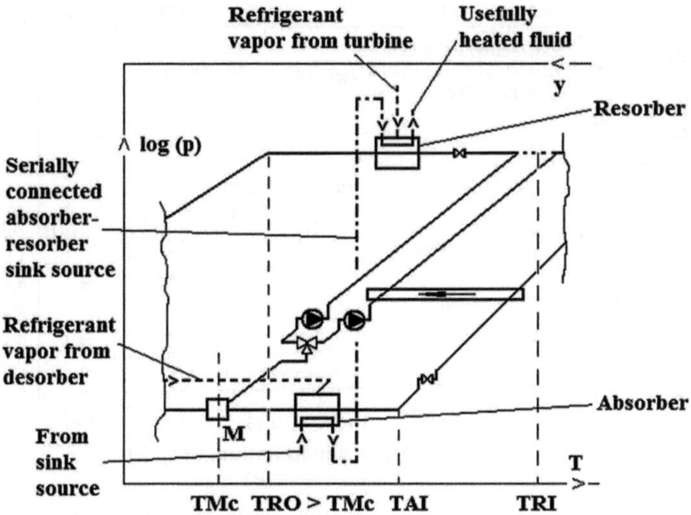

Fig. 3.53 Schematic of the absorber and resorber serially connected to a same sink source in order to ensure the cycle operation with four temperatures

for two sink temperatures, $T_M = 27$ °C and $T_M = 32$ °C. For a fixed T_{DI}, every figure indicates that COP_c and $T_{Rf,O}$ increase, while COP_h, η_w and η_{ex} decrease, when δY_{rctf} increases. Consequently, COP_{trgn} remains quasi-constant along the δY_{rctf} variation. Also, decreasing the desorber inlet temperature up to $T_{DI} = 25$ °C keeps quasi-constant the functions COP_h and COP_{trgn} values, while those of η_{ex} decrease and COP_c increase, as Fig. 3.52a–d show. Additionally, it is possible that lower T_{DI} operation with given T_M obliges the cycle be confronted with vapor threshold concentration problems. Solving them needs the Y_{dphlgm} increase, through δY_{rctf} increase over certain values, as for example here Fig. 3.52e, f indicates. Concluding for the three-temperature operation with up to $T_{DI} = -25$ °C, $T_{Rf,O} = (40–50)$ °C (values higher than 40 °C are satisfactory), moderate sink sources $T_M = (27–32)$ °C and chosen input data, the cycle could produce work, cooling and heating with effectiveness values of $\eta_w = 0.10–0.20$, $COP_c = 0.20–0.25$ and $COP_h = 0.60–0.80$, respectively, and a thermodynamic global effectiveness of $COP_{trgn} = 1.00–1.20$.

Four-Temperature Trigeneration Cycle

The second model results have been obtained considering that the cycle operation depends on four main temperatures, T_{DI} (and related T_{DO}), T_M, T_{RO} (and related $T_{Rf,O}$, $T_{Rf,O} > T_{RO} > T_M$) and T_{GO}. This type of operation can be achieved if the absorber and the resorber were serially connected to the same sink source, like in Fig. 3.53. The four-temperature working is proposed in order to increase $T_{Rf,O}$ without necessarily employ the dephlegmation, which has the drawbacks

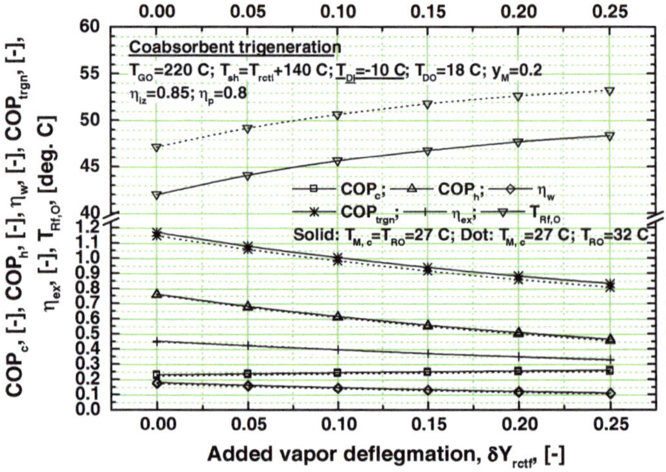

Fig. 3.54 COP_c, COP_h, η_w, COP_{trgn}, η_{ex}, $T_{Rf,O}$ results versus δY_{rctf}, for $T_{DI} = -10$ °C in case of the four-temperature operation

mentioned in the previous section. To this extent, in Fig. 3.54 the functions COP_c, COP_h, η_w, COP_{trgn}, η_{ex}, $T_{Rf,O}$ are plotted versus δY_{rctf}, for $T_{DI} = -10$ °C. This figure compares results of the three-temperature operation having $T_M = T_{RO} = 27$ °C, with those obtained according a four-temperature operation, when $T_M = 27$ °C < $T_{RO} = 32$ °C. The model results show in this case, that, if appropriately chosen the $T_{RO} - T_M$ temperature difference, the functions COP_c, COP_h, η_w, COP_{trgn} and η_{ex} have practically the same values in both, three- versus four-temperature operation modes and for the same input data, but the $T_{Rf,O}$ is sensibly higher, for a same δY_{rctf}.

Use of the Trigeneration Cycle Just as an Inverted Osenbrück Cycle

Producing just mechanical work (electricity) using the ammonia-water combination and depleted heat sources resulting from the industrial processes, is already applied in practice in order to save primary energy. One of the most known cycle, used to such circumstances, is the Kalina cycle, which is a modified Clausius-Rankine cycle (Ogriseck 2009). In our case, if certain applications required it, it is possible to maximize the work production of the cycle at hand, running it just as an inverted Osenbrück cycle (including only the absorber 4, generator 10, pump 8, HE 20 and the expansion valve 24, Fig. 3.50). In order to avoid any confusion, the inverted Osenbrück cycle is plotted alone in the $\log(p) - 1/T$ diagram in Fig. 3.55. Similar to the Kalina cycle, the inverted Osenbrück cycle is based as well on the direct Clausius-Rankine process which produces mechanical work through heat input. Its operation results from that of Fig. 3.50 cycle. In Fig. 3.56, we give

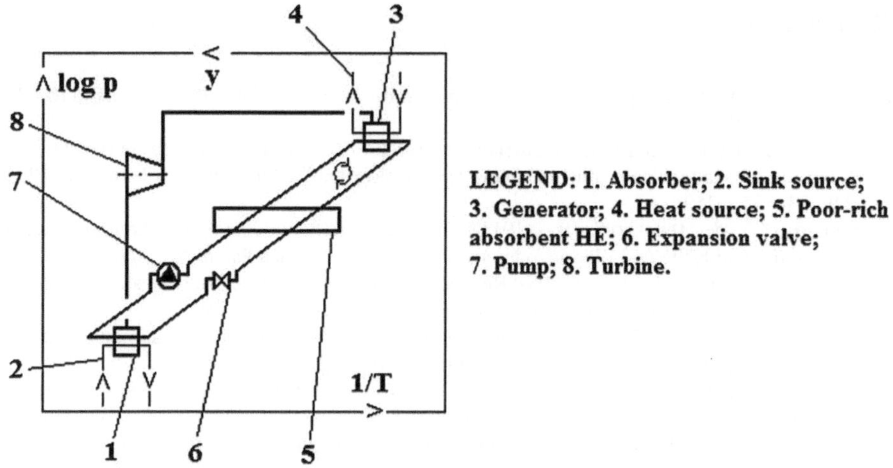

Fig. 3.55 The inverted Osenbrück cycle, used for mechanical work production, depicted in the $\log(p) - 1/T$ diagram

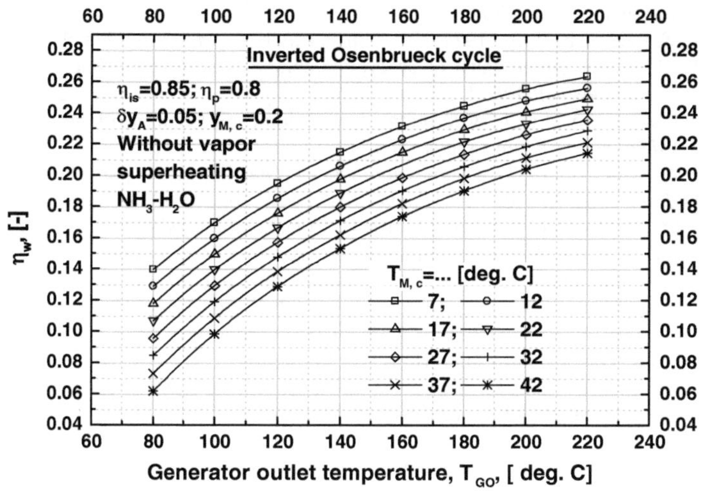

Fig. 3.56 NH_3–H_2O inverted Osenbrück cycle results: η_w versus T_{GO}, having as parameter $T_{M,c}$

NH_3–H_2O results of a work efficiency η_w family, having $T_{M,c}$ as parameter, plotted against the generator outlet temperature, T_{GO}, and without vapor superheating. The curves show that the inverted Osenbrück cycle answers with diminished returns to a continuous T_{GO} increase, but the cycle is very efficient, its theoretical relative to Carnot efficiency being within 0.66 and 0.73 for the model input data (see plot details).

References

Altenkirch, E. (1950). Kompressionskältemaschine mit Lösungslreislauf. *Kältetechnik 2*(10–12), 251–259, 279–284, 310–315.

Altenkirch, E. (1951). Der Einfluß endlicher Temperaturdifferenzen auf die Betriebskosten von Kompressionskälteanlagen mit und ohne Lösungslreislauf. *Kältetechnik 3*(8–10), 201–205, 229–234, 255–259.

Altenkirch, E., & Tenckhoff, B. (1911). Absorptionkältemaschine zur kontinuierlichen Erzeugung von Kälte und Wärme, oder auch von Arbeit. German Patent No. 278076.

Badescu, V., & Staicovici, M. D. (2005). Renewable energy for passive house heating. Model of the active solar heating system. *Energy and Buildings, 38*, 129–141.

Baehr, H. D. (1981). The COP of absorption and resorption heat pumps with ammonia-water as working fluids. *International Journal of Refrigeration, 4*(2), 83–86.

Belozerov, G., Mednikova, N., Pytchenko, V., & Serova, E. (2007). Cascade type refrigeration systems working on CO_2/NH_3 for technological processes of products freezing and storage. In *Proceedings of Ohrid 2007 IIR/IIF conference*, Macedonia, 19–21 April 2007.

Bliss, R. W. (1961). Athmospheric radiation near the surface of the ground. *Solar Energy, 5*, 103.

Brunt, D., & Quart, J. (1932). Notes on radiation in the atmosphere. *Royal Meteorological Society, 58*, 389–420.

Cox, N., Mazur, V., & Colbourne, D. (2009). The development of azeotropic ammonia refrigerant blends for industrial process applications. In *Proceedings of the Ohrid 2009 IIR/IIF conference*, Macedonia, 6–8 May 2009.

Duffie, J. A., & Beckman, W. A. (1974). *Solar energy thermal processes*. New York: Wiley.

Groll, E. A. (1997a). Current status of absorption/compression cycle technology. In *ASHRAE transactions symposia*, PH-97-1-1.

Groll E. A. (1997b). Modeling of absorption/compression cycles using working pair dioxide/acetone. In *ASHRAE transactions symposia*, PH-97-12-1.

Groll, E. A., & Kruse, H. (1992). *Kompressionskältemaschine mit Lösungslreislauf für umweltverträgliche Kältemittel. KK DIE KÄLTE und Klimatechnik* (pp. 206–218). Stuttgart: Gentner.

Itard, L. C. M., & Machielsen, C. H. M. (1994). Considerations when modelling compression/ resorption heat pumps. *Revue Internationale du Froid, 17*(7), 453–460.

Kashiwagi, T. (1980–1990). Advances in working fluids and cycles for absorption systems. Private communication, Tokyo University of Agriculture and Technology.

Kirilin, V. A., Sicev, V. V., & Seindlin, A. E. (1985). *Thermodynamics* (540 pp.). Bucharest: Scientific and Technical Publishing House (in Romanian).

Lund, J., Sanner, B., Rybach, L., Curtis, R., & Hellström, G. (2004, September). Geothermal (ground-source) heat pumps-a world review. *GHC Bulletin*.

Niebergall, W. (1959). *Handbuch der Kältetechnik, Band 7, Sorptions-Kältemaschinen* (Vol. 7). Berlin, Göttingen, Heidelberg: Springer.

Ogriseck, S. (2009). Integration of Kalina cycle in a combined heat and power plant, a case study. *Applied Thermal Engineering, 29*(14), 2483–2848.

Osenbrück, A. (1895). *Verfahren zur Kälteerzeugung bei Absorptionsmachinen*. Germany: Kaiserliches Patentamt.

Penzias, A. A., & Wilson, R. W. (1965). A measurement of excess antenna temperature at 4080 Mc/s. *Astrophysical Journal, 142*, 419–421.

Perez, B. H. (1993). Conceptual design of high-efficiency absorption cooling cycle. *International Journal of Refrigeration, 16*(6), 429–433.

Popa, B., & Vintila, C. (1977). *Thermotechnics and thermal machines*. Bucharest: The Didactic and Pedagogic Publishing House (in Romanian).

Prosper de Mulder. (2004). *Vertical freezers with* CO_2. Glasgow: Star Refrigeration. (Doncaster).

Radermacher, R., & Alefeld, G. (1982). Lithiumbromid-Wasser—Lösungen als Absorber für Ammoniak oder Methylamin. *Brennstoff-Wärme-Kraft, 34*(1), 31–38.

Radcenco, V., Florescu, A., Duicu, T., Burchiu, N., Dimitriu, S., Zdrenghea, P., Ghitulescu, M., Dobrovicescu, A., & Dragomir, T. (1985). *Heat pump installations*. Thermo-cooling series. Bucharest: Technical Publishing House.

Risberg, T., Horntvedt, B., Madland, D., Nordtvedt, S. R. (2004). Process dynamics in an industrial prototype compression/absorption heat pump. In *Proceedings of the 6th IIR Gustav Lorenzen natural working fluids conference*, Glasgow, UK, 29th August–1st September 2004.

Sloan, J., & Roncin, E. (1920). Procédé et dispositifs de production des frigories à cicle fermé. French Patent No. 537,438.

Staicovici, M. D. (2002). Further research zero CO_2 emission power production: The "COOLENERG" process. *Energy the International Journal, 27*(9), 831–844.

Staicovici, M. D. (2004). The "COOLENERG" clean power process—a challenge for the advanced absorption cooling technology. In *Proceedings of the 21st international congress of refrigeration,* Washington, August 2003.

Staicovici, M. D. (2006a). Coabsorbent cycles. In *Proceedings of the Gustav Lorentzen natural working fluids conference* (pp. 219–222), Trondheim, Norway, 28–31 May 2006.

Staicovici, M. D. (2006b). Heat pump. Patent Romania No. 123151/02.03.2006 (in Romanian).

Staicovici, M. D. (2006c). Cooling plant. Patent Romania No. 123152/06.07.2006 (in Romanian).

Staicovici, M. D. (2006d). Procedure of increasing the effectiveness and feasibility of a nontruncated coabsorbent cooling cycle and installation of application. Patent Romania No. 123082/26. 09.2006 (in Romanian).

Staicovici, M. D. (2006e). The coabsorbent cycle technology for heat pumping applications. *Presented to the confort, efficiency, energy conservation and environment protection, UTCB conference*, Bucharest, Romania, 29–30 November 2006.

Staicovici, M. D. (2007a). Coabsorbent heat pumping method by coupling with a thermal power station and application plant. Romanian Patent No. 123405/22.02.2007 (in Romanian).

Staicovici, M. D. (2007b). Coabsorbent cycle technology for ammonia/water heat pumping applications. In *Proceedings of the ammonia refrigeration technology for today and tomorrow IIR conference*, Ohrid, Macedonia, 19–21 April 2007.

Staicovici, M. D. (2007c). Coabsorbent heat pumps for the future. *IEA-Heat Pumps Newsletter, 25*(1), 29–32.

Staicovici, M. D. (2007d). Coabsorbent cycle technology for low grade sources thermal recovery. In *Proceedings of the technological innovations in air conditioning and refrigeration industry XII European conference* (Vol. I–II), Milan, June 2007.

Staicovici, M. D. (2007e). Coabsorbent technology for heat pumping applications. In *Proceedings of the IIRIIF 22 international congress of refrigeration, E1-2-1039*, Beijing.

Staicovici, M. D. (2007f). Coabsorbent technology for future cooling applications, Sent to Rajendra Shende, Director of the United Nations Environment Programme (UNEP), July 2007, and Didier Coulomb, Director of the International Institute of Refrigeration, November 2007.

Staicovici, M. D. (2007g). Coabsorbent cycles heat pumping and mechanical work producing procedure and applying installation. International patent file deposition No. PCT/RO 2007/000018/24.09.2007.

Staicovici, M. D. (2007h). A coabsorbent technology IPAE prospectus presented and offered by the author during a visit which this has paid to Lars Gulev from the Danish Board of the District Heating, in order to cooperate and possibly apply the mentioned technology in Copenhagen to district heating and cooling.

Staicovici, M. D. (2008a). Coabsorbent heat pumps. In *Proceedings of the 9th IEA heat pumps conference*, Zurich, May 2008.

Staicovici, M. D. (2008b). *Proposals for the nine sigma RPF 11087-1*: "Highly efficient low temperature heat driven cooling technologies", March 2008.

Staicovici, M. D. (2008c). Coabsorbent technology for heat pumping and power production. In *Proceedings of the 8th IIR Gustav Lorentzen conference on natural working fluids*, Copenhagen, 07–10 September 2008.

Staicovici, M. D. (2009). Further results in ammonia/water coabsorbent technology district tri-generation. In *Proceedings of Ohrid 2009 IIR/IIF conference*, Macedonia, 6–8 May 2009.
Staicovici, M. D. (2011). Heat exchange analysis of coabsorbent cycle absorption processes. In *Proceedings of ISHPC 2011 conference.*
Staicovici, M.-D., & Staicovici, A.-M. (2013). Hybrid coabsorbent heat pumping cycles. In *Proceeding of the 5th IIR conference: Ammonia refrigeration technology, Ohrid 2013.*
Swinbank, W. C., & Quart, J. (1963). Long-wave radiation from clear skies. *Royal Meteorological Society, 89.*
Whillier, A. (1967). "Design factors influencing solar collectors", Low temperature engineering applications of solar energy, ASHRAE, New York.

Chapter 4
A Few New Coabsorbent Cycle Configurations: The Internal Composition and the Coabsorbent Cycle Truncation

The coabsorbent cycles number is huge. For this reason, their selection with criteria such thermodynamics principles respect and low complexity, usefulness and feasibility accomplishment, is mandatory. The simplest cycles, with potential applications, are derived through the nontruncated cycles composition, as: (a) internal; (b) external, and (c) joined (skipped here). We start with the internal one, the truncation.

The truncation is an important and unique technique of the coabsorbent technology. It enables a nontruncated cooling coabsorbent cycle to work with a smaller temperature heat source for the generation process, $T_M < T_{GO} \leq T_{GO,nt}$, and a same sink source temperature, Fig. 4.1a. Alternatively, the truncation enables a nontruncated heating (heat transformer) coabsorbent cycle to work with a higher temperature sink source for the absorption processes, $T_M > T_{AO} \geq T_{AO,nt}$, and a same heat source temperature, Fig. 4.1b. On one side, the truncation is done upon a certain *heat (source) truncation line*, marking the end temperature of the endothermal generation and desorption processes. This line is placed to the right side of the cooling cycle mixing point, as Fig. 4.1a shows, and includes the mixing point of the heating cycle compulsorily. On the other side, the truncation is done upon a *sink (source) truncation line*, marking the end temperature of the exothermal absorption and resorption processes. This line is placed to the left side of the heating cycle mixing point, as Fig. 4.1b shows, and includes the mixing point of the cooling cycle compulsorily. The truncation lines can be designed in a free or in a pre-established way, according to the available sources characteristics, Fig. 4.1. When a cooling and a heating coabsorbent cycle are considered simultaneously, the generation process of the cooling cycle and the desorption process of the heating cycle have the same heat truncation line. Similarly, the resorption process of the cooling cycle and the absorption process of the heating cycle have the same sink truncation line. The two truncation lines, used simultaneously, define a *truncation column*, intrinsically related to a truncated coabsorbent cycle and referred to later in this chapter. The most part of the applications have quasi-constant temperature heat and sink sources, equal, within the accuracy limits of

M.-D. Staicovici, *Coabsorbent and Thermal Recovery Compression Heat Pumping Technologies*, Heat and Mass Transfer, DOI: 10.1007/978-3-642-54684-6_4,

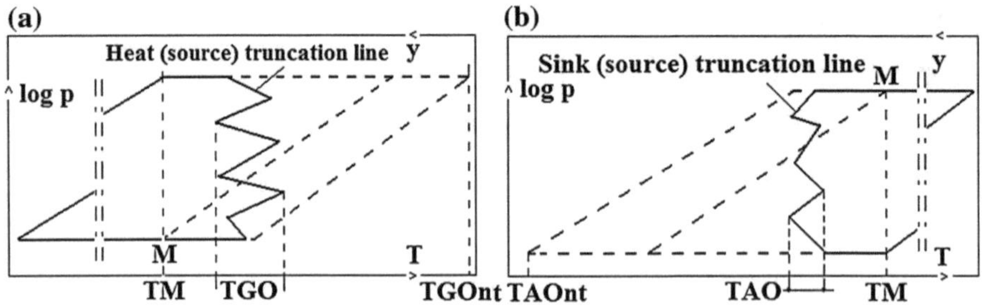

Fig. 4.1 The truncation of the coabsorbent cycles: **a** cooling cycle; **b** heating cycle

the necessary heat exchange temperature pinch, to the mixing points temperatures, $T_{GO} = T_{M,heating} = const.$, e.g. (40–70) °C, and $T_{RO} = T_{M,cooling} = const.$, e.g. (10–40) °C, respectively. The heat and sink lines are separated by a minimum temperature difference $\Delta T_M = T_{M,heating} - T_{M,cooling} \geq$ (8–10) °C, which is required in order that the cycle operation be possible. The higher this temperature difference, the better for the cycle COP increase and its complexity decrease. Unlike other known absorption cycles, the most important feature of a truncated cycle is that in theory it is capable to process low temperature supplying sources forming heat/sink pairs of small temperature differences, e.g. $\Delta T_M =$ (10–50) °C. A cycle provided with a truncation column might achieve high temperature lifts with small internal power consumption in both cooling and heating working modes ($COP_{w,cooling} \approx 10^1$–$10^2$; $COP_{w,heating} \approx 10^1$–$10^3$). This is why, despite a higher complexity, which is conveniently chosen by the designer according to the heat/sink sources availability, the truncated cycles can have important applications in the clean heat pumping and power production. The nature and human industrial activity create such pairs of heat/sink sources, the problem is to identify and use them appropriately. One of the most attractive is that offered by a thermal power plant. In this case, the condenser cooling water and the cooling tower cooled water could be the truncated cycle heat/sink sources, respectively. Some recent published works (Staicovici 2006a, b, 2007a, b, c, d, e, f, g, 2009a, b) emphasize these aspects, describing for instance a proposal of considering the pure absorption and hybrid truncated cycles with the ammonia/water in the coupling with the thermal power stations for efficient district power, heat and/or cooling in cities (mentioned throughout this book as applications).

Historically, the truncation idea relegates back to the industrial development, beginning with the early 20-th century. This faced the thermal engineer with a new challenge, that of recovering the low grade heat sources resulting from the various technological processes for heating (Bokelmann 1988) but especially for cooling (Niebergall 1959) production. The absorption technology matches the best a low grade heat source with the cooling task. A pioneering work performed in order to supply an absorbent cooling cycle with a low temperature heat source for cooling purposes was accomplished by Altenkirch about one century ago. He proposed the first the multi-stage absorption technique (Altenkirch 1914). A cooling cycle based

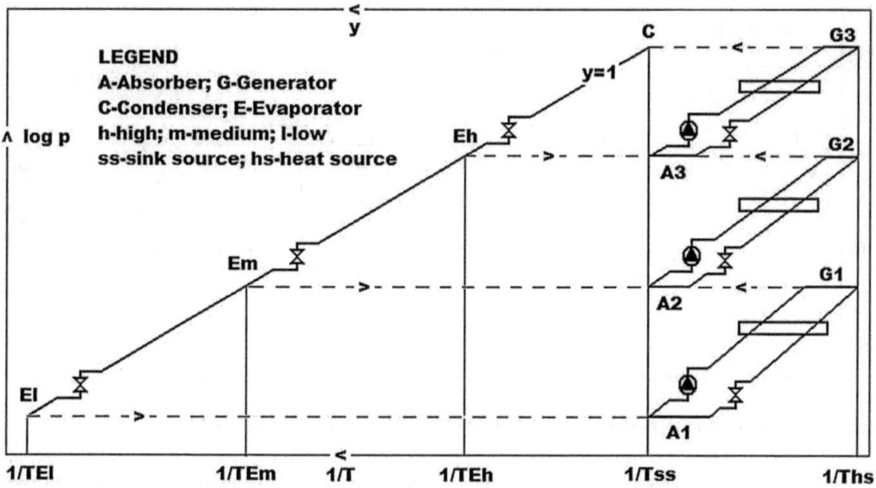

Fig. 4.2 A multi-stage absorption cooling cycle, as proposed by Altenkirch about one century ago

on it is schematically given in Fig. 4.2. It consists of a cascade of several absorber (A)–generator (G) loops. Two adjacent loops are internally isobar connected on the generated-absorbed vapor side and externally to the sink and heat sources (see the legend). Each generator is provided with vapor rectification when a volatile absorbent working combinations is used. The last generated vapor is condensed. Subsequently, it is expanded, evaporated to the stages pressures for the cooling be produced with different temperatures (here high, medium and low) and finally is absorbed in the absorbers. Each A–G loop has its own separate absorbent administration. When a rectified vapor is handled, its concentration differs from stage to stage and the cycle is suffering from the absorbent inventory problem (in this case, the loops are quasi-closed). This technical solution, although clever, is quite complex and was confined to refrigeration applications, only, those of heating being not disclosed by Altenkirch.

The coabsorbent technology is completing the multi-stage absorption, introducing three types of truncation columns. Each column type is provided with absorbents which are mixing to its every generation-resorption pressure stage. In this work, we define a *degree of truncation* of a coabsorbent cycle, equal to the number of pressure stages which a truncation column, no matter what type of, is interposing between the lowest and the highest pressure stages of the nontruncated cycle of origin. According to its degree of truncation, a column and the cycle using it, having one, two, …, five, etc., intermediary pressure stages, will be named simple, double, triple, quadruple, quintuple, etc. truncation column and truncated cycle, respectively. In Fig. 4.3 we present principally the three possible truncation columns with their absorbent mixing process. The first type is given schematically in Fig. 4.3a. Considering a *n-stage* column and the two consecutive resorber-generator loops, that is R_i–G_{i+1} and R_{i-1}–G_i, $2 < i < n - 1$, $n \in N$, the absorbent with the concentration $y_{GO,i+1}$ coming of the generator G_{i+1} is mixing up with the absorbent of concentration $y_{RO,i-1}$ coming of the resorber R_{i-1}. They generate

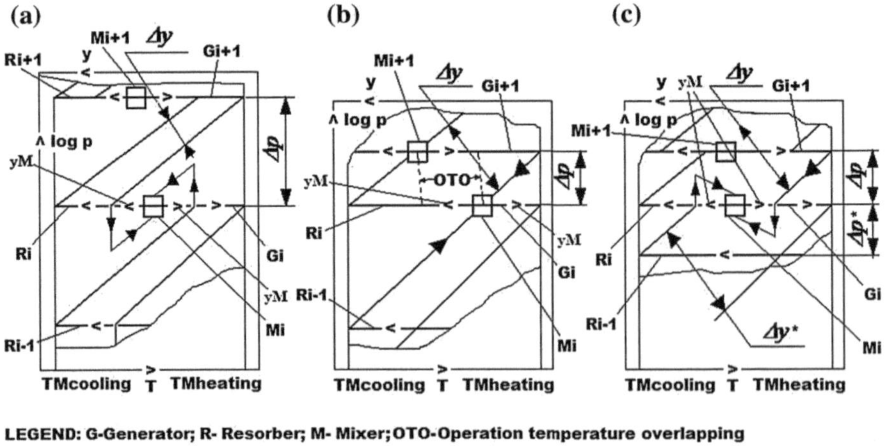

Fig. 4.3 Possible types of coabsorbent cycle truncation columns and their internal absorbent mixing. **a** Nonoverlapping. **b** Balance. **c** Overlapping

in the *i*-th stage mixer M_i the absorbent with the mean concentration y_{Mi}, which subsequently is distributed in a pre-established way to the stage G_i generator and R_i resorber. The $y_{M,i}$ mean concentration has as upper and lower limits the $y_{GO,i+1}$ and $y_{RO,i-1}$ concentration values, respectively:

$$y_{GO,i+1} > y_{M,i} > y_{RO,i-1}. \tag{4.1}$$

The two loops are graphically outdistanced in the log(*p*) − 1/*T* diagram, similar to the Altenkirch multi-stage cycle, Fig. 4.2, because $y_{GO,i+1} > y_{RO,i-1}$. In fact, the resemblance is apparent, only. Indeed, unlike the Altenkirch multi-stage, where the equalities $y_{RO,i-1} = y_{GI,i}$ and $y_{GO,i+1} = y_{RI,i}$ hold true, here we have $y_{RO,i-1} \neq y_{GI,i} = y_{M,i}$ and $y_{GO,i+1} \neq y_{RI,i} = y_{M,i}$. From these simple considerations and taking advantage of Eq. (4.1), it results that the absorbent mixing process of this column type resembles to that depicted in Fig. 3.3a, provided that we consider the generation replaces the desorption and the resorption replaces the absorption. For this reason, this type of column is named *nonoverlapping truncation column*.

The second column type is given schematically in Fig. 4.3b. The considerations made for Fig. 4.3a hold true in this case as well, except the fact that in the mixer M_i is generated the absorbent with the mean concentration y_{Mi}, which is characterized by the:

$$y_{GO,i+1} = y_{M,i} = y_{RO,i-1} \tag{4.2}$$

The absorbents mixing process resembles to that shown in Fig. 3.3b. For this reason, this type of column is named *balance (fractal) truncation column* or simpler the *truncation column*, known in the literature under this name through the author prior publications (e.g. Staicovici 2008a). Finally, the third column type is shown in Fig. 4.3c. Again, the considerations made for Fig. 4.3a hold true here as well. However, in this case, the absorbents coming of the generator G_{i+1} and resorber

R_{i-1} are mixing up in the mixer M_i and generate the absorbent with the mean concentration $y_{M,i}$, which has as upper and lower limits the $y_{RO,i-1}$ and $y_{GO,i+1}$ concentration values, respectively:

$$y_{RO,i-1} > y_{M,i} > y_{GO,i+1}. \tag{4.3}$$

From this point of view the mixing process resembles to that depicted in Fig. 3.3c, with concentrations overlapping, and the column will be named *overlapping truncation column.*

The cooling and heating cycles provided with nonoverlapping and overlapping truncation columns have been thoroughly analysed and modeled by the author. The results show that their heat pumping effectiveness is much less than that of the heat pumping cycles provided with balance columns. For this reason, they will not be presented further in this book, yielding the space to the balance (fractal) truncation of the coabsorbent cycle, analysed thoroughly next.

4.1 Balance (Fractal) Truncation of the Coabsorbent Cycle

The coabsorbent cycle truncated with the help of a balance column, was disclosed by the author priorly (e.g. Staicovici 2006a, b, 2007d, g, 2008a, b, 2009a). It is already known as the "truncated cycle". This name will be kept hereinafter. It can be considered the standard (base line) design in truncation. In comparison to a nonoverlapping column cycle, the balance column cycle needs more pressure stages (usually one or maximum two) for a same heating or cooling task and same sources. However, this type of cycle can compensate for this drawback, working principally with a better internal heat recovery in certain conditions, for the benefit of the heat pumping cycle effectiveness. Here we make reference mainly to the generator-absorber (resorber)-heat exchange (gax) use. A gax effect is possible but through the removal of concentrations of the adjacent pressure stages mixers M_i and M_{i+1}, $i = 1, \ldots, n-1$, in such a way that the higher this removal, the better for the gax effect stimulation. The removal effect, named shorter, the *adjacent concentrations removal* ("*acr*" abbreviation hereinafter) is met to a smaller or higher extent to all possible types of coabsorbent cycle truncation columns shown in Fig. 4.3, but the balance column maximizes it as an intrinsic property of this column type, only. Typical for *acr* is that it enables an operation temperature overlapping (OTO) of two adjacent devices, shown comparatively for the R_i and G_{i+1} devices in Fig. 4.3a–c. More about the gax effect will be written in Sect. 4.2.1. Continuing, the cycle is well adapted to the operation in off-design conditions, brought about more often than not by the sources temperature seasonal variation. This is why, next, we shall consider that Eq. (4.2) is fulfilled essentially with equality by all its i stages, $2 < i < n-1$, $n \in N$, although in practice it is very likely that the cycle operates with mixing concentrations having small deviations from Eq. (4.2), that is $y_{GO,i+1} = y_{M,i} \pm \delta y$ and $y_{RO,i-1} = y_{M,i} \pm \delta y$, with e.g. $\delta y = 0.005 - 0.01$.

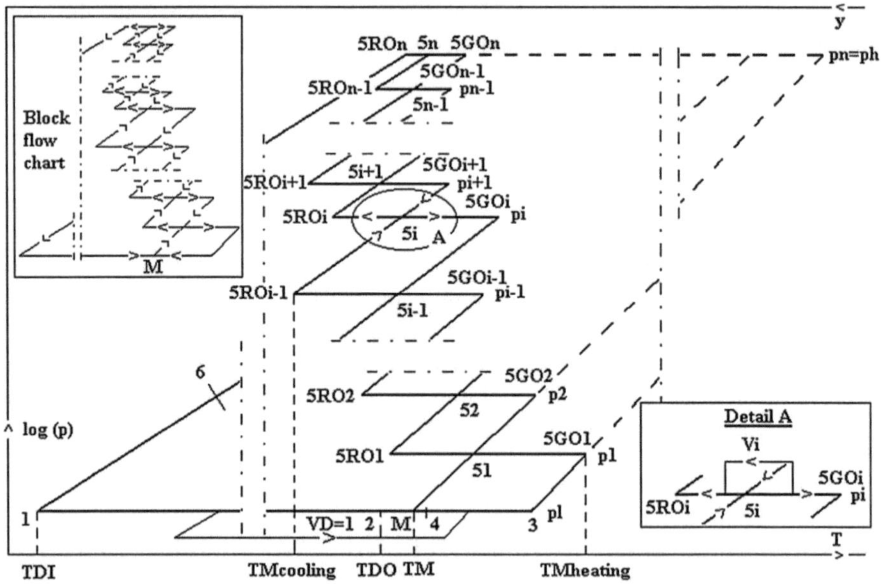

Fig. 4.4 The truncated cooling coabsorbent cycle

4.1.1 Cooling Cycle

The truncation technique is presented next, describing the cooling cycle using a balance truncation column (Staicovici 2006a, 2007g, 2009a). A truncated cooling cycle is plotted by the solid line for the absorbent circulation in the log p − 1/T diagram of Fig. 4.4 (see also the block flow chart). It comes of the nontruncated cooling cycle, Fig. 3.2, represented by the external contour with the solid line in the left side and with the dashed line in the right side of the diagram. It includes a low pressure desorption process (1–2), p_l, e.g. (0.1–2) bar, where the useful cooling effect occurs, e.g. $T_{DI} =$ (213,15–273,15)K and a truncation column having pressures between the low value and a high pressure p_n, e.g. (10–50) bar. The truncation column is made up of a low pressure absorption process (3–4) coupled on the vapor side with the low pressure desorption process (1–2), a low pressure mixing process of the absorbents coming of the low pressure absorption and the desorption processes in order to generate the y_M one mean concentration and $T_M(p_l, y_M) \geq T_{M,cooling}$ temperature absorbent, point M, and a series of i stages, $i = 1, \ldots, n$, $n \in N$, of isobar opposite intermediary generation $(5_i - 5_{GO,i})$ and resorption $(5_i - 5_{RO,i})$ processes, having increasing pressures, $p_n \geq p_{i+1} > p_i > p_l$ and temperatures between $T_{M,cooling}$ and $T_{M,heating}$ and being coupled on the vapor side. The vapor have essentially increasing mean concentrations $Y_{G,m,i} \leq Y_{G,m,i+1} \leq Y_{G,m,n-1}$ and fulfil the concentration threshold condition $y_{RO,i} \leq Y_{G,m,i}$, $i = 1, \ldots, n$. The stages are supplied in pre-established proportions by the uniform absorbents coming, for the first stage of the mixture of the one mean concentration absorbent with that coming of the second generation stage, point 5_1, for each of the following next i stages of the mixture of the absorbent coming of the first inferior resorption stage $i - 1$ with that coming of the first superior generation stage $i + 1$, points 5_i, $i = 2, \ldots, n - 1$ (see arrows and detail A), and for the high pressure stage of the last but one resorption process $n - 1$, point 5_n. The truncation column has also processes between its successive stages, of the one mean concentration absorbent pressure increasing from p_l to p_1 and of the absorbent coming of resorption processes, p_{i-1} to p_i, $i = 2, \ldots, n$, of pressure reduction of the absorbent coming of the generation processes, from p_1 to p_l and from p_i to p_{i-1}, $i = 2, \ldots, n$, and of heat recovery between the opposite processes of absorption-generation and of resorption-generation, of the gax and of the absorbent-absorbent type. Functionally, the rich absorbent leaves the truncation column last stage resorption process ($5_n - 5_{RO,n}$) at p_n, is subcooled by the desorbed vapor in a recovering way from $T_{RO,n} \geq T_{M,cooling}$ till a temperature closer to T_{DI}, portion ($5_{RO,n} - 6$), is expanded till p_l, it suffers the desorption process (1–2) where it extracts the heat from the medium which must be cooled and reaches the state parameters $T_{DO} \leq T_M$ and $y_{DO} \geq y_M$, and the absorbent coming of the generation process ($5_1 - 5_{GO,1}$) of the first truncation column stage is subcooled in a recovering way, is expanded till the low pressure, it suffers the absorption process (3–4) of the desorbed refrigerant vapor, is mixed up at the low pressure with the absorbent coming of the desorption process and it generates the one mean concentration absorbent which is covering the truncation column in the way described above until the last stage at p_n, in order to close the cycle.

Summarizing and formalizing, the truncation column of the cooling cycle in Fig. 4.4 is a "vertical" collection of "parallelograms" noted from the low pressure bottom to the high pressure top by $(5i) - (5RO_i) - (5i + 1) - (5GO_{i+1}) - (5i)$, $i = 1, 2, \ldots, n - 1$. The $i = 0$ notation is reserved for the low pressure bottom. The parallelograms notation is done in this way, because it starts from the place where the mixing point is located and conventionally it is considered an origin. Each parallelogram is covered clockwise in this case. Each two parallelograms are adjacent in a single point (i), $i = 1, 2, \ldots, n - 2$ and exchange mass (vapor and absorbent) and heat. A parallelogram "operates" between a low pressure resorption process and a high pressure generation process and two distinct isosteres. Each isostere has the concentration of the absorbent which exits the process it comes of. The parallelograms increase gradually their mean concentration from the bottom to the top. In this respect, the full truncation column, with its successive generation and condensation (resorption here) processes, is in fact a *truncation-rectification column* (or a *truncation-dephlegmation column*), as, besides truncation, the generated vapor increases naturally its concentration, approaching more often than not the unit in function of the particular application, when the absorbent covers the column from the bottom to the top and the cycle operates with volatile absorbent working combinations (see The Trigeneration Cycle Thermodynamic Model explanation). Moreover, because this time, besides generation, the rectification (dephlegmation) process can be accomplished as well intrinsically by the same heat source amount supplying just the cycle generation processes, it is expected that, comparative to

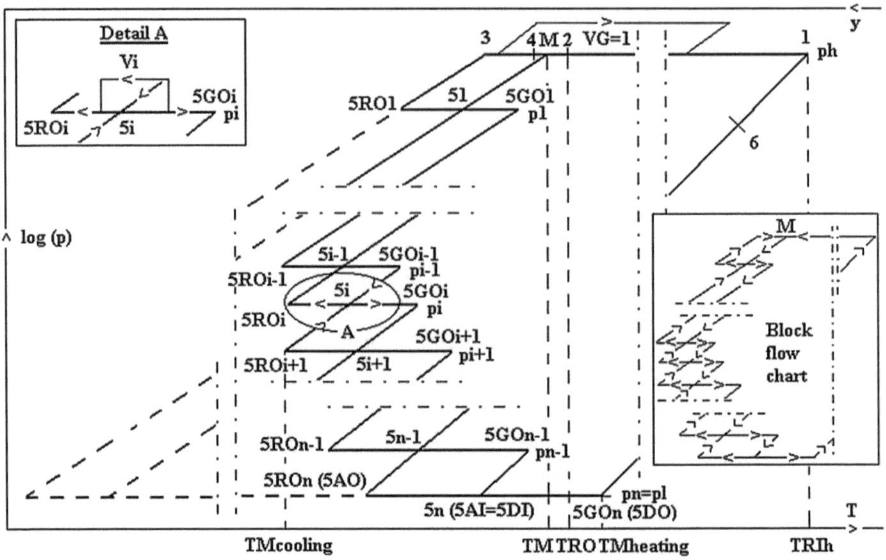

Fig. 4.5 The truncated heating (heat transformer) coabsorbent cycle

other cycles, the cycle at hand has a higher effectiveness with respect to the Carnot cycle, and this will be emphasized in the future applications. However, in quite rare cases, when the generated vapor mean concentration of the last generator is not satisfactory, producing a concentration threshold problem, a vapor rectification (dephlegmation) process must be performed, consuming in most cases a reduced additional rectification heat input. Continuing, in case of the parallelogram including the notation $i = 0$, the resorption process is replaced by an absorption process. The truncated cooling cycle is obtained, adding to the left side of the truncation column a pair of processes, namely a resorption [here the $(5_n - 5RO_n)$] and a desorption one [here the (1–2)], connected by the isostere $(5RO_n - 1)$. The block flow chart in Fig. 4.4 shows very clearly this joining between the truncation column and the left side pair of resorption-desorption processes.

4.1.2 Heating Cycle

A truncated heating cycle is plotted by the solid line in Fig. 4.5 in the log $(p) - 1/T$ diagram (see also the block flow chart). It comes of the nontruncated heating cycle, represented by the external contour with the solid line in the right side and with the dashed line in the left side of the diagram. Formally, it is obtained by adding to the right side of the truncation column a pair of processes, namely a resorption [here the (1–2)] and a desorption one [here the $(5DI - 5DO)$], interconnected by the isostere $(5DO - 1)$. In case of the heating cycle, the cooling cycle "bottom" is placed at the high pressure top, where the mixing point is located as well. For this reason, bearing in mind the previous convention, the truncation column starts the notation of its parallelograms from the high pressure stage to the low pressure stage. For the sake of computation simplifying, the last stage of the absorption ($5AO - 5AI(=5DI)$) is noted by ($5RO_n - 5_n$), like a resorption stage. Similarly, the stage of the desorption ($5DI(=5AI) - 5DO$) is noted by ($5_n - 5GO_n$), like a generation stage. In case of the heating cycle, the parallelograms mean concentration decreases gradually from the high pressure to the low pressure stages, and so does the generated vapor, as well. The operation of the truncated heating cycle has many similarities with the operation of the truncated cooling cycle. This is why we shall not explain it here. If necessary, the reader may find a complete description of this cycle in Staicovici (2007g, 2008a).

4.1.3 Truncation Theory

A truncated coabsorbent cycle has particular topologic and flow properties, which determine its thermal behaviour. We shall emphasize these basic properties next (Staicovici 2009a), starting with the following:

Lemma 1 *Formally, graphically, the representation of a (non)truncated heating (heat transformer) coabsorbent cycle in the log p − 1/T diagram can be obtained through an 180° plane rotation of the (non)truncated cooling coabsorbent cycle representation in the same diagram.*

Proof Indeed, rotating by 180° the graphs of the nontruncated and truncated cooling coabsorbent cycles, shown in Figs. 3.2 and 4.4, respectively, the representation of the nontruncated and truncated heating coabsorbent cycles, shown in Figs. 3.29 and 4.5, respectively, is obtained in the same diagram. This is possible because all cycle "parallelograms" of the truncation column have the same clockwise absorbent flow covering direction, which obviously is plane rotation free (q.e.d.). This result will be used later. The lemma 1 is valid also if the cycle "parallelograms" of the truncation column has the same counter-clockwise absorbent flow covering direction before and after rotation. Another useful result is referring to a truncated cycle "anatomy" as follows:

Lemma 2 *A truncated cooling and heating (heat transformer) coabsorbent cycle can be obtained through the overlapping of nontruncated coabsorbent cycles.*

Proof Next, we shall proof the lemma using truncated cooling coabsorbent cycles covered by nontruncated cooling cycles, but equally the lemma can be proved using truncated heating coabsorbent cycles covered by nontruncated heating cycles. Let us consider the truncated cooling coabsorbent cycle plotted in Fig. 4.6, which is a simpler noted version of Fig. 4.4, and the isosteres passing through the points

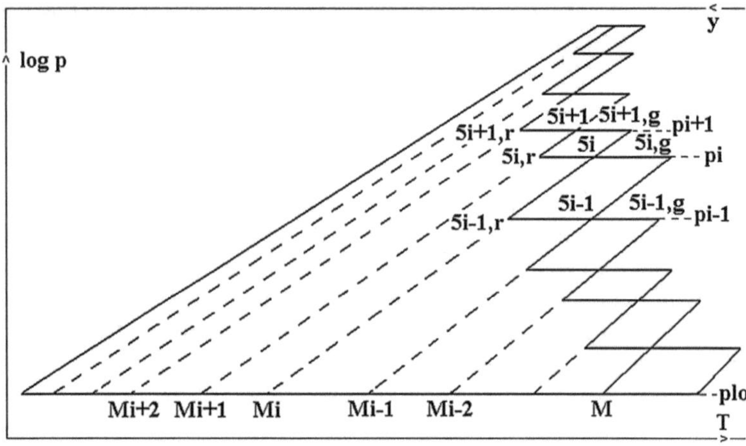

Fig. 4.6 A truncated cooling coabsorbent cycle build by non-truncated coabsorbent cycles overlapping

5_i, $i = 2, \ldots, n - 1$. With them, we construct a sequence of three nontruncated cooling coabsorbent cycles noted as, Fig. 4.6: $(i + 1)$: $M_{i+2}5_{i+1,r}5_{i+1}5_{i+1,g}M_i$; (i): $M_{i+1}5_{i,r}5_i5_{i,g}M_{i-1}$; $(i - 1)$: $M_i5_{i-1,r}5_{i-1}5_{i-1,g}M_{i-2}$, and having as mixing points the M_{i+1}, M_i and M_{i-1}, defined by the isosteres passing through the points 5_{i+1}, 5_i and 5_{i-1}, respectively. In the three-cycle sequence, the adjacent cycles are overlapping in such a way that the $(i - 1)$ cycle resorption outlet and the $(i + 1)$ cycle generation outlet isosteres, the $y_{i-1,r}$ and $y_{i+1,g}$, respectively, are overlapping on the (i) cycle mixing isostere y_{M_i}. As a result, given the different pressure the three cycles are running on top with, the p_{i-1}, p_i and p_{i+1}, with $p_{i-1} < p_i < p_{i+1}$, there will be three different portions of overlapping on the y_{M_i} isostere, namely the $M_i5_{i-1,r}$, with two overlappings, the $5_{i-1,r}5_i$, with a single overlapping, and the $5_i5_{i+1,g}$, with no overlapping, respectively. We make now the remark that a mixing isostere is covered by a double absorbent flow, while the extreme isosteres of a nontruncated cycle are covered by a single absorbent flow, with an opposite direction as compared to that of the double absorbent flow. Bearing this and the above mentioned in mind, we can now infer that the two-overlapping y_{M_i} isostere portion is vanishing, which reason for was represented by a dotted line in Fig. 4.6, and the single overlapping portion becomes single flowed. This is why the single overlapping and the no overlapping y_{M_i} isostere portions are operational and were consequently represented by solid lines. Extending our conclusion to the whole cycle, all two-overlapping portions of the y_{M_i}, $i = 2, \ldots, n - 1$, isosteres are vanishing, reason for which they were represented by dotted lines too. Concerning the y_{M_n} isostere, this is vanishing too, because here we don't have a two-overlapping portion, but given the flow balance, it is covered by a single absorbent flow, which is counterbalanced by the $y_{n-1,r}$ counterflow current. The y_M isostere is two-flow covered, as it results after a simple consideration, given the contribution of n-th desorber flow, which remains unbalanced because of the whole cycle p_{low} bottom vanishing, for the reasons outlined above. In this way, the whole cycle solid line has been proved, as resulting from the nontruncated cycles overlapping. Bearing in mind the lemma 1, we extend our conclusion to the truncated heating (heat transformer) coabsorbent cycles as well, and in this way the whole lemma has been proved (q.e.d.).

Not only a truncated coabsorbent cycle is topologic intrinsic related to a nontruncated cycle, as the lemma 2 showed, but these cycles have absorbent flow intrinsic similarities also, as it is stated by the following:

Lemma 3 *A truncated coabsorbent cycle, of cooling or heating (heat transformer) type, and its nontruncated coabsorbent cycle of origin, have the same typical equation, which is independent of the truncation way.*

Proof Let us first consider the truncated cooling coabsorbent cycle plotted in Fig. 4.4, which symbols of will be used hereinafter, together with those introduced for the nontruncated cycles in Sect. 3.2. We note the outlet absorbent mass flow factors and the each generation and resorption pressure stage vapor by,

$$g_{1,i} = \frac{Y_{G,i} - y_{GI,i}}{y_{GI,i} - y_{GO,i}}, \quad i = 1, \ldots, n, \tag{4.4}$$

$$r_{1,i} = \frac{Y_{G,i} - y_{RI,i}}{y_{RO,i} - y_{RI,i}}, \quad i = 1, \ldots, n, \tag{4.5}$$

and

$$V_i = \frac{V_{Gi}}{V_D}, \quad i = 1, \ldots, n, \tag{4.6}$$

respectively. The mass balance on different parts of the cycle gives:

$$V_n r_{1,n} = d, \tag{4.7}$$

$$V_1 g_{1,1} = a, \tag{4.8}$$

$$(a_1 + d_1) + g_{1,2} V_2 = \left(r_{1,1} + g_{1,1}\right) V_1, \tag{4.9}$$

$$r_{1,i-1} V_{i-1} + g_{1,i+1} V_{i+1} = \left(r_{1,i} + g_{1,i}\right) V_i, \quad i = 2, \ldots, n-1, \tag{4.10}$$

$$y_{M,i-1} r_{1,i-1} V_{i-1} + y_{M,i+1} g_{1,i+1} V_{i+1} = y_{M,i} \left(r_{1,i} + g_{1,i}\right) V_i, \quad i = 2, \ldots, n-1 \tag{4.11}$$

$$r_{1,n-1} V_{n-1} = \left(r_{1,n} + g_{1,n}\right) V_n. \tag{4.12}$$

Introducing Eqs. (4.7) and (4.8) in Eq. (4.9), using this result in the first Eq. (4.10) and further on successively introducing the result of a precedent equation into the next one, it is obtained the recurrent equation:

$$r_{1,i} V_i = g_{1,i+1} V_{i+1} + r_{1,n} V_n, \quad i = 1, \ldots, n-1. \tag{4.13}$$

With the help of the recurrent Eq. (4.13), we eliminate step by step the V_2, V_3, …, V_{i-1} reduced vapor quantities of each cycle stage, and obtain:

$$V_i \prod_{j=2}^{i} g_{1,j} = V_1 \prod_{j=1}^{i-1} r_{1,j} - r_{1,n} V_n \prod_{j=2}^{i-1} r_{1,j} \left(1 + \sum_{k=2}^{i-1} \prod_{j=2}^{k} s_j \right), \quad i = 2, \ldots, n-1. \tag{4.14}$$

In Eq. (4.14) the s_j symmetry factor of the coabsorbent "j" stage is given by,

$$s_j = \frac{g_{1,j}}{r_{1,j}} = \frac{y_{RO,j} - y_{M,j}}{y_{M,j} - y_{GO,j}}, \quad j = 1, \ldots, n, \tag{4.15}$$

where $y_{M,j}$ is the coabsorbent "j" stage mixer concentration. In Eq. (4.14) we make $i = n$ in order to obtain,

$$\frac{V_1}{V_n} = \frac{\prod_{j=2}^{n} r_{1,j}}{\prod_{j=1}^{n-1} r_{1,j}} \left(1 + \sum_{k=2}^{n} \prod_{j=2}^{k} s_j \right). \tag{4.16}$$

From Eqs. (4.7) and (4.8) we obtain,

$$\frac{a}{d} = \left(\frac{V_1}{V_n} \right) \frac{g_{1,1}}{r_{1,n}}. \tag{4.17}$$

Combining Eqs. (4.16) and (4.17), it results the *characteristic equation* of a n-stage balance (fractal) truncated cooling cycle, given by:

$$\frac{a}{d} = \sum_{k=1}^{n} \prod_{j=1}^{k} s_j. \tag{4.18}$$

In order to prove the first part of the lemma, we write Eq. (4.15) in the following form,

$$s_j = \frac{\Delta y_{R,j}}{\Delta y_{G,j}}, \quad j = 1, \ldots, n, \tag{4.19}$$

and remark that,

$$\Delta y_{G,j} = \Delta y_{R,j-1}, \quad j = 2, \ldots, n, \tag{4.20}$$

where from,

$$\prod_{j=1}^{k} s_j = \frac{\Delta y_{R,k}}{\Delta y_{G,1}}. \tag{4.21}$$

Using this result, Eq. (4.18) rewrites,

$$\frac{a}{d} = \sum_{k=1}^{n} \prod_{j=1}^{k} s_j = \sum_{k=1}^{n} \frac{\Delta y_{R,k}}{\Delta y_{G,1}} = \frac{y_{RO,n} - y_M}{y_M - y_{GO,1}} = S. \tag{4.22}$$

and the first part of the lemma has been proved. In order to prove that the property expressed by Eq. (4.22) holds true for a truncated heat transformer coabsorbent cycle as well, it is enough to remark that, according to Lemma 1, the heating cycle comes of the cooling one through its simple rotation by 180° and that for the latter cycle the theorem was already proved. In this way, taking this result into account and the fact that the truncation was chosen at random, the whole lemma is proved (q.e.d.).

The model and the design of a truncated cycle is requiring the knowledge of the reduced vapor quantities of each cycle stage V_i, $j = 1, \ldots, n$. They are calculated next. Unlike the nonoverlapping and overlapping column cycles, these vapors can be obtained in case of the balance column cycles in a closed form. As already discussed, from the (T_{DO}, y_{DO}) input data, the desorber flow factor d results and further the absorber flow factor a is computed, using the equations given in Tables 3.1 and 3.2. The vapors V_1 and V_n can be calculated now from Eqs. (4.8) and (4.7), respectively, resulting in:

$$V_1 = \frac{a}{g_{1,1}} \tag{4.23}$$

and

$$V_n = \frac{d}{r_{1,n}} \tag{4.24}$$

The rest of vapors are obtained considering Eqs. (4.14) and (4.23):

$$V_i = V_1 \left(\frac{g_{1,1}}{g_{1,i}}\right) \frac{\sum_{k=i}^{n} \prod_{j=i}^{k} s_j}{\sum_{k=1}^{n} \prod_{j=1}^{k} s_j} = V_1 \left(\frac{g_{1,1}}{g_{1,i}}\right) \frac{1}{S} \sum_{k=i}^{n} \prod_{j=i}^{k} s_j, \quad i = 2, \ldots, n-1. \tag{4.25}$$

In Eq. (4.25), the numerator, equal to the sum of products, can be expressed alternatively, following a similar way as in Eqs. (4.19)–(4.22):

$$\sum_{k=i}^{n} \prod_{j=i}^{k} s_j = \frac{y_{RO,n} - y_{Mi}}{y_{Mi} - y_{GOi}}. \tag{4.26}$$

In Eq. (4.26), each stage mean concentration, y_{Mi}, is known from Eq. (4.2). This is of great help, and the calculus of the vapors V_i, $i = 2, \ldots, n - 1$, becomes straightforward. It is not the case of the nonoverlapping and overlapping column cycles, when y_{Mi} values are not known and must be calculated through an iterative computation, considering additionally Eq. (4.11).

Using Eqs. (4.22), (4.23) and (4.26), Eq. (4.25) can be expressed in a simpler form:

$$V_i = V_1\left(\frac{g_{1,1}}{g_{1,i}}\right)\frac{\frac{y_{RO,n}-y_{Mi}}{y_{Mi}-y_{GOi}}}{\frac{y_{RO,n}-y_{M}}{y_{M}-y_{GO,1}}} = \frac{d}{g_{1,i}}\frac{y_{RO,n}-y_{Mi}}{y_{Mi}-y_{GOi}}, \quad i = 2,\ldots,n-1. \quad (4.27)$$

The algorithm given by Eqs. (4.4)–(4.27) is a complete tool for computing the cooling cycle (Fig. 4.4) absorbent (massic) flow. However, this must be modified correspondingly, in order to use it to the heating cycle (Fig. 4.5). The computation algorithm of the heating cycle reduced vapor quantities of each stage V_i, $i = 2, \ldots, n-1$, can be obtained elegantly bearing in mind the useful lemma 1 and utilizing the results obtained so far. Indeed, from this lemma it results that the change from the cooling to the heating cycle in the log(*p*) − 1/*T* diagram through an 180° plane rotation will replace the generation processes with the resorption ones and vice versa. Consequently, the ratio of the generators outlet flow factors in Eqs. (4.23) and (4.25) will be replaced in case of the heating cycle by the ratio of the resorbers outlet factors. Also, the heating cycle symmetry factor of the coabsorbent "*j*" stage, s_j, $j = 2, \ldots, n-1$, becomes the reverse of the s_j value known from the cooling cycle. In this way, anticipating, the V_i, $i = 2, \ldots, n-1$, generated vapor of the heating cycle will be calculated by Eq. (4.42). However, it is meaningful as well to adopt an independent way to calculate these vapor quantities, resembling to a good extent to that we used for the cooling cycle vapor assessment, which we shall traverse next. We consider the notations in Fig. 4.5. The equations of the $g_{1,i}$ and $r_{1,i}$ flow factors, given by Eqs. (4.4) and (4.5) for the cooling cycle, hold here true as well. Also, we note:

$$V_i = \frac{V_{G,i}}{V_G}, \quad i = 1,\ldots,n. \quad (4.28)$$

The mass balance on the first resorber and the last generator exits, gives:

$$r_{1,1}V_1 = g, \quad (4.29)$$

and

$$g_{1,n}V_n = r \quad (4.30)$$

respectively. From Eq. (3.22) of Table 3.2, we can obtain in case of the truncated heating cycle an equation similar to Eq. (4.22), using Eqs. (4.29) and (4.30):

$$S = \frac{g}{r} = \frac{r_{1,1}V_1}{g_{1,n}V_n} = \sum_{k=1}^{n}\prod_{j=1}^{k} s_j = \frac{y_M - y_{GO,n}}{y_{RO,1} - y_M}. \quad (4.31)$$

Equation (4.31) is the *characteristic equation* of a *n*-stage balance (fractal) truncated heating cycle. Continuing with the mass balance on the different parts of the heating cycle we can write:

$$(g_1 + r_1) + r_{1,2}V_2 = \left(r_{1,1} + g_{1,1}\right)V_1 \quad (4.32)$$

$$g_{1,i-1}V_{i-1} + r_{1,i-1}V_{i+1} = \left(g_{1,i} + r_{1,i}\right)V_i, \quad i = 2, \ldots, n-1 \tag{4.33}$$

$$y_{M,i-1}g_{1,i-1}V_{i-1} + y_{M,i+1}r_{1,i-1}V_{i+1} = y_{M,i}\left(g_{1,i} + r_{1,i}\right)V_i, \quad i = 2, \ldots, n-1 \tag{4.34}$$

$$g_{1,n-1}V_{n-1} = \left(g_{1,n} + r_{1,n}\right)V_n \tag{4.35}$$

Equations (4.29) and (4.30), written for the outlet flow factors, are introduced in Eq. (4.32). The obtained result is utilized in the first Eq. (4.33). Further on, the result of a precedent equation is introduced successively into the next one and the recurrent equation is obtained:

$$g_{1,i}V_i = r_{1,i+1}V_{i+1} + g_{1,n}V_n, \quad i = 1, \ldots, n-1 \tag{4.36}$$

With the help of the recurrent Eq. (4.36), the V_2, V_3, …, V_{i-1} reduced vapor quantities of each cycle stage are eliminated step by step. As a result, the following equation holds true:

$$V_i \prod_{j=2}^{i} r_{1,j} = V_1 \prod_{j=1}^{i-1} g_{1,j} - g_{1,n}V_n \left(1 + \sum_{k=2}^{i-1} \prod_{j=2}^{k} s_j\right) \prod_{j=2}^{i-1} g_{1,j}, \quad i = 2, \ldots, n-1 \tag{4.37}$$

In Eq. (4.37), the s_j "j" stage symmetry factor is given by:

$$s_j = \frac{r_{1,j}}{g_{1,j}} = \frac{y_{M,j} - y_{GO,j}}{y_{RO,j} - y_{M,j}}, \quad j = 1, \ldots, n, \tag{4.38}$$

where $y_{M,j}$ is the coabsorbent "j" stage mixer concentration. Equation (4.37) can be rewritten with the help of Eq. (4.38) in a more convenient form, after some calculus, as:

$$\frac{V_i}{V_1} = \frac{r_{1,1}}{g_{1,i}} \frac{1}{\prod_{j=1}^{i} s_j} \left[1 - \frac{g_{1,n}}{g_{1,1}} \frac{V_n}{V_1} \left(1 + \sum_{k=2}^{i-1} \prod_{j=2}^{k} s_j\right)\right], \quad i = 2, \ldots, n-1. \tag{4.39}$$

In Eq. (4.39), we take into account Eqs. (4.31) and (4.38), in order to obtain:

$$\frac{V_i}{V_1} = \frac{r_{1,1}}{g_{1,i}} \frac{\sum_{k=1}^{n} \prod_{j=1}^{k} s_j - \sum_{k=1}^{i-1} \prod_{j=1}^{k} s_j}{S \prod_{j=1}^{i} s_j}, \quad i = 2, \ldots, n-1. \tag{4.40}$$

The last ratio numerator of Eq. (4.40) right hand member may be written after a simple calculus, as:

$$\sum_{k=1}^{n} \prod_{j=1}^{k} s_j - \sum_{k=2}^{i-1} \prod_{j=2}^{k} s_j = \frac{1}{s_i} \left(\sum_{k=i}^{n} \prod_{j=i}^{k} s_j\right) \prod_{j=1}^{i} s_j, \quad i = 2, \ldots, n-1. \tag{4.41}$$

We take into account Eqs. (4.38) (4.29) and (4.31) and combine Eqs. (4.41) and (4.40), in order to obtain the final useful result:

$$V_i = V_1\left(\frac{r_{1,1}}{r_{1,i}}\right)\frac{\sum\limits_{k=i}^{n}\prod\limits_{j=i}^{k} s_j}{\sum\limits_{k=1}^{n}\prod\limits_{j=1}^{k} s_j} = V_1\left(\frac{r_{1,1}}{r_{1,i}}\right)\frac{\frac{(y_{M,i}-y_{GO,n})}{(y_{RO,i}-y_{M,i})}}{\frac{(y_M-y_{GO,n})}{(y_{RO,1}-y_M)}}$$
$$= \frac{r}{r_{1,i}}\frac{(y_{M,i}-y_{GO,n})}{(y_{RO,i}-y_{M,i})}, \quad i = 2,\ldots,n-1. \tag{4.42}$$

Equation (4.42) resembles to Eqs. (4.25) and (4.27), written for the cooling cycle, as we have for seen. It enables the calculation of the heating cycle reduced vapor quantities of each stage V_i, $i = 2, \ldots, n - 1$. The model is completed, calculating the vapors V_1 and V_n, from Eqs. (4.29) and (4.30), respectively. Prior to do this, of course we have to calculate first the g factor from the (T_{GO}, y_{GO}) input data and then the r factor, using Eq. (3.30) of Table 3.2.

The lemmas 2 and 3 are conducting us to emphasize a more comprehensive nontruncated coabsorbent cycle property, expressed by the following:

Theorem *The nontruncated cooling and heating (heat transformer) coabsorbent cycles are fractals from the topologic and the absorbent flow point of view.*

Proof According to the Benoit Mandelbrot definition, every fragmented geometric shape, that is the truncated cycle, of a nontruncated cooling and heating (heat transformer) coabsorbent cycle, considered as a whole (origin), can be subdivided in parts, each of which, namely the nontruncated cycle, being a reduced size of the whole from the geometric point of view, according to the lemma 2, and behaves as a whole from the absorbent flow point of view, according to the lemma 3, and these statements proof the theorem (q.e.d.).

Moreover, because a heat transport is always accompanying a mass transport, which in our case is the absorbent flow, the theorem entitles us to go even further and formulate the following:

Corollary *The nontruncated cooling and heating (heat transformer) coabsorbent cycles are mass and heat fractals.*

The theorem explains why the truncation column enjoying the property given by Eq. (4.2) was named a "fractal (balance) truncation column". Out of the three truncation types, depicted in Fig. 4.3, only that fulfilling Eqs. (4.2) and (4.20), makes a nontruncated cycle to enjoy the property of being a fractal when applied to it. The theorem suggests the introduction of new shorted names for the nontruncated cooling and heating coabsorbent cycles, such as *cooling* and *heating fractals*, respectively, used hereinafter.

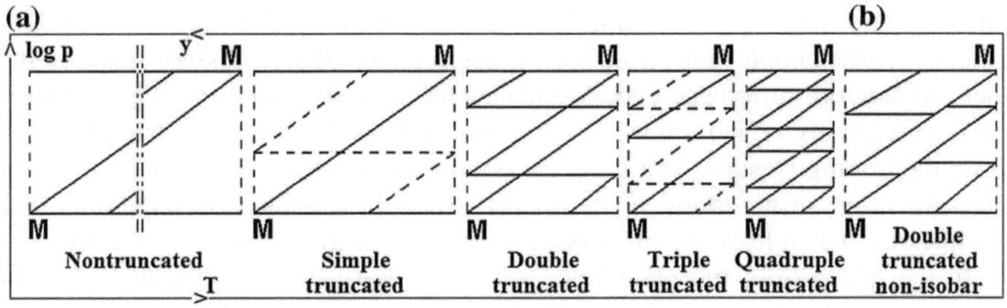

Fig. 4.7 **a** Isobar columns: the dependence of the degree of truncation on the cooling and heating fractals mixing points relative position; **b** Non-isobar truncation column

4.1.4 Truncation Columns, Common-Column Cycles, Column Cycles, Reverse Truncated Cycles and Fractals Symbolic Representation

4.1.4.1 Truncation Columns

Only the truncation columns made up by parallelograms covered in a clockwise direction have a technical application in the heat pumping, as the reader may easily prove. We could assert as a rule of thumb that there is at least one column of the type we analyse, which can be used to truncate a coabsorbent cycle in every cooling or heating application requiring this technique. Physically, the columns with even degrees of truncation have only one possible structure, while the odd ones can benefit of an infinite number of structures for a given application. This statement results from the $\log(p) - 1/T$ diagram plotting of the truncation columns, Fig. 4.7a. Indeed, in case of the even-degree plotting, the isosteres coming of the mixing points do not meet together neither directly, nor indirectly, through intermediary isobars or isobars and isosteres and cannot be moved at convenience for fixed input data. For this reason only one structure results and this configuration was marked by solid line. On the contrary, in case of the odd-degree plotting, the isosteres originating from the mixing points do meet together, directly or indirectly, so the remaining isosteres and isobars structure can be designed at convenience in order to generate an infinite family of columns, which reason for it was marked by dashed line. Of course, the input data of an application determine the degree of a truncation column. Out of them, the mixing points relative position is the most important. In Fig. 4.7a we have plotted a few ammonia/water columns with isobar stages and the end generation and resorption temperatures equal to those of the mixing points. The stages number increases with the mixing points temperature difference decrease (from the left to the right in the figure). The first left picture shows a nontruncated fractal column, where the temperature difference is high enough for the cooling mixing isostere be higher than that of the heating one. Decreasing the temperature difference so that the two mixing isosteres overlap, as in the second picture, the column results simple truncated, with a single intermediary pressure stage. A further temperature difference decrease makes the heating fractal mixing isostere to have increasingly higher values as compared to that of the cooling fractal. In this case, the mixing isosteres can have a common pressure interval, a single common pressure value, or no common pressure at all, so the columns are increasing their complexity to double, triple and quadruple-and higher truncation, respectively.

In the last period, the plate heat exchangers have been proposed also in the absorption technology, to replace the classic tube and shell devices playing the role of mass and heat exchangers (e.g. Risberg 2004, for the NH_3–H_2O hybrid heat pumps). Despite their high heat exchange coefficients and compactness for large heat exchange surfaces, the plate exchangers introduction is confronted with a problem, that of creating sensible vapour pressure losses between the generation and the absorption devices. In our case, this disadvantage could be technically overcome using non-isobar stage truncated columns, were the generation takes place at a higher pressure than that of the coupled resorption process, Fig. 4.7b.

A truncation column is not only a succession of parallelograms connected in the pressure stages mixing points, as we formally described the truncated fractals. It could be regarded from another point of view, too, that of a "two-flow column". Indeed, we could consider it as being represented by two zigzag lines covered in opposite directions by absorbent flows originating from the cooling and heating mixing points M, connecting all resorption and generation processes, respectively, and having mass and heat exchange on isobars. In this way, the left-side upstream of the resorption processes continuously increases the absorbent concentration from the $y_{M,cooling}$ till the $y_{RO,n}$, and the right-side downstream of the generation processes continuously decreases the absorbent concentration from the $y_{M,heating}$ till the $y_{GO,1}$. Concerning the cooling fractal, its truncation effect results in that the $y_{M,cooling}$ absorbent, having the flow f_M, is split in two complementary flows, the V_1g_1 and the V_nr_n, having the end concentrations the $y_{GO,1}$ and $y_{RO,n}$, respectively, similarly to the nontruncated cooling fractal case, described in Sect. 3.2.1. In this case, the Eqs. (3.14) and (3.15) hold true here as well, Table 3.2, because of the lemma 3. From these equations it results that the absorbent concentration increase till the $y_{RO,n}$ is done by sacrificing the absorbent flow amount, which is S times lower than that having the $y_{GO,1}$ low concentration. Indeed, the $V_nr_n = V_1g_1/S$, and typically $S \geq 2.0 - 3.0$. Both the $y_{RO,n}$ and the V_nr_n will influence the useful cooling effect capacity. Now, we can understand why the S values close to unit are preferred. A similar effect is met in case of the heating fractal column. Indeed, the most reduced concentration absorbent flow, related to the useful heating effect capacity, is $S \geq 2.0 - 3.0$ times lower than the highest concentration absorbent complementary flow. This assertion bases on Eqs. (3.25) and (3.26) practical evaluation, Table 3.2.

The "height" of a truncation column is obviously given by its parallelograms number, but a more analytical column index might be introduced, such as:

$$h_t = \frac{y_{RO,n} - y_{GO,1}}{T_{RO,n} - T_{GO,1}}. \tag{4.43}$$

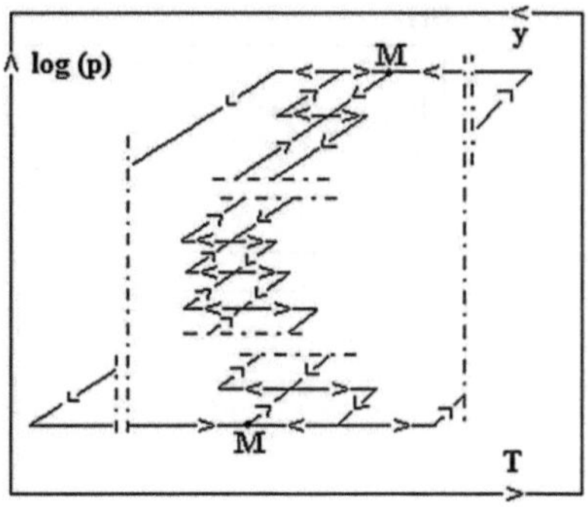

Fig. 4.8 The block flow-chart of the common-column cooling and heating cycle

The low generation temperatures combined with the high sink temperatures in the cooling operation and the high resorption temperatures combined with the low heat temperatures in the heating operation lead to high columns use. They require an increased number of parallelograms, typically three to five for the ammonia/water working combination. The column index normal values are in this case in the range of $h_t = 0.02–0.03\ \mathrm{K}^{-1}$.

4.1.4.2 Common-Column Cycles

The truncation columns of the cooling and heating cycles in Figs. 4.4 and 4.5, respectively, are constructively similar and theirs parallelograms have the same absorbent clockwise covering. Overlapping the cycles in Figs. 4.4 and 4.5, in order to form a common-column cooling and heating cycle, like in the schematic of Fig. 4.8, might be advantageous in some applications, from the constructive point of view, saving the manufacture of a column. The common-column cycle produces simultaneously cooling and heating. However, energetically, this has not the meaning of a cogeneration, similar to that met for example by the nontruncated cooling cycle (two useful effects for a single heat input). In this case, the absorbents and the vapor are overlapping, but the connections with the sink and especially with the heat sources of the two cycles are overlapping too, in an additive way. Principally, a common-column cycle cannot be equally constructed employing nonoverlapping or overlapping columns.

4.1.4.3 Column Cycles and Inverted Truncated Cycles

The integration of a truncation column in a closed circuit may result in a *direct (clockwise or power) column cycle*, Fig. 4.9a. Its parallelograms are clockwise covered. A turbo-generator is connected between the column top and the column bottom. The power column cycle has been analyzed elsewhere (Staicovici 2007d). The results show that, comparatively, the inverted Osenbrück cycle (see Sect. 3.2.5.2) performs better and is simpler constructively, so, in order to be more

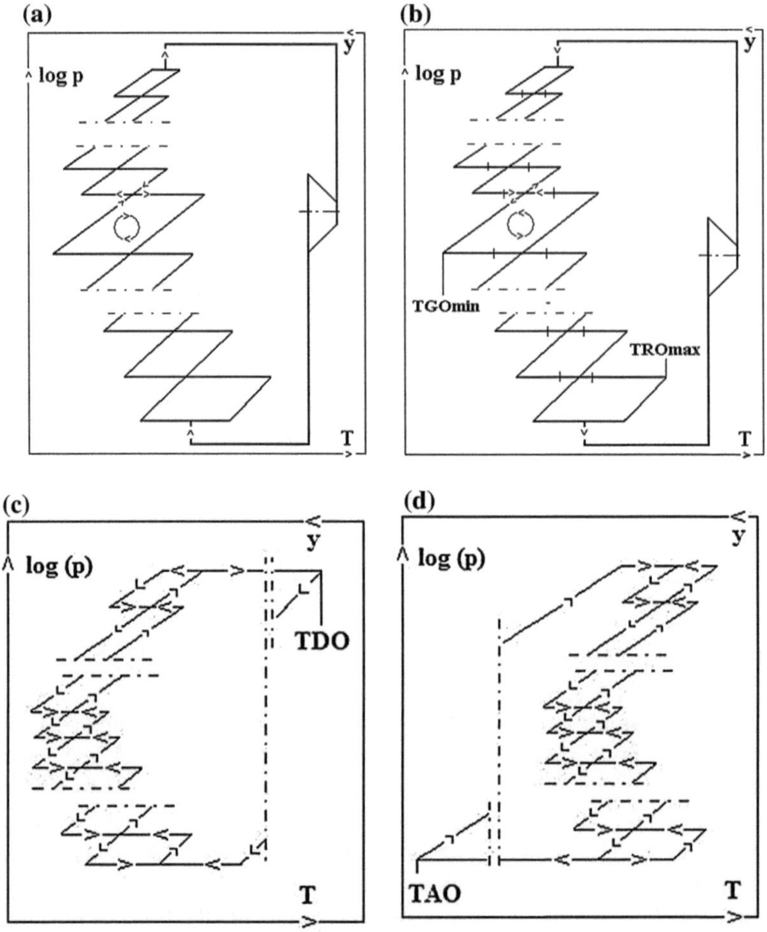

Fig. 4.9 Block flow-charts of: **a** direct column cycle; **b** inverted column cycle; **c** inverted truncated cooling cycle; **d** inverted truncated heating cycle

effective, a power column cycle must increase its internal gax and sensible heat recovery. Alternatively, a truncation column integration may result in an *inverted (counter-clockwise or heat pump) column cycle*, Fig. 4.9b. Its parallelograms are covered counter-clockwise in this case. According to lemma 2, this column comes of a truncated heating coabsorbent cycle covered by nontruncated heating cycles. A compressor is connected between the column bottom and the column top. No results of this cycle has been obtained so far.

Further on, using an inverted column, Fig. 4.9b, two types of cycles can be conceived, depicted in Figs. 4.9c, d. The inverted truncated cooling cycle of Fig. 4.9c bases on an inverted column which is connected to its right side with a desorber (top) and a resorber (bottom). Alternatively, the inverted truncated heating cycle of Fig. 4.9d comes of an inverted column which is connected to its left side with an absorber (bottom) and a generator (top). We can remark that, similar to the truncated cycles of Figs. 4.4 and 4.5, according to the Lemma 1, formally, graphically, the representation of the inverted truncated heating cycle in the log $p - 1/T$ diagram, Fig. 4.9d, can be obtained through an 180° plane rotation of the inverted truncated cooling cycle representation in the same diagram, Fig. 4.9c. Taking into account this fact and given the relative alike graphics, the reader could think that the group of cycles plotted in Figs. 4.4 and 4.5 on one side and that of those plotted in Fig. 4.9c, d, on the other side, have the same practical importance and application. In fact this is not at all true, conceptually the two groups differ fundamentally one another. Indeed, the cycles of Figs. 4.4 and 4.5 base on direct columns, fulfilling heat pumping tasks with unfavoring supply, i.e. with sink/heat sources outdistanced by small, limited, temperature gaps (see the sources related internal temperature difference $\Delta T_M = T_{M,heating} - T_{M,cooling}$, Fig. 4.4). On the contrary, the cycles of Fig. 4.9c, d base as mentioned on inverted columns, which, unlike the direct one, are not conditional upon the difference between the extreme temperatures of the resorber-generator devices running on the isobar stages (see $T_{RO,\max} - T_{GO,\min}$, Fig. 4.9b), but mainly upon the cycle output requirements. In this respect, an inverse column operation could be replaced by that of a single heat transformer cycle. The reasons to use an inverted column cycle instead of a single heat transformer cycle for same input data are therefore not mandatory, but recommendable, such as:

(i) the decrease of the cycle maximum operating pressure, which correspondingly decreases the solution pumping work and increases safety;
(ii) the permission of many, not all, out of the generation and resorption processes be designed in such a way that theirs operation intervals match better to sources;
(iii) the benefic limitation of the internal heat source temperature T_{DO}, Fig. 4.9c, and cooling source temperature T_{AO}, Fig. 4.9d;
(iv) the sensitivity of the heating and cooling effectivenesses with respect to the number of stages increase is less as compared to that of the direct column cycles.

The computation of an inverted column cycle is related to that of a direct column cycle by the following simple and useful:

Lemma 4 *A coabsorbent cooling (heating) cycle plotted in the log $p - 1/T$ diagram is computable with the algorithm of the heating (cooling) cycle resulting from the cycle to be computed which suffered successively an 180° plane rotation of its plot in the same diagram and an inversion of its absorbent flow covering.*

Proof Let us consider for instance an inverted column cooling cycle plotted in a log $p - 1/T$ diagram. According to the lemma 1, after an 180° plane rotation the cooling cycle plot becomes the plot of a heating cycle. During this process, the initial cycle devices role is inverted, that is for instance a generator becomes a resorber and vice versa. Further on, proceeding with the inversion of the absorbent flow covering, the heating cycle devices return to their initial role and the cycle becomes a direct column heating cycle. Because our choice of the initial cycle was arbitrary, the lemma is proved (q.e.d.).

The lemma 4 enables one to compute new cycles with the help of older cycles, which have already known computation algorithms. In our case, we can use the lemma 4 to compute the new cycles of Fig. 4.9c, d with the algorithms of the older cycles represented in Figs. 4.5 and 4.4, respectively, already given in Sect. 4.1.3. However, this work will not give any results of these cycles.

4.1.4.4 Truncated Cycles Symbolic Representation

Table 4.1 is giving a symbolic representation of the truncated cycles, truncation column, common-column cycles and hybrid truncated cycles. The pictures numbers 1 and 2 are a schematic of the already known heating and cooling fractals, respectively, the number 3 is the symbol of a truncation column, inspired by the two-flow column interpretation described above, and the numbers 4 and 5 are the symbols of the truncated cooling and heating fractals, respectively. The hybrid versions of the cycles 4 and 5 are those with the numbers 6 and 7. The common-column cycles are represented by the numbers 8–10, for the full and the hybrid truncation columns case.

4.2 Model of Cooling and Heating Truncated Cycles

Some important cooling and heating truncated cycles belonging to those presented above have been modelled and the results thereof are shown in this section. The model bases on Sect. 3.2.1.2 equations, written for the cycle devices, and on the gax heat exchange analysis, given in the next paragraph. Besides this, each stage generated vapor are calculated by the equations included in Sect. 4.1.3. The internal heat exchange has been considered carefully. In Fig. 4.10, we show several ways to perform it. All have been modeled in order to find the most effective one. In case of the "squat" parallelograms, provided with a small Δp and benefiting of *acr*, Fig. 4.10a, theirs resorption and generation processes have a large temperature overlapping in operation, so the gax latent heat recovery might be applicable. In our model we considered the gax recovery is valid whenever a good thermal match came out from a mandatory analysis done for each such a heat exchange type. Besides this, in the portions where there is not a temperature overlapping, the generation process can partially recover the sensible heat of the absorbent coming of it, and a part of the absorption process latent heat can be recovered by the absorbent coming of it. On the contrary, when the parallelograms have "middle" or "high" Δp values, Fig. 4.10b, the absorbents coming of the resorption and generation processes have a large temperature overlapping in operation, so they can exchange sensible heat. Again, in the portions where there is not a sensible heat

Table 4.1 Symbolic representation of fractals, truncation column, common-column cycles and hybrid truncated cycles

Cooling fractal	Heating fractal	Clockwise truncation column	Truncated cooling fractal	Truncated heating fractal
1 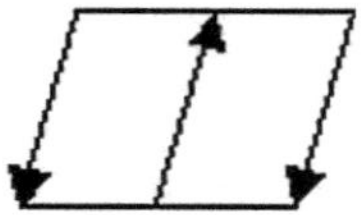	2 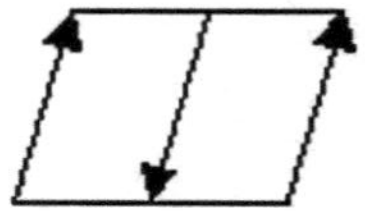	3 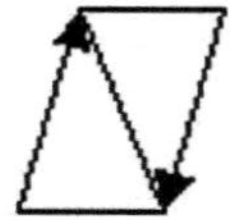	4 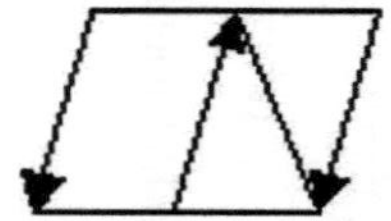	5 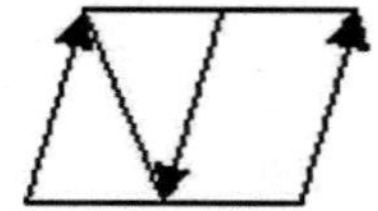
Hybrid cooling fractal	Hybrid heating fractal	Common-column cooling + heating fractal	Hybrid common-column cooling + heating fractals with equal columns	Hybrid common-column cooling + heating fractals with unequal columns
6 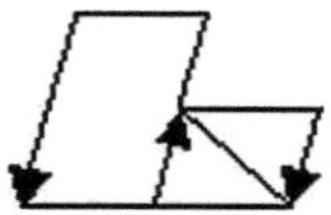	7 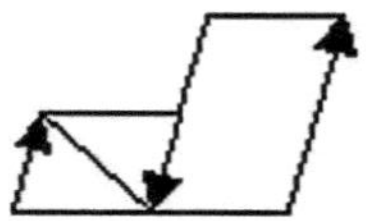	8 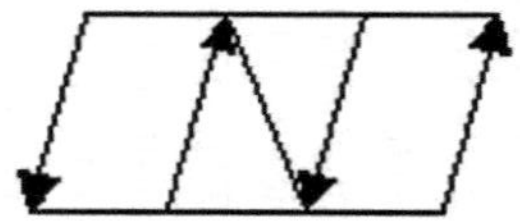	9 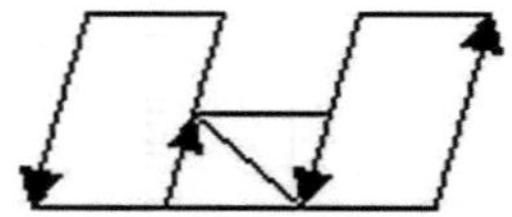	10 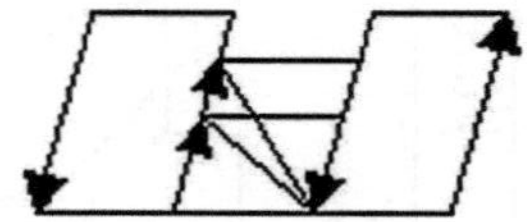

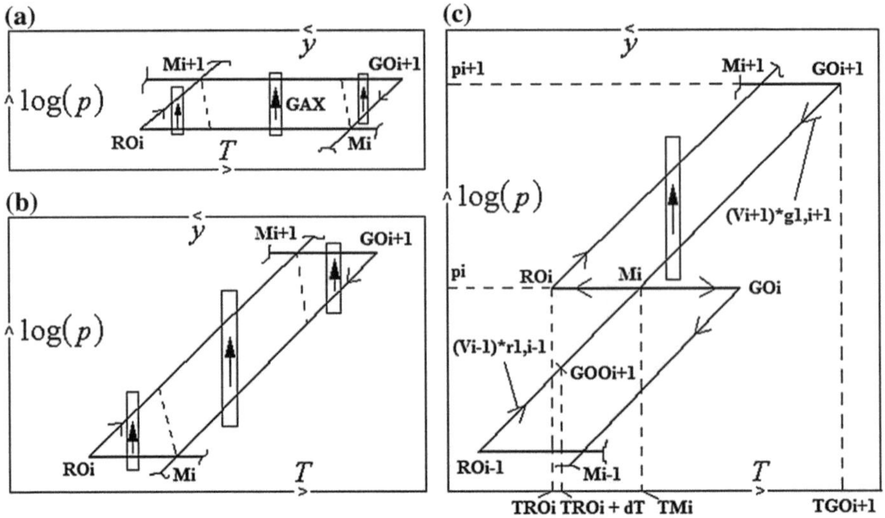

Figs. 4.10 Several internal heat exchange variants which could be applied in a truncation column operation

temperature overlapping, a heat recovery, similar to that described in Fig. 4.10a, can be practiced. Finally, a third type of internal heat recovery has been analysed, Fig. 4.10c. Here, two parallelograms, the M_i–RO_i–M_{i+1}–GO_{i+1}–M_i and the M_i–GO_i–M_{i-1}–RO_{i-1}–M_i, are shown in connection in the M_i stage mixing point. The absorbent coming of the G_{i+1} generator (see the point GO_{i+1}) is subcooled from the TGO_{i+1} temperature till the $TRO_i + \delta T$ one, where, $TRO_i < TRO_i + \delta T < TM_i$, $\delta T > 0$, in order to preheat the absorbent coming of the R_i resorber (see the point RO_i), prior to be mixed up in the M_i mixing point with the absorbent coming of the R_{i-1} resorber (see the point RO_{i-1}). The subcooling is beneficial, because it is avoiding an undesirable effect, that of the flash evaporation of the rich and superheated absorbent exiting the G_{i+1} generator when it is expanded in the mixer from p_{i+1} till p_i, with $p_{i+1} > p_i$, $i = 1, 2, \ldots, n - 1$, $i, n \in N$. There are also additional ways to perform the internal heat recovered, like the use of the R_i resorber rejected heat in order to preheat the absorbent exiting the R_{i-1} resorber or the recovery of the generated vapor sensible heat prior to be resorbed. In our model we preferred to use the heat recovery described in Fig. 4.10c. This is the simplest, provides a column with a right operation, it is one of the most effective and it is versatile, being the most adapted to the sink/heat sources temperature difference variation which a truncated coabsorbent plant can experience during for instance the transition from a season to another. Indeed, during such transitions, the $\Delta T_M = T_{M,heating} - T_{M,cooling}$ may vary to such extent that for instance a squat parallelogram of design may become a middle or a high one, and vice versa. In this case, the internal heat exchange depicted in Fig. 4.10a and b are not effective for the off-design operation situations, but that according to Fig. 4.10c is capable to ensure a column appropriate work in all cases.

4.2.1 Gax Use in "*Acr*" Provided Truncated Coabsorbent Cycles

4.2.1.1 Introductory Part

The cycle COP_c and COP_h could be improved further when operating with *acr*, performing an additional internal GAX heat recovery. The first attempts to improve the single-stage cooling and heating absorption cycle were made by Altenkirch, which proposed in 1913 the absorber heat exchange system (AHX). This system, called also "with absorbent return", capitalized the temperature difference which exists in an absorber and a generator between theirs colder and warmer ends, in such a way that the rich (high concentration) absorbent is preheated in the absorber it comes of prior to exit and enter the generator, and the poor (low concentration) absorbent is subcooled in the generator it comes of, prior to exit the generator and supply the absorber (see Fig. 4.10a, b). In this case, the absorber and generator operating temperatures are not superposing. The AHX was used starting with 1960 in the residential air conditioning, but the heating version was never commercially produced. Altenkirch also proposed in 1913 a better absorption heat recovery, with operating temperatures overlapping (Niebergall 1959), called later GAX (Generator Absorber heat eXchange), and noted as "gax" (see Sect. 4.1 introductory part). This recovered the high temperature heat released by the absorber in the temperature overlapping region with the generator, using an additional heat exchanger with an intermediary heat transfer fluid, as compared to the AHX system. Because of its high potential to increase the heat pumping cycles effectiveness, the gax become extensively studied and applied during the last 30 years (e.g. Scharfe et al. 1986; Herold et al. 1991; Erickson 1992; Rane and Erickson 1993; Bassols and Schneider 1993; Staicovici 1995a, b, 1998a, b). In this work we shall analyse mainly the gax processes where the latent exothermal resorption heat (most often, the resorption replaces the absorption in our case) is recovered by an opposite latent endothermal generation process in a temperature overlapping zone. The gax heat can be transferred directly, or indirectly, through an intermediary heat transfer fluid loop. The indirect way is more feasible, although obviously slightly less efficient. There are two important aspects which conditionate a gax performance: (a) the two processes take place on large concentration intervals with a significant temperature overlapping thereof; (b) there is a good match between the heat release and the heat receipt capacity of the two heat connected processes. The (a) condition can be fulfilled by the coabsorbent technology as this has intrinsic favoring properties of using the gax via *acr* to each of its truncated, multi-effect cycles, etc. (see e.g. Fig. 4.3b). The (b) condition has been studied by several authors mainly in case of the direct heat transfer (e.g. Scharfe et al. 1986; Erickson 1992; Staicovici 1995a). Its verification is mandatory for every practical gax application and for both heat transfer types, checking the (i) and (ii) conditions be fulfilled simultaneously (see Chap. 2). It must be done step by step along the gax devices, in order to ensure a minimum acceptable temperature

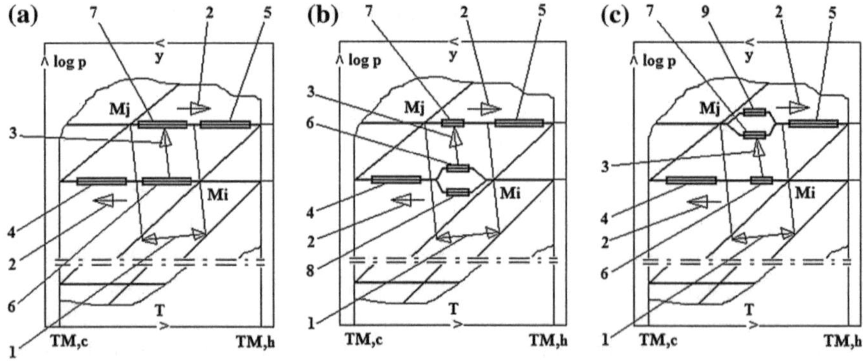

Legend: a) Full Match Gax; b) Partial Match Gax solution proposal for a Resorber Heat Excess (RHE) Gax problem; c) Partial Match Gax solution proposal for a Generator Heat Excess (GHE) Gax problem. (1): Operation Temperature Overlapping (OTO); (2): Absorbent Flow Direction; (3): Gax Heat Recovery; (4): i-th Gax Complementary Resorber (Ri,gax,c); (5): j-th Gax Complementary Generator (Gj,gax,c); (6): i-th Gax Resorber (Ri,gax); (7): j-th Gax Generator (Gj,gax); (8): i-th Gax Heat Excess Resorber (Ri,gax,e); (9): j-th Gax Heat Excess Generator (Gj,gax,e).

Figs. 4.11 Schematic of the three possible gax solutions applyable in a truncated column provided with *acr,* in order to respect the (i) condition, for: **a** Full match gax, enabling a single gax solution per application; **b** and **c** Partial match gax confronted with RHE-Gax and GHE-Gax problems, respectively, enabling an infinite number of gax solutions per application

pinch throughout the heat exchange. In our case, the *acr* enables an overlapping of temperatures in the operation of the 1st stage resorber and the 2nd stage generator, the 3rd stage resorber and the 4th stage generator, etc. There are three gax situations described shortly below, which could be solved, fully or partially, in a manner suggested for instance in Fig. 4.11a–c. According to the author experience gained so far in assessing a gax of a cooling cycle with full truncation column, the highest heat recovery is obtained between the 1st stage resorber and the 2nd stage generator. In this particular case, many times, not always, the condition (i) is fulfilled along the gax satisfactorily. This simplest operation of the full match gax is analyzed in the Sect. 4.2.1.2. In Fig. 4.11a it is shown how the full gax operation takes place between the resorber and generator sub-devices 6 and 7, respectively. The sub-devices 4 and 5 are complementary ones and finish the resorption and the generation processes, respectively. This is the classic gax configuration, the simplest one, which offers a single gax solution per application, only. The gax recovery between e.g. the 3rd stage resorber and the 4th stage generator of a full truncation column is usually partially satisfactory only, as the heat released by the resorber in the operation temperature overlapping (OTO, see Fig. 4.3b) is higher than that needed by the generator, that is the condition (i) is not satisfied. Possible solutions of this gax type confronted with a resorber heat excess (RHE), are proposed in Fig. 4.11a, b. According to Fig. 4.11b, the resorber is split in two isobar sub-devices, 6 and 8, and the full match gax will take place between the resorber 6 and the generator 7 sub-divices, while the excess resorber heat is eliminated through the gax heat excess resorber 8. Alternatively, the condition (i) cannot be fully satisfied as well, but only partially, if the heat released by resorber in the operation

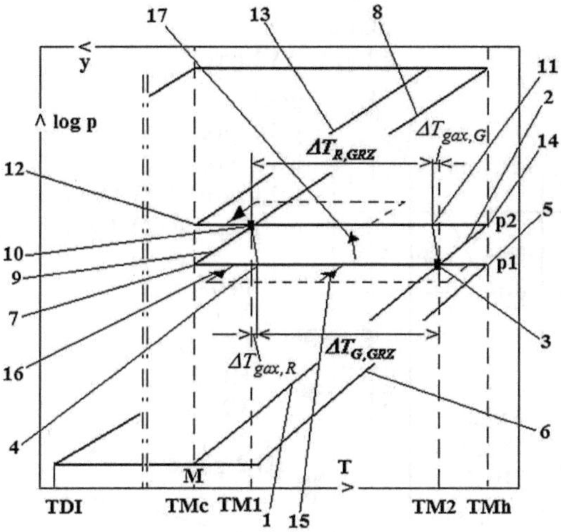

Point	p, [bar]	T, [C]	y, Y, [-]
1-(M)	0.4078	-	0.1719
2-(GO,2)	3.131	89.2	0.1719
3-(M,1)	2.342	79.65	0.1719
4-(R1,gax)	2.342	43.45	0.3456
5-(GO,1)	2.342	89.2	0.1313
6-(GO,1)	2.342	-	0.1313
7-(RO,1)	2.342	32.0	0.4088
8-(GO,3)	12.36	89.2	0.4088
9-(RO,1)	2.342	-	0.4088
10-(M,2)	3.131	40.45	0.4088
11-(G2,gax)	3.131	76.65	0.2272
12-(RO,2)	3.131	32.0	0.4585
13-(RO,2)	3.131	-	0.4585
14-(GO,2)	3.131	89.2	0.1719
15-	2.342	84.55	0.7864
$(V_{1,gax})$	$V_{1,gax} = 0.8609$		
16-	2.342	84.55	0.7864
$(V_{1,gax}^{c})$	$V_{1,gax}^{c} = 0.5096$		
17	$R_{1,gax}$-$G_{2,gax}$ gax HE		

Fig. 4.12 The full gax process plotted in the log $(p) - 1/T$ diagram and the joined data table for the double truncated cooling fractal cycle at hand, provided with *acr*

temperature overlapping were smaller than that needed by the generator. This type of gax, met especially to the heating cycles operation, is confronted with a generator heat excess (GHE). Its possible solutions are proposed in Fig. 4.11a, c. According to Fig. 4.11c, this time the generator is split in two isobar sub-devices, 7 and 9, and the full match gax will take place between the resorber 6 and the generator 7 sub-divices, while the excess generator heat is provided through the gax heat excess generator 9. The working with a partial match gax will be considered in Sect. 4.2.1.3. Here it will be shown that a partial match gax could be capitalized and included as well in the internal heat management of a truncated cycle, through simple yet effective technical solutions, such those introduced in Fig. 4.11a–c, in order to benefit additionally the cycle effectiveness. And last, but not the least, it will be shown that the gax configurations outlined in Fig. 4.11b, c enable an infinite number of gax solutions per application.

4.2.1.2 Full Match Gax

The analysis at hand is done next with the help of a deep cooling study case based on a double truncated cooling fractal. This is graphically represented in Fig. 4.12 (see also Fig. 4.11a). The 1st stage resorber portion implied in the gax recovery is noted by $R_{1,gax}$, and operates between the points 3 and 4 at p_1. The complementary portion of the first resorber, cooled externally, is noted by $R_{1,gax}^{c}$ (portion 4–7 at p_1). The vapors $V_{1,gax}$ (point 15) and $V_{1,gax}^{c}$ (point 16), generated by the first stage generator (portion 3–5 on the p_1 isobar), are resorbed by $R_{1,gax}$ and $R_{1,gax}^{c}$, respectively.

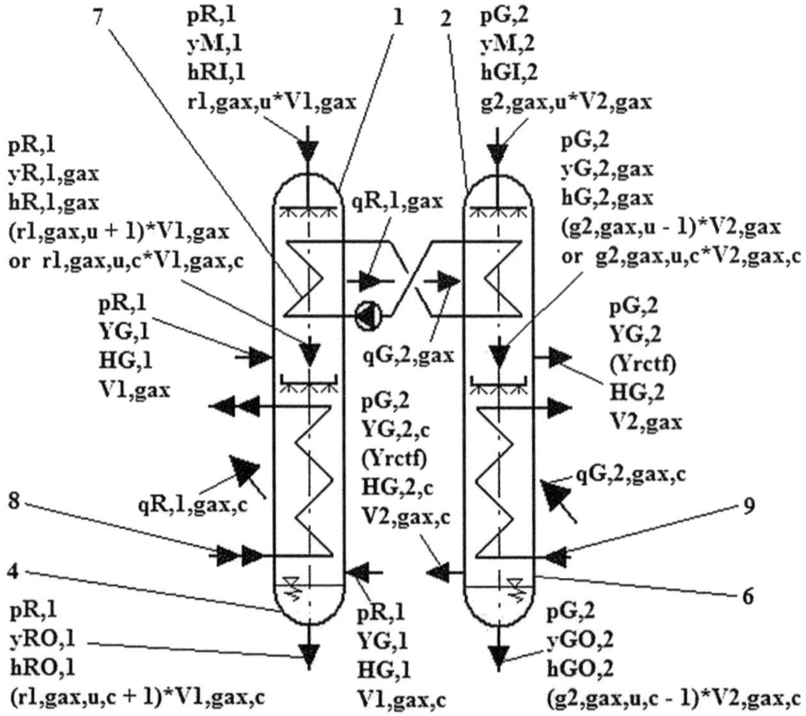

Legend: 1- Gax Resorber (R1gax in Fig. 4.12); 2-Gax Generator (G2gax in Fig. 4.12); 4- Gax Complementary Resorber (R1gax,c in Fig. 4.12); 6- Gax Complementary Generator (G2gax,c); 7- Gax Heat Transfer Loop; 8- Sink Source; 9- Heat Source.

Fig. 4.13 Mass and heat state parameters and data of a resorber ($R_{1,gax}$ and $R^c_{1,gax}$)—generator ($G_{2,gax}$ and $G^c_{2,gax}$) structure with full match gax, $q_{R,1,gax} \cong q_{G,2,gax}$, used in a truncation provided with *acr*. Note: the subscript "c" holds true for the superscript "c" in equations

The resorption heat released by $R_{1,gax}$, noted by 17, is transferred directly or through a heat transfer loop to the second stage generator first portion, noted by $G_{2,gax}$ (portion 10–11 at p_2). The remaining second stage generator (portion 11–14) is heated externally. The code of Fig. 4.12 cycle was completed with the gax assessment, given next and using the above nomenclature. The algorithm is a particular case of that which laid down the basis of the DD method presentation (see Chap. 2). Indeed, the $R_{1,gax}$ and $R^c_{1,gax}$ portions in Fig. 4.12, are the main (R^m) and the complementary (R^c) parts of the resorber shown in Fig. 2.3b, respectively. However, given the importance of the gax recovery in a truncated cycle, for the sake of clarity the reader is furnished with its complete calculus mode, useful to the beginning of the cycle design phase. Next, the mass and heat balances are referred to in conjuncture with Fig. 4.13.

$R_{1,gax}$ refrigerant mass balance:

$$Y_{G,1}V_{1,gax} + r_{1,gax}y_{M,1} = r_{R,1,gax}\left(r_{1,gax} + V_{1,gax}\right) \tag{4.44}$$

Resulting equation:

$$r_{1,gax} = V_{1,gax} r_{1,gax,u} \tag{4.45}$$

where:

$$r_{1,gax,u} = \frac{Y_{G,1} - y_{R,1,gax}}{y_{R,1,gax} - y_{M,1}} \tag{4.46}$$

$R_{1,gax}$ heat balance:

$$q_{R,1,gax} = H_{G,1} V_{1,gax} + h_{RI,1} r_{1,gax} - h_{R,1,gax}\left(r_{1,gax} + V_{1,gax}\right) \tag{4.47}$$

Resulting equation:

$$q_{R,1,gax} = V_{1,gax}\left[(H_{G,1} - h_{R,1,gax}) + r_{1,gax,u}(h_{RI,1} - h_{R,1,gax})\right] \tag{4.48}$$

${R_{1,gax}}^c$ refrigerant mass balance:

$$Y_{G,1} V^c_{1,gax} + y_{R,1,gax}\left(r_{1,gax} + V_{1,gax}\right) = y_{RO,1}\left[\left(r_{1,gax} + V_{1,gax}\right) + V^c_{1,gax}\right] \tag{4.49}$$

Resulting equation:

$$r_{1,gax} + V_{1,gax} = V^c_{1,gax} r^c_{1,gax,u}, \tag{4.50}$$

where:

$$r^c_{1,gax,u} = \frac{Y_{G,1} - y_{RO,1}}{y_{RO,1} - y_{R,1,gax}}. \tag{4.51}$$

$R^c_{1,gax}$ heat balance:

$$q^c_{R,1,gax} = H_{G,1} V^c_{1,gax} + h_{R,1,gax}\left(r_{1,gax} + V_{1,gax}\right) - h_{RO,1}\left[\left(r_{1,gax} + V_{1,gax}\right) + V^c_{1,gax}\right] \tag{4.52}$$

Resulting equation:

$$q^c_{R,1,gax} = V^c_{1,gax}\left[\left(H_{G,1} - h_{RO,1}\right) + r^c_{1,gax,u}\left(h_{R,1,gax} - h_{RO,1}\right)\right] \tag{4.53}$$

$V_{1,gax}$ and $V^c_{1,gax}$ calculus:

Mass continuity between the $R_{1,gax}$ and $R^c_{1,gax}$:

$$r_{1,gax,u} V_{1,gax} + V_{1,gax} = r^c_{1,gax,u} V^c_{1,gax}. \tag{4.54}$$

G_1 vapor mass balance:

$$V_{1,gax} + V^c_{1,gax} = V_1. \tag{4.55}$$

In Eq. (4.55), V_1 is the 1st generator vapor, Eq. (4.23). From Eqs. (4.54) and (4.55), it results:

$$V_{1,gax} = V_1 \frac{r^c_{1,gax,u}}{r^c_{1,gax,u} + r_{1,gax,u} + 1}, \tag{4.56}$$

$$V_{1,gax}^{c} = V_1 \frac{r_{1,gax,u} + 1}{r_{1,gax,u}^{c} + r_{1,gax,u} + 1}. \quad (4.57)$$

The gax is designed from the very beginning in such a way that, Fig. 4.12:

$$T_{G,2,gax} = T_{M,1} - \Delta T_{gax,G} \quad (4.58)$$

and

$$T_{R,1,gax} = T_{M,2} + \Delta T_{gax,R} \quad (4.59)$$

The positive items $\Delta T_{gax,G}$ and $\Delta T_{gax,R}$ in Eqs. (4.58) and (4.59), Fig. 4.12, must be chosen in order to ensure a simultaneous fulfilment of the (i) and (ii) conditions (see Chap. 2). Moreover, although these items are not appearing explicitly, they are key parameters in the models to follow. In this respect, the models will work with theirs minimum and maximum values, noted as $\Delta T_{gax,G(R),\min(\max)}$, and with theirs intermediate values. Continuing, from the two Eqs. (4.58) and (4.59) and according to Fig. 4.3b, the operation temperature overlapping (OTO) results in:

$$\Delta T_{OTO} = T_{G,2,gax} - T_{R,1,gax} = T_{M,1} - T_{M,2} - \Delta T_{gax,G} - \Delta T_{gax,R} \quad (4.60)$$

Also, from Fig. 4.12 we can emphasize two Gax Recovery temperature Zones (GRZ), where the gax takes place actually, one for the resorber, $\Delta T_{R,GRZ}$:

$$\Delta T_{R,GRZ} = \left[T_{R,1,gax}, T_{M,1}\right] \quad (4.61)$$

and one for the generator, $\Delta T_{G,GRZ}$:

$$\Delta T_{G,GRZ} = \left[T_{M,2}, T_{G,2,gax}\right] \quad (4.62)$$

marked further in figures of the gax study cases. The two temperature sets (zones) have in common, of course, the ΔT_{OTO} temperature set. First, the unit mass flow factors $r_{1,gax,u}$ and $r_{1,gax,u}^{c}$ are calculated with Eqs. (4.45) and (4.46). Then, the $V_{1,gax}$ and $V_{1,gax}^{c}$ vapors are calculated, using Eqs. (4.56) and (4.57). Second, the gax heat $q_{R,1,gax}$ is computed with Eq. (4.48), from the heat balance Eq. (4.47). Checking the calculus correctness requires, however, to compute the complementary resorption heat $q_{R,1,gax}^{c}$ as well, for comparing the results with the known total first resorption heat $q_{R,1} = q_{R,1,gax} + q_{R,1,gax}^{c}$. The benefit of improving the cycle internal heat management through gax recovery is generally expressed by the COP_c and COP_h equations, given below. They take into account the smaller need of generation heat for the same useful cooling or heating effects, as:
Cooling COP:

$$COP_c = \frac{Useful\_cooling\_effect}{Heat\_input} = \frac{q_D}{\sum_i \left(q_{Gi} - q_{Ri,gax}\right)} \quad (4.63)$$

Heating COP:

$$COP_h = \frac{Useful\_heating\_effect}{Heat\_input} = \frac{q_R}{\sum_i \left(q_{Gi} - q_{Ri,gax}\right)} \quad (4.64)$$

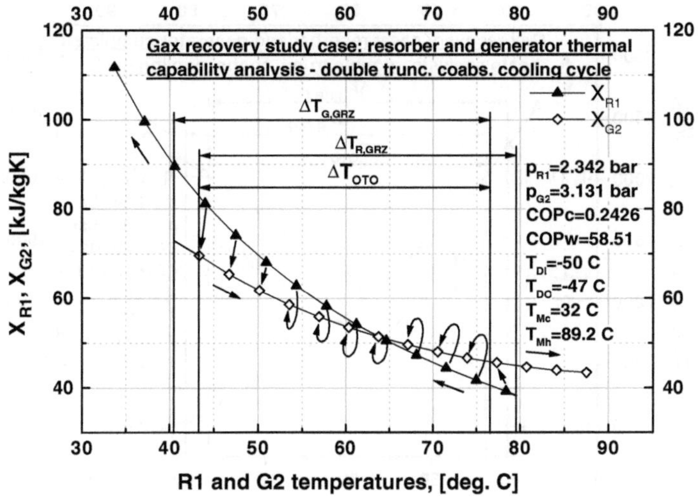

Fig. 4.14 The thermal capability of the 1st stage resorber and 2nd stage generator of Fig. 4.12 versus working temperature (X-plot)

The above algorithm must be completed compulsorily with a detailed analysis of the gax heat exchange. This can be done in the design phase, checking the fulfillment of the conditions (i) and (ii), mentioned previously, in each reasonable large interval of temperature throughout the coupled resorption and generation processes at hand. In our study case and everywhere throughout this book we have done this with the help of the DD method. The thermal capability of the 1st stage resorber and 2nd stage generator of Fig. 4.12 is plotted versus working temperature in Fig. 4.14 (noted as "X-plot" hereinafter). The operation temperature overlapping, marked as OTO, takes place between the points 4 and 11 values in Figs. 4.12, 4.14 and 4.15 that is $T_{R1,gax} = 43.45$ °C and $T_{G2,gax} = 76.65$ °C, respectively. The heat transfer from resorber to the generator in the gax recovery zone can be performed in a possible way, as that suggested by Fig. 4.14. The evolution of the generation and absolute resorption cumulated heat vs. temperature for same study case is plotted in Fig. 4.15 (noted as "q-plot" hereinafter). The gax recovery zone of the resorber, marked by Resorber GRZ, corresponds to points 3 and 4 in Fig. 4.12, and takes place between $T_{M,1} = 79.65$ °C and $T_{R1,gax} = 43.45$ °C. The Generator GRZ ranges between its points 10 and 11, which temperatures of are $T_{M,2} = 40.45$ °C and $T_{G2,gax} = 76.65$ °C, respectively. The results shown by this plot are particularly important, as they indicate to what extent the condition (i) is fulfilled globally. For a full match, the absolute heat released in the 1st stage Resorber GRZ must equal that received in the 2nd stage Generator GRZ, that is (see Fig. 4.15 notations):

$$\left[q_{R,1}\left(T_{R,1,gax}\right)\right]_{p_{R,1}} - \left[q_{R,1}\left(T_{M,1}\right)\right]_{p_{R,1}} = \left[q_{G,2}\left(T_{G,2,gax}\right)\right]_{p_{G,2}} - \left[q_{G,2}\left(T_{M,2}\right)\right]_{p_{G,2}} \tag{4.65}$$

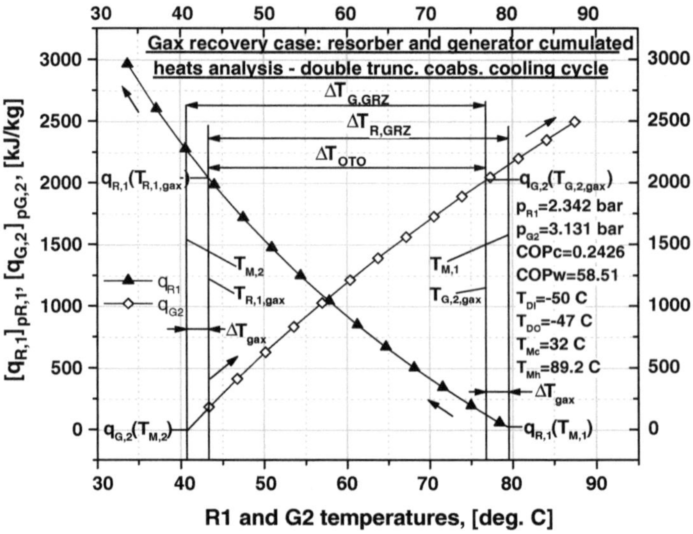

Fig. 4.15 The evolution of the 1st stage resorber and 2nd stage generator cumulated heat versus temperature (q-plot) for Fig. 4.12 study case

Moreover, in this respect, the two Figs. 4.14 and 4.15 must be correlated by the following equations:

$$\left|q_{R,i,gax}\right| = \int_{T_{R,i,gax}}^{T_{M,i}} d\left|q_{R,i,gax}\right| = \int_{T_{R,i,gax}}^{T_{M,i}} \left|X_{R,i,gax}\right| dT$$

$$= \int_{T_{M,j}}^{T_{G,j,gax}} X_{G,j,gax} dT = \int_{T_{M,j}}^{T_{G,j,gax}} dq_{G,j,gax} = q_{G,j,gax}$$

where $j = i \pm 1$ (see next Fig. 4.16 for cooling and heating cycles). In our case, Eq. (4.65) is fulfilled with an accuracy higher than 99.5 %, which is satisfactory. When Eq. (4.65) is not fulfilled satisfactorily, it is possible to match the gax splitting the absorbent flows and appropriately supplying with absorbent and vapor the heat exchanging devices, as suggested by Fig. 4.11b, c. The model is more complicated this time, but could be derived in a way similar to that we followed here, shown next.

4.2.1.3 Partial Match Gax and Capitalization Methods Thereof

A more detailed schematic of the partial match gax is given in Fig. 4.16. The mass and heat balances, which will follow, are done in conjuncture with this figure, for all sub-devices outlined in Fig. 4.11b, c, considered here on a same chart. They

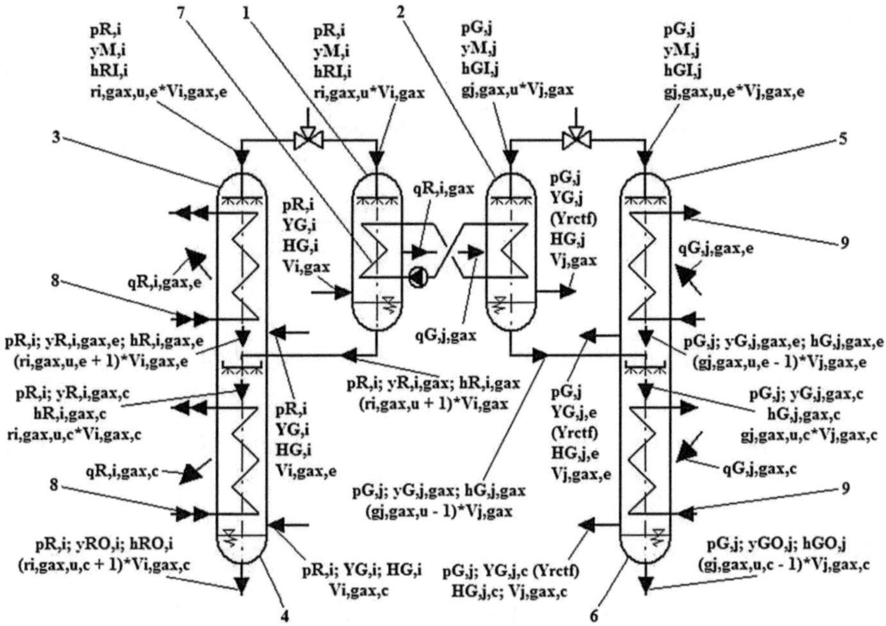

Legend: (1): i-th Gax Resorber (Ri,gax); (2): j-th Gax Generator (Gj,gax); (3): i-th Gax Heat Excess Resorber (Ri,gax,e); (4): i-th Gax Complementary Resorber (Ri,gax,c); (5): j-th Gax Heat Excess Generator (Gj,gax,e); (6): j-th Gax Complementary Generator (Gj,gax,c); (7): Gax Heat Transfer Loop; (8): Sink Source; (9): Heat Source. Note: j=i+1 in case of a refrigeration cycle and j=i-1 in case of a heating cycle.

Fig. 4.16 Mass and heat state parameters and data of the i-th resorber ($R_{i,gax}$, $R^e_{i,gax}$ and $R^c_{i,gax}$) coupled with the j-th generator ($G_{j,gax}$, $G^e_{j,gax}$ and $G^c_{j,gax}$), in a configuration enabling a partial match gax operation, $q_{R,i,gax} = q_{G,j,gax}$, used in the truncation provided with *acr*. *Note* the subscripts "c" and "e" hold true for the superscripts "c" and "e" in equations

will serve to model later separately both RHE and GHE problems, which the partial match gax is confronted with.

Gax resorber-pos. 1:

Mass balance:

$$y_{M,i} r_{i,gax,u} V_{i,gax} + Y_{G,i} V_{i,gax} = y_{R,i,gax}\left(r_{i,gax,u} + 1\right) V_{i,gax} \tag{4.66}$$

$$r_{i,gax,u} = \frac{Y_{G,i} - y_{R,i,gax}}{y_{R,i,gax} - y_{M,i}} \tag{4.67}$$

Heat balance:

$$q_{R,i,gax} + h_{R,i,gax}\left(r_{i,gax,u} + 1\right) V_{i,gax} = H_{G,i} V_{i,gax} + h_{RI,i} r_{i,gax,u} V_{i,gax} \tag{4.68}$$

$$q_{R,i,gax} = V_{i,gax}\left[\left(H_{G,i} - h_{R,i,gax}\right) + r_{i,gax,u}\left(h_{RI,i} - h_{R,i,gax}\right)\right] \tag{4.69}$$

Gax heat excess resorber-pos. 3:

Mass balance:

$$y_{M,i} r^e_{i,gax,u} V^e_{i,gax} + Y_{G,i} V^e_{i,gax} = y^e_{R,i,gax} \left(r^e_{i,gax,u} + 1 \right) V^e_{i,gax} \tag{4.70}$$

$$r^e_{i,gax,u} = \frac{Y_{G,i} - y^e_{R,i,gax}}{y^e_{R,i,gax} - y_{M,i}} \tag{4.71}$$

Heat balance:

$$q^e_{R,i,gax} + h^e_{R,i,gax} \left(r^e_{i,gax,u} + 1 \right) V^e_{i,gax} = H_{G,i} V^e_{i,gax} + h_{RI,i} r^e_{i,gax,u} V^e_{i,gax} \tag{4.72}$$

$$q^e_{R,i,gax} = V^e_{i,gax} \left[\left(H_{G,i} - h^e_{R,i,gax} \right) + r^e_{i,gax,u} \left(h_{RI,i} - h^e_{R,i,gax} \right) \right] \tag{4.73}$$

Gax complementary resorber-pos. 4:

Mass balance:

$$y^c_{R,i,gax} r^c_{i,gax,u} V^c_{i,gax} + Y_{G,i} V^c_{i,gax} = y_{RO,i} \left(r^c_{i,gax,u} + 1 \right) V^c_{i,gax} \tag{4.74}$$

$$r^c_{i,gax,u} = \frac{Y_{G,i} - y_{RO,i}}{y_{RO,i} - y^c_{R,i,gax}} \tag{4.75}$$

Heat balance on pos. 1 and 3 outlet and pos. 4 inlet:

$$\left(r^e_{i,gax,u} + 1 \right) V^e_{i,gax} h^e_{R,i,gax} + \left(r_{i,gax,u} + 1 \right) V_{i,gax} h_{R,i,gax} = r^c_{i,gax,u} V^c_{i,gax} h^c_{R,i,gax} \tag{4.76}$$

$$h^c_{R,i,gax} = \frac{\left(r^e_{i,gax,u} + 1 \right) V^e_{i,gax} h^e_{R,i,gax} + \left(r_{i,gax,u} + 1 \right) V_{i,gax} h_{R,i,gax}}{r^c_{i,gax,u} V^c_{i,gax}} \tag{4.77}$$

Heat balance on pos. 4:

$$q^c_{R,i,gax} + h_{RO,i} \left(r^c_{i,gax,u} + 1 \right) V^c_{i,gax} = H_{G,i} V^c_{i,gax} + h^c_{R,i,gax} r^c_{i,gax,u} V^c_{i,gax} \tag{4.78}$$

$$q^c_{R,i,gax} = V^c_{i,gax} \left[\left(H_{G,i} - h_{RO,i} \right) + r^c_{i,gax,u} \left(h^c_{R,i,gax} - h_{RO,i} \right) \right] \tag{4.79}$$

Gax generator-pos. 2:

Mass balance:

$$y_{M,j} g_{j,gax,u} V_{j,gax} = Y_{G,j} V_{j,gax} + y_{G,j,gax} \left(g_{j,gax,u} - 1 \right) V_{j,gax} \tag{4.80}$$

$$g_{j,gax,u} = \frac{Y_{G,j} - y_{G,j,gax}}{y_{M,j} - y_{G,j,gax}} \tag{4.81}$$

Heat balance:

$$q_{G,j,gax} + h_{GI,j} g_{j,gax,u} V_{j,gax} = H_{G,j} V_{j,gax} + h_{G,j,gax} \left(g_{j,gax,u} - 1 \right) V_{j,gax} \quad (4.82)$$

$$q_{G,j,gax} = V_{j,gax} \left[\left(H_{G,j} - h_{G,j,gax} \right) + g_{j,gax,u} \left(h_{G,j,gax} - h_{GI,j} \right) \right] \quad (4.83)$$

Gax excess heat generator-pos. 5:

Mass balance:

$$y_{M,j} g^e_{j,gax,u} V^e_{j,gax} = Y^e_{G,j} V^e_{j,gax} + y^e_{G,j,gax} \left(g^e_{j,gax,u} - 1 \right) V^e_{j,gax} \quad (4.84)$$

$$g^e_{j,gax,u} = \frac{Y^e_{G,j} - y^e_{G,j,gax}}{y_{M,j} - y^e_{G,j,gax}} \quad (4.85)$$

Heat balance:

$$q^e_{G,j,gax} + h_{GI,j} g^e_{j,gax,u} V^e_{j,gax} = H^e_{G,j} V^e_{j,gax} + h^e_{G,j,gax} \left(g^e_{j,gax,u} - 1 \right) V^e_{j,gax} \quad (4.86)$$

$$q^e_{G,j,gax} = V^e_{j,gax} \left[\left(H^e_{G,j} - h^e_{G,j,gax} \right) + g^e_{j,gax,u} \left(h^e_{G,j,gax} - h_{GI,j} \right) \right] \quad (4.87)$$

Gax complementary generator-pos. 6:

Mass balance:

$$y^c_{G,j,gax} g^c_{j,gax,u} V^c_{j,gax} = Y_{G,j} V^c_{j,gax} + y_{GO,j} \left(g^c_{j,gax,u} - 1 \right) V^c_{j,gax} \quad (4.88)$$

$$g^c_{j,gax,u} = \frac{Y^c_{G,j} - y_{GO,j}}{y^c_{G,j,gax} - y_{GO,j}} \quad (4.89)$$

Heat balance on pos. 2 and 5 outlets and pos. 6 inlet:

$$\left(g^e_{j,gax,u} - 1 \right) V^e_{j,gax} h^e_{G,j,gax} + \left(g_{j,gax,u} - 1 \right) V_{j,gax} h_{G,j,gax} = g^c_{j,gax,u} V^c_{j,gax} h^c_{G,j,gax} \quad (4.90)$$

$$h^c_{G,j,gax} = \frac{\left(g^e_{j,gax,u} - 1 \right) V^e_{j,gax} h^e_{G,j,gax} + \left(g_{j,gax,u} - 1 \right) V_{j,gax} h_{G,j,gax}}{g^c_{j,gax,u} V^c_{j,gax}} \quad (4.91)$$

Heat balance on pos. 6:

$$q^c_{G,j,gax} + h^c_{G,j,gax} g^c_{j,gax,u} V^c_{j,gax} = H^c_{G,j} V^c_{j,gax} + h_{GO,j} \left(g^c_{j,gax,u} - 1 \right) V^c_{j,gax} \quad (4.92)$$

$$q^c_{G,j,gax} = V^c_{j,gax} \left[\left(H^c_{G,j} - h_{GO,j} \right) + g^c_{j,gax,u} \left(h_{GO,j} - h^c_{G,j,gax} \right) \right] \quad (4.93)$$

The cycle COP_c and COP_h, calculated by Eqs. (4.63) and (4.64) hold true in case of the cycle with RHE (or GHE) gax, provided that $q_{R,i,gax} = q_{G,j,gax}$, according to Eq. (4.96).

Note: If the j-th stage generated vapor are rectified up to an Y_{rctf} value, then in equations above the $Y_{G,j}$, $Y^e_{G,j}$ and $Y^c_{G,j}$ vapor concentrations values are replaced by Y_{rctf}.

Vapor calculation
Mass continuity between pos. 1 and 3 outlets and pos. 4 inlet:

$$\left(r^e_{i,gax,u} + 1\right)V^e_{i,gax} + \left(r_{i,gax,u} + 1\right)V_{i,gax} = r^c_{i,gax,u}V^c_{i,gax} \tag{4.94}$$

Refrigerant species mass continuity between pos. 1 and 3 outlets and pos. 4 inlet:

$$\left(r^e_{i,gax,u} + 1\right)V^e_{i,gax}y^e_{R,i,gax} + \left(r_{i,gax,u} + 1\right)V_{i,gax}y_{R,i,gax} = r^c_{i,gax,u}V^c_{i,gax}y^c_{R,i,gax} \tag{4.95}$$

Partial match gax capitalization condition in order that the gax resorber pos. 1 be designed to deliver exactly the maximum amount of heat acceptable by the gax generator pos. 2, in case of RHE or for the gax generator pos. 2 be designed to need exactly the maximum amount of heat deliverable by the gax resorber pos. 1, in case of GHE:

$$q_{R,i,gax} = q_{G,j,gax} \tag{4.96}$$

or

$$V_{i,gax}\left[q_{R,i,gax}\right] = V_{j,gax}\left[q_{G,j,gax}\right] \tag{4.97}$$

where

$$\left[q_{R,i,gax}\right] = \left(H_{G,i} - h_{R,i,gax}\right) + r_{i,gax,u}\left(h_{RI,i} - h_{R,i,gax}\right) \tag{4.98}$$

and

$$\left[q_{G,j,gax}\right] = \left(H_{G,j} - h_{G,j,gax}\right) + g_{j,gax,u}\left(h_{G,j,gax} - h_{GI,j}\right) \tag{4.99}$$

G_i vapor mass balance:

$$V_{i,gax} + V^e_{i,gax} + V^c_{i,gax} = V_i \tag{4.100}$$

Mass continuity between pos. 2 and 5 outlets and pos. 6 inlet:

$$\left(g^e_{j,gax,u} - 1\right)V^e_{j,gax} + \left(g_{j,gax,u} - 1\right)V_{j,gax} = g^c_{j,gax,u}V^c_{j,gax} \tag{4.101}$$

Refrigerant species mass continuity of between pos. 2 and 5 outlets and pos. 6 inlet:

$$\left(g^e_{j,gax,u} - 1\right)V^e_{j,gax}y^e_{G,j,gax} + \left(g_{j,gax,u} - 1\right)V_{j,gax}y_{G,j,gax} = g^c_{j,gax,u}V^c_{j,gax}y^c_{G,j,gax} \tag{4.102}$$

G_j vapor mass balance:

$$V_{j,gax} + V^e_{j,gax} + V^c_{j,gax} = V_j \tag{4.103}$$

Resorber Heat Excess (RHE) Gax Operation Model

In this case, the i-th resorber is capable to eliminate in the gax recovery zone more heat than the j-th generator needs in the same zone. Theoretically, this intrinsic problem, specific mostly to the refrigeration cycles ($j = i + 1$), could be partially overcome, splitting the gax resorber in two paralel devices, $R_{i,gax}$ and $R^e_{i,gax}$ (see Fig. 4.11b). In this configuration, the $R_{i,gax}$ device is designed in such a way that it delivers in a recovering way an amount of heat equal to that needed by the gax generator $G_{j,gax}$, that is $q_{R,i,gax} = q_{G,j,gax}$, according to Eq. (4.96). The useful gax resorption heat $q_{R,i,gax}$ is rejected during the resorption of the $V_{i,gax}$ vapor amount, while the $q_{G,j,gax}$ heat input is generating the $V_{j,gax}$ vapor amount. The gax heat transfer is considered to be done similarly to the full match gax case, Fig. 4.13, with the help of a heat transfer closed loop (pos. 7, Fig. 4.16). The extra-heat of the i-th gax resorber, $q^e_{R,i,gax}$, generated through the absorption of the $V^e_{i,gax}$ vapor amount, is eliminated in the gax heat excess resorber, $R^e_{i,gax}$ (pos. 3, Fig. 4.16), by means of the sink source (pos. 8, Fig. 4.16). The absorbents coming of $R_{i,gax}$ and $R^e_{i,gax}$ are mixed up and supply together the rest of the resorber sub-device, $R^c_{i,gax}$ (pos. 4, Fig. 4.16), named similarly as that in Fig. 4.13, pos. 4, the i-th gax complementary resorber and playing the role of finishing the i-th resorption process, through the $V^c_{i,gax}$ vapor amount resorption. Figure 4.16 includes also the position 4, representing the j-th gax heat excess generator, $G^e_{j,gax}$. Given the case at hand, the Eqs. (4.94)–(4.103) are solved together with the obvious Eq. $q^e_{G,j,gax} = 0$, therefore:

$$V^e_{j,gax} = 0 \tag{4.104}$$

and consequently the $G^e_{j,gax}$ sub-device vanishes. The j-th generation process is ending in our case in the $G^c_{j,gax}$ sub-device, that is the j-th gax complementary generator, pos. 6, Fig. 4.16, heated externally by the heat source 9 and generating the vapor amount $V^c_{j,gax}$. Introducing Eq. (4.104) in Eqs. (4.101) and (4.102), it results:

$$y_{G,j,gax} = y^c_{G,j,gax} \tag{4.105}$$

Taking into account Eq. (4.105), the independent equations remaining out of Eqs. (4.94)–(4.103) form a nonlinear implicit system having as solution the unknown variables $V_{i,gax}$, $V^e_{i,gax}$, $V^c_{i,gax}$, $V_{j,gax}$, $V^c_{i,gax}$ and $y^e_{R,i,gax} = y^e_{R,i,gax}\left(V^e_{i,gax}\right)$. The solution depends on the unknown absorbent concentration, $y^e_{R,i,gax}$, recorded to the outlet of the i-th gax heat excess resorber (pos. 3, Fig. 4.16). It can be chosen at convenience, $y^e_{R,i,gax} \in \left(y_{M,i}, y_{RO,i}\right)$. However, the computation is considerably simplified if we considered that:

$$y^e_{R,i,gax} = y_{R,i,gax} = y^c_{R,i,gax} \tag{4.106}$$

without to influence our results for the case at hand. The first equality in Eq. (4.106) leads to the following result [see Eqs. (4.67) and (4.71)]:

$$r^e_{i,gax,u} = r_{i,gax,u} \tag{4.107}$$

Next we shall work with Eq. (4.107), and applying Eqs. (4.104)–(4.107) to the system of Eqs. (4.94)–(4.103), this becomes simpler:

$$\left(r^e_{i,gax,u} + 1\right) V^e_{i,gax} + \left(r_{i,gax,u} + 1\right) V_{i,gax} = r^c_{i,gax,u} V^c_{i,gax} \tag{4.108}$$

$$V_{i,gax}\left[q_{R,i,gax}\right] = V_{j,gax}\left[q_{G,j,gax}\right] \tag{4.109}$$

$$V_{i,gax} + V^e_{i,gax} + V^c_{i,gax} = V_i \tag{4.110}$$

$$\left(g_{j,gax,u} - 1\right) V_{j,gax} = g^c_{j,gax,u} V^c_{j,gax} \tag{4.111}$$

$$V_{j,gax} + V^c_{j,gax} = V_j \tag{4.112}$$

Solving the system of Eqs. (4.108)–(4.112), it is obtained a closed form solution, the only one out of the partial match gax analysis solutions:

$$V_{i,gax} = V_j \frac{g^c_{j,gax,u}}{g_{j,gax,u} + g^c_{j,gax,u} - 1} \frac{\left[q_{G,j,gax}\right]}{\left[q_{R,i,gax}\right]} \tag{4.113}$$

$$V^e_{i,gax} = V_i \frac{r^c_{i,gax,u}}{r_{i,gax,u} + r^c_{i,gax,u} + 1} - V_{i,gax} \tag{4.114}$$

$$V^c_{i,gax} = V_i \frac{r_{i,gax,u} + 1}{r_{i,gax,u} + r^c_{i,gax,u} + 1} \tag{4.115}$$

$$V_{j,gax} = V_j \frac{g^c_{j,gax,u}}{g_{j,gax,u} + g^c_{j,gax,u} - 1} \tag{4.116}$$

$$V^c_{j,gax} = V_j \frac{g_{j,gax,u} - 1}{g_{j,gax,u} + g^c_{j,gax,u} - 1} \tag{4.117}$$

RHE-Gax problem study cases. Next, we shall present comparatively a first set of three partial match gax study cases. These are numbered by 1st to 3rd, and are included in the further coming Results of the RHE-Gax Model Run with $\Delta T_{gax,R,min}$–Infinite Equivalent Solutions to a RHE-Gax Problem, respectively. All study cases are facing a same RHE-Gax problem, met for example between the 1st

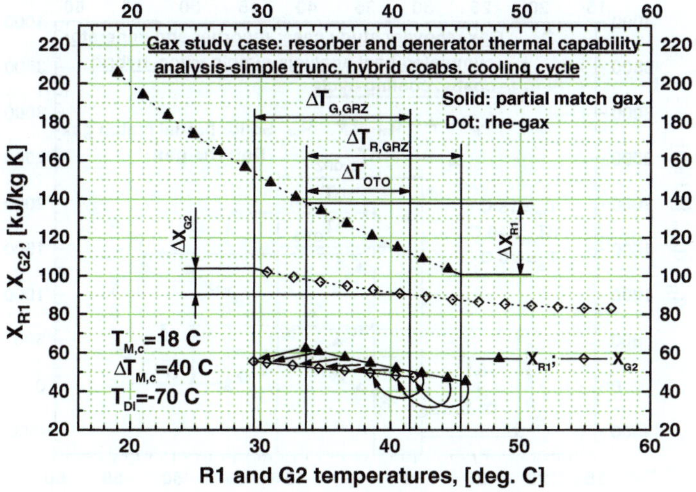

Fig. 4.17 The X-plot of the 1st stage resorber, 2nd stage generator and of the gax thereof belonging to a partial match gax study case of a RHE problem, met to a simple truncated hybrid coabsorbent cooling cycle, Fig. 4.44, and analyzed with the RHE-Gax model run with $\Delta T_{gax,R} = \Delta T_{gax,R,\min}$

($i = 1$), resorber, $R_{1,gax}$ and the 2nd ($j = 2$), generator, $G_{2,gax}$ of the simple truncated hybrid coabsorbent cooling cycle (see Fig. 4.44 of Sect. 4.3.2). The main input data of all study cases are: T_{DI}/T_{DO}=−70 °C/−67 °C, $T_{M,c}$ = 18 °C, $T_{M,h}$ = 58 °C, $T_{M,1}$ = 45.4 °C, $T_{M,2}$ = 29.46 °C. The cycle uses rectification, Y_{rctf} = 0.998, and the effectivenesses are COP_w = 1.558, COP_c = 0.2625 and COP_{mvc} = 1.065.

Results of the RHE-Gax Model Run with $\Delta T_{gax,R,min}$

The 1st study case analyzes the solution of the RHE-Gax problem using the model given in the *Resorber Heat Excess (RHE) Gax* Operation Model (see Fig. 4.11b also). As we expect, in this case the model is run with $\Delta T_{gax,R} = \Delta T_{gax,R,\min}$, therefore $\Delta T_{R,GRZ}$ is maximum, $\Delta T_{R,GRZ} = \Delta T_{R,GRZ,\max}$, Eq. (4.61). Figures 4.17 and 4.18 show the X-plot and q-plot, respectively, before (dotted line plot) and after (solid line plot) applying the partial match gax method. The operation temperature overlapping is small, it takes place between $T_{R1,gax}$ = 33.46 °C and $T_{G2,gax}$ = 41.4 °C, only, considering that $\Delta T_{gax,R,\min} = \Delta T_{gax,G}$=4 °C. Without applying the method, the X-plot and q-plot reveal the comparative values, $X_{R1,gax} \in \Delta X_{R1}$ = [137, 102] versus $X_{G2,gax} \in \Delta X_{G2}$ = [113, 91], and $q_{R1,rhe\text{-}gax} \cong$ 1,430 − 30 = 1,400 versus $q_{G2,rhe\text{-}gax} \cong$ 630 − (− 530) = 1,160, respectively. These values show that the resorber and the generator sub-devices have a visible mutual thermal mismatch, which entitles us to appeal to the gax method at hand. The gax resorber (pos. 6 in Fig. 4.11b) will absorb a vapor amount of $V_{1,gax}$ = 0.2503 [Eq. (4.113)], only, the excess amount, $V^{e}_{1,gax}$ = 0.3014 [Eq. (4.114)] being absorbed in the gax heat excess resorber (pos. 8 in Fig. 4.11b). In this case, the heat delivered

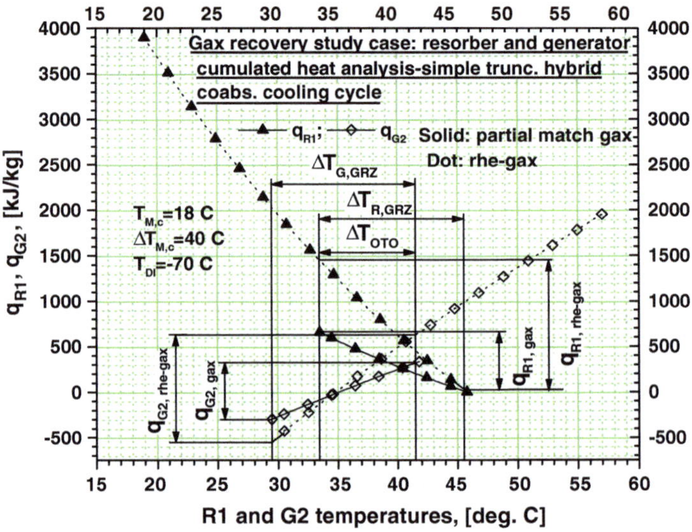

Fig. 4.18 The q-plot for same as in Fig. 4.17

by the gax resorber $R_{1,gax}$ will diminish to a value equal to that needed by the gax generator $G_{2,gax}$, that is $q_{R,1,gax} = q_{G,2,gax} \cong 657.2$ [Eq. (4.96) and Fig. 4.18], which generates the $V_{2,gax} = 0.4898$ vapor amount [Eq. (4.116)], while $V^e_{2,gax} = 0.0$ [Eq. (4.104)]. The heat of the gax heat excess resorber, $q^e_{R,1,gax} = 791.3$, is the highest out of all configurations analyzed and is eliminated in $R^e_{1,gax}$, by the sink source. The gax complementary resorber $R^c_{1,gax}$ will absorb the rest of vapor $V^c_{1,gax} = 1.064$ [Eq. (4.115)], out of the 1st stage generated vapor amount of $V_1 = 1.616$, and rejects to the sink a heat amount of $q^c_{R,i,gax} = 2656$. The gax generator $G^c_{2,gax}$ will generate the rest of vapor $V^c_{2,gax} = 0.5202$ [Eq. (4.117)], out of the 2st stage generated vapor amount of $V_2 = 1.01$, consuming a heat source amount of $q^c_{G,2,gax} = 1501.8$.

Summarizing, the method application enables in theory to rectify partially a RHE-Gax problem, recovering the gax resorber heat for the benefit of the cooling cycle effectiveness. Here, this increased by more than 12 %, i.e. from $COP_c = 0.234$ to $COP_c = 0.2625$.

Results of the RHE-Gax Model Run with $\Delta T_{gax,R,max}$

A 2nd study case analyzes the solution of the RHE-Gax problem mentioned priorly, making use of the full match gax configuration, outlined in Fig. 4.11a. It can be obtained from the partial match gax configuration, Fig. 4.16, making $V^e_{i,gax} = 0$ and dimensioning correspondingly the $y_{GO,1}$ temperature item and further $\Delta T_{R,GRZ}$ [see Fig. 4.12 and Eq. (4.61)], in order that Eq. (4.96) be fulfilled. In this particular case, one might expect that $\Delta T_{gax,R}$ takes its highest value, $\Delta T_{gax,R} = \Delta T_{gax,R,\max}$, therefore $\Delta T_{R,GRZ}$ will have a minimum value, $\Delta T_{R,GRZ} = \Delta T_{R,GRZ,\min}$. The solvers of our study case are equally the models of the full match gax model [see

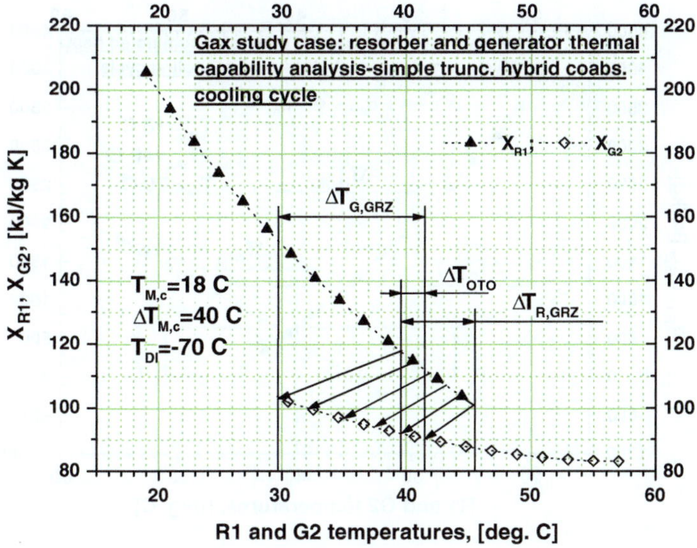

Fig. 4.19 The X-plot of the study case of Fig. 4.17, solving the RHE-Gax problem when considering that $\Delta T_{gax,R} = \Delta T_{gax,R,\max}$

Eqs. (4.44)–(4.65)] and the model given to the *Resorber Heat Excess (RHE) Gax Operation Model*, because the latter reduces to the former when $V^e_{i,gax} = 0$ and Eqs. (4.113) and (4.114) become Eqs. (4.96) and (4.56), respectively. This time, the solution of the problem at hand cannot be obtained in a closed form. Indeed, the $\Delta T_{gax,R,\max}$ value can be found through a trial-and-error calculus, only. Each chosen $\Delta T_{gax,R}$ proposes an $\Delta T_{R,i,gax}(\Delta T_{gax,R,\max})$ concentration variable value, and the trial-and-error calculus ends when the heat of the gax resorber, $q_{R,i,gax}$, computed with $h_{R,i,gax} = \left[h_{R,i,gax}\left(y_{R,i,gax}\right)\right]_{PR}$ and $r_{i,gax,u} = \left[r_{i,gax,u}\left(y_{R,i,gax}\right)\right]_{PR}$ functions [see Eq. (4.48) where $i = 1$], satisfies Eq. (4.96). As one can remark, this type of gax with $\Delta T_{gax,R,\max}$ is nothing else but the single gax solution, which the classic configuration of Fig. 4.11a is capable to offer in an RHE-Gax working mode.

The X-plot and q-plot of Figs. 4.17 and 4.18 are repeated in Figs. 4.19 and 4.20, respectively. The diminution of the $R_{1,gax}$ resorber heat, in order to match the heat needed by the generator $G_{2,gax}$, has been done by restricting the $\Delta T_{R,GRZ}$ domain (set) from $\Delta T_{R,GRZ,\max}$=[29.46, 45.4] (see Figs. 4.17 and 4.18) to $\Delta T_{R,GRZ,\min}$ = [39.66,45.4] (see Figs. 4.19 and 4.20). The restriction was achieved choosing a higher value for $\Delta T_{gax,R}$, comparatively, while keeping the same 1st study case $\Delta T_{gax,G} = 4$ °C value. The trial-and-error calculus ended when the guessed values, $\Delta T_{gax,R,\max}$=10.2 °C and $y_{R,1,gax} = 0.1264$, resulted in Eq. (4.96) fulfiling, $q_{R,1,gax} = q_{G,2,gax} \cong 657.2$. The accepted value, $\Delta T_{gax,R,\max} = 10.2 > 4$ °C, leads to both (i) and (ii) heat transfer conditions fulfilling, as priorly stated and proved by Figs. 4.20 and 4.19 plots, respectively.

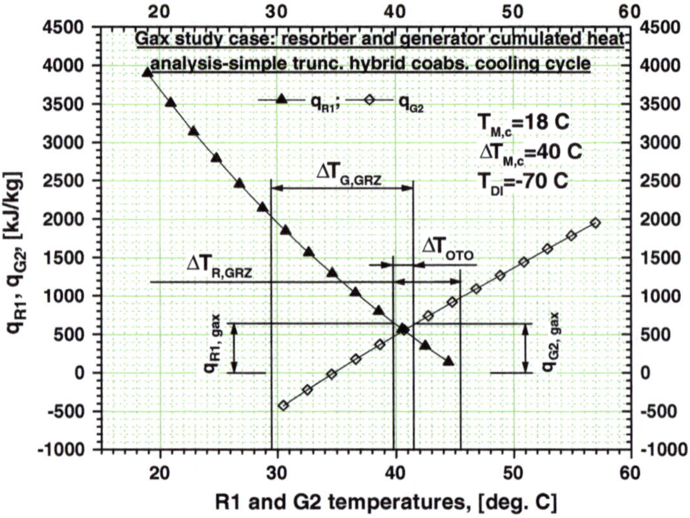

Fig. 4.20 The same as in Fig. 4.19, for the q-plot

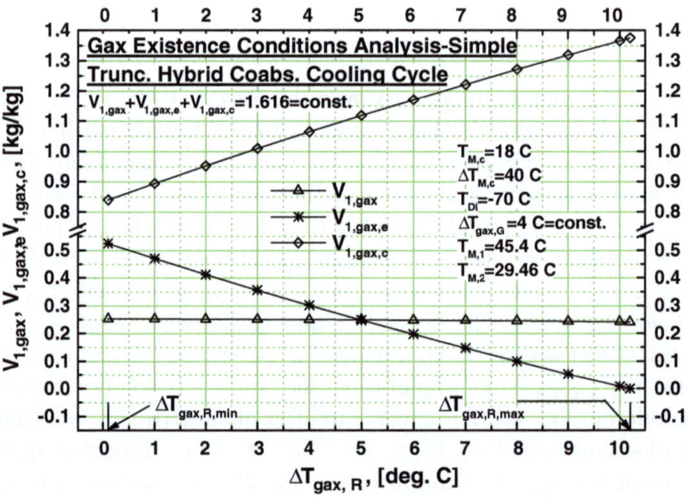

Fig. 4.21 Partial results plot of infinite equivalent solutions to the RHE-Gax problem mentioned priorly, shown for all resorber sub-device's vapor versus $\Delta T_{gax,R}$

Infinite Equivalent Solutions to a RHE-Gax Problem

According to a 3rd study case, running the *Resorber Heat Excess (RHE) Gax* Operation Model model for the configuration outlined in Fig. 4.11b and with intermediate $\Delta T_{gax,R}$ values, $\Delta T_{gax,R,\max} \leq \Delta T_{gax,R} \leq \Delta T_{gax,R,\max}$, an infinite number of equivalent solutions for the same RHE-Gax problem mentioned priorly will result. Partial results thereof are given in Figs. 4.21 and 4.22 for the resorber vapor

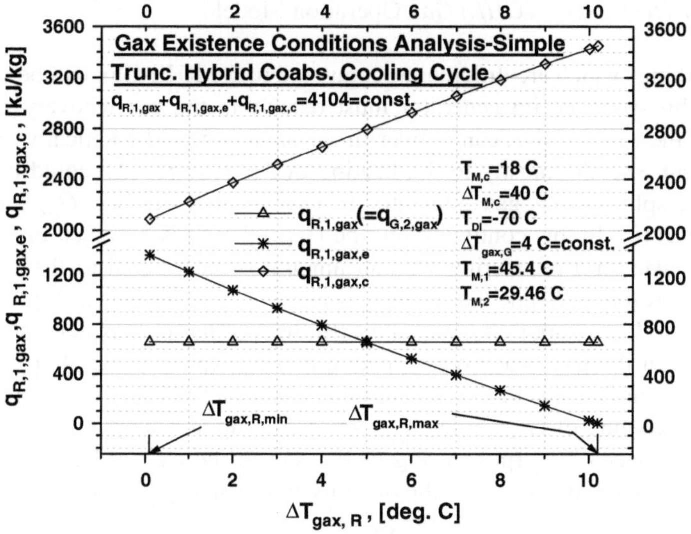

Fig. 4.22 Partial results of infinite equivalent solutions to the RHE-Gax problem mentioned priorly, shown for all resorber sub-device's heat versus $\Delta T_{gax,R}$

and heat, respectively and $\Delta T_{gax,R,\min} = 1$ °C and $\Delta T_{gax,R,\max} = 10.2$ °C. All solutions respect the (ii) heat transfer condition (provided that the $\Delta T_{gax,R,\min}$ value is acceptable). Moreover, Fig. 4.22 proves that the (i) condition (Eq. (4.96)) is fulfilled as well, because the heat of the gax heat excess resorber and gax complementary resorber sub-devices are continuous decreasing and increasing complementary functions, respectively, enabling the gax resorber heat be kept constant and equal to that of the gax generator, for each $\Delta T_{gax,R}$ variable value. The results obtained above are not confined to the particular RHE-Gax problem at hand, they are general. Expressed otherwise, these results emphasize that, for a given $q_{G,j,gax}$ generation heat and using the configuration outlined in Fig. 4.11b, it is always possible to find an infinite number of gax solutions for a RHE-Gax problem, having $q_{R,i,gax} = q_{G,j,gax}$ and being characterized by Gax Recovery Zones, $\Delta T_{R,GRZ}$, which respect the double inequality below:

$$\begin{aligned}\left[T_{R,i,gax,\max}, T_{M,i}\right] = \Delta T_{R,GRZ,\min} \le \Delta T_{R,GRZ} \le \Delta T_{R,GRZ,\max} \\ = \left[T_{R,i,gax,\min}, T_{M,i}\right]\end{aligned} \tag{4.118}$$

where

$$T_{R,i,gax,max} = T_{M,j} + \Delta T_{gax,R,\max} \tag{4.119}$$

and

$$T_{R,i,gax,\min} = T_{M,j} + \Delta T_{gax,R,\min} \tag{4.120}$$

Generator Heat Excess (GHE) Gax Operation Model

Only the things which are different as compared to the RHE-Gax will be considered here. In this case, the jth generator must receive in the gax recovery zone more heat than the i-th resorber can deliver in the same zone. Theoretically, this intrinsic problem, specific mostly to the heating cycles (j = i − 1), could be partially overcome, splitting the gax generator in two parallel devices, $G_{j,gax}$ and $G^e_{j,gax}$ (see Fig. 4.11c). In this configuration, the $G_{j,gax}$ device is designed in such a way that it receives in a recovering way an amount of heat equal to that deliverable by the gax resorber $R_{i,gax}$, that is $q_{R,i,gax} = q_{G,j,gax}$, according to Eq. (4.96), and produces the $V_{j,gax}$ vapor amount. The extra-heat of the jth gax generator, $q^e_{G,j,gax}$, is received in the gax heat excess generator, $G^e_{j,gax}$ (pos. 5, Fig. 4.16), by means of the heat source (pos. 9, Fig. 4.16), and generates the $V^e_{j,gax}$ vapor amount. The absorbents coming of $G_{j,gax}$ and $G^e_{j,gax}$ are mixed up and supply together the generator sub-device $G^c_{j,gax}$ (pos. 6, Fig. 4.16). Given the case at hand, Eqs. (4.94)–(4.103) are solved together with the obvious equation $q^e_{R,i,gax} = 0$, therefore:

$$V^e_{i,gax} = 0 \tag{4.121}$$

and consequently the $R^e_{i,gax}$ sub-device vanishes. Introducing Eq. (4.121) in Eqs. (4.94) and (4.95), it results:

$$y_{R,i,gax} = y^c_{R,i,gax} \tag{4.122}$$

Taking into account Eq. (4.122), the independent equations remaining out of Eqs. (4.94)–(4.103) form a nonlinear implicit system having as solution the unknown variables $V_{i,gax}$, $V^c_{i,gax}$, $V_{j,gax}$, $V^e_{j,gax}$, $V^c_{i,gax}$ and $y^e_{G,j,gax} = y^e_{G,j,gax}\left(V^e_{j,gax}\right)$. The solution depends on the unknown absorbent concentration, $y_{G,j,gax}{}^e$, recorded to the outlet of the j-th gax heat excess generator (pos. 5, Fig. 4.16). It can be chosen at convenience, $y^e_{G,j,gax} \in (y_{M,j}, y_{GO,i})$. However, again, similar to the RHE-Gax case, the computation is considerably simplified if we considered this time that:

$$y^e_{G,j,gax} = y_{G,j,gax} = y^c_{G,j,gax} \tag{4.123}$$

without to influence our further results. The first equality in Eq. (4.123) leads to the following result [see Eqs. (4.81) and (4.85)]:

$$g^e_{j,gax,u} = g_{j,gax,u} \tag{4.124}$$

Next we shall work with Eq. (4.124). Applying Eqs. (4.121)–(4.124) to the system of Eqs. (4.94)–(4.103), this becomes simpler and the solution writes in a closed form in case of GHE gax as:

$$V_{i,gax} = V_i \frac{r^c_{i,gax,u}}{r_{i,gax,u} + r^c_{i,gax,u} + 1} \tag{4.125}$$

$$V^c_{i,gax} = V_i \frac{r_{i,gax,u} + 1}{r_{i,gax,u} + r^c_{i,gax,u} + 1} \tag{4.126}$$

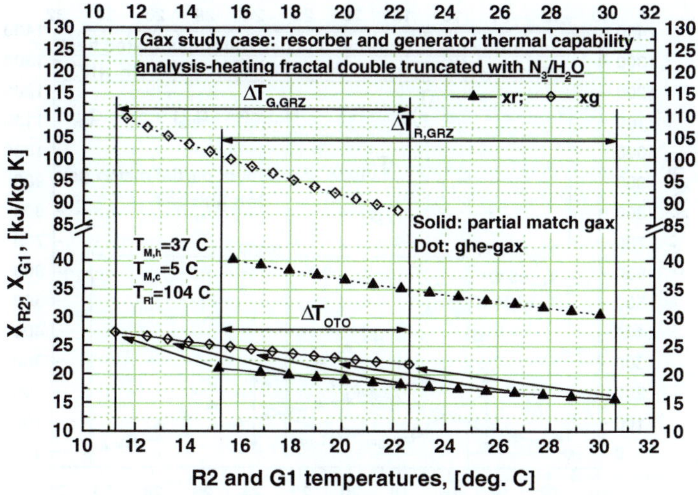

Fig. 4.23 The X-plot of the 1st stage generator, 2nd stage resorber and of the gax thereof belonging to a partial match gax study case of a GHE problem, met to an NH_3–H_2O heating double truncated fractal and run with $\Delta T_{gax,G} = \Delta T_{gax,G,\min}$

$$V_{j,gax} = V_i \frac{r^c_{i,gax,u}}{r_{i,gax,u} + r^c_{i,gax,u} + 1} \frac{[q_{R,i,gax}]}{[q_{G,j,gax}]} \tag{4.127}$$

$$V^e_{j,gax} = V_j \frac{g^c_{j,gax,u}}{g_{j,gax,u} + g^c_{j,gax,u} - 1} - V_{j,gax} \tag{4.128}$$

$$V^c_{j,gax} = V_j \frac{g_{j,gax,u} - 1}{g_{j,gax,u} + g^c_{j,gax,u} - 1} \tag{4.129}$$

GHE-Gax problem study cases. Next, we shall present comparatively a second set of three partial match gax study cases. These are numbered by 4th–6th, and are included in the further coming Results of the RHE-Gax Model Run with $\Delta T_{gax,R,min}$–Infinite Equivalent Solutions to A GHE-Gax Problem, respectively. All study cases are facing a same GHE-Gax problem, met for example between the 1st ($j = i - 1 = 1$), generator, $G_{1,gax}$ and the 2nd ($i = 2$), resorber, $R_{2,gax}$ of the double truncated coabsorbent heating cycle (see Fig. 4.37a of Sect. 4.2.3). The main input data of all study cases are: $T_{RI}/T_{Rf,O}$ = (104/74) °C, $T_{M,c}$ = 5 °C, $T_{M,h}$ = 37 °C, $T_{M,1}$ = 11.23 °C, $T_{M,2}$ = 30.44 °C. The effectivenesses are COP_w = 414.6, COP_h = 0.3376.

Results of the GHE-Gax Model Run with an Intermediate $\Delta T_{gax,G}$

The 4th study case analyzes the solution of the GHE-Gax problem using the model given in the *Resorber Heat Excess (RHE) Gax* Operation Model (see Fig. 4.11c also). The model is run with an intermediate $\Delta T_{gax,G}$ value, $\Delta T_{gax,G}$ = 7.885 °C > $\Delta T_{gax,G,min}$ = (0.1 – 4) °C. Figures 4.23 and 4.24 show

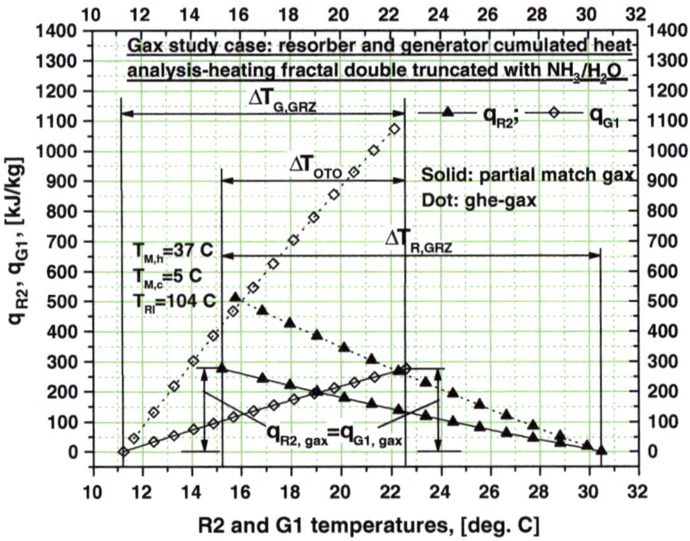

Fig. 4.24 The same as in Fig. 4.23, for the q-plot

the X-plot and q-plot, respectively, before (dotted line plot) and after (solid line plot) applying the partial match gax method. The operation temperature overlapping takes place between $T_{R2,gax} = 15.23$ °C and $T_{G1,gax} = 22.55$ °C, therefore $\Delta T_{OTO} = 7.319$ °C. The X and q plots show that the generator sub-device needs much more heat than the resorber sub-device can supply. A partial correction of this thermal mismatch is done splitting the generator in two sub-devices, according to Fig. 4.11c. Further on, the gax generator (pos. 7 in Fig. 4.11c) and the gax resorber (pos. 6 in Fig. 4.11c) are thermally coupled in such a way that the (i) condition is fulfilled, $q_{G,1,gax} = q_{R,2,gax} = 532.8$, Fig. 4.24. During this process, the former generates the vapor amount $V_{1,gax} = 0.2548$ (Eq. (4.127)), while the latter resorbs the amount $V_{2,gax} = 0.231$ (Eq. (4.125)). The gax heat excess generator (pos. 9 in Fig. 4.11c) receives the external heat $q_{G,\bar{1},\overline{gax}}^{e} = 574.6$, generating the vapor amount $V_{\bar{1},\overline{gax}}^{e} = 0.2748$ [Eq. (4.128)]. More results of this case can be extracted from Figs. 4.27 and 4.28.

Summarizing, this time the method application enables in theory to rectify partially a GHE-Gax problem, recovering the gax resorber heat for the benefit of the heating cycle effectiveness. Here, this increased by more than 10 percent, i.e. from $COP_h = 0.3057$ to $COP_h = 0.3376$.

Results of GHE-Gax Model Run with $\Delta T_{gax,G,max}$

The 5th study case makes use again of the full match gax configuration, outlined in Fig. 4.11a (see the Results of the RHE-Gax Model Run with $\Delta T_{gax,R,\max}$), but this time it analyzes the GHE-Gax problem mentioned priorly. The full gax configuration can be obtained from the partial match gax configuration, Fig. 4.16, making $V_{j,gax}^{e} = 0$ and dimensioning correspondingly the $\Delta T_{gax,G}$ temperature

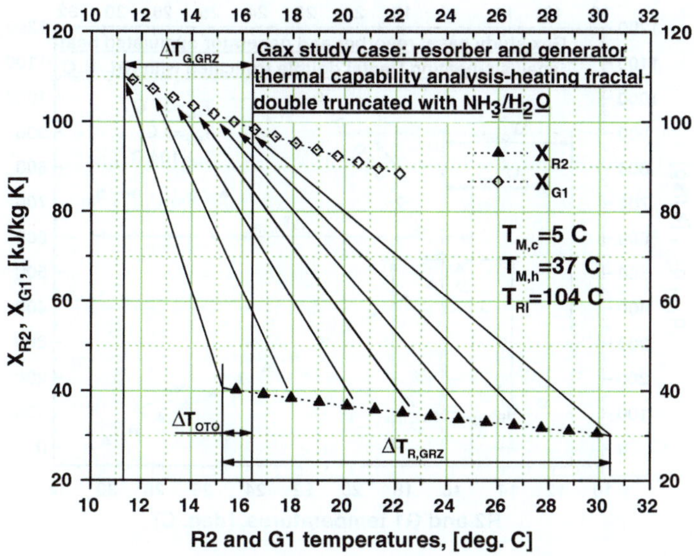

Fig. 4.25 The X-plot of the study case of Fig. 4.23, solving the GHE-Gax problem when considering that $\Delta T_{gax,G} = \Delta T_{gax,G,\max}$

item and further $\Delta T_{G,GRZ}$ [see Fig. 4.12 and Eq. (4.62)], in order that Eq. (4.96) be fulfilled. In this particular case, one might expect that $\Delta T_{gax,G}$ takes its highest value, $\Delta T_{gax,G} = \Delta T_{gax,G,\max}$, therefore $\Delta T_{G,GRZ}$ will have a minimum value, $\Delta T_{G,GRZ} = \Delta T_{G,GRZ,\min}$. The solver of our study case is the model given to the *Generator Heat Excess (GHE) Gax* Operation Model. The $\Delta T_{gax,R,\max}$ value can be found through a trial-and-error calculus, only. Each chosen $\Delta T_{gax,G}$ proposes an $y_{G,j,gax} = y_{G,j,gax}(\Delta T_{gax,G,\max})$ concentration variable value, and the trial-and-error calculus ends when the heat of the gax resorber, $q_{G,j,gax}$, computed with $h_{G,j,gax} = \left[h_{G,j,gax}\left(y_{G,j,gax}\right)\right]_{PG}$ and $g_{j,gax,u} = \left[g_{j,gax,u}\left(y_{G,j,gax}\right)\right]_{PG}$ functions [see Eq. (4.83) where $j = 1$], satisfies Eq. (4.96). As one can remark, similar to Results of the RHE-Gax Model Run with $\Delta T_{gax,R,\max}$, this type of gax with $\Delta T_{gax,G,\max}$ is nothing else but the single gax solution, which the classic configuration of Fig. 4.11a is capable to offer in an GHE-Gax working mode. The X-plot and q-plot of Figs. 4.23 and 4.24 are repeated in Figs. 4.25 and 4.26, respectively. The diminution of the $G_{1,gax}$ generator heat, in order to match the heat released by the resorber $R_{2,gax}$, has been done by restricting the $\Delta T_{G,GRZ}$ domain from $\Delta T_{G,GRZ,\max} = [11.23, 26.44]$ (valid for $\Delta T_{gax,G} = 4$ °C), or $\Delta T_{G,GRZ,\max} = [11.23, 22.55]$ (valid for $\Delta T_{gax,G} = 7.885$ °C, see Figs. 4.23 and 4.24) to $\Delta T_{G,GRZ,\min} = [11.23, 16.37]$. The restriction was achieved choosing a higher value for $\Delta T_{gax,G}$, comparatively, while keeping the same $\Delta T_{gax,R} = 4$ °C value. The trial-and-error calculus ended when the guessed values, $\Delta T_{gax,G,\max} = 14.0718$ °C and $y_{G,1,gax} = 0.379$, resulted in Eq. (4.96) fulfilling, $q_{R,2,gax} = q_{G,1,gax} \cong 532.8$. The accepted value, $\Delta T_{gax,G,\max} = 14.0\,718$ °C > 4 °C, leads to both (i) and (ii) heat transfer conditions fulfilling, as priorly stated and proved by Figs. 4.25 and 4.26 plots, respectively.

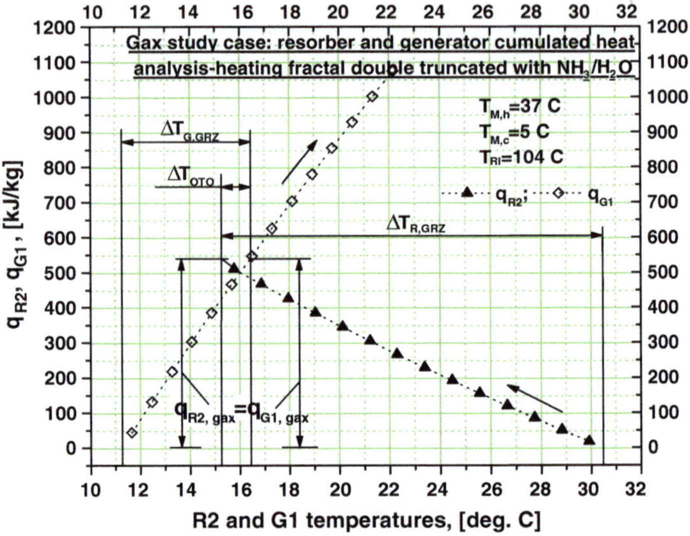

Fig. 4.26 The same as in Fig. 4.25, for the q-plot

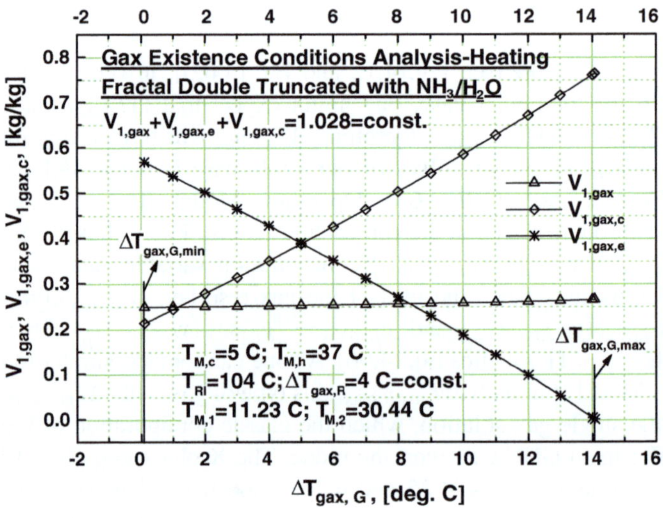

Fig. 4.27 Partial results of infinite equivalent solutions to the GHE-Gax problem mentioned priorly, shown for all generator sub-device's vapor versus $\Delta T_{gax,G}$

Infinite Equivalent Solutions to A GHE-Gax Problem

According to the 6th study case, running the 4.2.1.3.2. sub-sub-paragraph model for the configuration outlined in Fig. 4.11c and with intermediate $\Delta T_{gax,G}$ values, $\Delta T_{gax,G,\min} \leq \Delta T_{gax,G} \leq \Delta T_{gax,G,\max}$, will result this time in an infinite number of equivalent solutions for the same GHE-Gax problem mentioned priorly. Partial results thereof are given in Figs. 4.27 and 4.28 for the resorber vapor and

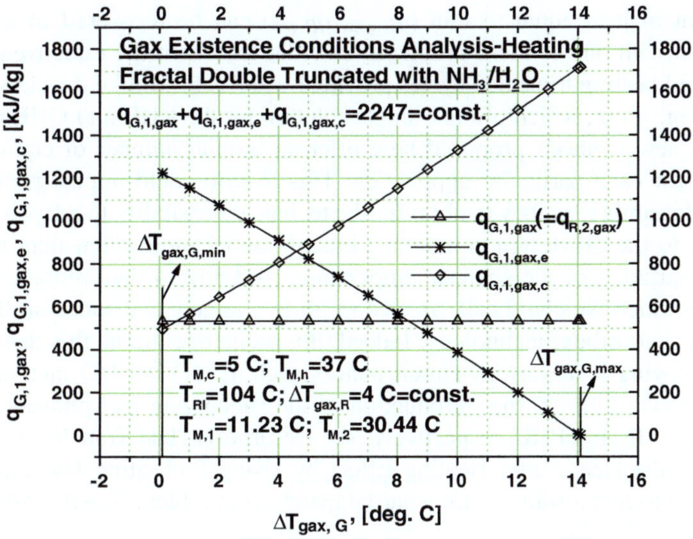

Fig. 4.28 Partial results of infinite equivalent solutions to the GHE-Gax problem mentioned priorly, shown for all resorber sub-device's heat versus $\Delta T_{gax,G}$

heat, respectively and $\Delta T_{gax,G,\min} = 0.1$ °C and $\Delta T_{gax,R,\max} = 10.2$ °C. All solutions respect the (ii) heat transfer condition (provided that the $\Delta T_{gax,G,\min}$ value is acceptable). Moreover, Fig. 4.28 proves that the (i) condition [Eq. (4.96)] is fulfilled as well, because, similar to RHE-Gax problem, the heat of the gax heat excess generator and gax complementary generator sub-devices are continuous decreasing and increasing complementary functions, respectively, enabling the gax generator heat be kept constant and equal to that of the gax resorber, for each $\Delta T_{gax,G}$ variable value. The results obtained above are not confined to the particular GHE-Gax problem at hand, they are general. Expressed otherwise, these results emphasize that, for a given $q_{R,i,gax}$ resorber heat and using the configuration outlined in Fig. 4.11c, it is always possible to find an infinite number of gax solutions for a GHE-Gax problem, having $q_{R,i,gax} = q_{G,j,gax}$ and being characterized by Gax Recovery Zones, $\Delta T_{G,GRZ}$, which respect the double inequality below:

$$\left[T_{M,j}, T_{G,j,gax,min}\right] = \Delta T_{G,GRZ,min} \leq \Delta T_{G,GRZ} \leq \Delta T_{G,GRZ,max} = \left[T_{M,j}, T_{G,j,gax,max}\right] \tag{4.130}$$

where

$$T_{G,j,gax,max} = T_{M,i} - \Delta T_{gax,G,min} \tag{4.131}$$

and

$$T_{G,j,gax,min} = T_{M,i} - \Delta T_{gax,G,max} \tag{4.132}$$

Final conclusions concerning the gax recovery technique reveal the fact that, principally, both RHE-and GHE-Gax problems can be partially capitalized, within the current absorption technological limits. To this extent, it has been shown that an amount of heat equal to min ($q_{G,gax}$, $q_{R,gax}$) can be recovered in a gax zone confronted with one of these problems, in order to benefit the effectiveness of the cooling and heating truncated cycles provided with "acr". Unlike the classic configuration, Fig. 4.11a, which offers a single solution for the RHE-and GHE-Gax problems, the new methods proposed here offer an infinite number of equivalent gax solutions for each particular application. This feature could impact positively the methods feasibility, as they could offer the technical possibility to adapt a cycle gax operation to the off-design conditions, brought about more often than not by the seasonal variation of the sources temperature. And now, a final remark, about the technique which we could call *"device splitting"*, largely and successfully used so far in the coabsorbent technology. Indeed, this technique merits this denomination after employing it in various circumstances such as: (1) the DD method conception (see Sect. 2.2); (2) the introduction of the subcooling and preheating hysteresis loops, SHL and PHL, respectively (see Desorption Heat Transfer Analysis and Enhancing the Heat Pump Heating Effect by Using Preheating Hysteresis Loops (PHL)); (3) to find a solution for a partial match gax problem (see this Sect. 4.2.1).

4.2.2 Model Results of Cooling Truncated Coabsorbent Cycles

The model of the truncated cycles (see this section for cooling and the next one for heating) was done according to a more realistic approach, following next. Having in mind a certain cooling or heating application, a most appropriate truncated cycle will result. Its truncated column must be compulsorily selected in order to ensure the cycle functioning for all working conditions met to the particular case at hand, while keeping the same configuration, that is the same design number of stages in operation (e.g. see for completion the introductory part of the Sect. 4.2). In order to avoid too high mass flow factors, which increase the pumping work consumption and decrease therefore the cycle COP_w, a certain truncation column device was considered operationally well designed throughout this book if its inlet/outlet concentration variation is higher than 0.03 in case of the H_2O–LiBr and NH_3–H_2O working combinations, during a transient work. Bearing in mind the above, our model was conducted to study on several examples what is the behavior of a truncated cycle when applying selection operation criteria such as (cycle internal temperature values are meant, only):

(a) The cycle sink or heat sources are changing separately their temperature within certain intervals during operation, that is $T_{M,c,\min} \leq T_{M,c} \leq T_{M,c,\max}$ or $T_{M,h,\min} \leq T_{M,h} \leq T_{M,h,\max}$;
(b) Both $T_{M,c}$ and $T_{M,h}$, differing one another with the cycle temperature difference, $\Delta T_M = T_{M,h} - T_{M,c}$, change their value simultaneously during the entire transition operation;
(c) The cycle T_{DI} of the cooling cycle or T_{RI} of the heating cycle are changing their temperature within certain intervals during operation, that is $T_{DI,\min} \leq T_{DI} \leq T_{DI,\max}$ or $T_{Rf,O,\min} \leq T_{Rf,O} \leq T_{Rf,O,\max}$, while keeping $\Delta T_M = const.$, or $\Delta T_M \neq const.$;

(d) The temperatures $T_{M,c}$ and $T_{M,h}$, forming the cycle $\Delta T_M = T_{M,h} - T_{M,c}$, and T_{DI} or T_{RI} change their value simultaneously, while the cycle absorbent mean concentration, y_M, and the characteristic equation, S, are kept constant during the entire transition operation.

Other selection operation criteria can be added to the A–D ones, if necessary. Many of the sink/heat sources interacting with a truncated cycle and included in Table 3.6 can be referred to when applying the above selection criteria. The main operation criterion used to select a truncated cycle and column depends on a certain application. However, besides this, an energetic analysis, seeking to maximize the cycle effectiveness (i.e. COP_c, COP_h, COP_w), must be had in view mandatory to each application. In this respect, the operation criteria emphasizing the application of the gax internal heat recovery to a maximum extent should be preferred. According to the A, B and C criteria, a cycle gax recovery is not favored during the transitory functioning, so these cannot be used in a first place as an effective operation strategy. The last one only, the D criterion, could be utilized mostly in an energetic selection. This will be described next. It is outlined in Fig. 4.29a and b, for the cooling and heating cycles, respectively. Let us consider a truncated cycle supposed to satisfy a certain cooling application and its nontruncated cycle it comes of, Fig. 4.29a. The nontruncated cycle can glide up and down along the constant isosteres $y_M = const.$, $y_{GO} = const.$ and $y_{RO} = const.$ (see left down corner arrows), simulating the transient operation with different T_{DI}, ΔT_M and T_{GO} values. According to Eq. (3.12), this cycle has a constant characteristic equation, $S = const.$, for all operating conditions. When the nontruncated cycle covers the transient operation domain, its associated cycle, the nontruncated one, will do the same, and, because of the lemma 3, it will keep the $y_M = const.$ and $S = const.$ as well. Expressed in other words, the truncated cycle is gliding through a plane-parallel movement along the constant y_M, $y_{GO,1}$, $y_{GO,i+1} = y_{RO,i-1}$, $y_{RO,n-1}$ and $y_{RO,n}$, isosteres, $i = 1, 2, \ldots, n - 1$, during the transient operation, while it preserves its initial configuration (see the dotted line isosteres of Fig. 4.29a). The explanation given so far for the cooling cycle holds true for the heating cycle, Fig. 4.29b, as well.

The first model results have been obtained with NH_3–H_2O quadruple truncated cooling fractals, Fig. 4.30. Theirs COP_c can be improved if the gax recovery and the last generator vapor rectification were practiced. However, for the moment, none of these methods have been considered in Fig. 4.30 results, but the bare cycle configuration plotted in Fig. 4.4. Besides this, the model worked with constant $\Delta T_M = T_{M,h} - T_{M,c} = 40$ °C, $y_{R5,O} = 0.8$, $\eta_{pump} = 0.6$ and "2/3" desorption interval, as input data (criterion C). The $COP_c \cong 0.11$ quasi-constant value obtained vs. T_{DI} enables the achievement of a high COP_w effectiveness (see plot). The results show the potential of the truncated cycle, which is capable to capitalize low temperature heat sources (see $T_{M,h}$ plot) in order to produce deep (industrial) cooling with enough high temperature sink sources (see $T_{M,c}$ plot). The cycle sensibility check with respect to T_{DI} variation, Fig. 4.31, shows there is a certain optimum T_{DI} value for the COP_w records its highest value when operating with constant cycle temperature gap and sink temperatures, ΔT_M and $T_{M,c}$, respectively (criterion C). Also, the working with fixed $T_{M,c}$ and T_{DI} emphasizes that COP_w reaches a maximum value

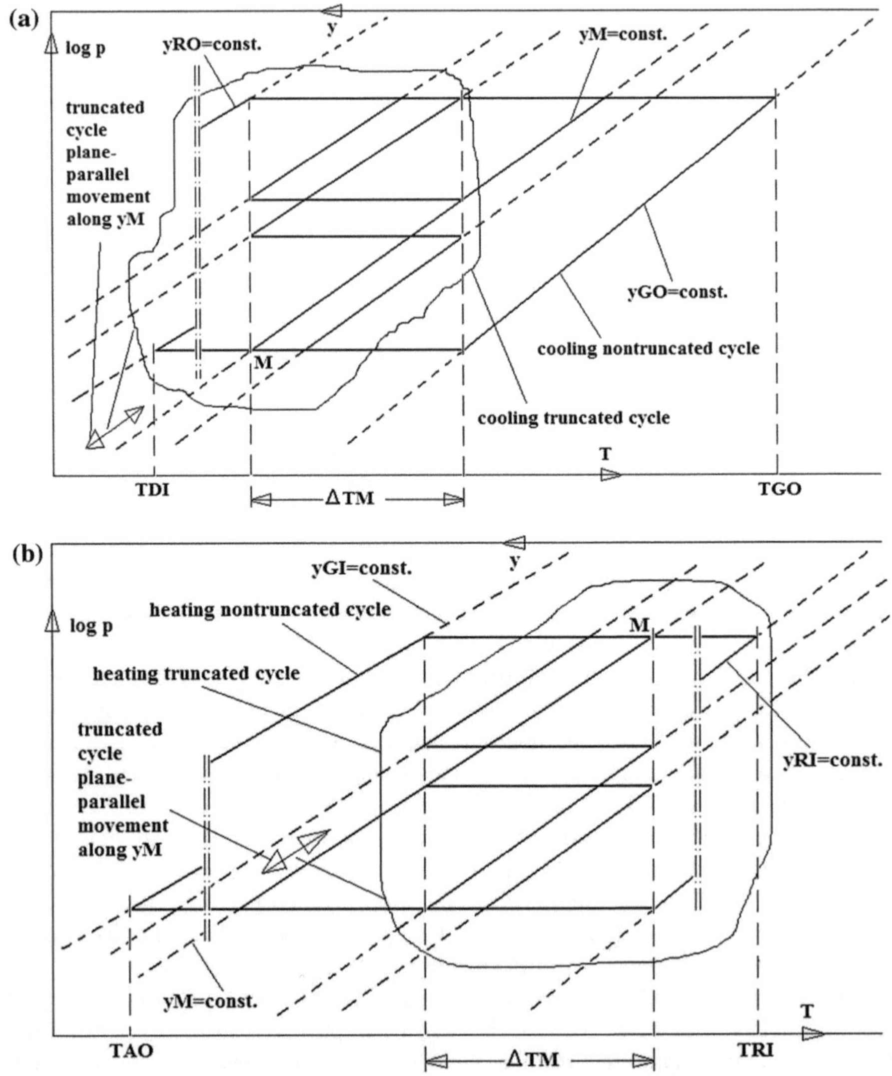

Fig. 4.29 The D criterion outline for: **a** the cooling cycles; **b** the heating cycles

for an optimum ΔT_M value, Fig. 4.32 (criterion B). The results of Figs. 4.31 and 4.32 indicate that large deviations from the design operation keeps the COP_c quasi-constant, while the COP_w is drastically penalized, as foreseen.

The second set of results have been obtained for NH_3/H_2O quadruple and double cooling fractals, shown in Figs. 4.33 and 4.12, respectively. They benefit of the gax recovery and a rectification of the last generator vapor ($\eta_{rectif.} = 0.82$). Referring to the quadruple cooling fractal, Fig. 4.33, the (i) and (ii) conditions fulfillment is analyzed for the gax recovery in a study case. The results of this case,

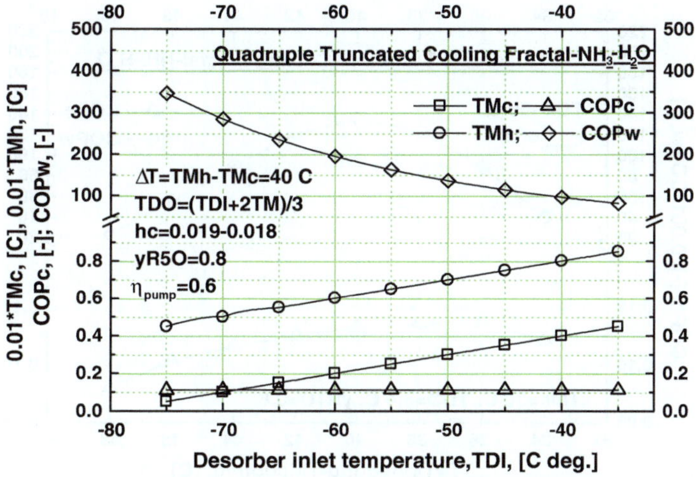

Fig. 4.30 NH_3–H_2O quadruple truncated cooling fractal: COP_c, COP_w, $T_{M,c}$ and $T_{M,h}$ functions versus T_{DI}

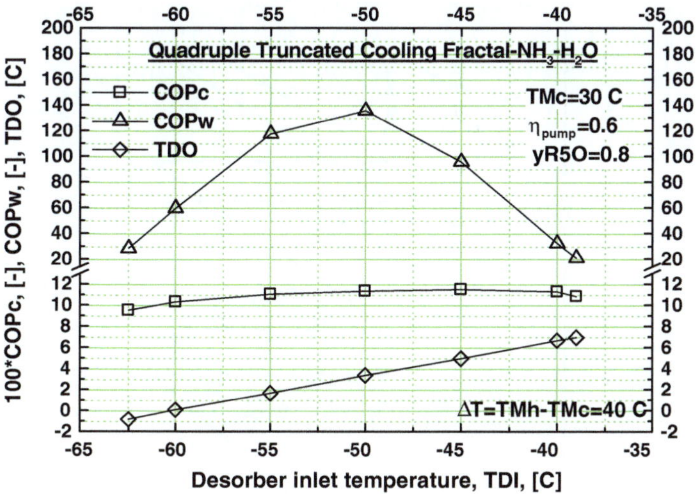

Fig. 4.31 NH_3–H_2O quadruple truncated cooling fractal: COP_c, COP_w and T_{DO} functions sensibility against T_{DI}, for fixed $T_{M,c} = 30$ °C and $\Delta T_M = 40$ °C

given in Figs 4.34 and 4.35 for the X-plot and q-plot, respectively, show that they are satisfactorily fulfilled as well. Moreover, these conditions have been checked with consistency and were found to be fulfilled for a larger range of operating parameters. In this respect, Figs. 4.36 and 4.37 plot the second set of deep cooling results for $T_{M,c} = 32$ °C and $T_{M,c} = 29$ °C ($\eta_{pump} = 0.8$), respectively (criterion C). They have as comparison term the NH_3/H_2O condensing deep cooling cycle (Niebergall 1959, p. 71), Table 4.2.

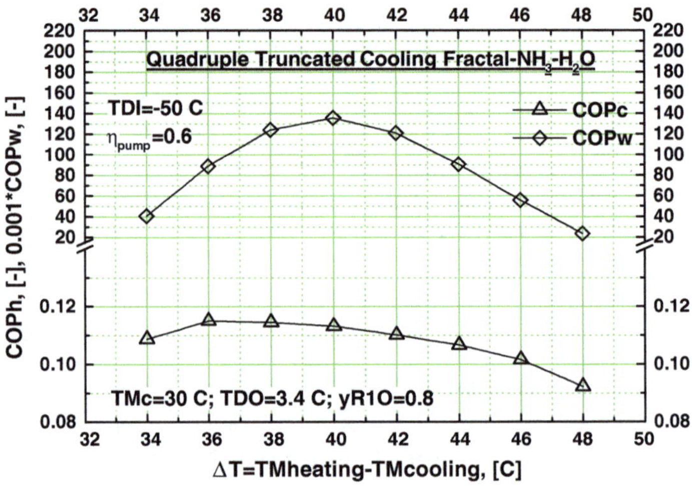

Fig. 4.32 NH_3–H_2O quadruple truncated cooling fractal: COP_c and COP_w functions sensibility against ΔT_M, for fixed $T_{M,c} = 30$ °C and $T_{DI} = -50$ °C

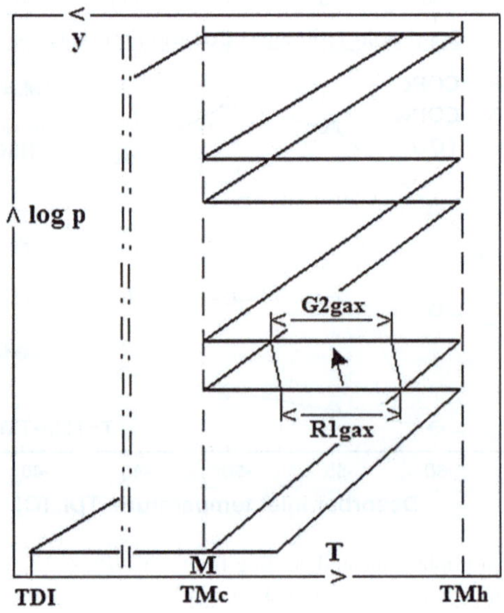

Fig. 4.33 Quadruple truncated cooling fractal with gax recovery between its 1st stage resorber and 2nd stage generator

Table 4.2 NH_3– H_2O condensing single-stage deep cooling cycle (Niebergall 1959, p. 71)

T_E (°C)	$T_{M,c}$ (°C)	$T_{M,h}$ (°C)	COP_c (−)	$COP_{c,rel.Carnot}$ (−)
(−50)–(−55)	15–18[a]	(~152)–(~159)[b]	0.26–0.22	0.36–0.31

[a]External sink
[b]External heat

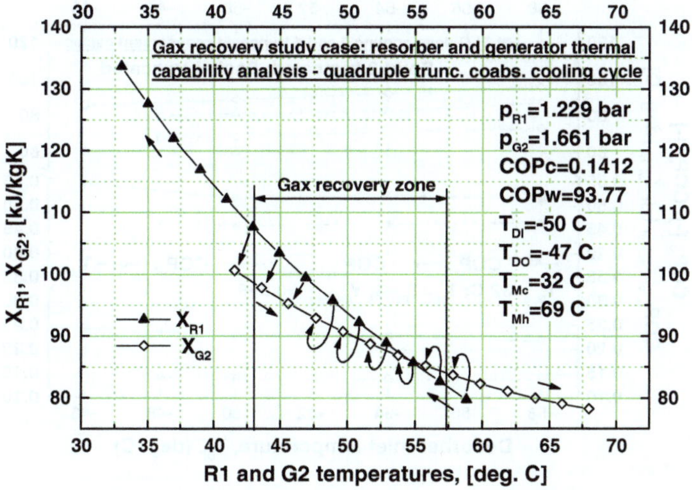

Fig. 4.34 The X-plot of the 1st stage resorber and 2nd stage generator of Fig. 4.32 cycle versus theirs working temperatures

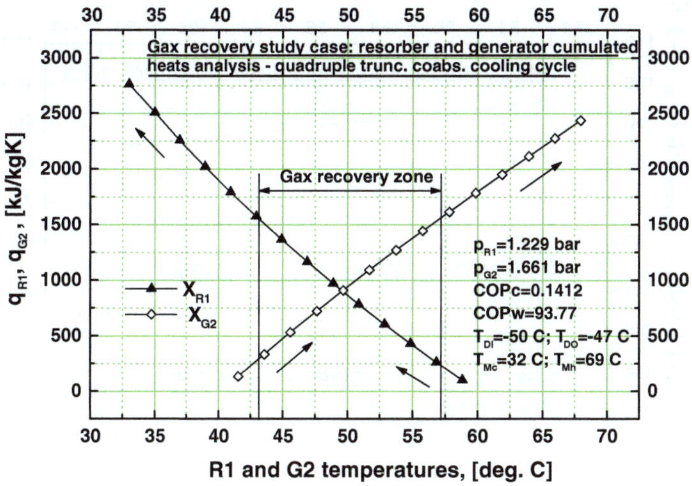

Fig. 4.35 The q-plot of the 1st stage resorber and 2nd stage generator of Fig. 4.32 cycle versus theirs working temperatures

The COP_c values of the double truncated cooling fractal are similar to those of Table 4.2, but comparatively they have been obtained with sink and heat source temperatures which are by cca. 10 °C higher and 60 °C lower, respectively. The comparative much higher performance of the double truncated cooling fractal is explained by its better internal heat management, typical for the coabsorbent technology. Indeed, in our particular case this is expressed by the $COP_{c,rel.Carnot}$,

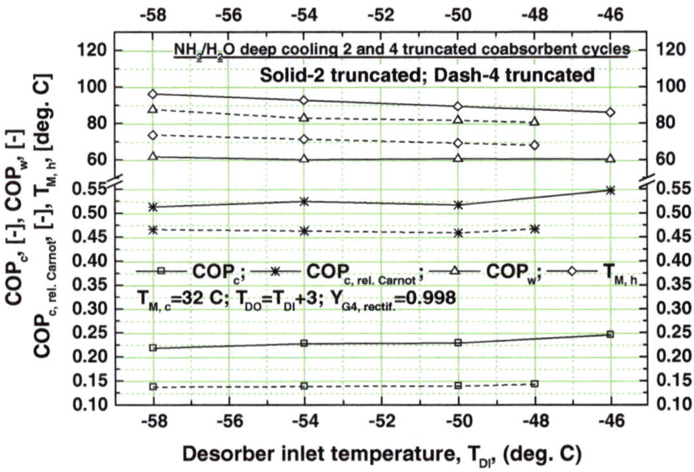

Fig. 4.36 NH_3–H_2O quadruple and double truncated cooling fractal having gax recovery and last generator vapor rectification: COP_c, COP_w, $COP_{c,rel.Carnot}$ and $T_{M,h}$ versus T_{DI}

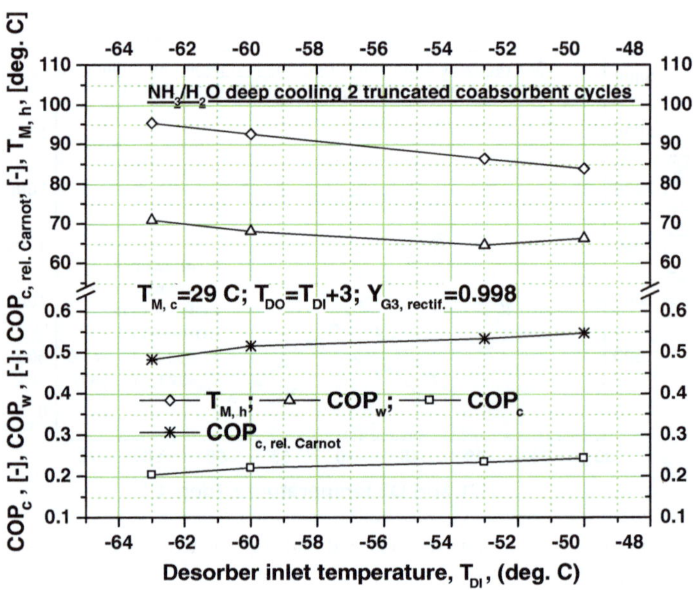

Fig. 4.37 NH_3–H_2O double truncated cooling fractal provided with gax recovery and last generator vapor rectification: COP_c, COP_w, $COP_{c,rel.Carnot}$ and $T_{M,h}$ versus T_{DI}

which is by cca. 57 % better than that of the classical condensing cooling cycle (see Table 4.2). Concerning the quadruple truncated cooling fractal, the gax recovery introduction increases its COP_c by about 27 %, from ~0.11 to about 0.14 (see Figs. 4.30 and 4.36 results). It is possible to increase further its COP_c, up to ~0.16,

Table 4.3 Model results of NH_3/H_2O double truncated deep cooling operation according to the criterion D ($y_{M,c} = 0.1719 = const.$, $S = 26.96 = const.$)

T_{DI}	$T_{M,c}$	$T_{M,h}$	COP_c	$COP_{c,rel.Carnot}$	COP_w	$V_1/V_2/V_3$
−50	32	88	0.2374	0.5485	43.86	1.349/1.148/1.012
−55	25.14	79.77	0.2404	0.5561	54.37	1.317/1.134/1.012
−60	18.26	71.52	0.2437	0.5639	67.98	1.287/1.121/1.012

if the gax taking place between the 3rd stage resorber and 4th stage generator (not represented in Fig. 4.33), were partially capitalized. It is true, this cycle COP_c is smaller than that of the baseline cycle, Table 4.2. However, comparatively it has been run with sink and heat source temperatures which are by cca. 10 °C higher and 80 °C lower, respectively. Moreover, its higher $COP_{c,rel.Carnot}$, by ~45 %, indicates this coabsorbent cycle has as well a much better internal heat management than that of the baseline cycle.

Summarizing, generally, the $COP_{c,rel.Carnot}$ figure of merit is much in favor of the coabsorbent cycle and this speaks a lot about the improvement the truncated cycles could bring in the absorption deep cooling, comparatively. In case of NH_3/H_2O, the explanation comes of the fact that a truncation column is a rectification-dephlegmation column as well (see Sect. 4.1.1 end).

The last results have been obtained applying the criterion D to an NH_3/H_2O double truncated deep cooling cycle operation, Table 4.3. The truncated cycle benefits of vapor rectification and of the highest gax recovery. The results hold true for $y_{M,c} = 0.1719 = const.$ and $S = 26.96 = const.$. Despite the unfavourable input data, the cycle works with very high effectiveness all the time, emphasized by the COP_c and especially by the $COP_{c,rel.Carnot}$ figures of merit. However, to put in work order a cycle operation according to the criterion D is not that simple. Indeed, the sudden temperature change of a single supplying source implies the checking of the rest of sources temperatures, and the sudden temperature change of two supplying sources may result more often than not in an inadequate operation of the cycle according to the criterion D.

4.2.3 Model Results of Heating Truncated Coabsorbent Cycles

NH_3–H_2O double and quadruple truncated heating fractals, plotted in the log $p-1/T$ diagram, Fig. 4.38a, b, respectively, have been modelled. A careful analysis showed that the gax recovery between the R2–G1 of the cycle plotted in Fig. 4.38a and between R2–G1 and R4–G3 of the cycle plotted in Fig. 4.38b is confronted with a GHE-Gax problem. Indeed, in this case the generation heat capability is constantly much higher than that of resorption, throughout the investigated domains with gax temperature overlapping, therefore the (i) heat transfer condition is not fulfilled and a partial match gax recovery technique must be applied (see the *Generator Heat Excess (GHE) Gax* Operation Model). Except the first results, given in Fig. 4.39a–c, all the other results plotted in Fig. 4.40a, b and Fig. 4.41a, b considered that a

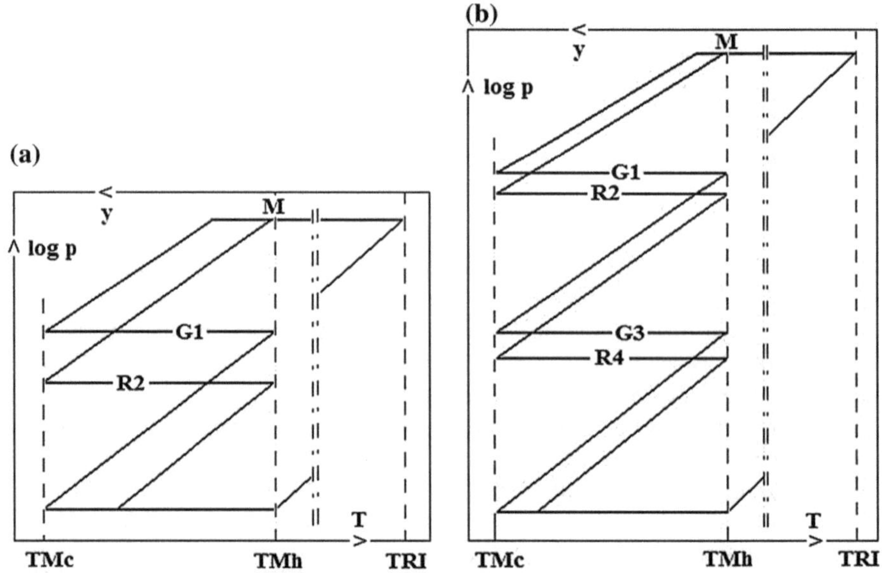

Figs. 4.38 Modeled NH_3–H_2O double (**a**) and quadruple (**b**) truncated heating fractals plotted in the log $p - 1/T$ diagram

GHE-Gax recovery is applied to the heating truncated cycles. Two district heating applications have been analyzed and results thereof are given next. The idea which lays down the basis of our model consists in coupling a coabsorbent heat pump with a steam Rankine cycle (SRC) power plant in such a way that the SRC condenser cooling source is the heat source of the fractal. This source has a quasi-constant temperature during the whole year (our input data is $T_{M,h} = 37$ °C). As fractal sink sources we considered a seasonal one, typical for winter time only, with $T_{M,c} = (-5)$–$(+5)$, Fig. 4.39a–c, a winter-summer transition period one, supplied for example by low temperature seas, oceans, lakes, rivers, etc., with $T_{M,c} = (5$–$15)$ °C, Fig. 4.40a, b and a whole year sink source resulted from the combination of the two first sources, with $T_{M,c} = (-5)$–$(+15)$, Fig. 4.41a, b. Such kind of heating applications were already suggested in Table 3.6. The resorber heating effect was enhanced considering that the cycles benefit of PHL (see Enhancing the Heat Pump Heating Effect by Using Preheating Hysteresis Loops (PHL).). The solution pump effectiveness input data was $\eta_{pump} = 0.8$. The model also considered that the relative position of the high pressure generator and resorber operation temperature and concentration intervals have a single point (balance) overlapping (T, y) interval, that is $(T_{GO}, y_{GO}) = (T_{RO}, y_{RO}) = (T_{M,h}, y_{M,h})$ (see Fig. 3.3b). A ratio evaluating the heating temperature effectiveness:

$$e_{ht} = \frac{\sum \delta T_{Rf,i}}{T_{RI} - T_{M,h}} \tag{4.133}$$

was introduced through Eq. (4.133). In this equation, the useful heating temperature delivered to the customer, $T_{Rf,O}$, is calculated by Eq. (3.128), and $0 \leq e_{ht} \leq 1$.

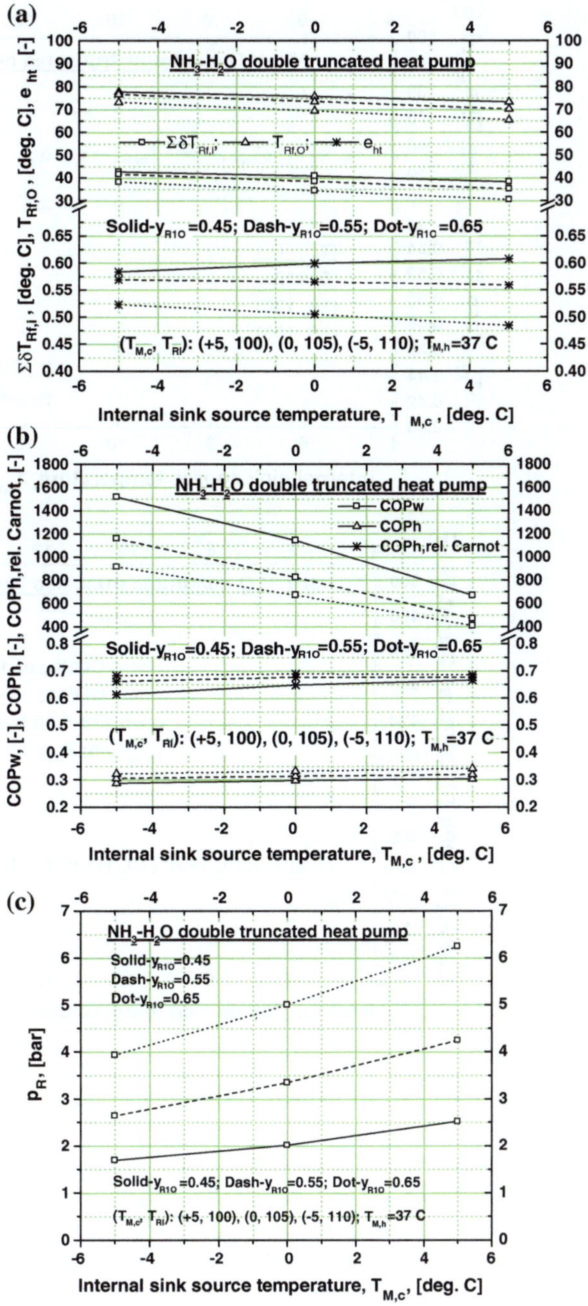

Figs. 4.39 functions **a** $\sum \delta T_{Rf,i}$, $T_{Rf,O}$, e_{ht}; **b** COP_w, COP_h, $COP_{h,rel.Carnot}$ and **c** p_R, plotted versus sink temperature, $T_{M,c}$ for the NH_3–H_2O double truncated heat pumps

The functions $\sum \delta T_{Rf,i}$, $T_{Rf,O}$, e_{ht}, COP_w, COP_h, $COP_{h,rel.Carnot}$ and p_R were plotted vs. cycle sink $T_{M,c}$, for the NH_3/H_2O double and quadruple truncated heat pumps in Figs. 4.39a–c and 4.40a, b, respectively (criterion C). The plot of Fig. 4.39a–c

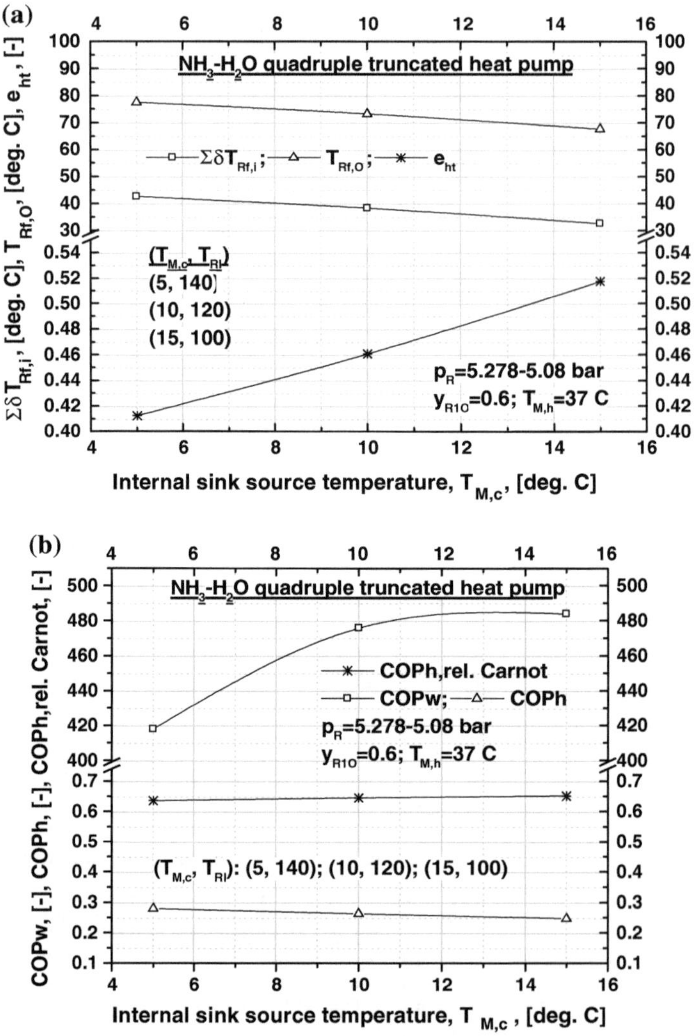

Figs. 4.40 Functions **a** $\sum \delta T_{Rf,i}$, $T_{Rf,O}$, e_{ht} and **b** COP_w, COP_h, $COP_{h,rel.Carnot}$ plotted versus sink temperature, $T_{M,c}$, for the NH_3–H_2O quadruple truncated heat pumps

has the y_{R1O} concentration as parameter. This indicates that the lower y_{R1O} values benefit the $\sum\delta T_{Rf,i}$, $T_{Rf,O}$, e_{ht}, COP_w and p_R functions, while the higher ones benefit the rest of functions, COP_h and $COP_{h,rel.Carnot}$. The results of Fig. 4.41a, b are given for the double truncated heating fractal (criterion C). Here, the same functions $\sum\delta T_{Rf,i}$, $T_{Rf,O}$, e_{ht}, COP_w, COP_h and $COP_{h,rel.Carnot}$ are plotted vs. $T_{M,c} = ((-5) - (+15))$ °C for $T_{M,h} = 37$ °C and a single $y_{R1O} = 0.45$ value. As expected, the GHE-Gax recovery increases COP_h as compared to the cycle which is not using it [here by $(17 - 7)\%$]. Similarly it happens to the quadruple truncated heating fractal, Fig. 4.40b, which COP_h of increases by cca. 10 % when applying

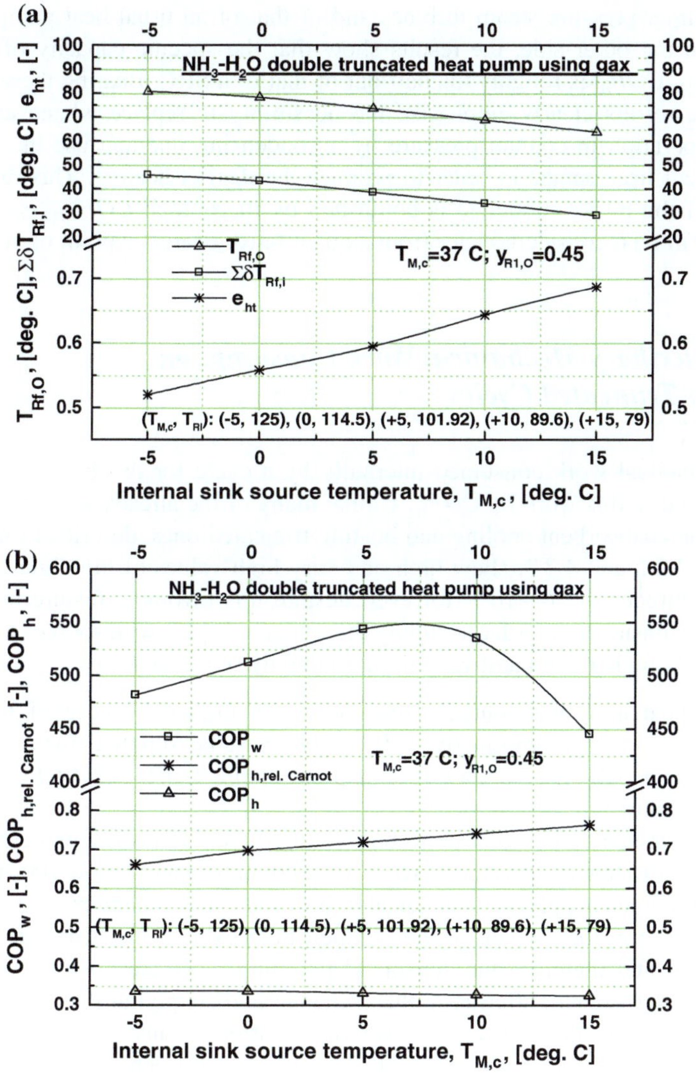

Figs. 4.41 Functions a $\sum \delta T_{Rf,i}$, $T_{Rf,O}$, e_{ht} and **b** COP_w, COP_h, $COP_{h,rel.Carnot}$, plotted versus sink temperature, $T_{M,c}$, for the NH_3–H_2O double truncated heat pump

the GHE-Gax recovery. Also, this recovery increases the e_{ht} and $COP_{h,rel.Carnot}$ functions values, but diminishes those of COP_w, comparatively.

Concluding, all figures results show that it would be possible to supply a district with warm water of a satisfactory temperature level, $T_{Rf,O} = (64\text{–}81)$ °C, necessary for house heating and domestic hot water production. On one side, a benefic property of the heating cycles at hand is their very small pumping mechanical work consumption, here $1{,}500 > COP_w > 400$. This COP_w figure of merit is by one and two orders of magnitude higher than that of a today SRC which cogenerates heat using counter-pressure steam turbines and of that of an usual heat pump, respectively. On the other side, the results show that the specific capacity of a fractal heat pump, COP_h, is roughly the half out of that of a SRC. Bearing these in mind, the above results clearly emphasize the possibility to replace whenever feasible a SRC operation in counter-pressure by a condensing one and use the truncated coabsorbent heat pumps in order to produce the district heating. This remarkable feature of the cycles modelled is confirmed by theirs high $COP_{h,rel.Carnot}$. These aspects will be reconsidered later from a quantitative energetic point of view.

4.2.4 Auxiliary Mechanical Work Consumption in Truncated Cycles

The mechanical work consumed internally by a cycle for its absorbent pumping is expressed in this work by COP_w. Unlike many of the already known absorption cycles, the coabsorbent cooling and heating truncated ones, described for instance in Sects. 4.2.2 and 4.2.3, show high and very high values of this figure of merit, with magnitudes of 10^1–10^3. However, despite this favorable feature their auxiliary work consumed in order to supply the truncated cycle with sources tends to be intrinsically higher comparatively, because mainly of two reasons:

(i) The heat and sink sources consumption is higher, because of the higher number of stages (normally 3–5) which must be supplied for a same useful effect output, or otherwise expressed, theirs COP_c and COP_h are lower, comparatively;
(ii) The heat source is supposed to be recovered from industrial processes rejecting heat to low temperatures. The source could be water below 100 °C in most applications, supplying with sensible heat the cycle generators and desorbers. In this case, the amount of the pumped fluid exchanging heat without phase change and therefore the consumed pumping work are obviously higher as compared to that suffering a phase change and supplying the classic absorption units. In this respect, here we can mention for instance the single-stage deep cooling ammonia-water plants using steam for generation to a considerable higher temperature (120–150) °C [(2–6) bar].

Bearing these in mind, the high COP_w favorable feature emphasized above becomes effectual when besides the internal work the auxiliary one needed for supplying the truncated cycle with sources is small as well, only. Assuming a supplying heat/sink source, with the density ρ_s [kg m^{-3}] and the mass m_s [kg], has to overcome a pressure loss Δp_s [N m^{-2}] during its pumping, then the consumed auxiliary work, w_{aux} [J], is calculated simply by:

$$w_{aux} = \frac{\Delta p_s m_s}{\rho \eta_p} \tag{4.134}$$

where η_p is the pump effectiveness. The mass m_s is calculated in case of a heat transfer without phase change with the help of the heat and sink source capacities $\sum_i q_{G,i}$ and/or $\sum_i q_{R,i}$, as:

$$m_s = \frac{\sum_i q_{G(R),i}}{c_{p,s} \Delta t_s} \quad (4.135)$$

where $c_{p,s}$ [J*kg^{-1}*K^{-1}] and Δt_s [°C] is the source specific heat and source temperature variation during the heat exchange process, respectively. The Δt_s's are computed for the heat and sink sources by:

$$\Delta t_s = T_{hs} - \left(T_{M,h} + \delta T_{M,h}\right) \quad (4.136)$$

and

$$\Delta t_s = \left(T_{M,c} - \delta T_{M,c}\right) - T_{ss} \quad (4.137)$$

respectively. In Eqs. (4.136) and (4.137), $T_{M,h}$ and $T_{M,c}$ hold for the already mentioned cycle internal mixing temperatures, T_{hs} and T_{ss} hold for the initial heat and sink external sources temperatures, and $\delta T_{M,h}$ and $\delta T_{M,c}$ hold for a necessary minimum temperature pinch, which should exist during the heat exchange process. Introducing Eq. (4.135) in Eq. (4.134), w_{aux} becomes:

$$w_{aux} = \frac{\Delta p_s}{\rho \eta_p} \frac{\sum_i q_{G(R),i}}{c_{p,s} \Delta t_s} \quad (4.136)$$

From Eq. (4.136) it results that the decrease of w_{aux} is principally conditional upon the decrease of Δp_s and $\sum_i q_{G(R),i}$ and the increase of ρ, η_p, $c_{p,s}$, and Δt_s functions. However, out of these, Δp_s and Δt_s could be used for the w_{aux} decrease only, because ρ and $c_{p,s}$ are physical properties that cannot be chosen at convenience to a given application, η_p depends on the pumps technological limits and $\sum_i q_{G(R),i}$ result from a cycle internal heat management optimization. Concerning Δp_s, minimizing the pressure loss of a truncated cycle-source interaction should make compulsorily the object of a careful design, because this is one of the most important way to capitalize these cycles thermodynamic advantage. Moreover, this way to proceed enables to keep Δt_s's, the second item that could be used in w_{aux} minimization, with reasonable small values, in order to benefit of the truncated cycle main quality, that of being able to process most effectively heat/sink sources with small (T_{hs}–T_{ss}) temperature gaps, or otherwise expressed from Eqs. (4.136) and (4.137), with small $\Delta T_M = (T_{M,h}$–$T_{M,c})$ values.

As study case we consider the quadruple truncated heat pump, Fig. 4.38b, thought to be coupled with a SRC for district heating and which few results of are given in Fig. 4.40a, b. Here it is considered that the heat source, i.e. the SRC steam, is heating the heat pump generators and desorber ($T_{M,h}$ = 39 °C) without an intermediary heat exchange fluid, so the heat source pumping work consumed is negligible. However, it is not the case of the sink source cooling the cycle resorbers, which pumping work consumption of is taken into account applying

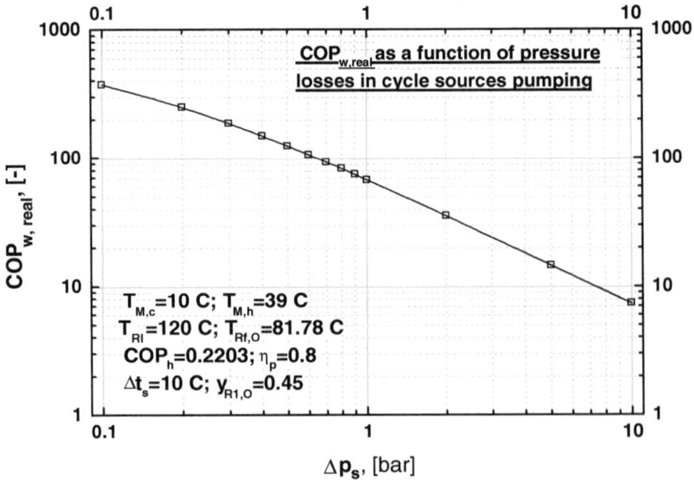

Figs. 4.42 Results of the $COP_{w,real}$ plotted versus Δp_s in a double logarithmic scale for the above study case of the district heating NH_3–H_2O quadruple truncated heat pump coupled to a SRC

the simple model given above. This auxiliary work is added to the denominator of Eq. (3.55) COP_w in order to calculate this figure of merit. All cycle internal and external pumping work consumption effectiveness is considered $\eta_p = 0.8$, giving the COP_w a real dimension, $COP_{w,real}$. The sink source input data are those of water. The values of $COP_{w,real}$ are plotted in Fig. 4.42 versus Δp_s, which includes the cycle input data, also. From this figure it results clearly that the auxiliary work consumption reduction is very important in order to keep the $COP_{w,real}$ with high values (in our case, Δp_s should not exceed 1 bar).

4.3 Hybrid Truncation of the Coabsorbent Cycle

The height of a full truncation column can be diminished in at least three practical cases. According to the first, the sink/heat sources have a limited availability (amount and/or temperature level) for a given task, even they are low-grade, or waste, etc., ones. According to the second, it is intended to diminish the column constructive complexity. And third, bearing in mind the remarks of Sect. 4.2.4, in order to diminish the auxiliary work consumption. Next we shall consider the truncation columns having its parallelograms covered clockwise, like in Fig. 4.9a. In all cases, a truncation column with a diminished height raises its generated vapor pressure till a final value, which is smaller as compared to the pressure achieved by a full truncation column, for the same sink/heat column supplying sources and input data. More often than not, the compensation for the vapor pressure difference till the full column pressure value is achieved in practice with the help of a mechanical vapor compression unit. The column using such a unit

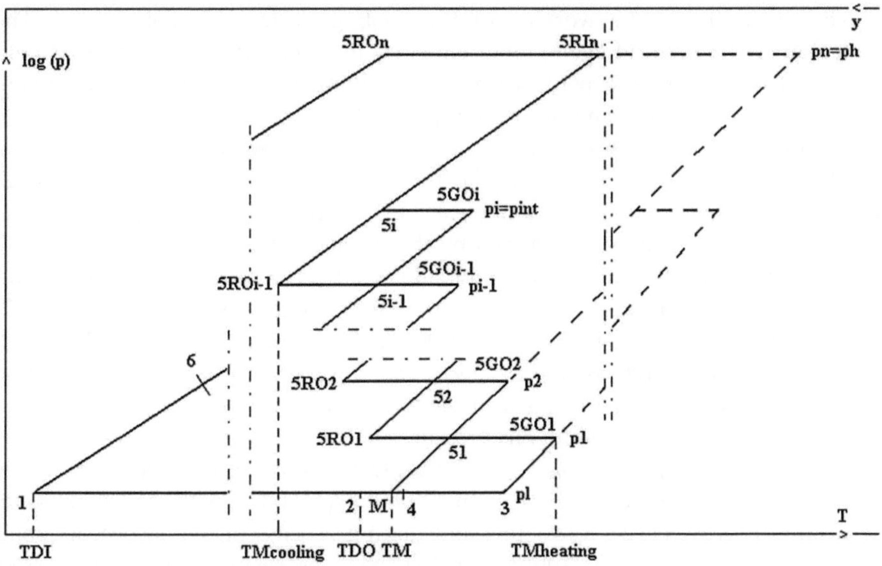

Fig. 4.43 The general diagram of the hybrid truncated cooling fractal

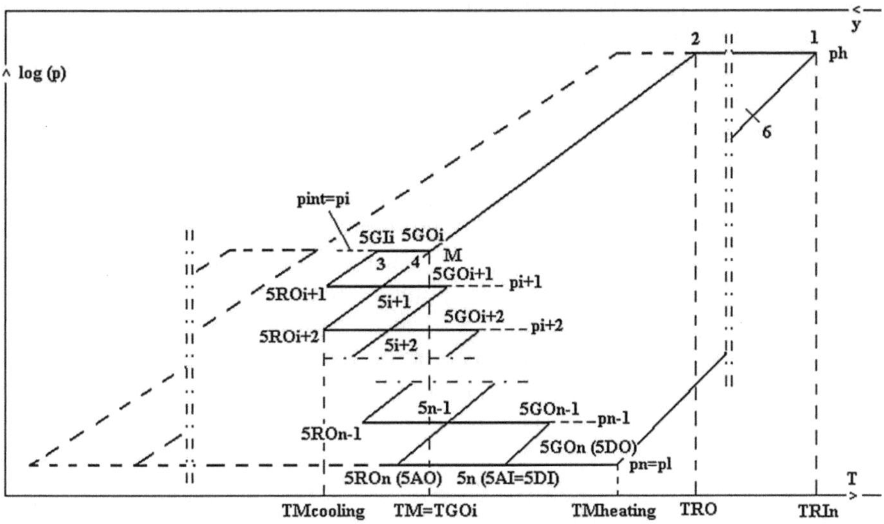

Fig. 4.44 The general diagram of the hybrid truncated heating fractal

is a *hybrid truncation column*. The hybrid truncation columns are most indicated in heat pumping applications when the truncation degree is small (presumably max. simple or double). Although, principally, they can be either of nonoverlapping, balance or of overlapping type, here we shall analyze the balance type, only (Staicovici 2007g, 2008a). The general log $(p) - 1/T$ hybrid truncated cooling and heating cycle (fractal) diagrams are given in Figs. 4.43 and 4.44, respectively.

The diagram and the working of the cooling cycle is described next. The truncated cooling hybrid cycle, represented by the solid line in Fig. 4.43, is connected to heat sources capable to provide heating to a maximum temperature $T_{M,heating}$ for generation and desorption processes, e.g. (40–70) °C and the cooling to a minimum temperature $T_{M,cooling}$ for resorption and absorption processes, e.g. (10–40) °C, in such a way that the $\Delta T_M = T_{M,heating} - T_{M,cooling} \geq (8–10)$ °C. The cycle at hand is coming of a nontruncated cooling non-isobar (hybrid) cycle, represented in the same very diagram by the outside solid line contour on the left side and by the dashed line on the right side (see Sect. 3.2.5). The truncated cooling hybrid cycle is including a low pressure desorption process (1–2) p_l, e.g. (0.1–2) bar, where the useful cooling effect occurs, e.g. $T_{DI} = (213, 15–273, 15)$K, a high pressure resorption process ($5_{RI,n} - 5_{RO,n}$), p_h, e.g. (10–50) bar, and a truncation hybrid column, used as the nontruncated non-isobar (hybrid) cycle truncation process. The truncation column has pressures between the low value and an intermediary pressure p_{int}, $p_h > p_{int} > p_l$. The truncation column is made up by a low pressure absorption process (3–4), coupled on the vapor side with the low pressure desorption process (1–2), a low pressure mixing process of the absorbents coming of the low pressure absorption and desorption processes in order to generate the one mean concentration absorbent y_M and temperature $T_M(p_l, y_M) \geq T_{M,cooling}$, point M, a series of i–1 stages, $i < n$, i, $n \in N$ of isobar opposite intermediary generation($5_k - 5_{GO,k}$) and resorption processes ($5_k - 5_{RO,k}$), $k = 1, \ldots, i - 1$, and a final i stage with an intermediary pressure generation process, $p_{int}=p_i$. The stages pressure increases gradually from the low value to the intermediary one, $p_{int}>p_{i+1} > p_i \geq p_l$, and theirs temperatures range within the $T_{M,cooling}$ and the $T_{M,heating}$ values. The stages are coupled on the vapor side. The vapor have essentially increasing mean concentrations $Y_{G,m,k} \leq Y_{G,m,k+1}$, $k \leq i - 1$, $k \in N$, fulfilling the concentration threshold condition $y_{RO,k} \leq Y_{G,m,k}$, $k \leq i - 1$, $y_{RO,n} \leq Y_{G,m,i}$. The stages are supplied in pre-established proportions by uniform absorbents coming, for the first stage of the mixture of the one mean concentration absorbent with that coming of the second generation stage, point 5_1, for each of the following next k, $k = 2, \ldots, i - 1$, stages of the mixture of the absorbent coming of the first inferior resorption $k - 1$ stage with that coming of the first superior generation stage $k + 1$, points 5_k, $k = 2, \ldots, i - 1$, and for the last i stage at p_{int} and the resorption process at p_h by the absorbent coming of the resorption process of the last but one resorption process, $i - 1$, point 5_i. As previously mentioned, the column benefits of a process of increasing the generated vapor pressure from the intermediary pressure to the high one. The column benefits also of processes between the successive stages, namely of pumping the one mean concentration absorbent y_M from p_l to p_1 and of the absorbent coming of the resorption processes p_{k-1} to p_k, $k = 2, \ldots, i - 1$. Besides this, the column has an inter-stage pressure reduction of the absorbent coming of the generation processes p_1 to p_l and from p_k to p_{k-1}, $k = 2, \ldots, i - 1$, and it is provided as well with the heat recovery between the opposite processes of absorption-generation and resorption-generation, of gax and of the absorbent to absorbent type (see for example the internal heat recovery mode of Fig. 4.10).

The working is as follows: the absorbent coming of the $i-1$ stage resorption process ($5_{i-1} - 5_{RO,i-1}$), is supplying in pre-established quantities the generation process at p_{int} ($5_i - 5_{GO,i}$) and the resorption at p_h ($5_{RI,n} - 5_{RO,n}$), the vapor generated at p_{int} is suffering first a process of pressure increasing from the p_{int} to p_h and then a resorption one at p_h, the rich absorbent leaves the resorption process ($5_{RI,n} - 5_{RO,n}$) at the p_h, it is subcooled in a recovering way with the desorbed vapor coming of the desorption process (1–2) at p_l, from $T_{RO,n} \geq T_{M,cooling}$ to a temperature closer to the T_{DI}, portion ($5_{RO,n} - 6$), it is expanded till the p_l, it suffers the desorption process (1–2) where it extracts the heat from the medium which must be cooled and reaches the state parameters $T_{DO} \leq T_M$ si $y_{DO} \geq y_M$, and the absorbent coming of the generation process of the first truncation column stage ($5_1 - 5_{GO,1}$) is subcooled in a recovering way, it is expanded until the low pressure, it suffers the absorption process (3–4) of the desorbed refrigerant vapor, it is mixing at the low pressure with the absorbent coming of the desorption process and it generates the one mean concentration absorbent which is covering the truncation column in the way described above until the last i stage, in order to close the cycle. A truncated cooling hybrid cycle can benefit of SHL (see Desorption Heat Transfer Analysis).

The truncated heating hybrid cycle is represented by the solid line in Fig. 4.44. It is coming of the nontruncated heating non-isobar (hybrid) cycle, represented in the same very diagram by the external solid line contour to the right side and by the dashed line to the left side. It is including a high pressure resorption process (1–2), p_h, e.g. (10–50) bar, where the useful heating effect occurs, e.g. $T_{RI,n} = (353, 15–453, 15)K$, a truncation hybrid column, which is used as the nontruncated non-isobar (hybrid) cycle truncation process and is similar to that of the cooling cycle, and a low pressure desorption processes ($5_{AI} \equiv 5_{DI} - 5_{DO}$), which for the sake of computation simplicity is considered the *n-th* generation process of the truncation column ($5n - 5GOn$). The truncation column has pressures between an intermediary value p_{int}, $p_{int}<p_h$, and a low pressure value p_l, e.g. (0.1–2) bar, and it is made up by an intermediary pressure generation process coupled on the vapor side with the high pressure resorption process, an intermediary pressure mixing process of absorbents coming of the high pressure resorption and the intermediary pressure generation processes in order to generate the one mean concentration absorbent y_M and temperature $T_M(p_{int}, y_M) \leq T_{M,heating}$, point M, a stage of low pressure absorption ($5_{AI} \equiv 5_{DI}–5_{AO}$), which for the sake of computation simplicity is considered the n - th resorption process of the truncation column ($5ROn - 5n$), coupled on the vapor side with the *n-th* low pressure generation process ($5n - 5GOn$), and respecting the concentration threshold condition $y_{RO,n} \leq Y_{G,m,n}$, and a series of $n - i$ stages of isobar opposite intermediary generation ($5_{i+k} - 5_{GO,i+k}$) and resorption processes (5_{i+k}–$5_{RO,i+k}$), $k = 1, 2, \ldots, n - i - 1$, noted by $i + 1, i + 2, \ldots, n - 1$, $i, k, n \in N$, with stages pressures decreasing from the intermediary pressure to the low one $p_{int} \geq p_{i+k} > p_{i+k+1} > p_l$, $k \in N$ and temperatures between $T_{M,cooling}$ and $T_{M,heating}$, being coupled on vapor side having essentially decreasing mean concentrations $Y_{G,m,i+k-1} \geq Y_{G,m,i+k} \geq Y_{G,m,n}$, and fulfilling the concentration threshold condition $y_{RO,i+k} \leq Y_{G,m,i+k}$, $k \leq n - i - 1$, $i, k, n \in N$. The stages are supplied in pre-established proportions by uniform absorbents coming, for the second stage of the mixture of the one mean concentration absorbent with that coming of the third stage of resorption $i + 2$, point 5_{i+1}, for each of the following next $i + k$ stages, $2 \leq k \leq n - i - 2$, of the mixture of the absorbent coming of the first superior resorption stage (5_{i+k+1}–$5_{RO,i+k+1}$) with that coming of the first inferior generation stage ($5_{i+k-1} - 5_{GO,i+k-1}$), points 5_{i+k}, for the $n - 1$ stage by the mixture of the absorbent coming of the $i = n - 2$ generation process with that of the low pressure resorption process, point 5_{n-1}, and for the low pressure processes stage of the absorbent coming of the $n - 1$ generation process. The truncation column includes also the inter-stage processes already mentioned to the cooling cycle, of absorbent pressure increase and decrease, as well as of heat recovery, which will not be repeated here.

The working is as follows: the absorbent coming of the resorption process at p_{i+1} ($5_{i+1} - 5_{RO,i+1}$), is pumped till $p_i = p_{int}$, it is preheated in a recovering way, it is supplying the generation process (3–4) at p_{int}, the refrigerant vapor generated is compressed from p_{int} to p_h where it is suffering the resorption process (1–2), the poor absorbent leaves the n-th low pressure generation process ($5n$ – $5GOn$) at p_l, it is pumped till p_h, it is preheated in a recovering way subcooling the absorbent coming of the resorption process (1–2) at p_h, it is suffering the resorption process where it is yielding the resorption heat to the medium which must be heated as the useful effect, and the absorbent coming of the resorption process at p_h is p_{int} generating the one mean concentration absorbent which is covering the truncation column until the last stage at p_l, in the way described above, for closing the cycle. Finally, two completing remarks are done. First, a truncated heating hybrid cycle can benefit of PHL (see Enhancing the Heat Pump Heating Effect by Using Preheating Hysteresis Loops (PHL)). Second, the useful resorption process of our hybrid heating cycle is favored in comparison to that of the heating cycle plotted in Fig. 4.5. Indeed, looking at the resorption domains (1–2) of the two cycles, it results that, comparatively, the hybrid cycle resorption process takes place on a smaller temperature interval T_{RI}–T_{RO} and to lower absorbent concentrations, enabling to solve more effectively certain heating applications characterized by a smaller temperature glide (e.g. domestic hot water and house heating). These favorable features become additional reasons to use the hybrid heating cycles provided with the smallest degrees of truncation, completing those referring generally to a full truncation column height diminution, exposed to the beginning of Sect. 4.3.

4.3.1 Hybrid Simple Truncated Cooling Cycles

A schematic of a hybrid simple truncated cooling cycle is depicted in Fig. 4.45, in the log p – $1/T$ diagram. The construction and operation of this cycle are particular cases of those described to the hybrid truncated cooling fractal given in Fig. 4.43. Hence, these will not be repeated here. We have modeled the operation of the cycle

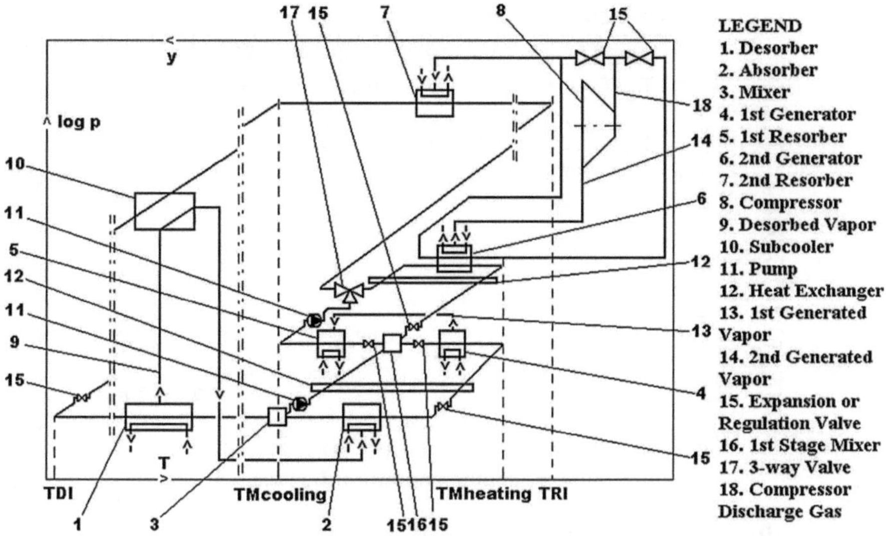

Fig. 4.45 Schematic of a hybrid simple truncated cooling cycle

at hand with the code consisting of those given in Sects. 3.2.1.2 and 3.2.5.1 and Sect. 4.2.1. The model considered the following internal heat management details: (i) out of the absorbent exiting the 1st resorber, only that supplying the 2nd generator is preheated (see the heat exchanger 12); (ii) whenever this was physically possible, the gax recovery has been considered between the 1st resorber (pos. 5) and the 2nd generator (pos. 6) (see also the study cases included in the Results of the RHE-Gax Model Run with $\Delta T_{gax,R,min}$–Infinite Equivalent Solutions to a RHE-Gax Problem); (iii) the vapor generated by the 2nd generator is rectified till a concentration equal to $Y_{rectif.} = 0.998$; (iv) the compressor discharge gas (pos. 18) superheat is recovered by the 2nd generator, as we did to the hybrid truncated cooling cycle of Sect. 3.2.5.1. The model results are given in Fig. 4.46a–f for COP_c, COP_w and COP_{mvc} and the NH_3–H_2O working combination. In order to compare these results with those of the hybrid truncated cooling cycle of Sect. 3.2.5.1, we have used here essentially the same values for the main cycle temperatures input data, that is $T_{M,c} = (12–42)$ °C, $T_{DI} = -10, \ldots, -60$ °C and $T_{DO} = T_{DI} + 3$ °C, and $\Delta T_M = T_{M,h} – T_{M,c} = 20; 30; 40$ °C (Staicovici and Staicovici 2013). Commenting upon the results, first, the cycle at hand extends comparatively the efficient cooling range towards lower temperatures by about (25–30) °C. Second, the cycle at hand increases comparatively the COP_w up to 30 %, but decreases the COP_h by about (40–70) %. Third, referring to gax recovery, this can be applied when ΔT_M takes higher values, that is $\Delta T_M = (30–40)$ °C. Concluding for this cycle, prior to its use, a comparative feasibility study should be accomplished, putting in balance its higher complexity and auxiliary pumping work consumption (see Sect. 4.2.4) on one side and the smaller compressor work consumption on the other side.

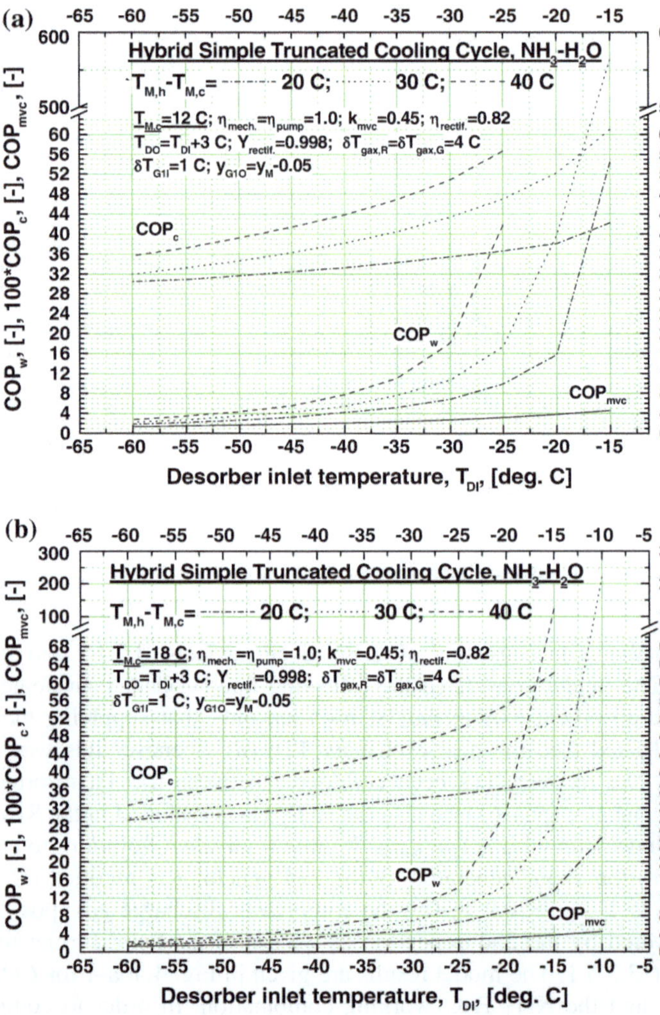

Fig. 4.46 **a** Hybrid simple truncated coabsorbent cooling cycle model results, $T_{M,c} = 12$ °C. **b** Hybrid simple truncated coabsorbent cooling cycle model results, $T_{M,c} = 18$ °C. **c** Hybrid simple truncated coabsorbent cooling cycle model results, $T_{M,c} = 24$ °C. **d** Hybrid simple truncated coabsorbent cooling cycle model results, $T_{M,c} = 30$ °C. **e** Hybrid simple truncated coabsorbent cooling cycle model results, $T_{M,c} = 36$ °C. **f** Hybrid simple truncated coabsorbent cooling cycle model results, $T_{M,c} = 42$ °C

4.3.2 Hybrid Simple Truncated Heating Cycles

The hybrid simple truncated heating cycle is represented schematically in the log $(p) - 1/T$ diagram of Fig. 4.47. Its construction and operation are particular cases of those described in a general way in conjunction with Fig. 4.44 and shall not be

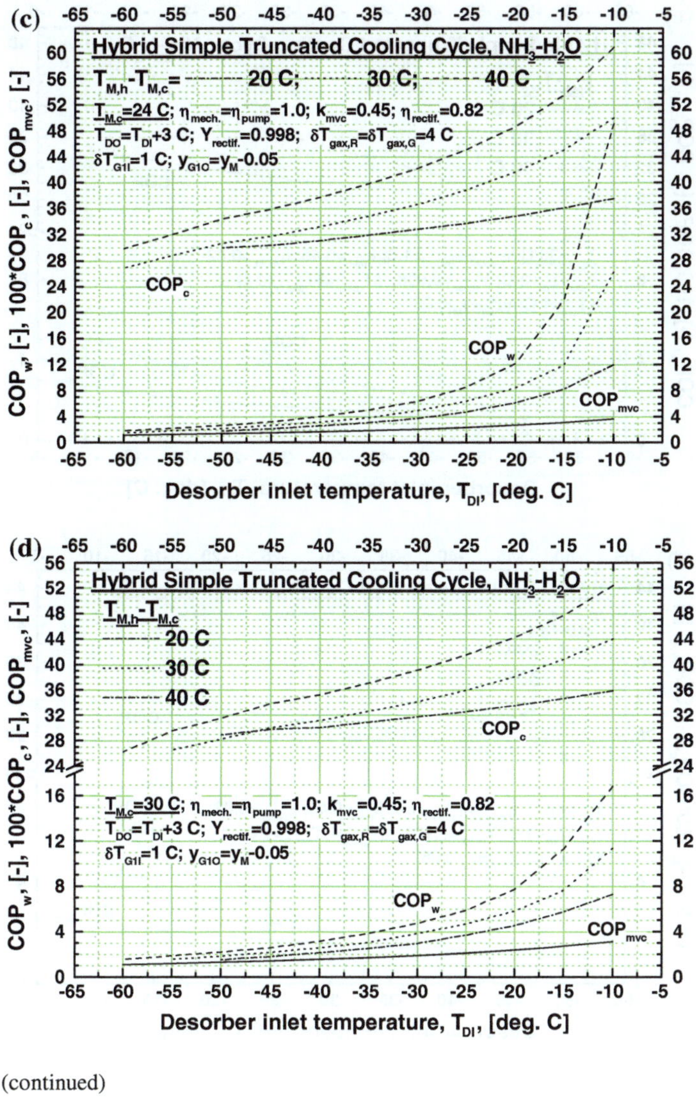

Fig. 4.46 (continued)

repeated here. We have modeled the operation of the cycle at hand again with the code consisting of those given in Sects. 3.2.1.2 and 3.2.5.1 and 4.2.1. The model considered the next internal heat management details: (i) the absorbent 18 is preheated prior to enter the resorber 8, recovering in the resorber absorbent preheater 15 an amount of the sensible heat of the absorbent exiting the resorber 8; (ii) whenever this was physically possible, the gax recovery of GHE-Gax problem type has been considered between the 1st resorber (pos. 4) and the generator (pos. 7); (iii) the resorber heating effect was enhanced considering that the cycle benefits of PHL (see Enhancing the Heat Pump Heating Effect by Using Preheating Hysteresis Loops

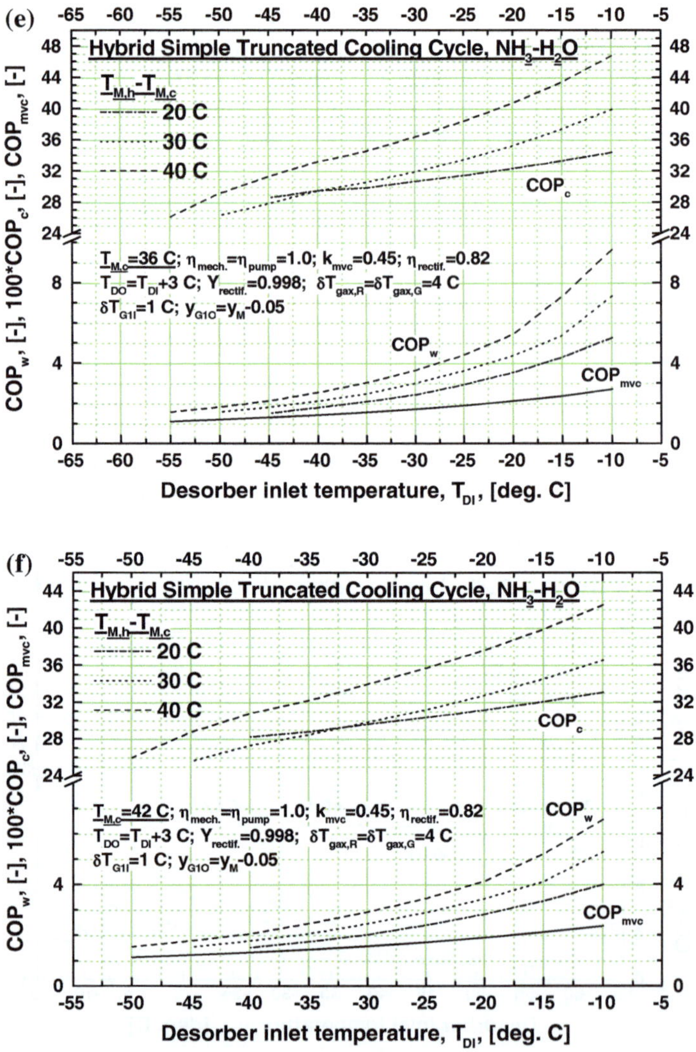

Fig. 4.46 continued

(PHL)); (iv) unlike the hybrid heat pump (Sect. 3.2.5.3), the cycle at hand cannot operate just as an Osenbrück cycle when ΔT_M becomes too small. The solution pump and compressor mechanical effectiveness input data were $\eta_{pump} = \eta_{comp.} = 1.0$. The model also considered that the generator and resorber outlet concentrations are equal (one another), $y_{GO} = y_{RO}$. The ratio evaluating the heating temperature effectiveness [see Eq. (4.133)] is expressed this time by:

$$e_{ht} = \frac{\sum \delta T_{Rf,i}}{T_{RI} - T_{RO}} \tag{4.137}$$

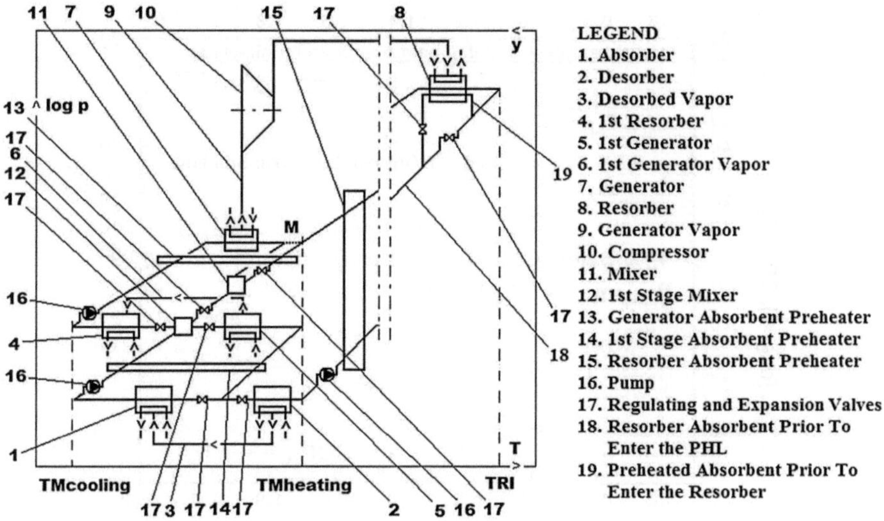

Fig. 4.47 The hybrid simple balance truncated heating cycle

The model results are plotted in Figs. 4.48a, 4.49a, 4.50a for $T_{M,c}$, COP_w and $T_{Rf,O}$ vs. p_R and in Figs. 4.48b, 4.49b, 4.50b for COP_h, e_{ht} and $\sum_{i=1}^{nR} \Delta T_{Rf,i}$ vs. p_R (Staicovici and Staicovici 2013). In order to make easier the comparison with the hybrid heat pump, the model input data for y_M, $T_{M,h}$, $T_{M,c}$ and T_{RI} are essentially the same for both cycles. The results confirm that the COP_w values of the cycle at hand are by (25–33) % higher for same resorber inlet and heating effect temperatures, T_{RI} and $T_{Rf,O}$, respectively. Also, when the gax recovery is possible, the heating effectiveness results in the highest values, $COP_h \cong 0.40$–0.45, which are quite close to those of the hybrid heat pump. Concluding, our cycle can be a hybrid heat pump challenger in house and industrial heating applications, despite its more complex structure, because of its higher COP_w and still reasonable COP_h.

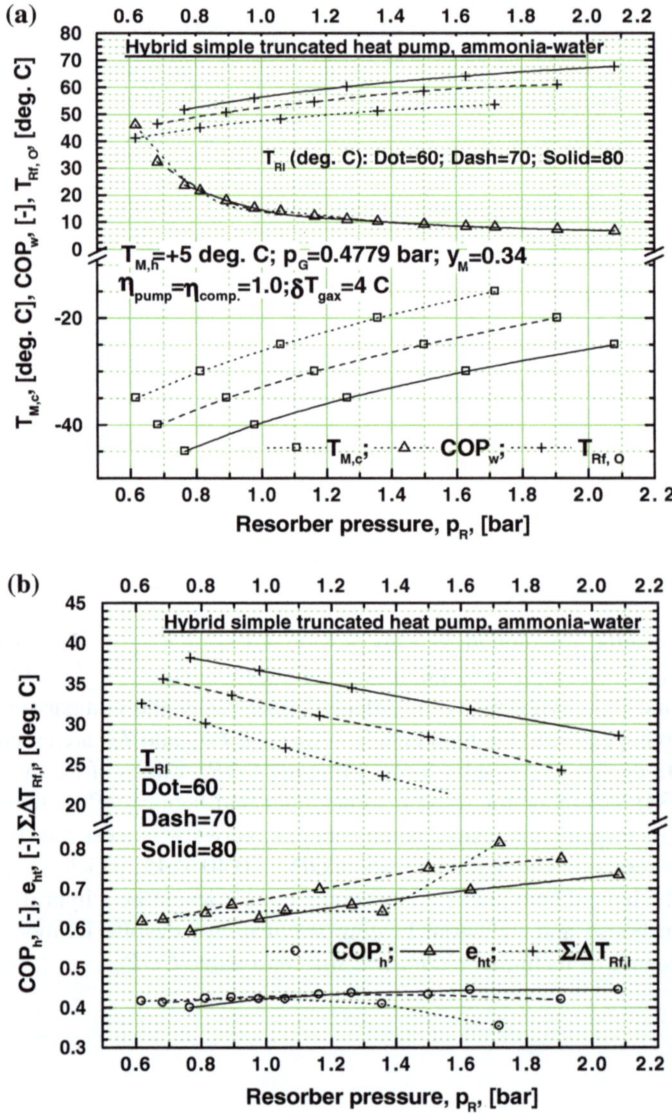

Fig. 4.48 **a** Results of $T_{M,c}$, COP_w and $T_{Rf,O}$ versus p_R, for $T_{M,h} = 5$ °C, $y_M = 0.34$ and $T_{RI} = (60–80)$ °C. **b** COP_h, e_{ht} and $\sum_{i=1}^{n_R} \Delta T_{Rf,i}$ versus p_R, for $T_{M,h} = 5°C$, $y_M = 0.34$ and $T_{RI} = (60–80)$ °C

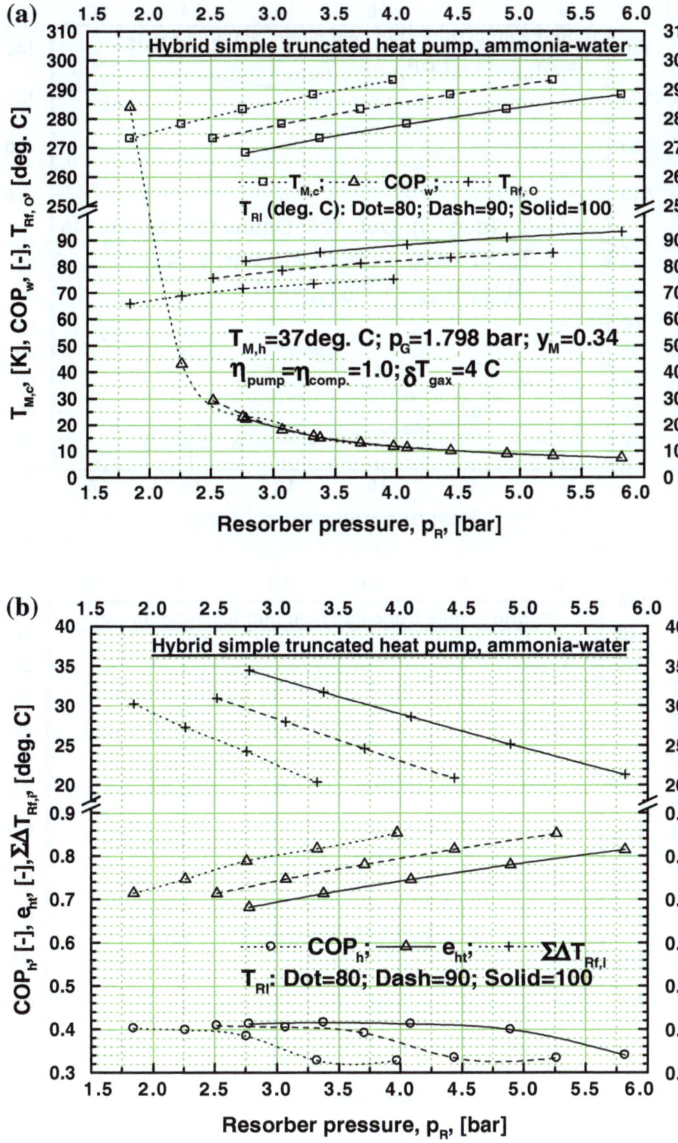

Fig. 4.49 **a** Results of $T_{M,c}$, COP_w and $T_{Rf,O}$ versus p_R, for $T_{M,h}$ = 37 °C, y_M = 0.34 and T_{RI} = (80–100) °C. **b** COP_h, e_{ht} and $\sum_{i=1}^{n_R} \Delta T_{Rf,i}$ versus p_R, for $T_{M,h}$ = 37 °C, y_M = 0.34 and T_{RI} = (80–100) °C

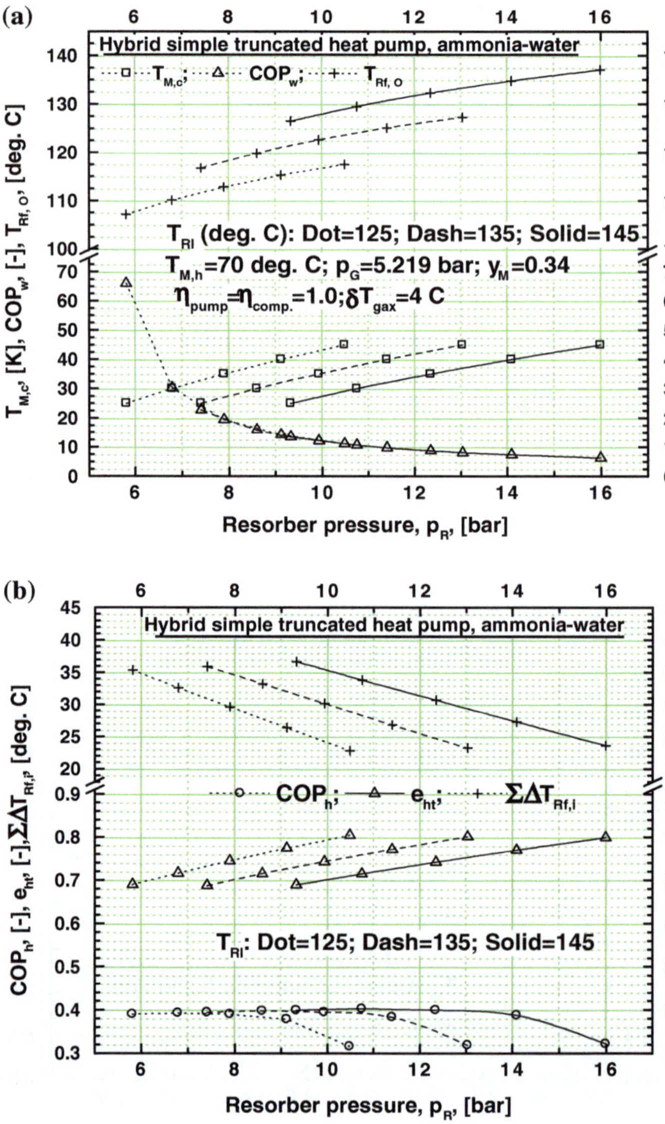

Fig. 4.50 **a** Results of $T_{M,c}$, COP_w and $T_{Rf,O}$ versus p_R, for $T_{M,h}$ = 70 °C, y_M = 0.34 and T_{RI} = (125–145) °C. **b** COP_h, e_{ht} and $\sum_{i=1}^{n_R} \Delta T_{Rf,i}$ versus p_R, for $T_{M,h}$ = 70 °C, y_M = 0.34 and T_{RI} = (125–145) °C

References

Altenkirch, E. (1914). Z. ges. Kälteind. Bd. 21, S. 7.

Bassols, J., & Schneider, R. (1993). First results of the operation of a gas-fired 250 kW absorption heat pump. In *Heat pump for energy efficiency environmental progress* (pp. 447–452).

Bokelmann, H. (1988). Industrielle Anwendung der Absorptionswärmepumpe. *BWK* Bd. 40 (1988) Nr. 6–Juni.

Erickson, D.C. (1992). Branched GAX absorption vapor compressor. US Patent 5 024 063. Vapor exchange duplex GAX absorption cycle. US Patent 5 097 676.

Herold, K. E., He, X., Erickson D. C., & Rane, M. V. (1991). The branched GAX absorption heat pump cycle. In *Japanese Association of Refrigeration Absorption Heat Pump Conference, Tokyo* (pp. 27–33).

Niebergall, W. (1959). *Handbuch der Kältetechnik, Band 7, Sorptions-Kältemaschinen*. Berlin, Göttingen, Heidelberg: Springer.

Rane, M.V., & Erickson D.C. (1993). Advanced absorption cycle: Vapor exchange GAX. In *International Absorption Heat Pump Conference. AES, ASME* (Vol. 31, pp. 25–32).

Risberg, T., Horntvedt, B., Madland, D., & Nordtvedt, S.R. (2004, August 29–September 1). Process dynamics in an industrial prototype compression/absorption heat pump. In *Proceedings of the 6th IIR Gustav Lorenzen Natural Working Fluids Conference*, Glasgow, UK.

Scharfe, J., Ziegler, F., & Radermacher, R. (1986). Analysis of advantages and limitations of absorber-generator heat exchange. *International Journal of Refrigeration, 9*, 326–333.

Staicovici, M. D. (1995a). Polybranched regenerative GAX cooling cycles. *International Journal of Refrigeration, 18*, 318–329.

Staicovici, M. D. (1995b). Polybranched regenerative GAX cycles for combined power and cooling production. In *Proceedings of 19th International Congress Refrigeration, Hague, Netherlands* (IVa, pp. 542–551).

Staicovici, M. D. (1998a). Contributions to two-phase flow with heat transfer in solar cooling plants (in Romanian). Ph.D. thesis, Universitatea Politehnica Bucuresti.

Staicovici, M. D. (1998b). Branching procedure of a regenerative GAX cycle and application plant. Patent Romania nr. 113 902/30.11.1998.

Staicovici, M. D. (2006a). Procedure of increasing the effectiveness and feasibility of a nontruncated coabsorbent cooling cycle and installation of application (in Romanian). Patent Romania No. 123082/26. 09.2006.

Staicovici, M. D. (2006b, November, 29–30). The coabsorbent cycle technology for heat pumping applications. In *Presented to the "Confort, efficiency, energy conservation and environment protection", UTCB Conference*, Bucharest, Romania.

Staicovici, M. D. (2007a). Coabsorbent heat pumping method by coupling with a thermal power station and application plant (in Romanian). Romanian patent No. 123405 /22.02.2007.

Staicovici, M. D. (2007b, April, 19–21). Coabsorbent cycle technology for ammonia/water heat pumping applications. In *Proceedings of the "Ammonia Refrigeration Technology for Today and Tomorrow" IIR Conference*, Ohrid, Macedonia.

Staicovici, M. D. (2007c). Coabsorbent heat pumps for the future. *IEA-Heat Pumps Newsletter*, 25(1), 29–32.

Staicovici, M. D. (2007d, June). Coabsorbent cycle technology for low grade sources thermal recovery. In *Proceedings of the "Technological Innovations in Air Conditioning and Refrigeration Industry" XII European Conference*, Milan (Vol. I–II).

Staicovici, M. D. (2007e). Coabsorbent technology for heat pumping applications. In *Proceedings of the IIRIIF 22 International Congress of Refrigeration, E1-2-1039*, Beijing.

Staicovici, M. D. (2007f). Coabsorbent technology for future cooling applications, Sent to Rajendra Shende, Director of the United Nations Environment Programme (UNEP), July 2007, and Didier Coulomb, Director of the International Institute of Refrigeration, November 2007.

Staicovici, M. D. (2007g). Coabsorbent cycles heat pumping and mechanical work producing procedure and applying installation. International patent file deposition no. PCT/RO 2007/000018/24.09.2007.

Staicovici, M. D. (2008a, May). Coabsorbent heat pumps. In: *Proceedings of the 9th IEA Heat Pumps Conference*, Zurich.
Staicovici, M. D. (2008b, March). Proposals for the Nine Sigma RPF 11087-1: Highly efficient low temperature heat driven cooling technologies.
Staicovici, M. D. (2009a). Coabsorbent cycles. Part one: Theory. *International Journal of Thermal Sciences*, *48*, 626–644. See also in http://dx.doi.org/10.1016/j.ijthermalsci.2008.05.004.
Staicovici, M. D. (2009b, May, 6–8). Further Results in Ammonia/Water Coabsorbent Technology District Trigeneration. In *Proceedings of Ohrid 2009 IIR/IIF Conference*, Macedonia.
Staicovici, M. D., & Staicovici, A. M. (2013). Hybrid coabsorbent heat pumping cycles. In *Proceeding of the 5th IIR Conference: Ammonia Refrigeration Technology, Ohrid.*

Chapter 5
Effectiveness of Coabsorbent Cycles and Cascades According to First and Second Principles of Thermodynamics

The effectiveness according to the first and second principles of thermodynamics is assessed in this chapter for the coabsorbent cycles presented so far, that is for the cooling and heating cycles of nontruncated, truncated and hybrid type (Staicovici 2007a, 2009). Finally, the first principle COP of the hybrid cooling and heating fractal cascades is analyzed. The ideal effectiveness is calculated in a two-step approach. During the first, an ideal mechanical work is produced by a direct Carnot cycle, calculated with the outlet temperatures of the opposite processes, generation and resorption or desorption and absorption, while during the second the work produces the useful effect by means of an inverted Carnot cycle, computed using the inlet and outlet temperatures of the desorption and absorption processes in case of cooling and of the generation and resorption processes in case of heating useful effects, respectively (see also Sect. 1.4.3). Next, the work and heat energy forms are ascribed with positive values, either these are cycle inputs or outputs.

5.1 Cooling Fractal (Nontruncated Cycle) COP

First Principle COP

The COP estimate is done in relation to Fig. 3.2 and Table 3.3. The COP referring to cooling, heating and to the cycle mechanical work consumption are expressed by Eqs. (3.53), (3.54) and (3.55), respectively. For the sake of completeness, these are repeated here by the new Eqs. (5.1)–(5.3), respectively:

$$COP_{c,cf} = \frac{q_D}{q_G} \tag{5.1}$$

$$COP_{h,cf} = \frac{q_R}{q_G} \tag{5.2}$$

$$COP_{w,cf} = \frac{q_D}{w_p} \tag{5.3}$$

M.-D. Staicovici, *Coabsorbent and Thermal Recovery Compression Heat Pumping Technologies*, Heat and Mass Transfer, DOI: 10.1007/978-3-642-54684-6_5,

Second Principle COP

The ideal coabsorbent cycle COP estimate is similar to that of an absorption cycle (Niebergall 1959). The energy balances on generation-resorption $(G - R)$ and desorption-absorption $(D - A)$ are $q_R + w = q_G$ and $q_D + w = q_A$, respectively. For the Carnot cooling COP it is considered first that the isobar $(G - R)$ processes produce the w ideal maximum mechanical work, in such a way that:

$$\eta_w = \frac{w}{q_G} = \frac{q_G - q_R}{q_G} = 1 - \frac{T_{RO}}{T_{GO}} \tag{5.4}$$

Subsequently, the work produced powers the ideal isobar $(D - A)$ processes in order to obtain a maximum efficiency of the useful cooling effect q_D, as:

$$COP_q = \frac{q_D}{w} = \frac{q_D}{q_A - q_D} = \left(\frac{T_{AO}}{T_{D,m}} - 1\right)^{-1} \tag{5.5}$$

where $T_{D,m} = \left(T_{DI,actual} + T_{DO}\right)/2$ and $T_{AO} \cong T_{RO}$. Combining Eqs. (5.4) and (5.5), the ideal cooling COP results in:

$$COP_{C,cf} = \eta_w COP_q = \frac{q_D}{q_G} = \left(1 - \frac{T_{RO}}{T_{GO}}\right)\left(\frac{T_{AO}}{T_{D,m}} - 1\right)^{-1} \tag{5.6}$$

Equations (5.6) and (1.214) are different only because $T_{D,m}$ has been used instead of T_{DI}. Equations (5.5) and (5.6) depend on the mean desorption temperature $T_{D,m}$, which is a function of the actual desorption inlet temperature, $T_{DI,actual}$. In case of the operation with non-volatile absorbent working pairs (e.g. H_2O–LiBr, NH_3–$LiNO_3$, NH_3–NaSCN), $T_{DI,actual}$ equals the design value, $T_{DI,actual} \cong T_{DI,design}$, and the cycle desorption works effectively. However, when the cycle is run by a volatile absorbent working combination (e.g., NH_3–H_2 O, NH_3–CH_3 NH_4), $T_{DI,actual}$ increases as compared to the design value due to a natural heat and temperature double deficiency, $T_{DO} > T_{DI,actual} \geq T_{DI,design}$ (see Desorption Heat Transfer Analysis). The desorption process becomes critical, with negative impact on cycle operation. Indeed, according to Eq. (5.6), the $COP_{C,cf}$ increases comparatively, while the $COP_{c,cf}$, Eq. (5.1), decreases, so the relative to Carnot cooling effectiveness, $COP_{c,rel.Carnot}$, will decrease. In order to avoid this situation, it is necessary to rectify the generated vapor. Alternatively, the desorption could be improved qualitatively if the technique using a subcooling hysteresis loop (SHL) were applied (see Desorption Heat Transfer Analysis).

5.2 Heating Fractal (Nontruncated Cycle) COP

First Principle COP

The heating COP of the cycle at hand is given by, Fig. 3.29 and Table 3.3:

$$COP_{h,hf} = \frac{q_R}{q_G + q_D} \tag{5.7}$$

The COP with respect to the cycle mechanical work consumption is expressed by:

$$COP_{w,hf} = \frac{q_R}{w_p} \tag{5.8}$$

Second Principle COP

The ideal heating COP of the classic heat transformer absorption cycle has been estimated long before (e.g. Radcenco et al. 1985; Keith 1995; Gutkowski 1987). Referring to our cycle, the energy balances on $(G - R)$ and $(D - A)$ are $q_R = q_G + w$ and $q_D = q_A + w$, respectively. In order to estimate the heat transformer ideal COP we consider first that the isobar $(D - A)$ processes produce the w ideal maximum mechanical work, in such a way that:

$$\eta_w = \frac{w}{q_D} = \frac{q_D - q_A}{q_D} = 1 - \frac{T_{AO}}{T_{DO}} \tag{5.9}$$

Subsequently, the work produced powers the ideal isobar $(G - R)$ heating processes in order to obtain a maximum efficiency useful effect q_R, upgrading the q_G heat extracted in the generation process:

$$COP_q = \frac{q_R}{w} = \frac{q_R}{q_R - q_G} = \frac{T_{Rf,O} + \Delta T_{M,h}}{\left(T_{Rf,O} + \Delta T_{M,h}\right) - T_{GI}} \tag{5.10}$$

The Carnot COP of the ideal heat transformer is given by:

$$COP_{C,hf} = \frac{q_R}{q_G + q_D} = \left(\frac{q_G}{q_R} + \frac{q_D}{q_R}\right)^{-1} \tag{5.11}$$

The first and second ratios in Eq. (5.11) result from (5.10) and Eqs. (5.9) and (5.10) as:

$$\frac{q_G}{q_R} = \frac{T_{GI}}{T_{Rf,O} + \Delta T_{M,h}} \tag{5.12}$$

and

$$\frac{q_D}{q_R} = \left(\eta_w COP_q\right)^{-1} = \left(1 - \frac{T_{AO}}{T_{DO}}\right)^{-1}\left(1 - \frac{T_{GI}}{T_{Rf,O} + \Delta T_{M,h}}\right) \tag{5.13}$$

respectively. Introducing Eqs. (5.12) and (5.13) in Eq. (5.11), the heating fractal Carnot efficiency results:

$$COP_{C,hf} = \left[\left(\frac{T_{GI}}{T_{Rf,O} + \Delta T_{M,h}}\right) + \left(1 - \frac{T_{AO}}{T_{DO}}\right)^{-1}\left(1 - \frac{T_{GI}}{T_{Rf,O} + \Delta T_{M,h}}\right)\right]^{-1} \tag{5.14}$$

Equation (5.14) is identical with Eq. (1.217), because $T_{DO} \equiv T_{GI}$. Making a parallel with the cooling cycle, the heat transformer has the resorption, responsible for the useful heating yield, as critical process, because it suffers of the preheating double deficiency problem (see the Sect. 3.2.5.2.1). The first negative consequence of this problem is that the $COP_{h,hf}$, Eq. (5.7), will decrease. The second one is that the design resorber inlet temperature, $T_{RI,design}$, is perpetually higher than the actual one, $T_{RI,actual}$, that is $T_{RO} < T_{RI,actual} < T_{RI,design}$. Consequently, the $COP_{C,hf}$ will increase, because in Eq. (5.14) $T_{RI} = T_{RI,design}$ is replaced by the actual value, $T_{RI,actual} = T_{Rf,O} + \Delta T_{M,h}$, where $\Delta T_{M,h} > 0$ [see Eq. (3.128)]. Combining the two last assertions, it results that, similarly to the cooling cycle, the relative to Carnot efficiency, $COP_{h,rel.Carnot}$, decreases. A simple method to improve qualitatively the resorption process, reducing the $COP_{h,rel.Carnot}$ decrease, is to use a preheating hysteresis loop (PHL), according to Enhancing the Heat Pump Heating Effect by Using Preheating Hysteresis Loops (PHL).

5.3 Truncated Cooling Fractal COP

First Principle COP

The COP estimate is done in relation to Fig. 4.4. The COP referring to cooling and to the cycle mechanical work consumption are expressed by Eqs. (5.15) and (5.16), respectively:

$$COP_{c,tcf} = q_D\left(1 - \frac{T_{RO}}{T_{GO}}\right)^{-1}_{\max}\left[\sum_{i=1}^{n} q_{G,i}\left(1 - \frac{T_{RO,i}}{T_{GO,i}}\right)\right]^{-1} \tag{5.15}$$

$$COP_{w,tcf} = \frac{q_D}{\sum\limits_{i=1}^{n} w_{p,i}} \tag{5.16}$$

The $COP_{c,tcf}$ of Eq. (5.15) was calculated considering the general case of a multi-temperature source generation supply, in such a way that not only the quantity, but the quality of each individual generation heat input, $q_{G,i}$, $i = 1, 2, \ldots, n$, has been taken into account as well.

Second Principle COP

Unlike the Sect. 1.4.3.2, here we confine the 2nd principle COP calculation only for the cooling effect. The energy balances on the ith generation–resorption $(G_i - R_i)$ and desorption–absorption $(D_i - A_i)$ processes are $q_{R,i} + w_i = q_{G,i}$ and $q_{D,i} + w_i = q_{A,i}$, $i = 1, 2, \ldots, n$, respectively. In order to assess the Carnot COP for the cycle at hand it is considered first that the ith isobar $(G_i - R_i)$ processes produce the w_i, $i = 1, 2, \ldots, n$, ideal maximum mechanical work, in such a way that:

$$\eta_{w,i} = \frac{w_i}{q_{G,i}} = \frac{q_{G,i} - q_{R,i}}{q_{G,i}} = 1 - \frac{T_{RO,i}}{T_{GO,i}} \tag{5.17}$$

Subsequently, the work produced powers the ith ideal isobar $(D_i - A_i)$ cooling processes for a maximum efficiency useful effect $q_{D,i}$, as

$$COP_{q,i} = \frac{q_{D,i}}{w_i} = \frac{q_{D,i}}{q_{A,i} - q_{D,i}} = \left(\frac{T_{AO,i}}{T_{D,m,i}} - 1\right)^{-1} \tag{5.18}$$

where $T_{D,m,i} = \left(T_{DI,i,actual} + T_{DO,i}\right)/2$. The thermal Carnot effectiveness of the ith shear is

$$COP_{C,i} = \frac{q_{D,i}}{q_{G,i}} = \eta_{w,i} COP_{q,i} = \left(1 - \frac{T_{RO,i}}{T_{GO,i}}\right)\left(\frac{T_{AO,i}}{T_{D,m,i}} - 1\right)^{-1} \tag{5.19}$$

The Carnot effectiveness of the truncated cooling fractal can be calculated using Eq. (5.19), as:

$$COP_{C,tcf} = \frac{\sum_{i=1}^{n} q_{D,i}}{\sum_{i=1}^{n} q_{G,i}} = \frac{\sum_{i=1}^{n} q_{G,i}\left(1 - \frac{T_{RO,i}}{T_{GO,i}}\right)\left(\frac{T_{AO,i}}{T_{D,m,i}} - 1\right)^{-1}}{\sum_{i=1}^{n} q_{G,i}} \tag{5.20}$$

The comments of the cooling fractal ideal COP regarding the desorption process hold true in the case at hand as well. An important particular case consists in considering $T_{RO,i} = T_{RO} = const.$, $T_{GO,i} = T_{GO} = const.$, $T_{AO,i} = T_{AO} = const.$ and $T_{D,m,i} = T_{D,m} = const.$, $i = 1, 2, \ldots, n$, when Eq. (5.20) becomes Eq. (5.6). Equation (5.20) was applied in its form of Eq. (5.6) to the double and quadruple truncated cooling cycles, Figs. 4.36, 4.37 and Table 4.3.

5.4 Truncated Heating Fractal COP

First Principle COP

The COP estimate is done in relation to Fig. 4.5, where the desorber-absorber stage is considered the last stage of generators-absorbers, i.e. labeled with $i = n$. The COP referring to heating and to the cycle mechanical work consumption are expressed by Eqs. (5.21) and (5.22), respectively:

$$COP_{h,thf} = q_{R,1-2}\left[\sum_{i=1}^{n} q_{G,i}\left(1 - \frac{T_{RO,i}}{T_{GO,i}}\right)\left(1 - \frac{T_{RO}}{T_{GO}}\right)_{\max}^{-1} + q_{G,3-4}\right]^{-1} \tag{5.21}$$

$$COP_{w,thf} = \frac{q_{R,1-2}}{\sum_{i=1}^{n} w_{p,i}} \tag{5.22}$$

The $COP_{h,thf}$ of Eq. (5.21) was calculated considering the general case of a multi-temperature source generation, similarly to Eq. (5.15).

Second Principle COP

The energy balances on the ith generation–resorption ($G_i - R_i$) and generation–resorption useful heating processes ($G_{3-4} - R_{1-2}$) are $q_{R,i} + w_i = q_{G,i}$, $i = 1, 2, \ldots, n$, and $q_{R,1-2} = q_{G,3-4} + \sum_{i=1}^{n} w_i$, respectively. In order to assess the Carnot COP for the cycle at hand it is considered first that the ith isobar ($G_i - R_i$) processes produce the w_i ideal maximum mechanical work, $i = 1, 2, \ldots, n$, in such a way that:

$$\eta_{w,i} = \frac{w_i}{q_{G,i}} = \frac{q_{G,i} - q_{R,i}}{q_{G,i}} = 1 - \frac{T_{RO,i}}{T_{GO,i}} \tag{5.23}$$

Subsequently, the ideal mechanical work w_i, $i = 1, 2, \ldots, n$, is used to power the ideal isobar $\left(G_{3-4,i} - R_{1-2,i}\right)$ process, when the heat $q_{R,3-4,i}$ (state 3–4), is upgraded, producing with maximum effectiveness the $q_{R,1-2,i}$ heat (state 1–2), in such a way that:

$$COP_{q,i} = \frac{q_{R,1-2,i}}{w_i} = \frac{q_{R,1-2,i}}{q_{R,1-2,i} - q_{G,3-4,i}} = \left(1 - \frac{T_{GI,3-4,i}}{T_{Rf,O,1-2,i} + \Delta T_{M,h,i}}\right)^{-1} \tag{5.24}$$

The Carnot COP is calculated for the ideal heat transformer. Using the ($G_{3-4} - R_{1-2}$) energy balance it results:

$$COP_{C,thf} = \frac{\sum\limits_{i=1}^{n} q_{R,1-2,i}}{\sum\limits_{i=1}^{n} q_{G,i} + \sum\limits_{i=1}^{n} q_{G,3-4,i}} = \frac{\sum\limits_{i=1}^{n} q_{R,1-2,i}}{\sum\limits_{i=1}^{n} q_{G,i} + \sum\limits_{i=1}^{n} q_{R,1-2,i} - \sum\limits_{i=1}^{n} w_i} \tag{5.25}$$

We express the last term of Eq. (5.25) denominator with the help of Eq. (5.23), in order to obtain:

$$\sum_{i=1}^{n} w_i = \sum_{i=1}^{n} q_{G,i}\left(1 - \frac{T_{RO,i}}{T_{GO,i}}\right) \tag{5.26}$$

Also, the numerator of Eq. (5.25) can be expressed with the help of Eqs. (5.24) and (5.26), resulting in:

$$\sum_{i=1}^{n} q_{R,1-2,i} = \sum_{i=1}^{n} q_{G,i}\left(1 - \frac{T_{RO,i}}{T_{GO,i}}\right)\left(1 - \frac{T_{GI,3-4,i}}{T_{Rf,O,1-2,i} + \Delta T_{M,h,i}}\right)^{-1} \tag{5.27}$$

Introducing Eqs. (5.26) and (5.27) in Eq. (5.25), the ideal COP of the truncated heating fractal is expressed by:

$$COP_{C,thf} = \left[1 + \frac{\sum\limits_{i=1}^{n} \frac{T_{RO,i}}{T_{GO,i}} q_{G,i}}{\sum\limits_{i=1}^{n} q_{G,i}\left(1 - \frac{T_{RO,i}}{T_{GO,i}}\right)\left(1 - \frac{T_{GI,3-4,i}}{T_{Rf,O,1-2,i}+\Delta T_{M,h,i}}\right)^{-1}}\right]^{-1} \tag{5.28}$$

The comments of the heating fractal ideal COP regarding the resorption process hold true in the case at hand as well. An important particular case consists in considering $T_{RO,i} = T_{RO} = const.$, $T_{GO,i} = T_{GO} = const.$, $T_{GI,3-4,i} = T_{GI,3-4} = const.$, $T_{Rf,O,1-2,i} = T_{Rf,O} = const.$, and $\Delta T_{M,h,i} = \Delta T_{M,h} = const.$, $i = 1, 2, \ldots, n$, when Eq. (5.28) becomes Eq. (5.29). Equation (5.28) was applied in its form of Eq. (5.29) to the double and quadruple truncated heating cycles, Figs. 4.39b, 4.40b and 4.41b.

$$COP_{C,thf} = \left[1 + \left(\frac{T_{GO}}{T_{RO}} - 1\right)^{-1}\left(1 - \frac{T_{GI,3-4}}{T_{Rf,O,1-2} + \Delta T_{M,h}}\right)\right]^{-1}. \quad (5.29)$$

5.5 Hybrid Cooling Fractal COP

First Principle COP

The COP estimate is done in relation to Fig. 4.43. Suppose a full truncated cooling fractal is equipped with a n stage truncation column. In our case, the hybrid cooling fractal is provided with a truncation column including $j = 1, 2, \ldots, i-1$ isobar stages, $i - 1 < n$, supplied by sink/heat sources. Suppose, also, the ith stage is non-isobar and the compensation for the generated vapor pressure difference till the full column pressure value is achieved with the help of a mechanical vapor compression unit which is consuming the work w_c. The cycle COP referring to the heat input and to the cycle mechanical work consumption (compressor and pumps) are expressed by Eqs. (5.30) and (5.31), respectively:

$$COP_{c,hcf} = \frac{q_D}{\sum_{j=1}^{i} q_{G,i}} \quad (5.30)$$

$$COP_{w,hcf} = \frac{q_D}{w_c + \sum_{j=1}^{i} w_{p,i}} \quad (5.31)$$

Second Principle COP

The energy balances on the jth generation-resorption (G_j − R_j) and desorption-absorption (D_j − A_j) processes are $q_{R,j} + w_j = q_{G,j}$ and $q_{D,j} + w_j = q_{A,j}$, $j = 1, 2, \ldots, i-1$, respectively. For the last, ith, non-isobar stage, the following heat balance holds true: $q_{R,i} + w_i = q_{G,i} + w_c$. In order to assess the Carnot COP for the cycle at hand, it is considered first that the jth isobar ($G_j - R_j$) processes produce the w_j, $j = 1, 2, \ldots, i$, ideal (maximum) mechanical work, given by Eq. (5.17). With its help, the total ideal work produced by the $i - 1$ stage truncation column is then written as:

$$\sum_{j=1}^{i} w_j = \sum_{j=1}^{i} q_{G,j}\left(1 - \frac{T_{RO,j}}{T_{GO,j}}\right) \quad (5.32)$$

Subsequently, an external minimum (ideal) mechanical vapor compression work, w_c, is added to this ideal work amount. Together, they are powering the inverted Carnot cycle in order to run the $(D - A)$ processes and to extract from desorber the maximum cooling heat, in such a way that:

$$COP_{C,c,mvc} = \frac{q_D}{w_c + \sum_{j=1}^{i} w_j} = \left(\frac{T_{AO}}{T_{D,m}} - 1\right)^{-1} \quad (5.33)$$

where, similar to the cooling fractal, $T_{D,m} = \left(T_{DI,actual} + T_{DO}\right)/2$ and $T_{AO} \cong T_{RO}$. Introducing Eq. (5.32) in Eq. (5.33) and solving it for w_c, we obtain the positively defined mvc work:

$$w_c = q_D\left(\frac{T_{AO}}{T_{D,m}} - 1\right) - \sum_{j=1}^{i} q_{G,j}\left(1 - \frac{T_{RO,j}}{T_{GO,j}}\right) > 0 \quad (5.34)$$

Equation (5.34) right side first term can be written obviously as:

$$q_D\left(\frac{T_{AO}}{T_{D,m}} - 1\right) = \sum_{j=1}^{n} w_j = \sum_{j=1}^{n} q_{G,j}\left(1 - \frac{T_{RO,j}}{T_{GO,j}}\right) \quad (5.35)$$

Introducing Eq. (5.35) in Eq. (5.34), it results that:

$$w_c = \sum_{j=1}^{n} w_j - \sum_{j=1}^{i} w_j \leq \sum_{j=1}^{n} w_j \quad (5.36)$$

because $\sum_{j=1}^{i} w_j \geq 0$. Equation (5.36) expresses quantitatively and ideally a truthfulness we already knew. Indeed, for the same cooling effect the hybrid cycle consumes less mechanical vapor compression work than a cooling Osenbrück (absorption mechanical vapor compression) cycle, therefore consumes less high quality primary energy (e.g. fossil fuel), but depleted (low grade) heat sources, only. Taking into account Eq. (5.34), the ideal effectiveness of the hybrid cycle with respect to the external work consumption is expressed by:

$$COP_{C,hcf,w} = \frac{q_D}{w_c} = \left[\left(\frac{T_{AO}}{T_{D,m}} - 1\right) - \frac{1}{q_D}\sum_{j=1}^{i} q_{G,j}\left(1 - \frac{T_{RO,j}}{T_{GO,j}}\right)\right]^{-1} > 0 \quad (5.37)$$

Another way to express the hybrid ideal effectiveness is to take into account all the energy input (mechanical vapor compression work + heat input). This figure of merit is less meaningful, but for the sake of completeness, it is given below:

$$COP_{C,hcf,w+q} = \frac{q_D}{w_c + \sum_{j=1}^{i} q_{G,j}} = \left[\left(\frac{T_{AO}}{T_{D,m}} - 1\right) + \frac{1}{q_D}\sum_{j=1}^{i} \frac{T_{RO,j}}{T_{GO,j}} q_{G,j}\right]^{-1} \quad (5.38)$$

An important particular case consists in considering that $T_{RO,j} = T_{RO} = const.$ and $T_{GO,j} = T_{GO} = const.$, $j = 1, 2, \ldots, i$. Keeping the frame of this particular case, with the help of Eqs. (5.30), (5.37) and (5.38) become Eqs. (5.39) and (5.40),

$$COP_{C,hcf,w} = \frac{q_D}{w_C} = \left[\left(\frac{T_{AO}}{T_{D,m}} - 1\right) - \left(1 - \frac{T_{RO}}{T_{GO}}\right) COP_{c,hcf}^{-1}\right]^{-1} \quad (5.39)$$

$$COP_{C,hcf,w+q} = \frac{q_D}{w_C + \sum_{j=1}^{i} q_{G,j}} = \left[\left(\frac{T_{AO}}{T_{D,m}} - 1\right) + \frac{T_{RO}}{T_{GO}} COP_{c,hcf}^{-1}\right]^{-1} \quad (5.40)$$

respectively. Keeping the frame of this particular case, it is worth paying to find the relative gain in the ideal COP when producing cooling by a hybrid operation instead of mechanical vapor compression, that is we are interested to find the positive defined ratio:

$$COP_{C,hcf.rel.mvc} = \frac{COP_{C,hcf,w} - COP_{C,c,mvc}}{COP_{C,c,mvc}} > 0 \quad (5.41)$$

Equation (5.41) can be rewritten as Eq. (5.42):

$$COP_{C,hcf.rel.mvc} = \frac{COP_{C,hcf,w}}{COP_{C,c,mvc}} \frac{COP_{C,hcf,w} - COP_{C,c,mvc}}{COP_{C,hcf,w}} > 0 \quad (5.42)$$

The last ratio of Eq. (5.42) is obviously less than unit, but positive. With these remarks, and considering Eqs. (5.39), (5.33), (5.30) and (5.6), this ratio becomes:

$$0 < \frac{COP_{C,hcf,w} - COP_{C,c,mvc}}{COP_{C,hcf,w}} = \frac{COP_{C,cf}}{COP_{c,hcf}} = \frac{COP_{C,tcf}}{COP_{c,hcf}} < 1,$$

wherefrom the ineqn. (5.43) holds true:

$$COP_{c,hcf} > COP_{C,cf} \quad (5.43)$$

and Eq. (5.42) rewrites:

$$COP_{C,hcf.rel.mvc} = \frac{COP_{C,hcf,w}}{COP_{C,c,mvc}} \frac{COP_{C,cf}}{COP_{c,hcf}} = \frac{COP_{C,hcf,w}}{COP_{C,c,mvc}} \frac{COP_{C,tcf}}{COP_{c,hcf}} > 0 \quad (5.44)$$

The same result would be obtained if we had considered the ineqns. $w_C > 0$ and $COP_{C,hcf,w} > 0$ within the frame of our particular case. The ineqn. (5.43) is conditional upon the 2nd principle point of view, but not upon the 1st one, when investigating the hybrid cycle effectiveness. Otherwise expressed, the ineqn. (5.43) is a sufficient condition but not necessary for the hybrid cooling operation be more effective than that of full mechanical vapor compression. In this respect, not all results of the hybrid nontruncated and simple truncated cycles plotted in Figs. 3.37a–f and 4.46a–f, respectively, are fulfilling ineqn. (5.43). Generally, the simple truncated, but especially the nontruncated hybrid cycle, characterized by small sources temperature gap, $\Delta T_M = (20-30)$ °C, are verifying it for quite large

Table 5.1 Comparative 1st and 2nd principle effectiveness cooling case for middle/northern areas

T_{DI}	$COP_{c,hcf}$ Eq. (5.30)	COP_{mvc} Eq. (3.108)	$COP_{w,hcf}$ Eq. (5.31)	$COP_{C,tcf}$ Eq. (5.6)	$COP_{C,c,mvc}$ Eq. (5.33)	$COP_{C,hcf,w}$ Eq. (5.39)	$COP_{C,hcf.rel.mvc}$ Eq. (5.41)	$\frac{COP_{C,tcf}}{COP_{c,hcf}}$
−60	0.2968	1.297	1.676	0.1951	2.805	8.183	1.918	0.6573
−50	0.309	1.568	2.391	0.2358	3.381	14.28	3.224	0.7632
−40	0.3246	1.938	3.583	0.2915	4.163	40.83	8.806	0.898
−30	0.3451	2.475	5.473	0.3722	5.286	−67.16	−13.71	1.079

desorber $T_{D,m}$ values. Next, a study case is considering the hybrid simple truncated cooling cycle of the Sect. 4.3.1, producing industrial and district cooling. The application is located to middle and northern geographical zones (Staicovici 2007b) which can supply the cycle during the whole year with sink/heat sources capable to ensure the internal temperatures of $T_{RO} = 289.15\,\mathrm{K}$, $T_{GO} = 310.15\,\mathrm{K}$ (e.g. rivers or sea water and thermal power station condensate heat, respectively). The results are included in Table 5.1 for a cycle which uses vapor rectification.

As long as ineqn. (5.43) is fulfilled, $T_{DI} = (-60) - (-40)$, the 2nd principle shows that $COP_{C,hcf,w}$ is much higher than $COP_{C,c,mvc}$. The relationship between these two figures of merit is extended naturally, although to a different scale, between the 1st principle $COP_{w,hcf}$ and COP_{mvc}, too. However, when ineqn. (5.43) is no more fulfilled, $T_{DI} \geq -30$, the 2nd principle information cannot be useful any longer, but that according to the 1st one holds still true, proving the advantage of the hybrid configuration use. Also, the big difference between $COP_{C,hcf,w}$ and $COP_{w,hcf}$ shows that there is much room for the hybrid cycle be improved.

5.6 Hybrid Heating Fractal COP

First Principle COP

The COP estimate is done in relation to Fig. 4.44. Suppose a full truncated heating fractal is equipped with a n stage truncation column. In our case, the hybrid heating fractal is provided with a truncation column, including $j = i+1, i+2, \ldots, n$ isobar stages, supplied by sink/heat sources. On account of simplicity, the desorption–absorption processes of the nth isobar stage have been assimilated with the rest of the generation–resorption processes. Suppose, also, the ith stage is non-isobar and the compensation for the generated vapor pressure difference till the full column pressure value is achieved with the help of a mechanical vapor compression unit which is consuming the work w_c. The cycle COP referring to the heat input and to the cycle mechanical work consumption (compressor and pumps) are expressed by Eqs. (5.45) and (5.46), respectively:

$$COP_{h,hhf} = q_{R,1-2}\left[\sum_{j=i+1}^{n} q_{G,j}\left(1-\frac{T_{RO,j}}{T_{GO,j}}\right)\left(1-\frac{T_{RO}}{T_{GO}}\right)_{\max}^{-1} + q_{G,3-4}\right]^{-1} \tag{5.45}$$

$$COP_{w,hhf} = \frac{q_{R,1-2}}{w_c + \sum\limits_{j=i+1}^{n} w_{p,j}} \tag{5.46}$$

Second Principle COP

The energy balances on the *j*th generation–resorption ($G_j - R_j$) and generation–resorption useful heating process ($G_{3-4} - R_{1-2}$) are $q_{R,j} + w_j = q_{G,j}$ and $q_{R,1-2} = q_{G,3-4} + w_c + \sum\limits_{j=i+1}^{n} w_j$, respectively. In order to assess the Carnot COP for the cycle at hand, it is considered first that the *j*th isobar ($G_j - R_j$) processes produce the w_j ideal maximum mechanical work, $j = i+1, i+2, \ldots, n$, in such a way that:

$$\eta_{w,j} = \frac{w_j}{q_{G,j}} = \frac{q_{G,j} - q_{R,j}}{q_{G,j}} = 1 - \frac{T_{RO,j}}{T_{GO,j}} \tag{5.47}$$

The total ideal work produced by the ($G_j - R_j$) processes, $j = i+1, i+2, \ldots, n$, becomes with Eq. (5.47) as:

$$\sum_{j=i+1}^{n} w_j = \sum_{j=i+1}^{n} q_{G,j}\left(1 - \frac{T_{RO,j}}{T_{GO,j}}\right) \tag{5.48}$$

Subsequently, an external minimum (ideal) mechanical vapor compression work, w_c, is added to this total ideal work amount, Eq. (5.48). Together, they are powering the inverted Carnot cycle in order to run the $\left(G_{3-4,i} - R_{1-2,i}\right)$ processes and to produce in the resorber the maximum heating effect, in such a way that:

$$COP_{C,h,mvc} = \frac{q_{R,1-2}}{w_c + \sum\limits_{j=i+1}^{n} w_j} = \left(1 - \frac{T_{GI,3-4}}{T_{Rf,O,1-2} + \Delta T_{M,h}}\right)^{-1} \tag{5.49}$$

Introducing Eq. (5.48) in Eq. (5.49) and solving it for w_c, we obtain the positively defined mvc work:

$$w_c = q_{R,1-2}\left(1 - \frac{T_{GI,3-4}}{T_{Rf,O,1-2} + \Delta T_{M,h}}\right) - \sum_{j=i+1}^{n} q_{G,j}\left(1 - \frac{T_{RO,j}}{T_{GO,j}}\right) > 0 \tag{5.50}$$

Equation (5.50) right side first term can be written obviously as:

$$q_{R,1-2}\left(1 - \frac{T_{GI,3-4}}{T_{Rf,O,1-2} + \Delta T_{M,h}}\right) = \sum_{j=1}^{n} w_j = \sum_{j=1}^{n} q_{G,j}\left(1 - \frac{T_{RO,j}}{T_{GO,j}}\right) \tag{5.51}$$

Introducing Eq. (5.51) in Eq. (5.50), it results that:

$$w_c = \sum_{j=1}^{n} w_j - \sum_{j=i+1}^{n} w_j \le \sum_{j=1}^{n} w_j \tag{5.52}$$

because $\sum_{j=i+1}^{n} w_j \geq 0$. Similarly to the hybrid cooling fractal, Eq. (5.52) shows in an ideal manner that for the same heating effect the hybrid heating cycle at hand consumes less mechanical vapor compression work than a heating Osenbrück cycle, therefore consumes less high quality primary energy. Taking into account Eq. (5.34), the ideal effectiveness of the hybrid cycle with respect to the external work consumption is expressed by:

$$COP_{C,hhf,w} = \frac{q_{R,1-2}}{w_C} = \left[\left(1 - \frac{T_{GI,3-4}}{T_{Rf,O,1-2} + \Delta T_{M,h}}\right) - \frac{1}{q_{R,1-2}} \sum_{j=i+1}^{n} q_{G,j}\left(1 - \frac{T_{RO,j}}{T_{GO,j}}\right)\right]^{-1} > 0 \tag{5.53}$$

Another way to express the hybrid ideal effectiveness is to take into account all the energy input (mechanical vapor compression work + heat input). This figure of merit is less meaningful, but for the sake of completeness, it is given below:

$$\begin{aligned} COP_{C,hhf,w+q} &= \frac{q_{R,1-2}}{w_C + \sum_{j=i+1}^{n} q_{G,j} + q_{G,3-4}} \\ &= \left\{ \begin{aligned} &\left(1 - \frac{T_{GI,3-4}}{T_{Rf,O,1-2} + \Delta T_{M,h}}\right) - \\ &- \frac{1}{q_{R,1-2}}\left[\sum_{j=i+1}^{n} q_{G,j}\left(1 - \frac{T_{RO,j}}{T_{GO,j}}\right) - \left(\sum_{j=i+1}^{n} q_{G,j} + q_{G,3-4}\right)\right] \end{aligned} \right\}^{-1} \end{aligned} \tag{5.54}$$

An important particular case consists in considering that $T_{RO,j} = T_{RO} = const.$ and $T_{GO,j} = T_{GO} = const.$, $j = i, i+1, \ldots, n$. Taking into account this case and Eq. (5.45) and bearing in mind that:

$$\frac{q_{G,3-4}}{q_{R,1-2}} = \frac{T_{GI,3-4}}{T_{Rf,O,1-2} + \Delta T_{M,h}}$$

Equations (5.53) and (5.54) become:

$$COP_{C,hhf,w} = \left[\frac{T_{RO}}{T_{GO}}\left(COP_{h,hhf}^{-1} - \frac{T_{GI,3-4}}{T_{Rf,O,1-2} + \Delta T_{M,h}}\right) - \left(COP_{h,hhf}^{-1} - 1\right)\right]^{-1} \tag{5.55}$$

and

$$COP_{C,hhf,w+q} = \left[\frac{T_{RO}}{T_{GO}}\left(COP_{h,hhf}^{-1} - \frac{T_{GI,3-4}}{T_{Rf,O,1-2} + \Delta T_{M,h}}\right) + 1\right]^{-1} \tag{5.56}$$

respectively. Further on, we are looking for the existence conditions of $COP_{C,hhf,w}$. In this respect, the ineqns. $COP_{C,hhf,w} > 0$ and $COP_{C,hhf,w} > COP_{C,h,mvc}$ are considered simultaneously, where:

$$COP_{C,h,mvc} = \left(1 - \frac{T_{GO}}{T_{Rf,O,1-2} + \Delta T_{M,h}}\right)^{-1} \tag{5.57}$$

is the Carnot COP of the mechanical vapor compression heating cycle, our hybrid heating coabsorbent cycle challenger, operating between T_{GO} and $T_{Rf,O,1-2} + \Delta T_{M,h}$. Taking into account Eqs. (5.55) and (5.57), it results that $COP_{h,hhf}$ must fulfil the double ineqns. (5.58) in order that $COP_{C,hhf,w}$ be useful in the ideal effectiveness assessment of the cycle at hand:

$$COP_{h,hhf,\min} < COP_{h,hhf} < COP_{h,hhf,\max} \tag{5.58}$$

where

$$COP_{h,hhf,\min} = \frac{T_{RO} - T_{RO}}{T_{GO}\left(T_{Rf,O,1-2} + \Delta T_{M,h}\right) - T_{RO}T_{GI}}\left(T_{Rf,O,1-2} + \Delta T_{M,h}\right) \tag{5.59}$$

and

$$COP_{h,hhf,\max} = \frac{T_{RO} - T_{RO}}{T_{GO}^2 - T_{RO}T_{GI}}\left(T_{Rf,O,1-2} + \Delta T_{M,h}\right) \tag{5.60}$$

Next, a short study case having as object the hybrid simple truncated heating cycle of Sect. 4.3.2 is presented. The heating application, $T_{Rf,O,1-2} = (48-63)$ °C, is considered to take place in a cold region, with sink temperatures of $T_{RO} = (-40, -20)$ °C and a heat sink source of $T_{GO} = T_{M,h} = T_{DO} = +5$ °C (see the dashed lines in Fig. 4.48a, b). The results have been completed here with the ideal effectiveness evaluation and are given in Table 5.2. The results show that as long as $COP_{h,hhf}$ respects the ineqn. (5.58), that is when $T_{RO} = -35$ °C, ..., -20 °C, the $COP_{C,hhf,w}$ values are considerable higher than those of $COP_{w,hhf}$, which, to theirs turn, are well above those of $COP_{C,h,mvc}$. Another remark we can make is that there is much room for improvements in the $COP_{w,hhf}$ effectiveness. When $COP_{h,hhf}$ is outside the $(COP_{h,hhf,\min}, COP_{h,hhf,\max})$ interval, as in case of $T_{RO} = -40$ °C, the $COP_{C,hhf,w}$ is negative, therefore cannot be applied, despite the fact that $COP_{w,hhf} >> COP_{C,h,mvc}$. In this respect, similar in a way to the hybrid cooling fractal, the ineqn. (5.58) express a sufficient condition but not necessary for the hybrid heating operation be more effective than that of full mechanical vapor compression.

5.7 COP of Hybrid Cooling and Heating Fractals Cascades

As shown so far, the coabsorbent technology is capable to process alone free sink/heat sources pairs in order to satisfy a cooling and/or heating task. There are, however, situations when the coupling of a coabsorbent cycle with a mechanical vapor compression unit in a cascade might benefit additionally the effectiveness of a heat pumping unit. The main objective of this section is to present simple algorithms which can be used in the preliminary design phase for a quick effectiveness estimate of these cascades COP, calculated with respect to the mechanical work consumption and depending on subcycles COP's, solely.

Table 5.2 A few results of the 1st and 2nd principle effectiveness evaluation in case of the hybrid simple truncated heating cycle, Sect. 4.3.2

T_{RO}	T_{GI}	$T_{Rf,O,1-2}$	$\frac{COP_{h,hhf,\min}}{COP_{h,hhf,\max}}$ Eqs. (5.59)/(5.60)	$COP_{h,hhf}$ Eq. (5.45)	$COP_{w,hhf}$ Eq. (5.46)	$COP_{C,h,mvc}$ Eq. (5.49)	$COP_{C,hhf,w}$ Eq. (5.55)	$COP_{C,hhf,w+q}$ Eq. (5.56)
−40	−32.73	48.44	0.4333 0.679	0.4121	32.27	7.402	−51.91	0.4154
−35	−27.6	52.66	0.4054 0.6899	0.4251	17.91	6.837	60.81	0.4222
−30	−22.47	56.61	0.3751 0.7032	0.4324	12.19	6.389	22.51	0.4242
−25	−17.31	60.58	0.3412 0.7212	0.4319	9.181	6.004	15.06	0.4199
−20	−12.19	62.93	0.3064 0.7432	0.42	7.342	5.802	12.61	0.4065

We analyze to the beginning the cooling cascade consisting in a first coabsorbent truncated cooling ccycle, Fig. 4.4, coupled with a second mechanical vapor compression cooling cycle, Fig. 3.36a, b. In Fig. 5.1, this case is represented by the configuration (a), for i = 0. Theirs heat balances can be written as:

$$\sum_{i=1}^{n} q_{R,i} + q_A = \sum_{i=1}^{n} q_{G,i} + q_D + w_p \tag{5.61}$$

and

$$q_E + w_c = q_C \tag{5.62}$$

respectively. The cascade consists in thermally coupling the coabsorbent cycle desorber with the mvc cycle condenser in such a way that:

$$q_D = q_C \tag{5.63}$$

The effectiveness of the two cycles with respect to the mechanical work consumption is given by Eqs. (3.55) and (3.108), noted here as:

$$I_c = \frac{q_D}{w_p} \tag{5.64}$$

and

$$J_c = \frac{q_E}{w_c} \tag{5.65}$$

respectively. The cascade COP is calculated by:

$$COP_{c,IJ} = \frac{q_E}{w_p + w_c} \tag{5.66}$$

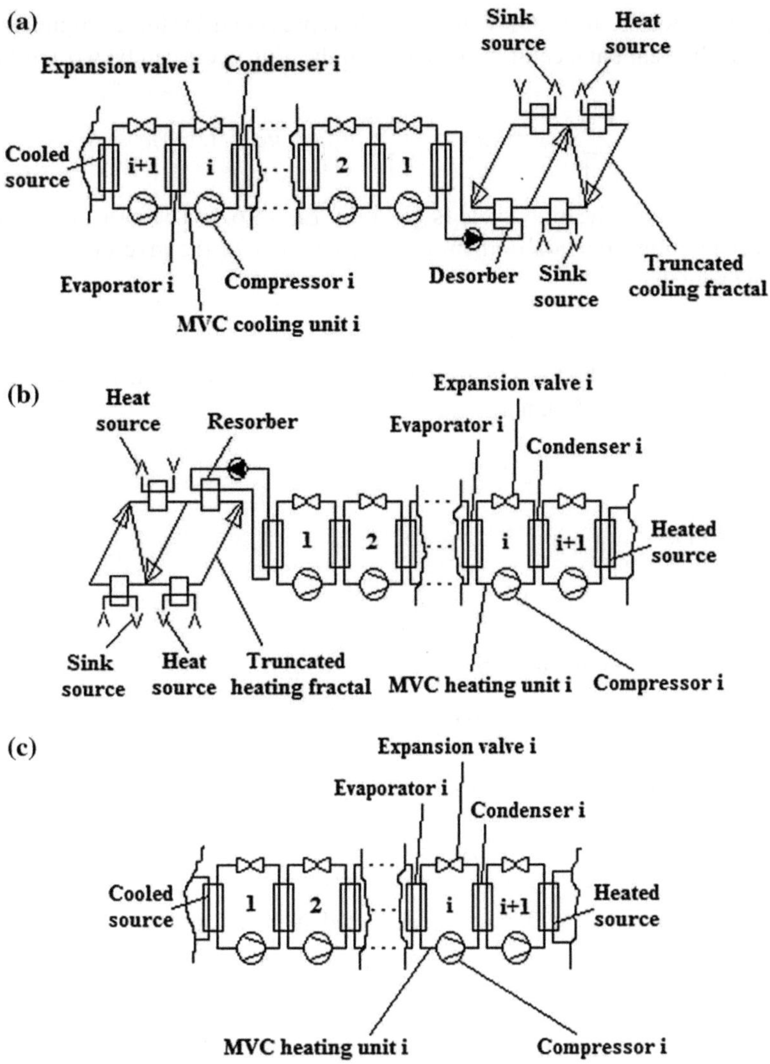

Fig. 5.1 Three cascade configurations analyzed in Sect. 5.7: **a** cooling cascade consisting in (i + 1) mvc cooling units + a truncated cooling cycle; **b** heating cascade consisting in a truncated heating cycle + (i + 1) mvc heating units; **c** (i + 1) mvc cooling and/or heating units

which, with the help of Eqs. (5.61)–(5.65), becomes:

$$COP_{c,IJ} = \frac{I_c J_c}{I_c + J_c + 1} \tag{5.67}$$

Suppose now we have a heating cascade consisting of a first coabsorbent truncated heating cycle, Fig. 4.5, coupled with a second mechanical vapor compression cycle, Fig. 3.36a, b. In Fig. 5.1, this case is represented by the configuration (b), for i = 0. The heat balance of the coabsorbent heating cycle can be written as:

$$\sum_{i=1}^{n} q_{R,i} + q_{R,1-2} = \sum_{i=1}^{n} q_{G,i} + q_{G,3-4} + w_p \tag{5.68}$$

and that of the mvc heating cycle is given by Eq. (5.62). The cascade consists in thermally coupling the coabsorbent cycle resorber with the mvc cycle evaporator in such a way that:

$$q_{R,1-2} = q_E \tag{5.69}$$

The effectiveness of the two cycles with respect to the mechanical work consumption is given by Eq. (3.55), noted as:

$$I_h = \frac{q_{R,1-2}}{w_p} \tag{5.70}$$

and Eq. (3.71):

$$J_h = \frac{q_C}{w_c} \tag{5.71}$$

The cascade COP is calculated by:

$$COP_{h,IJ} = \frac{q_C}{w_p + w_c} \tag{5.72}$$

which, with the help of Eqs. (5.68)–(5.71), becomes:

$$COP_{h,IJ} = \frac{I_h J_h}{I_h + J_h - 1} \tag{5.73}$$

Equations (5.67) and (5.73) can be extended to three-subcycle cascades, if for instance the "J" effectiveness resulted from a cascade having subcycles COP's equal to "*K*" and "*L*". For the sake of consistency, the older noting is kept for the three subcycles COP's, that is "*I*", "*J*" and "*K*", where from the following three-subcycle cascades COP's hold true for cooling and heating:

$$COP_{c,IJK} = \frac{I_c J_c K_c}{I_c J_c + J_c K_c + K_c I_c + I_c + J_c + K_c + 1} \tag{5.74}$$

and

$$COP_{h,IJK} = \frac{I_h J_h K_h}{I_h J_h + J_h K_h + K_h I_h - I_h - J_h - K_h + 1}, \tag{5.75}$$

respectively. Equations (5.67) and (5.73) on one side and Eqs. (5.74) and (5.75) on the other side, can be checked for correctness considering that one of the subcycles COP is very high. In this case, equations. mentioned above should confirm that the cascade subcycles number reduces practically by a unit. Indeed, supposing for example that the "*I*" effectiveness is very high and dividing these equations by its value, the terms having "*I*" as denominator can be neglected with respect to the others and Eqs. (5.74) and (5.75) become to the limit Eqs. (5.67) and (5.73), respectively, while the letters become Eq. (3.55). Important properties of the cascades at hand is emphasized next. Let us note by $COP_{c(h),i}$ a cooling (c) or heating (h) cascade effectiveness consisting of $i \in N$, $i = 1, 2, \ldots, m$, subcycles, having the cooling (heating) effectivnesses equal to $I_{c(h)}, J_{c(h)}, \ldots, M_{c(h)}$, respectively. It is easy to show from Eqs. (5.67) to (5.73), (5.74) and (5.75) that two successive cooling and heating cascades effectivenesses are characterized by the followings recurrent equations:

$$COP_{c,i+1} = L_{c,i+1} \frac{COP_{c,i}}{COP_{c,i} + L_{c,i+1} + 1} \tag{5.76}$$

and

$$COP_{h,i+1} = L_{h,i+1} \frac{COP_{h,i}}{COP_{h,i} + L_{h.i+1} - 1} \tag{5.77}$$

respectively, where $L_{c,i+1}$ and $L_{h,i+1}$ are the known effectivenesses of the $(i + 1)$th subcycle added to the ith cascade in order to construct the $(i + 1)$th cascade. As a first cascades property, from Eqs. (5.76) and (5.77) it results that *a cascade effectiveness is smaller than the effectiveness of each component subcycle*. Indeed, this results if we divided each member of Eqs. (5.76) and (5.77) by $L_{c,i+1}$ and $L_{h,i+1}$, respectively and then by $COP_{c(h)i}$ effectivenesses and remarked that the right member of each equation is each time smaller than unit. A second property is emphasized as follows. To this extent, we consider that the $L_{c,i+1}$ and $L_{h,i+1}$ effectivenesses are not known this time. Theirs values can be found from Eqs. (5.76) and (5.77) as functions of $COP_{c,i}$ and $COP_{c,i\,+\,1}$ on one side and of $COP_{h,i}$ and $COP_{h,i+1}$ on the other side, respectively, considered to be known, resulting in:

$$L_{c,i+1} = \frac{\left(1 + COP_{c,i}\right) COP_{c,i+1}}{COP_{c,i} - COP_{c,i+1}} \tag{5.78}$$

and

$$L_{h,i+1} = \frac{\left(COP_{h,i} - 1\right) COP_{h,i+1}}{COP_{h,i} - COP_{h,i+1}} \tag{5.79}$$

respectively. Bearing in mind that in each case $L_{c,i+1}, L_{h,i+1} > 0$, it results that $COP_{c,i} - COP_{c,i+1} > 0$ and $COP_{h,i} - COP_{h,i+1} > 0$, because ineqn. (5.79) the ineqn $COP_{h,i} - 1 > 0$ is obviously true, or:

$$COP_{c,i} > COP_{c,i+1} \quad \text{and} \quad COP_{h,i} > COP_{h,i+1},\ i = 1, 2, \ldots, m - 1 \tag{5.80}$$

Ineqns. (5.80) show the known fact, that *the series of both cooling and heating cascades effectiveness is strictly decreasing*. A third property, is emphasized for cascades consisting exclusively of mvc cooling and heating cycles, Fig. 5.1c. Bearing in mind Eq. (1.65), from Eqs. (5.65) and (5.71) it results in our case:

$$J_{c,j} + 1 = J_{h,j},\ j = 1, \ldots, i + 1 \tag{5.81}$$

This known property of the mvc cooling and heating cycles, expressed by Eq. (1.81), is translated further to the mvc cascades, as well. Indeed, from Eqs. (5.67), (5.81) and (5.73), it results:

$$COP_{c,IJ} + 1 = \frac{I_c J_c}{I_c + J_c + 1} + 1 = \frac{(I_c + 1)(J_c + 1)}{(I_c + 1) + (J_c + 1) - 1} = \frac{I_h J_h}{I_h + J_h - 1} = COP_{h,IJ} \tag{5.82}$$

Equation (5.82) is, obviously, valid for the higher order cascades, as well. Further on, with the help of Eqs. (5.81), (5.82), (5.76) and (5.77) can be rewritten as Eqs. (5.83) and (5.84)

$$COP_{c,i+1} = \frac{L_{c,i+1} COP_{c,i}}{COP_{h,i} + L_{c,i+1}} \tag{5.83}$$

and

$$COP_{h,i+1} = \frac{L_{h,i+1} COP_{h,i}}{COP_{h,i} + L_{c,i+1}} \tag{5.84}$$

respectively. Dividing Eqs. (5.83) and (5.84) term by term, the following recurrent equation results:

$$\frac{COP_{c,i+1}}{COP_{h,i+1}} = \frac{L_{c,i+1} COP_{c,i}}{L_{h,i+1} COP_{h,i}}. \tag{5.85}$$

We apply the recurrent equation in itself, in order to obtain:

$$\frac{COP_{c,i+1}}{COP_{h,i+1}} = \frac{L_{c,i+1} COP_{c,i}}{L_{h,i+1} COP_{h,i}} = \frac{L_{c,i+1} L_{c,i} COP_{c,i-1}}{L_{h,i+1} L_{h,i} COP_{h,i-1}} = \cdots = \frac{\prod_{j=1}^{i+1} L_{c,j}}{\prod_{j=1}^{i+1} L_{h,j}} \tag{5.86}$$

where $L_{c,1} = COP_{c,1}$ and $L_{h,1} = COP_{h,1}$. In Eq. (5.86), we use Eq. (5.82), and this becomes:

$$COP_{c,i+1} = \left(COP_{c,i+1} + 1\right) \frac{\prod_{j=1}^{i+1} L_{c,j}}{\prod_{j=1}^{i+1} L_{h,j}} \tag{5.87}$$

Equation (5.87) is solved for $COP_{c,i+1}$. Considering just an i cascade, instead of $i + 1$, and taking into account Eqs. (5.81) and (5.82), it results:

$$COP_{c,i} = \frac{\prod_{j=1}^{i} L_{c,j}}{\prod_{j=1}^{i} (L_{c,j} + 1) - \prod_{j=1}^{i} L_{c,j}} = \left[\prod_{j=1}^{i} \left(1 + \frac{1}{L_{c,j}}\right) - 1\right]^{-1} \tag{5.88}$$

Alternatively, Eq. (5.86) is used in order to find $COP_{h,i+1}$. Again, working with i instead of $i+1$, and introducing Eqs. (5.81) and (5.82) results, this is expressed by:

$$COP_{h,i} = \frac{\prod_{j=1}^{i} L_{h,j}}{\prod_{j=1}^{i} L_{h,j} - \prod_{j=1}^{i} (L_{h,j} - 1)} = \left[1 - \prod_{j=1}^{i} \left(1 - \frac{1}{L_{h,j}}\right)\right]^{-1} \tag{5.89}$$

The ideal cascades effectiveness is obtained simply replacing in Eqs. (5.88) and (5.89) the effectivenesses $L_{c,j}$ and $L_{h,j}$, by the Carnot effectiveness, given by Eqs. (1.208) and (1.209), respectively. Performing this replacement, the ideal mvc cooling and heating cascades effectiveness is calculated by:

$$COP_{c,C,i} = \left(\prod_{j=1}^{i} \frac{T_{C,j}}{T_{E,j}} - 1\right)^{-1} \tag{5.90}$$

and

$$COP_{h,C,i} = \left(1 - \prod_{j=1}^{i} \frac{T_{E,j}}{T_{C,j}}\right)^{-1} \tag{5.91}$$

respectively. In the ideal cascade, the condensation and evaporation temperatures of two adjacent mvc ideal cycles are characterized by the following relationship:

$$T_{C,j-1} = T_{E,j}, \quad j = 2, \ldots, i \tag{5.92}$$

Introducing Eqs. (5.92) in Eqs. (5.90) and (5.91) and performing the products in the brackets, the following results are obtained:

$$COP_{c,C,i} = \left(\frac{T_{C,i}}{T_{E,1}} - 1\right)^{-1} \tag{5.93}$$

and

$$COP_{h,C,i} = \left(1 - \frac{T_{E,1}}{T_{C,i}}\right)^{-1} \tag{5.94}$$

According to Eqs. (1.208) and (1.209), Eqs. (5.93) and (5.94) are nothing else but the ideal cooling and heating effectivenesses of a single cycle operating within the first cascade cycle evaporation temperature, $T_{E,1}$, and the last cascade cycle condensation temperature, $T_{C,i}$, respectively. In this way it has been proved the following:

Lemma *The cooling and heating effectivenesses of an ideal cascade, consisting of i ideal mvc cycles, numbered by* $1, \ldots, j, \ldots, i$*, equal those of a single ideal equivalent*

Table 5.3 Results of the deep cooling cascade study case

COP_w	COP_c	$COP_{h,cg}$	I_c Eq. (5.95)	J_c Eq. (5.65)	$COP_{c,IJ}$ Eq. (5.67)	$COP_{c,mvc}$
130.4	0.2375	17.0	3.920	4.674	1.910	1.260

mvc cycle operating within the evaporation and condensation temperatures of the first and last cascade ideal cycles, respectively.

Equations (5.88) and (5.89) generate Eqs. (5.67), (5.73)–(5.75), and others similar, if effectivenesses $I_{c(h)}$, $J_{c(h)}$, $K_{c(h)}, \ldots,$ were considered those noted here by $L_{c(h),1}$, $L_{c(h),2}$, $L_{c(h),3}, \ldots,$ etc. Although Eq. (5.81) is not valid for absorption heat pumping cycles, Eqs. (5.88) and (5.89) can be used formally to calculate a cascade effectiveness, including absorption cooling and heating cycles, as well. Additionally, Eqs. (5.88) and (5.89) confirm also clearly that if a cycle had high cooling or heating effectiveness, it can be eliminated practically from the cascade effectiveness computation. The truncated cycles, analyzed in Chap. 4 and mentioned in Fig. 5.1, belong to this category, enjoying the advantage of a small mechanical work consumption, $COP_{w,tcf(thf)} \approx (10^2 - 10^3)$. Two study cases, first of cooling and then of heating, will be analyzed below as effective cascades capitalizing this absorption cycles advantage.

5.7.1 Deep Cooling Cascade Study Case

We consider a cascade consisting of a first cycle, of ammonia-water double truncated cooling coabsorbent type, thermally coupled with an ammonia mvc cooling second cycle in such a way that the desorber of the coabsorbent cycle is providing the condensing for the mvc cycle, that is $q_D = q_C$, when $T_{DI} = -20$ °C and $T_C = -15$ °C. A CO_2 mvc cooling cycle is a better option in our case, but was not analyzed here. Additionally, the deep cooling useful effect is obtained in the mvc evaporator at $T_E = -50$ °C, while the internal coabsorbent cycle sink and heat sources temperatures are $T_{M,c} = 32$ °C and $T_{M,h} = 70$ °C, respectively. The heat source is supplied by a Steam Rankine Cycle (SRC) through a heat and power cogeneration process. This process penalizes the SRC power output, as already known and discussed. This effect appears in our case as an "additional work" consumed by the first cycle for its operation. These aspects will be presented in Chap. 7 to follow. Here, we take them into account, calculating Eq. (5.64) with:

$$I_c = \frac{q_D}{w_p + \Delta w_{SRC}} = \left(COP_w^{-1} + COP_c^{-1} COP_{h,cg}^{-1}\right)^{-1} \tag{5.95}$$

In Eq. (5.95), Δw_{SRC} is the "additional work" mentioned and COP_w, COP_c and $COP_{h,cg}$ hold for Eqs. (5.16), (5.15) and for the cogeneration effectiveness, Eq. (7.17). The results are given in Table 5.3. They show that the deep cooling cascade effectiveness, Eq. (5.67), is by more than 50 % higher than that of a system where mvc is used exclusively (last column, the CO_2–NH_3 known cascade). Table 5.3.

Table 5.4 The results of the cold region heating cascade study case, $T_C = 61$ °C

T_{RO}/T_{RI} (°C)	COP_h Eq. (5.7)	I_h Eq. (5.70)	J_h Eq. (5.71)	$COP_{h,IJ}$ Eq. (5.73)	$COP_{h,mvc}$
22.75/26.05	0.4627	195.1	6.738	6.545[a]	4.599[b]

[a] No auxiliary pumping work has been taken into account
[b] Two-stage compression heat pump

Results confirm that $I_c > COP_{c,IJ}$ and $J_c > COP_{c,IJ}$, but $COP_{c,IJ} > COP_{c,mvc}$. The last ineqn. is explained by the much lower internal work consumption of the coabsorbent cycle, $COP_w = 130.4$, as compared to that of the mvc cycle, $COP_{c,mvc} = 1.26$.

5.7.2 Cold Region Heating Cascade Study Case

We consider a cold region heating cascade. Such regions most available free sink heat sources are the ground water and the ground. These might ensure a coabsorbent cycle internal heat temperature of about $T_{M,h} = +5$ °C. The sink sources are mainly the ambient air and the sky, characterized by negative values during up to 6–8 months per year. This entitles us to rely on an internal cycle sink of about $T_{M,c} = -15$ °C. The above sink/heat temperatures can ensure the operation of the truncated cycles, but the heating effect level is not satisfactory, $T_{Rf,O} = (30 - 40)$ °C, and the resorber heat is delivered on a large temperature interval, $T_{M,h}$–$T_{Rf,O}$. The coupling to a mvc heat pump [see Eq. (5.69)], having, on the contrary, a punctual evaporation temperature, is therefore not well matched, so the more appropriate heat transformer (condensing) cycle will be considered further on. The results are given in Table 5.4, for an ammonia/water heat transformer and an ammonia heat pump.

The heat transformer resorber is preheating the heat pump working fluid from $T_E \cong 22.5$ °C up to $T_{RI} \cong 26$ °C, diminishing in this way the mvc device burden which, otherwise, would had been obliged to extract the heat from a well lower temperature heat source, of $T_{M,h} = +5$ °C. The energetic solution of choosing a cascade proves theoretically to be beneficial in this case, as $COP_{h,IJ} > COP_{h,mvc}$. Again, similar to the Sect. 5.7.1 study case, this result is possible because of the low internal work consumption of the absorption cycle, $I_h = 195.1$, as compared to that of the mvc cycle, $COP_{h,mvc} = 4.599$.

References

Gutkowski, K.M. (1987). Application of a heat transformer to solar energy systems. In *Proceedings of the 16th International Congress of Refrigeration*, B1.

Keith, E. H., Radermacher, R., & Klein, S. A. (1995). *Absorption chillers and heat pumps*. Cleveland: CRC Press.

Niebergall, W. (1959). *Handbuch der Kältetechnik Sorptions-Kältemaschinen* (Vol. 7). Berlin: Springer-Verlag.

Radcenco, V., Florescu, A., Duicu, T., Burchiu, N., Dimitriu, S., Zdrenghea, P., et al. (1985). *Heat pump installations*. Bucharest: Technical Publishing House. (in Romanian).

Staicovici, M. D. (2007a). Coabsorbent cycles heat pumping and mechanical work producing procedure and applying installation. *International patent file deposition No. PCT/RO 2007/000018*/24.09.2007.

Staicovici, M. D. (2007b). A coabsorbent technology IPAE prospectus presented and offered by the author during a visit which this has paid to Lars Gulev from the Danish Board of the District Heating, in order to cooperate and possibly apply the mentioned technology in Copenhagen to district heating and cooling.

Staicovici, M. D. (2009). Coabsorbent cycles part one: Theory. *International Journal of Thermal Sciences, 48*, 626–644.

Chapter 6
External Coabsorbent Cycle Composition

The basic idea leading to the coabsorbent cycle elaboration can be extended to other known or new cycles, more complex than the single-stage one, provided that the thermodynamics principles be respected, and new cycle usefulness and feasibility be had in view. The new, simplest cycles, obtained through an external nontruncated cycle composition, are derived here by adding a convenient number of pressure (Fig. 6.1a, b, c) or concentration (Fig. 6.1e, f) stages to a basic fractal. The possible mixing points are noted by M and upset M for cooling and heating fractals, respectively (Staicovici 2009a).

One of the external composition application is to transpose advanced (multi-effect) cycles (Alefeld 1982) in coabsorbent technology (Figs. 6.2, 6.3, 6.4), made according to the:

Rule of transposition in coabsorbent cycles (Staicovici 2006): *A separate absorbent condensing cycle of origin can be transposed in a coabsorbent cycle, if the condensing cycle subcycles of the cascade were replaced each by a nontruncated coabsorbent cycle (fractal), and then interconnected, more often than not through their superposition, provided that at least a mixing point exists.*

The transposition into coabsorbent cycles favors in certain conditions the absorption technology feasibility through COP increase and cycle complexity decrease [in Figs. 6.2 and 6.4, the cycles (b) have only 6 main devices instead of 8 in (a)].

6.1 The Pressure-Stages Multi-Effect Coabsorbent Cooling Cycle (PSMECCC) Thermal Analysis

Out of the advanced (multi-effect) cycles transposed into coabsorbent cycles, the cooling cycle with pressure stages, Figs. 6.1a, b, c, and 6.2a, b, will be analyzed here. It can operate with both broad and narrow solubility field working combinations, has simpler flow (require one mixing point only), is less complex and is efficient. The two last aspects will be analyzed in more detail next: (I) We consider

M.-D. Staicovici, *Coabsorbent and Thermal Recovery Compression Heat Pumping Technologies*, Heat and Mass Transfer, DOI: 10.1007/978-3-642-54684-6_6,

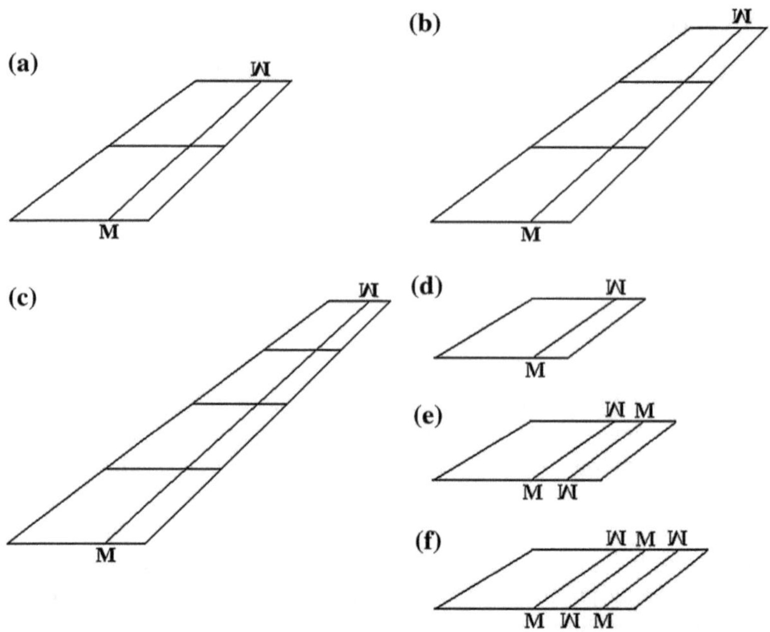

Fig. 6.1 New coabsorbent cycles with double (**a**), triple (**b**) and quadruple (**c**) pressure stages and with simple (**e**) and double (**f**) concentration stages, $\log(p) - 1/T$ diagram

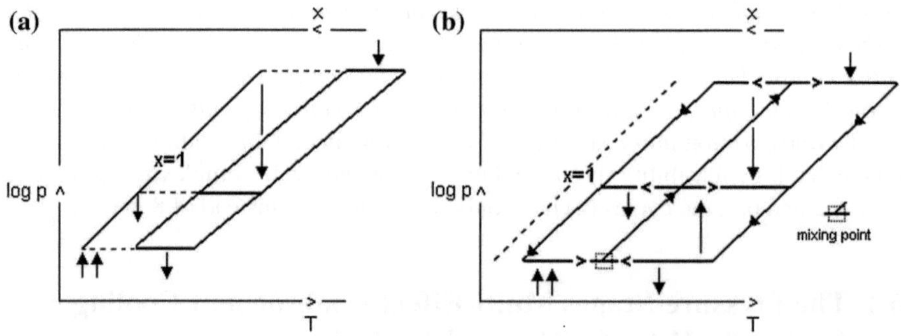

Fig. 6.2 Double-effect cycle with condenser heat recovery (**a**), transposed in coabsorbent cycle (**b**)

the *n* pressure stages cooling cycle of origin, versus the transposed. The origin advanced cycle COP has the expression:

$$COP_{acc,cond} = \sum_{i=k}^{n} \prod_{j=1}^{k} \eta_j \tag{6.1}$$

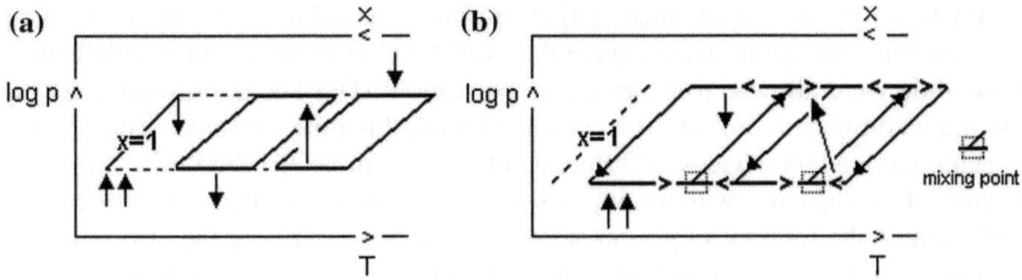

Fig. 6.3 Double-effect cycle with absorber heat recovery (**a**), transposed in coabsorbent cycle (**b**)

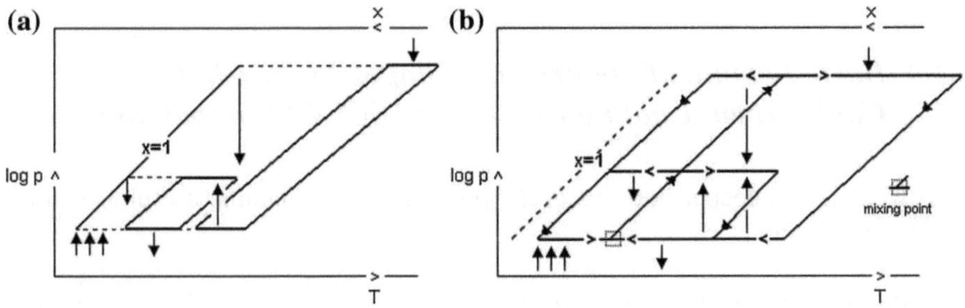

Fig. 6.4 Triple-effect cycle with condenser and absorber heat recovery (**a**), transposed in coabsorbent cycle (**b**)

In Eq. (6.1), obtained first by Staicovici (1992) for up to triple-effect and extended later for higher-order effect by Ziegler et al. (1993), η_j are the cooling COP's of the cascade subcycles, which normally have different values, as the cycles work with different parameters. Equation (6.1) shows that an investment for COP increase, obtained by adding new effects to the cascade, has diminished thermal returns, and the complexity increases. On the contrary, the transposed advanced cycle shows that the investment for COP increase will have close to proportional expected thermal returns, because, when replacing the condensing by resorption, the rejected available heat is higher than the condensing heat, comparable to that of its vapor coupled generation for symmetry factors sensibly equal to unit, that is we shall have rather a linear expression in η_j,

$$COP_{acc,coabs} \cong \sum_{i=1}^{n} \eta_i \tag{6.2}$$

as cycle COP than that given by Eq. (6.1), where η_i, $i = 1, 2, \ldots, n$, are the COP's of each cooling subcycle of the cascade; (II) The large temperature overlapping, met to resorption/generation gax recovery use, enables the comparative heat source temperature decrease and cycle exergy COP increase. This is why, a transposed cycle can usually have one more added effect for the same maximum operating temperature, as compared to its origin cycle (Staicovici 2007); (III) The complete elimination of the vapor rectification during the volatile absorbent working combinations operation favours again the cycle COP, if the desorption with temperature glide were allowed to a particular application; (IV) The transposed cycle has less main component devices as its origin. Indeed, this time the same role and same working parameters devices of different subcycles can be replaced by a single one of equivalent capacity, without any absorbent inventory problem risk. For a n_{effect} cycle, out of the total of $4n_{effect}$ main devices $2(n_{effect} - 1)$ have the same function (absorbers and desorbers) and the same working parameters and can be replaced by those belonging to the basic fractal with corresponding capacity, so $2(n_{effect} + 1)$ main devices remain to be manufactured, only.

6.1.1 Basic Lemma of the Pressure-Stages Multi-Effect Coabsorbent Cooling Cycle (PSMECCC) Computation

Next we shall go deeper into this cycle, proving a basic lemma, useful to its thermal computation:

Lemma *A pressure-stages coabsorbent cycle, resulted from the transposition of a n-effect advanced absorption cycle recovering the condensing heat, has the absorption and desorption absorbent flow factors equal to*

$$a = V_1 \sum_{i=1}^{n} g_{1,i} \prod_{j=1}^{i-1} r_{gax,j} \tag{6.3}$$

and

$$d = V_1 \sum_{i=1}^{n} r_{1,i} \prod_{j=1}^{i-1} r_{gax,j} \tag{6.4}$$

respectively, and the same characteristic equation as that of the basic cooling fractal, where $r_{gax,i}$, $i = 1, \ldots, n - 1$, is the gax factor of the $(i + 1)$th resorption process yielding heat to the ith *generation internal heat recovering process.*

Proof Let us consider the coabsorbent cycle, plotted in the $\log(p) - 1/T$ diagram in Fig. 6.5. Constructively, it results by overlapping and connecting through mass and heat exchange i, $i = 1, 2, \ldots, n$, coabsorbent cycles, Fig. 3.2, in a cascade, so that the$(i + 1)$th subcycle resorber heat is recovered by the ith subcycle generator. It works between p_0 and p_n pressures and $y_{RO,1}$ and $y_{GO,1}$ extreme concentrations, with y_M as mixing concentration. The whole resulting advanced cycle has a common bottom with the same desorber, externally heated by a source usefully cooled and the same absorber, externally cooled, respectively. The first subcycle resorber, $R(i = 1)$, is externally cooled, producing useful heat. The nth cycle generator is the only connected externally to a high temperature heat source. The absorbents coming of the desorber and absorber are mixing up in the mixer M generating the

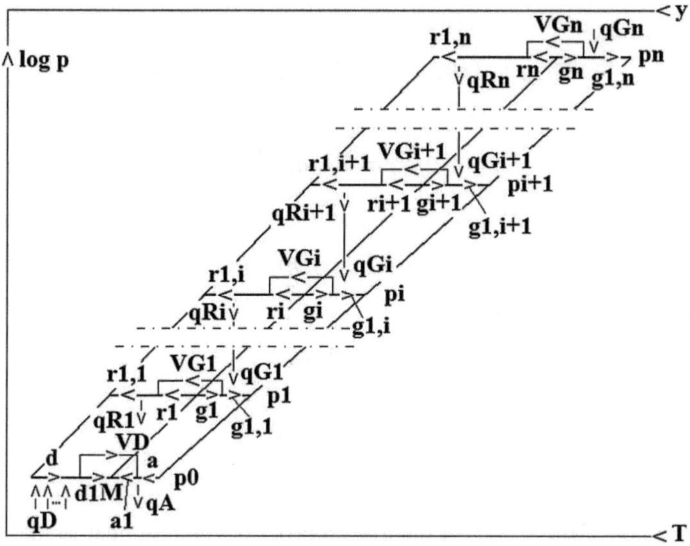

Fig. 6.5 A *n*-effect cooling cycle with condenser heat recovery, transposed into a pressure-stages multi-effect coabsorbent cooling cycle (PSMECCC)

absorbent of mean concentration. The y_M absorbent supplies further all the i pressure stages generators and resorbers, $i = 1, 2, \ldots, n$. The absorbents coming of generators and resorbers have the same $y_{RO,1}$ and $y_{GO,1}$ extreme concentrations. They are collected from each pressure stage for supplying the absorber and desorber, respectively, and closing the cycle. Thermally, the cycle produces i cooling effects, $i = 2, \ldots, n$, with a single heat input. The mass balances of absorption and desorption processes, as well as the heat balance of the $(i + 1)$th and ith pressure stage gax heat exchange, $i = 1, \ldots, n - 1$, results in:

$$a = \sum_{i=1}^{n} V_i g_{1,i} \tag{6.5}$$

and

$$d = \sum_{i=1}^{n} V_i r_{1,i} \tag{6.6}$$

and

$$q_{R,i+1} = q_{G,i} \tag{6.7}$$

respectively, where V_i, $i = 1, \ldots, n - 1$, is given by Eq. (4.6). Also, in Eq. (6.7) we have:

$$q_{R,i} = V_{G,i} q_{R,u,i} \tag{6.8}$$

with

$$q_{R,u,i} = \left(H_{G,i} - h_{RI,i}\right) + r_{1,i}\left(h_{RI,i} - h_{RO,i}\right) \tag{6.9}$$

and

$$q_{G,i} = V_{G,i} q_{G,u,i} \tag{6.10}$$

with

$$q_{G,u,i} = \left(H_{G,i} - h_{GI,i}\right) + g_{1,i}\left(h_{GO,i} - h_{GI,i}\right) \tag{6.11}$$

From Eqs. (6.7–6.11), we obtain:

$$V_{j+1} = V_j r_{gax,j} \tag{6.12}$$

where

$$r_{gax,j} = \frac{q_{G,u,j}}{q_{R,u,j+1}} \tag{6.13}$$

is the gax factor of the $(j+1)$th resorption process yielding heat to the jth generation internal heat recovering process, $j = 1, \ldots, n-1$. In the recurrent Eq. (6.12) we make successively $j = 1, 2, \ldots, i-1$, in order to find:

$$V_i = V_1 \prod_{j=1}^{i-1} r_{gax,j}, \; i = 1, \ldots, n-1 \tag{6.14}$$

In Eq. (6.14) we choose V_1 and remark that $V_i < V_{i+1}$, $i = 1, \ldots, n-1$, because $r_{gax,j} > 1$, for all j, where from it results that V_1 has the smallest value amongst all other stages reduced vapor. With Eq. (6.14), Eqs. (6.5) and (6.6) become Eqs. (6.3) and (6.4), and the first part of the lemma has been proved. In order to prove its second part, it suffices to multiply and divide $g_{1,i}$ in Eq. (6.3) by $r_{1,i}$, and remark that

$$\frac{g_{1,i}}{r_{1,i}} r_{1,i} = \frac{g_{1,1}}{r_{1,1}} r_{1,i} = S r_{1,i}, \quad i = 1, \ldots, n \tag{6.15}$$

wherefrom, introducing this result in Eq. (6.3), we obtain

$$a = S V_1 \sum_{i=1}^{n} r_{1,i} \prod_{j=1}^{i-1} r_{gax,j} = Sd \tag{6.16}$$

or

$$\frac{a}{d} = S \tag{6.17}$$

which proofs the whole lemma (*q.e.d.*). The cycle first principle cooling COP is given by

$$COP_{psccac} = \frac{q_D}{q_{G,n}} = \left(V_1 q_{G,u,n} \prod_{j=1}^{n-1} r_{gax,j}\right)^{-1} q_D \tag{6.18}$$

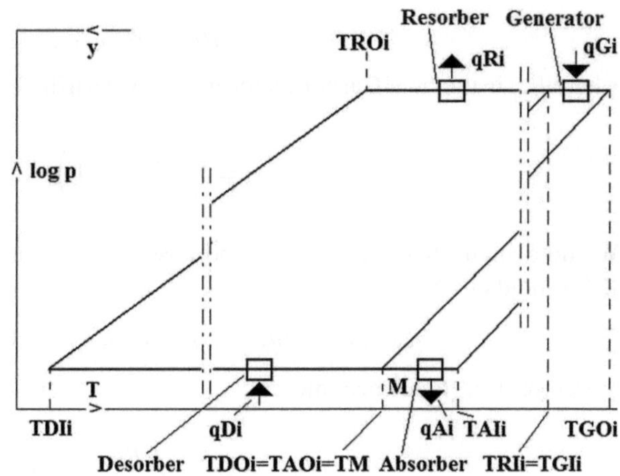

Fig. 6.6 The *i*th subcycle of the pressure-stages multi-effect coabsorbent cooling cycle (PSMECCC), showing the temperatures the heat inputs and outputs are exchanged with

where q_D is given by Eq. (3.42). The ideal cooling effectiveness of a multi-effect coabsorbent cycle can be calculated with the help of a theorem, proved to the next section.

6.1.2 Carnot COP Theorem of the Pressure-Stages Multi-Effect Coabsorbent Cooling Cycle (PSMECCC)

The *n*-effect cycle structure, Fig. 6.5, is completed by Fig. 6.6. Here are noted the temperatures and the heats involved in our approach for the *i*th, $i = 1, 2, \ldots, n$, thermal cascade subcycle. Next we shall obtain an internal Carnot COP expression of the cooling cycle of Fig. 6.5, considering the internal temperatures, only (see the motivation of Chap. 5 introductory part). The way to proceed in order to obtain the ideal effectiveness is identical to that followed in Chap. 5, derived in two known steps and repeated here for the sake of clarity. In the first step, it is evaluated the ideal work w_i of the *i*th cascade subcycle, $i = 1, 2, \ldots, n$. This is produced by a direct Carnot cycle, running between the *i*th generator, which has the heat input $q_{G,i}$ at $T_{GO,i}$, and the *i*th resorber, which rejects the heat $q_{R,i}$ at $T_{RO,i}$, Fig. 6.6. The ideal work w_i is expressed by

$$w_i = q_{G,i}\theta_{w,i}, \tag{6.19}$$

where $\theta_{w,i}$ is the mechanical work Carnot factor given by, (Kirilin et al. 1985):

$$\theta_{w,i} = 1 - \frac{T_{RO,i}}{T_{GO,i}}. \tag{6.20}$$

In the second step, the ideal work w_i of the *i*th cascade subcycle, $i = 1, 2, \ldots, n$, is utilized to extract the ideal heat $q_{D,i}$ from the sink heat source. This is done by a reverse Carnot cycle running between the *i*th desorber, which has the heat input $q_{D,i}$ at $T_{DI,i}$, and the *i*th absorber, which rejects the heat $q_{A,i}$ at $T_{AO,i}$, Fig. 6.6. The ideal heat $q_{D,i}$ is expressed by

$$q_{D,i} = w_i \theta_{q,i}, \tag{6.21}$$

where $\theta_{q,i}$ is the heat Carnot factor given by, (Kirilin et al. 1985):

$$\theta_{q,i} = \left(\frac{T_{AO,i}}{T_{DI,i}} - 1 \right)^{-1}. \tag{6.22}$$

We note again that the desorber and absorber are common devices, for which Eq. (6.23) holds true

$$T_{DI,i} = T_{DI},\ T_{AO,i} = T_{AO}, \quad i = 1, 2, \ldots, n. \tag{6.23}$$

therefore, Eq. (6.22) becomes:

$$\theta_{q,i} = \theta_q = \left(\frac{T_{AO}}{T_{DI}} - 1 \right)^{-1}. \tag{6.24}$$

The Carnot cooling COP of the ith absorption cycle, $i = 1, 2, \ldots, n$, is defined now by:

$$COP_{C,i} = \frac{q_{D,i}}{q_{G,i}} = \frac{q_{D,i}}{w_i} \frac{w_i}{q_{G,i}}. \tag{6.25}$$

With Eqs. (6.19)–(6.21) and (6.24), Eq. (6.25) becomes (Niebergall 1959):

$$COP_{C,i} = \theta_{w,i} \theta_q = \left(1 - \frac{T_{RO,i}}{T_{GO,i}} \right) \left(\frac{T_{AO}}{T_{DI}} - 1 \right)^{-1}. \tag{6.26}$$

With this preliminary summarizing knowledge we can now formulate the following:

Theorem *The Carnot COP of a n-effect pressure-stages coabsorbent cooling cycle is given by:*

$$COP_{C,n-effect} = \left(1 - \frac{T_{RO,1}}{T_{GO,n}} \right) \left(\frac{T_{AO}}{T_{DI}} - 1 \right)^{-1}, \tag{6.27}$$

where $T_{RO,1}$, $T_{GO,n}$, T_{AO} and T_{DI} hold for the outlet temperatures of the first resorber, last generator and absorber and the desorber inlet temperature producing the useful effect, respectively.

Proof We write the thermal balance of the $(i + 1)$th fractal

$$q_{G,i+1} + q_{D,i+1} = q_{R,i+1} + q_{A,i+1}, \tag{6.28}$$

we divide it by $q_{D,i+1}$ and take into account the thermal cascade condition

$$q_{R,i+1} = q_{G,i}, \tag{6.29}$$

in order to obtain:

$$1 + \frac{q_{D,i+1}}{q_{G,i+1}} = \frac{q_{G,i}}{q_{G,i+1}} + \frac{q_{A,i+1}}{q_{G,i+1}}, \quad i = 1, 2, \ldots, n. \tag{6.30}$$

In Eq. (6.30) we note that

$$\frac{q_{D,i+1}}{q_{G,i+1}} = COP_{C,i+1} = \theta_{w,i+1}\theta_q, \tag{6.31}$$

and

$$\frac{q_{G,i}}{q_{G,i+1}} = \frac{q_{D,i}}{q_{D,i+1}} \frac{\theta_{w,i+1}}{\theta_{w,i}}. \tag{6.32}$$

We consider now the energy balance of the $(i + 1)$th reverse Carnot cycle. This is running with its $(i + 1)$th desorber, having the heat input $q_{D,i+1}$ at $T_{DI,i+1}$, and the $(i + 1)$th absorber, rejecting the heat $q_{A,i+1}$ at $T_{AO,i+1}$. Dividing its energy balance equation by $q_{G,i+1}$, we obtain:

$$\frac{q_{A,i+1}}{q_{G,i+1}} = \frac{q_{D,i+1}}{q_{G,i+1}} + \frac{w_{i+1}}{q_{G,i+1}}. \tag{6.33}$$

In Eq. (6.33), the right member first term is given by Eq. (6.31). Also, the last one results from Eq. (6.19), wherefrom, Eq. (6.33) rewrites as:

$$\frac{q_{A,i+1}}{q_{G,i+1}} = COP_{C,i+1} + \theta_{w,i+1}. \tag{6.34}$$

We introduce Eqs. (6.31), (6.32) and (6.34) in Eq. (6.30) and, simplifying, we obtain the following useful iterative equation:

$$q_{D,i} = q_{D,i+1}\theta_{i,i+1}, \quad i = 1, 2, \ldots, n-1 \tag{6.35}$$

where:

$$\theta_{i,i+1} = \theta_{w,i}\left(\theta_{w,i+1}^{-1} - 1\right). \tag{6.36}$$

We write Eq. (6.35) for $i = 1, 2, \ldots, n - 1$ and multiply term by term the resulted equations, obtaining

$$\prod_{i=1}^{n-1} q_{D,i} = \prod_{i=2}^{n} q_{D,i}\theta_{i-1,i}, \tag{6.37}$$

wherefrom, we have:

$$q_{D,1} = q_{D,n}\prod_{i=2}^{n}\theta_{i-1,i}. \tag{6.38}$$

Eq. (6.38) is introduced in Eq. (6.35) and, making successively $i = 1, 2, \ldots, n - 1$, we can write the desorber heats as:

$$q_{D,i} = q_{D,n} \prod_{j=i+1}^{n} \theta_{j-1,j}, \quad i = 1, 2, \ldots, n-1 \tag{6.39}$$

Next we need to tot the desorber heats. With Eq. (6.39), this sum is written by:

$$\sum_{i=1}^{n} q_{D,i} = q_{D,n} \left(1 + \sum_{i=1}^{n-1} \prod_{j=i+1}^{n} \theta_{j-1,j} \right). \tag{6.40}$$

The 2-nd principle COP of the pressure-stage n-effect coabsorbent cooling cycle is calculated by

$$COP_{C,n-effect} = \frac{1}{q_{G,n}} \sum_{i=1}^{n} q_{D,i}, \tag{6.41}$$

or, with Eq. (6.40), Eq. (6.41) rewrites:

$$COP_{C,n-effect} = \frac{q_{D,n}}{q_{G,n}} \left(1 + \sum_{i=1}^{n-1} \prod_{j=i+1}^{n} \theta_{j-1,j} \right) = COP_{C,n} \left(1 + \sum_{i=1}^{n-1} \prod_{j=i+1}^{n} \theta_{j-1,j} \right). \tag{6.42}$$

Further on we need to solve the bracket in Eq. (6.42). From Eqs. (6.20) and (6.36) we can write

$$\theta_{j-1,j} = \frac{T_{GO,j-1} - T_{RO,j-1}}{T_{GO,j-1}} \frac{T_{RO,j}}{T_{GO,j} - T_{RO,j}}, \quad j = 2, 3, \ldots, n. \tag{6.43}$$

Remarking that in our ideal cascade the heat exchange is characterized by

$$T_{RO,j} = T_{GO,j-1}, \tag{6.44}$$

Eq. (6.43) becomes

$$\theta_{j-1,j} = \frac{T_{GO,j-1} - T_{RO,j-1}}{T_{GO,j} - T_{RO,j}}, \tag{6.45}$$

wherefrom, the bracket takes the form:

$$1 + \sum_{i=1}^{n-1} \prod_{j=i+1}^{n} \theta_{j-1,j} = \sum_{i=1}^{n} \frac{T_{GO,i} - T_{RO,i}}{T_{GO,n} - T_{RO,n}} = \frac{T_{GO,n} - T_{RO,1}}{T_{GO,n} - T_{RO,n}} \tag{6.46}$$

Finally, bearing in mind that

$$COP_{C,n} = \frac{T_{GO,n} - T_{RO,n}}{T_{GO,n}} \left(\frac{T_{AO}}{T_{DI}} - 1 \right)^{-1}, \tag{6.47}$$

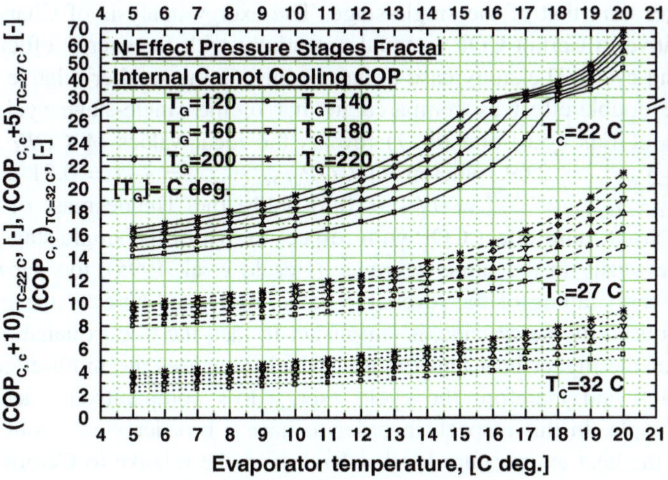

Fig. 6.7 The Carnot efficiency of the *n*-effect pressure-stages condensing cooling cycle versus evaporator temperature, Eq. (6.27)

we introduce Eqs. (6.46) and (6.47) in Eq. (6.42) and hence Eq. (6.27) and the theorem proof result (*q.e.d.*).

The theorem applies to the coabsorbent and particularly condensing cycles at hand, run by four temperatures. Equation (6.27) was plotted in Fig. 6.7 in function of the $T_{GO,n}$ and T_C parameters. Here we considered the condensing advanced cycle, with the resorber and absorber temperatures equal to the condensing temperature, $T_{RO,1} = T_{AO} = T_C$, and the desorber temperature equal to that of the evaporator, $T_{DI} = T_E$.

The theorem is completing a gap in the theory of the multi-effect cooling cycles based on condensing/resorption heat recovery. It is proved for the first time here, although attempts to disclose it publicly in this final form date several years back (Staicovici 2009e). In a past work (Herold et al. 1995), the authors, lacking the results of this theorem, used the three temperature ideal effectiveness of the single stage absorption cooling cycle (see Eq. 5.6) in order to assess the ideal effectiveness of the similar type double-effect condensing cycle. Their result is numerically correct, but the way to obtain it is not rigorous, thermodynamically.

The increase of the generation temperature benefits, on one side, the advanced cycle quantitatively, that is its actual COP, calculated according to the 1st principle, is significantly higher than that of the single-stage COP. This is shown for instance by the next study case of a H_2O–LiBr cycle. Indeed, operated as single-stage, by $T_G \cong 80$ °C, $T_C = 32$ °C and $T_E = 5$ °C, the cycle achieves a chilling figure of merit of about $COP_c \cong 0.72$, while running as double-effect, by $T_G \cong 140$ °C and keeping constant T_C and T_E, it achieves an effectiveness of up to $COP_c \cong 1.20$. On the other side, however, according to the 2nd principle, obtaining the same heat pumping effect ($T_E = 5$ °C for $T_C = 32$ °C) but at the expense of a higher generating temperature, is poorer qualitatively. In other words, the actual relative to Carnot double-effect

COP is lower than that of the single-stage. The exergy analysis of Chap. 7 shows that the relative to Carnot COP is equal essentially with the exergy efficiency of a heat pumping cycle. The study case at hand confirms the smaller relative to Carnot COP of the double-effect as compared to that of the single-stage cycle. Indeed, the calculus of Eqs. (5.6) and (6.27) reveals the ideal values of $COP_{C,cf} = 1.40$ and $COP_{C,c,2\text{-}effect} = 2.69$, wherefrom $COP_{cf,rel.Carnot} = \frac{0.72}{1.40} = 0.51$ and in Eq. $COP_{2-effect,c,rel.Carnot} = \frac{1.20}{2.69} = 0.44 < 0.51$ holds true. The decrease of the actual relative to Carnot multi-effect COP with respect to that of the single-stage has been analyzed previously in the same reference (Herold et al. 1995), too. However, the actual COP short falling of the reversible value, motivated here by the irreversible effects introduced by the cascade, is simplistic. In fact, the true explanation is given by the use of the higher heat temperature, which increases the multi-effect source-task mismatch and therefore the cycle exergy loss comparatively, according to the 2nd principle. In this respect, in order to have a broader view about the influence which the heat temperature level is having on the relative to Carnot effectiveness, as compared to that of the single-stage, the truncated cooling cycle, discussed to Chap. 4, is mentioned in completion. This time, unlike multi-effect, its operation is ensured by generation temperatures below those of the single-stage and, according to the 1st and 2nd principles, it should have comparatively actual and relative to Carnot COP's with lower and higher values, respectively. The theoretical results of Tables 4.2 and 4.3 and Figs. 4.36 and 4.37 already confirmed our assessment. In the light of the above presented, the much better relative to Carnot COP and therefore the higher exergy efficiency, is motivated in this case by the better source-task match of the truncated cycle as compared to that of the single-stage one. Given the fact that exergy loss costs money and that, according to the conclusion of our simple analysis, the multi-effect cycle has a higher loss of exergy of as compared to that of a single-stage cycle performing the same task, it is recommended to perform a feasibility study prior each heat pumping application involving multi-effect cycles.

The pressure-stages condensing advanced cycles are mostly used today in the air conditioning (e.g. the H_2O–LiBr double- and treble-effect machines). The favourable prerequisites show that their transposition into the coabsorbent technology opens a practical way towards an increased COP in the future, analyzed to the next section.

6.2 Use Analysis of Water–Lithium Bromide Pressure-Stages Multi-Effect Coabsorbent Cycle (PSMECCC) in Air Conditioning

The global warming strongly motivates an increased interest of the coabsorbent technology (e.g. Staicovici 2006, 2009a, b, c, d) for a much more efficient air conditioning supplied mainly by renewable sources (e.g. solar energy, biomass, geothermal). Taking these into account and given preliminary results (Staicovici 2009a), this section is studying the possibility of integrating, from thermodynamic and heat transfer viewpoint, the pressure-stages multi-effect coabsorbent cooling cycle (PSMECCC) in a classic air conditioning system (ACS), for its benefit (Staicovici 2011). The operation of cycles of 4- and 5-effect type with H_2O–LiBr integrated in a common flow chart is modeled. The simulation results, accounting for different fresh air fractions and typical dry-warm and humid–warm climates, are compared to those of air conditioning produced nowadays with H_2O–LiBr advanced absorption chillers working in condensation.

6.2.1 Structure and Heat Exchange Analysis of PSMECCC

The n-stage PSMECCC of Fig. 6.5 has been conceived in two variants, i.e. as just absorption (Staicovici 2009b) and as hybrid absorption (Staicovici 2009c) cycle, respectively. The hybrid operation has been chosen as a solution to increase the cycle effectiveness, through the increase of the number of effects, n, for the same heat/sink and cooling temperatures. This can be done by applying the hybrid operation to the cycle bottom. In this case, the desorber and absorber become non-isobar and a mechanical vapour compression means (e.g. compressor or fan) can be used to transport the desorbed vapour from the lower pressure desorber to the higher pressure absorber (Staicovici 2009c). This solution cannot be applied comfortably to the condensing advanced cycles, but in this case it could be feasible. Indeed, unlike the separate evaporators and absorbers of a condensing advanced cycle, here, as mentioned, the desorber and absorber are common devices of all cascade subcycles (see cycle constructive details to the proof of Sect. 6.1.1 basic lemma), so a single equivalent mechanical vapour compression unit can be utilized, instead of n. The solution is depicted in the $\log(p) - 1/T$ diagram of Fig. 6.8. The number of effects is increased, for the same temperature of sources, by decreasing the distance between the mean concentration isostere and that of the resorbers outlet in the hybrid cycle, as compared to the absorption one, and this is possible introducing the non-isobar (hybrid) operation of desorber and absorber in the cycle bottom. The constructive difference between the absorption and hybrid absorption variants is minimum. Here only the hybrid version is depicted in the $\log(p) - 1/T$ diagram, Fig. 6.9. Briefly describing the operation, the mean concentration y_M absorbent 6, generated in the mixer 5 by mixing up the absorbents coming of the absorber 4 and desorber 3, is pumped to each pressure stage by the pumps 7_i and valves 11, where the generators 9_i and resorbers 10_i are supplied, $i = 1, 2, \ldots, n$. The absorbents with concentrations y_{GO} and y_{RO}, coming of the generators and resorbers, respectively, are expanded and gathered in order to supply the absorber and desorber. The desorbed vapor is sucked by the compressor 23, compressed from p_D to p_A, superheated in the HE 22 and absorbed and the absorbent coming of desorber is pumped in the mixer, in order to close the cycle. The absorbents coming of the $(i + 1)$th resorber and the ith generator are subcooled in the HE 16_i in order to preheat the absorbent supplying the ith stage generator and resorber, and the $(i + 1)$th resorber heat is transferred to the

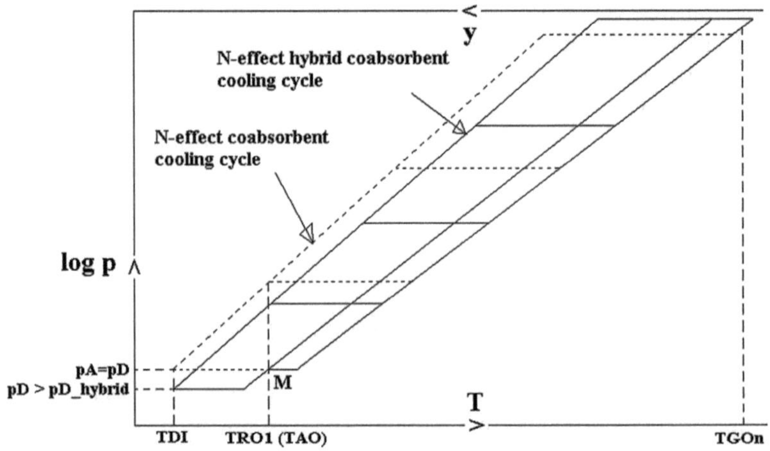

Fig. 6.8 Schematic of the coabsorbent versus hybrid coabsorbent PSMECCC plot

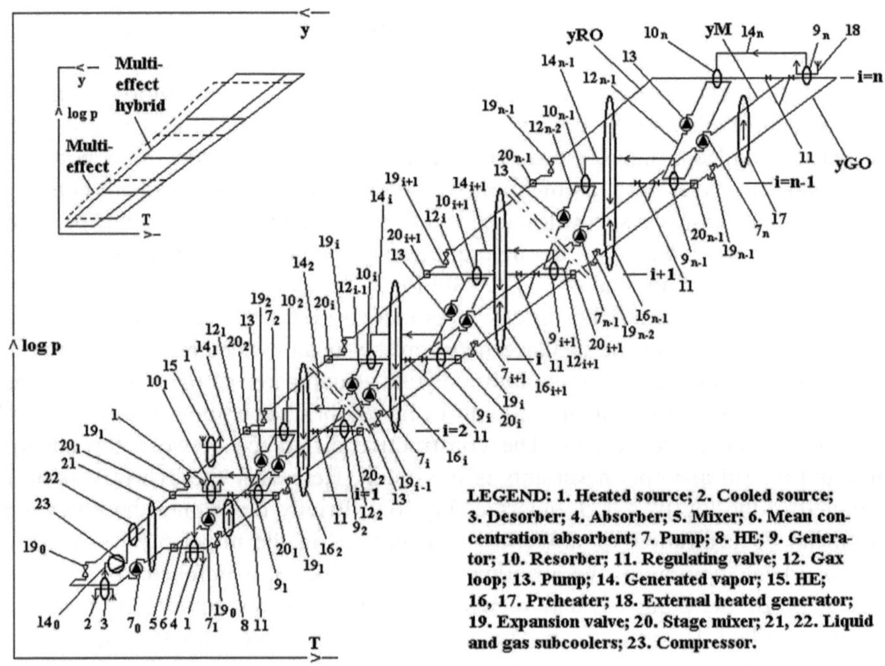

Fig. 6.9 PSMECCC hybrid cycle (see also effects number increase in the *left upper corner* figure)

ith generator by the gax loop 12_i, $i = 2, \ldots, n - 1$. The desorber usefully cools the source 3, the 1st resorber and the absorber are cooled by the sink source 1 and the nth generator is heated by the heat source 18.

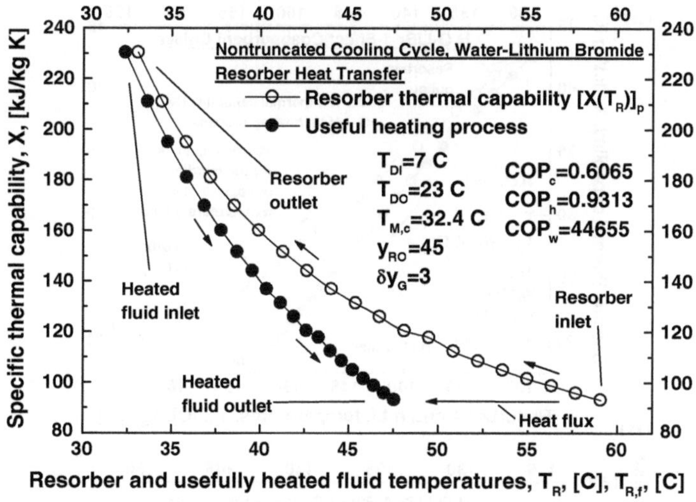

Fig. 6.10 Typical 1st resorber thermal capability versus resorption temperature of a H_2O–LiBr PSMECCC

Bearing in mind the air conditioning application we analyze, a natural question is put forward: is it possible to replace the classic condensing subcycle of a pressure-stages cascade by a nontruncated coabsorbent subcycle? Two main issues derive further from this question: (i) is it possible to replace the classic evaporator, providing a single evaporation temperature value for water chilling (e.g. 7 °C), by a desorber operating on a large desorption interval with gliding temperature (e.g. 7–23 °C) for same water chilling? and (ii) is it possible to supply the generator of an inferior cascade subcycle with the heat delivered on a large gliding temperature interval by the resorber of the adjacent superior cascade subcycle with the help of a gax process instead of doing the same with a condenser characterized by a single value condensing temperature? The (i) and (ii) questions answer implies first a task-source match analysis from the heat exchange point of view and second, a thermodynamic a analysis. Concerning the heat exchange analysis of both (i) and (ii) questions, the answer is positive, there is a satisfactory task-source match in all study cases considered, either we are referring to desorbtion, Fig. 3.18, and resorption processes, Fig. 6.10, or to a gax process, Fig. 6.11a, b, c (Staicovici 2011). The thermodynamic analysis is done to the Sect. 6.2.2.

6.2.2 PSMECCC-Classic Air Conditioning System Link

The H_2O–LiBr PSMECCC cycle was integrated in a classic air conditioning system (ACS), Fig. 6.12. The outdoor fresh air fraction f mixes up in an air mixer (AM) with the depleted air fraction $(1 - f)$. The air mixture is cooled in an air conditioning

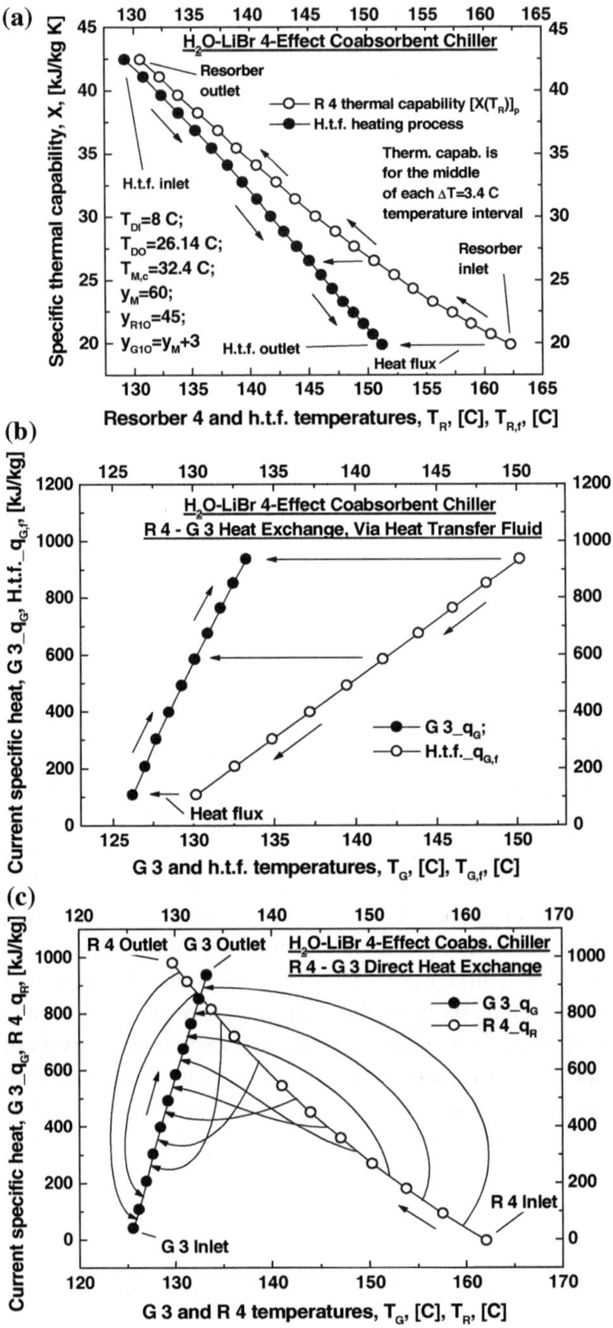

Fig. 6.11 Typical resorber-generator (gax) heat exchange of a H_2O–LiBr PSMECCC: **a** and **b** intermediary gax loop (see position 12_i, Fig. 6.9) using a heat transfer fluid (h.t.f.), which first receives the resorber heat (**a**) and than it releases it to the generator (**b**); **c** a direct heat transfer gax

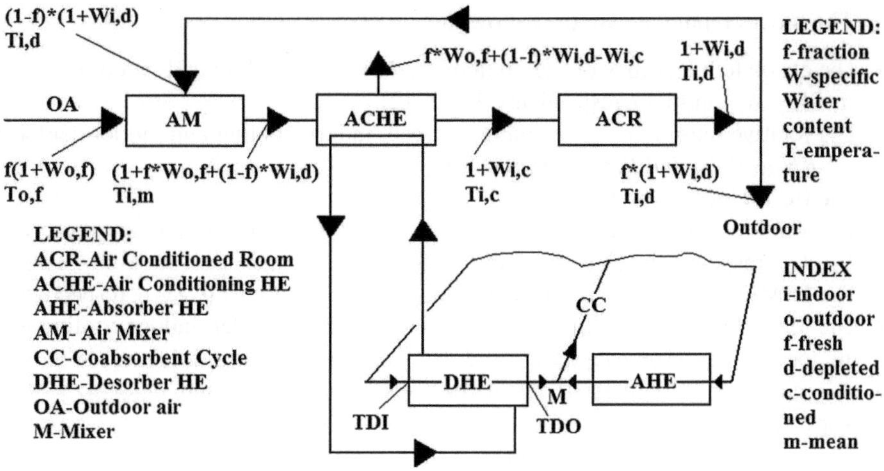

Fig. 6.12 PSMECCC-classic ACS

Fig. 6.13 Air conditioning temperatures

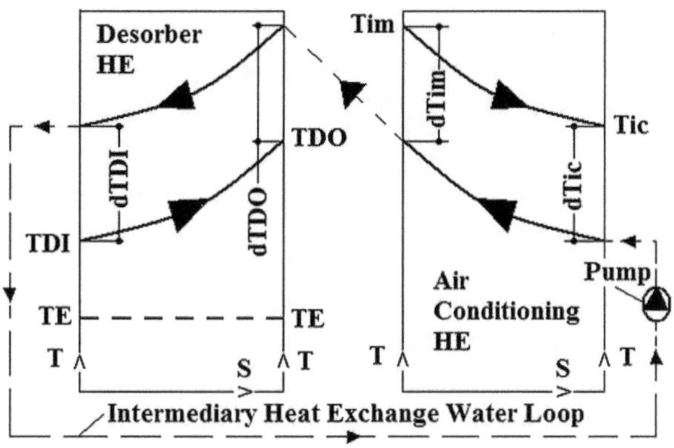

heat exchanger (ACHE), where a moisture amount is eliminated. Further, the conditioned air is introduced in the air conditioned room (ACR). Finally, the resulting depleted air is evacuated from the room, the depleted air fraction *f* is rejected in the atmosphere (see "Outdoor") and the remaining $(1 - f)$ fraction joins the fresh air fraction *f* in the air mixer, in order to close the open air conditioning cycle. The inlet and outlet air and the moisture quantities and temperatures of each device are shown in Fig. 6.12, as well. The air conditioning heat exchanger (ACHE) is cooled by the PSMECCC cycle desorber (DHE) using an intermediary heat transfer fluid loop. The temperature details of this heat transfer are given in Fig. 6.13. The system of Fig. 6.12 was modeled and model details are given next.

The model input data are the followings:

- The absolute humidities of the outdoor fresh, indoor cooling and indoor depleted air: $W_{o,f}$, $W_{i,c}$ and $W_{i,d}$, respectively, Fig. 6.12;
- The relative humidities of the outdoor fresh, indoor cooling and indoor depleted air: $RH_{o,f}$, $RH_{i,c}$ and $RH_{i,d}$, respectively;
- The fresh air fraction, f, Fig. 6.12;
- The outdoor fresh, indoor cooling and indoor depleted air temperatures: $T_{o,f}$, $T_{i,c}$, $T_{i,d}$, respectively, Figs. 6.12 and 6.13;
- The PSMECCC mixer and 1st resorber outlet temperatures: T_M, T_{R1O}, respectively;
- The temperature pinch of the desorber inlet, desorber outlet, indoor cooling and indoor mixing: dT_{DI}, dT_{DO}, $dT_{i,c}$ and $dT_{i,m}$, respectively, Fig. 6.13;
- The pump and compressor effectivenesses: $\eta_{pump} = 0.6$ and $\eta_{compress} = 0.9$, respectively;
- The generators 9_i degassing interval: $y_{GO,i} - y_{GI,i} = 3$, $i = 1, 2, \ldots, n$;

The model output data are the followings:

- The indoor mixer temperature, $T_{i,m}$;
- The desorber inlet, desorber outlet and generator 9_n outlet temperatures: T_{DI}, T_{DO} and $T_{G,n}$, respectively;
- The cooling, Eq. (6.65), heating, Eq. (6.66), with respect to mechanical work consumption, Eq. (6.67), and with respect to the Carnot cycle effectivenesses: COP_c, COP_h, COP_w and $COP_{rel.Carnot}$, respectively.

The model equations are given in Table 6.1 and explained next. From Fig. 6.13, the temperature analysis shows that Eqs. (6.48) and (6.50) hold true, wherefrom Eqs. (6.49) and (6.51) result, respectively. In order to avoid the skin stress, the air conditioning standard recommends Eq. (6.52) be respected (Harris et al. 1985). The outdoor standard conditions typical temperature is given by Eq. (6.53) for a humid–warm climate (Harris et al. 1985), wherefrom, with Eq. (6.52), a threshold for the minimum air conditioned temperature is expressed by Eq. (6.54). The chillers manufacturers do not respect it, in practice it is rather accepted Eq. (6.55) value (Staicovici 2009d), adopted here as well. The standard recommendations for the indoor dry bulb temperature, $T_{i,c}$ (Harris et al. 1985), are for the short-term and long-term convenience given by Eqs. (6.56) and (6.57), respectively. The specific enthalpies of the outdoor fresh, indoor depleted and indoor conditioned air are given by the Eqs. (6.58)–(6.60) (Kirilin et al. 1985), while the air mixer enthalpy, humidity, temperature and specific cooling load are calculated by Eqs. (6.61)–(6.64), respectively. The effectiveness of cooling, heating and with respect to the mechanical work consumption are written by Eqs. (6.65)–(6.67), respectively. As an important remark, from Eqs. (6.49) and (6.51) it results that decreasing the pinch temperatures (TP's) of the desorber, dT_{DI} and dT_{DO}, and of the air conditioning HE, $dT_{i,c}$ and $dT_{i,m}$, the T_{DI} and T_{DO} temperatures will increase, Fig. 6.13 and therefore the COP_c is favored. In case of the desorber, the TP's can be decreased, here $dT_{DI} = dT_{DO} =$ (2–3) °C, because a small TP favors the heat transfer between water and H_2O–LiBr mixture, characterized by a coupled mass and heat transfer (Staicovici 2000). However, decreasing the TP's of ACHE doesn't fully comply with its satisfactory operation,

Table 6.1 Model main equations

$T_{DI} + dT_{DI} = T_{i,c} - dT_{i,c}$	(6.48)	$I_{o,f} = T_{o,f} + W_{o,f}\left(2501 + 1.93 * T_{o,f}\right)$	(6.58)
$T_{DI} = T_{i,c} - \left(dT_{i,c} + dT_{DI}\right)$	(6.49)	$I_{i,d} = T_{i,d} + W_{i,d}\left(2501 + 1.93 * T_{i,d}\right)$	(6.59)
$T_{DO} + dT_{DO} = T_{i,m} - dT_{i.m}$	(6.50)	$I_{i,c} = T_{i,c} + W_{i,c}\left(2501 + 1.93 * T_{i,c}\right)$	(6.60)
$T_{DO} = T_{i,m} - (dT_{i.m} + dT_{DO})$	(6.51)	$I_{i,m} = f * I_{o,f} + (1-f)I_{i,d}$	(6.61)
$T_{o,f} - T_{i,c} < 20\,°\mathrm{F} = 11.11\,°\mathrm{C}$	(6.52)	$W_{i,m} = f * W_{o,f} + (1-f)W_{i,d}$	(6.62)
$T_{o,f} = 95\,°\mathrm{F} = 35\,°\mathrm{C}$	(6.53)	$T_{i,m} = \left(I_{i,m} - 2501 * W_{i,m}\right)/\left(1 + 1.93 * W_{i,m}\right)$	(6.63)
$T_{i,c} > 35 - 11.11 = 23.89\,°\mathrm{C}$	(6.54)	$q_D = I_{i,m} - I_{i,c}$	(6.64)
$T_{i,c} = 18\,°\mathrm{C}$	(6.55)	$COP_c = \sum_{i=1}^{n} q_{D,i}/q_{G,n}$	(6.65)
$T_{i,c,st} = (78\text{–}80)\,°\mathrm{F} = (25.6\text{–}26.7)\,°\mathrm{C}$	(6.56)	$COP_h = q_{R,1}/q_{G,n}$	(6.66)
$T_{i,c,lt} = 75\,°\mathrm{F} = 23.88\,°\mathrm{C}$	(6.57)	$COP_w = \sum_{i=1}^{n} q_{D,i}/w_p(+w_c)$	(6.67)

Table 6.2 H_2O–LiBr PSMECCC air conditioning model

Item	Humid–warm climate (close to Shanghai, China)		
	Small ACHE TP's	Average ACHE TP's	Average-high ACHE TP's
Indoor depleted air conditions	$\left[T_{i,d}\right]_{f=0} = T_{i,d,st} + dT_{i,d,st} = 28.9\,°\mathrm{C}$	$\left[T_{i,d}\right]_{f=0} = 76\,°\mathrm{F}\,(24.44\,°\mathrm{C})$	$\left[T_{i,d}\right]_{f=0} = 76\,°\mathrm{F}\,(24.44\,°\mathrm{C})$
Inlet/outlet ACTP	$dT_{i,c} = 2.5\,°\mathrm{C}$ $dT_{i,m} = 3.0\,°\mathrm{C}$	$dT_{i,c} = 5\,°\mathrm{C}$ $dT_{i,m} = 5\,°\mathrm{C}$	$dT_{i,c} = 8\,°\mathrm{C}$ $dT_{i,m} = 8\,°\mathrm{C}$
T_{DI}	$T_{DI} = 13.5\ °\mathrm{C}$	$T_{DI} = 11\ °\mathrm{C}$	$T_{DI} = 8\ °\mathrm{C}$
Absorber TP's	$dT_{AI}/dT_{AO} = 3.0\,°\mathrm{C}$	$dT_{AI}/dT_{AO} = 2.0\,°\mathrm{C}$	$dT_{AI}/dT_{AO} = 3.0\,°\mathrm{C}$

$T_{o,f} = 95\,°\mathrm{F}\,(35\,°\mathrm{C});\ RH_{o,f} = 70\,\%;\ T_{i,c} = 64.4\,°\mathrm{F}\,(18\,°\mathrm{C});\ RH_{i,c} = 50\,\%;\ T_M = 32.4\,°\mathrm{C};$
$dT_{DI}/dT_{DO} = 2.0\,°\mathrm{C};\ T_{M,e} = 29.4\,°\mathrm{C};\ y_{R1O} = 46;\ T_{Rf,O} \cong 48\,°\mathrm{C};\ RH_{i,d} = 52\,\%$

Study cases I–III, humid–warm climate. Table bottom is listing the input data

Table 6.3 H_2O–LiBr PSMECCC air conditioning model

Items	Dry-warm climate (Riyadh, Saudi Arabia)	
	Average ACHE TP's	Average-high ACHE TP's
Inlet/outlet ACHE TP's	$dT_{i,c}/dT_{i,m} = 5\,°C$	$dT_{i,c}/dT_{i,m} = 8\,°C$
Desorber internal inlet temperature	$T_{DI} = 11\,°C$	$T_{DI} = 8\,°C$
$T_{o,f} = 107.5\,°F\,(42\,°C)$; $T_{WB,o} = 25\,°C$; $RH_{o,f} = 25\,\%$; $T_{i,c} = 64.4\,°F\,(18\,°C)$; $RH_{i,c} = 50\,\%$; $[T_{i,d}]_{f=0} = 76\,°F\,(24.44\,°C)$; $RH_{i,d} = 52\,\%$; $y_{R1O} = 46$; $dT_{DI}/dT_{DO} = (2-3)\,°C$; $T_{Rf,O} \cong 45\,°C$; $T_M = 29.4\,°C$; $T_{M,e} = T_{WB,o} + 1 = 26\,°C$; $dT_{AI} = dT_{AO} = 2\,°C$		

Study cases II–III, dry-warm climate. Table bottom is listing the input data

because $dT_{i,c}$ and $dT_{i,m}$ decrease favors T_{DO} increase, but the (air + water vapor) to water heat transfer is diminished, even water vapor is condensing. This contradictory situation can be solved by increasing the effects number for the same input data, using the hybrid solution, Figs. 6.8 and 6.9.

The model results (Staicovici 2009d) were obtained for two climate types (humid–warm and dry-warm) and three ACHE TP's (small, average and average-high). The PSMECCC cycles were considered those with 4-effect, 5-effect and 5-effect hybrid. The climate input data are included in Tables 6.2 and 6.3, while the output data are given by Tables 6.4, 6.5, and 6.6. The COP_c results, included in Tables 6.4, 6.5, and 6.6, are plotted in Figs. 6.14, 6.15, and 6.16, respectively, in order to enable one to analyze them at a glance.

Results comments follow: (I) ACHE TP's are key parameters. The COP_c increases with TP's decrease, from 8 °C (case III) to 2.5–3.0 °C (case I), for diminishing the heat exchange exergy loss increases the effectiveness. Practically, the coabsorbent technology could be applied successfully keeping the TP's within 5–8 °C. The H_2O–LiBr 2- and 3-effect condensing cycles have the $COP_c = 1.24$–1.35 and $COP_c = 1.62$ for similar input data (Yabase et al. 2005). Comparatively, the PSMECCC COP_c could be better by cca. 50–100 % (and more), provided the desorber higher gliding temperature operation be promoted; (II) Fig. 6.17a shows a nowadays typical ACS diagram with 2- and 3-effect condensing chiller (Staicovici 2009d). Figure 6.17b gives results obtained for case III modeled with the worst working conditions ($f < 0.3$). Despite this, Fig. 6.17b results show that PSMECCC could meet closely the actual air conditioning heat exchange conditions if integrated in a classic ACS; (III)$T_{GO,n}$ is less than 170 °C in all cases, so one of the heat sources could be the solar energy, increasingly cheaper and environmentally friendly; (IV) The PSMECCC-ACS yields warm water still useful ($COP_h \cong 0.9$, $T_{Rf,O} \cong 48$ °C).

6.2.3 PSMECCC (Heat Source) Energy Savings in Air Conditioning

Figures 6.13 and 6.17b indicate that the coabsorbent technology presence makes the ACS more flexible when $T_{i,m}$ increases. Indeed, the $T_{i,m} - dT_{i,m}$ and $T_{DO} + dT_{DO}$

Table 6.4 H_2O–LiBr PSMECCC air conditioning model

Item	$dT_{i,c} = 2.5$ °C; $dT_{i,m} = 3.0$ °C, humid–warm climate							Machine type
f	0.1	0.2	0.3	0.4	0.5	0.6	0.7	4-effect absorption, 10 main devices
COP_c	–	–	–	2.002	2.072	2.137	2.202	$T_{GO,n} = 144.6\,°C$
COP_w	–	–	–	942.3	974.6	1006	1037	
$COP_{rel.Carnot}$	–	–	–	0.349	0.351	0.353	0.354	
T_{DO} (°C)	–	–	–	26.37	26.98	27.59	28.2	
$T_{i,m}$ (°C)	–	–	–	31.37	31.98	32.59	33.2	
COP_c	1.975	2.067	2.157	2.246	2.334	2.415	2.50	5-effect absorption, 12 main devices
COP_w	1,327	1,389	1,451	1,511	1,571	1,626	1,683	$T_{GO,n} = 165.6\,°C$
$COP_{rel.Carnot}$	0.323	0.329	0.335	0.340	0.344	0.347	0.350	
T_{DO} (°C)	24.51	25.13	25.75	26.37	26.98	27.59	28.2	
$T_{i,m}$ (°C)	29.51	30.13	30.75	31.37	31.98	32.59	33.2	
COP_c	2.741	2.984	3.087	3.225	3.356	3.586	3.722	5-effect hybrid, 12 main devices + 1 fan
T_{DO} (°C)	24.51	25.13	25.75	26.37	26.98	27.59	28.2	$T_{GO,n} = (118-123)\,°C$
$T_{i,m}$ (°C)	29.51	30.13	30.75	31.37	31.98	32.59	33.2	$COP_w = 60-70$

Study case I-small ACHE TP's

Table 6.5 H_2O–LiBr PSMECCC air conditioning model

Item	$dT_{i,c} = dT_{i,m} = 5$ °C, 5-effect hybrid, 12 main devices + 1 fan										Climate type
f	0.15	0.2	0.3	0.4	0.5	0.6	0.7	0.8	0.9	1.0	Humid-warm
COP_c	–	–	1.828	1.995	2.156	2.309	2.458	2.648	2.738	2.872	$T_{GO,n} = 162.4$ °C
COP_w	–	–	68.6	69.88	70.99	71.95	72.81	73.57	74.24	74.87	
T_{DO} (°C)	–	–	20.68	21.74	22.8	23.85	24.9	25.94	26.97	28.0	
$T_{i,m}$ (°C)	–	–	27.68	28.74	29.8	30.85	31.9	32.94	33.97	35.0	
COP_c	2.042	2.216	2.548	2.865	3.166	3.453	3.731	4.001	4.259	4.512	Dry-warm
COP_w	64.71	65.55	66.93	68.03	68.94	69.72	70.39	70.99	71.51	71.99	$T_{GO,n} = 134.2$ °C
T_{DO} (°C)	20.09	20.97	22.73	24.49	26.25	28.0	29.76	31.51	33.25	35.0	
$T_{i,m}$ (°C)	27.09	27.97	29.73	31.49	33.25	35.0	36.76	38.51	40.25	42.0	

Study case II-average ACHE TP's

Table 6.6 H_2O–LiBr PSMECCC air conditioning model

Item	$dT_{i,c} = dT_{i,m} = 8$ °C, 5-effect hybrid, 12 main devices + 1 fan										Climate type
f	0.1	0.273	0.3	0.4	0.5	0.6	0.7	0.8	0.9	1.0	Humid-warm
COP_c	–	1.774	1.822	1.995	2.162	2.321	2.477	2.626	2.77	2.91	$T_{GO,n} = 158.5\,°C$
COP_w	–	33.23	33.39	33.91	34.46	34.74	35.08	35.38	35.65	35.9	
T_{DO} (°C)	–	17.39	17.68	18.74	19.8	20.85	21.9	22.94	23.97	25.0	
$T_{i,m}$ (°C)	–	27.39	27.68	28.74	29.8	30.85	31.9	32.94	33.97	35.0	
f	0.154	0.2	0.3	0.4	0.5	0.6	0.7	0.8	0.9	1.0	Dry-warm
COP_c	2.062	2.217	2.543	2.853	3.148	3.43	3.704	3.975	4.218	4.463	$T_{GO,n} = 135.4\,°C$
COP_w	34.78	35.14	35.8	36.32	36.76	37.13	37.46	37.75	38.0	38.23	
T_{DO} (°C)	17.17	17.97	19.73	21.49	23.25	25.0	26.76	28.51	30.25	32.0	
$T_{i,m}$ (°C)	27.17	27.97	29.73	31.49	33.25	35.0	36.76	38.51	40.25	42.0	

Study case III-average-high ACHE TP's

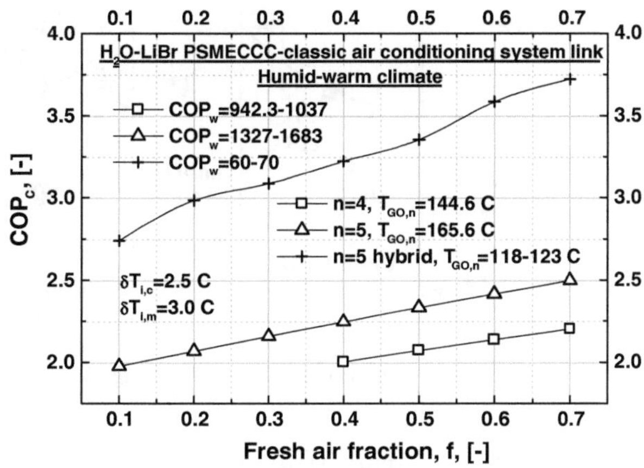

Fig. 6.14 Study case I-small ACHE TP's. Cooling effectiveness of 4-, 5- and 5 hybrid-effect H_2O–LiBr PSMECCC cycles plotted versus fresh air fraction, for humid–warm climate

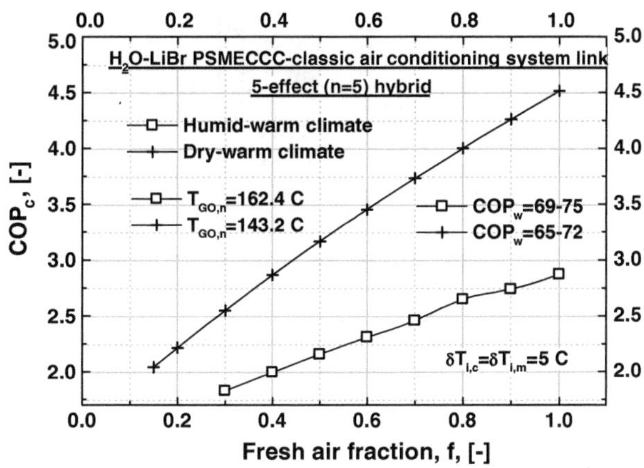

Fig. 6.15 Study case II-average ACHE TP's. Cooling effectiveness of 5 hybrid-effect H_2O–LiBr PSMECCC cycles plotted versus fresh air fraction, for humid–warm and dry-warm climate

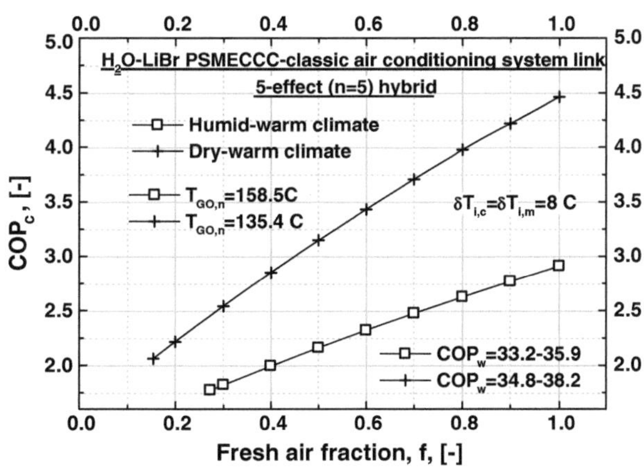

Fig. 6.16 Study case III-average ACHE TP's. Cooling effectiveness of 5 hybrid-effect H_2O–LiBr PSMECCC cycles plotted versus fresh air fraction, for humid–warm and dry-warm climate

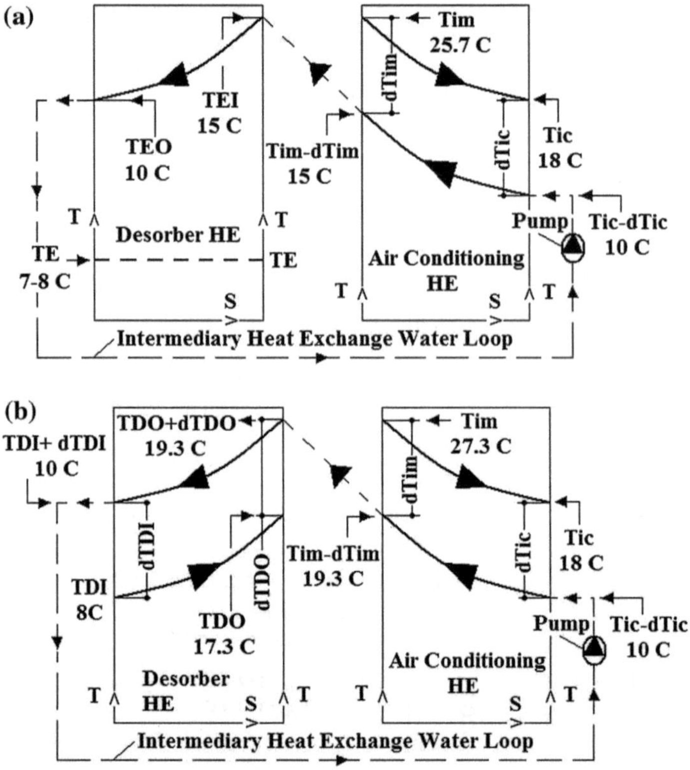

Fig. 6.17 H_2O–LiBr air conditioning, T_M = (29.4–32.4) °C: **a** Thermax unit (f = 0.1; COP_c = 1.62–1.24); **b** PSMECCC unit (f = 0.15–0.27; COP_c = 2.06–1.77)

Table 6.7 PSMECCC-ACS energy savings model

$COP_{c,\min} = \frac{q_{D,\min}}{q_{G,n,\min}}$ (6.68)	$COP_c = \frac{q_D}{q_{G,n}}$ (6.69)	$q_{G,n,rel} = \frac{q_{G,n}}{q_{G,n,\min}} = q_{D,rel} COP_{c,rel}$ (6.70)
$q_{D,rel} = \frac{q_D}{q_{D,\min}}$ (6.71)	$COP_{c,rel} = \frac{COP_{c,\min}}{COP_c}$ (6.72)	

temperatures are able to follow with a constant TP the $T_{i,m}$ increase, keeping, unlike classic ACS, a relative constant minimum exergy loss. This PSMECCC-ACS better cooling source-task (cooling effect) match is translated in important heat source energy savings, comparatively, evaluated next (see Table 6.7 model). The current COP_c increases with f increase and covers the range $COP_{c,\min} \leq COP_c \leq COP_{c,\max}$ when f covers the interval $f_{\min} \leq f \leq f_{\max}$. The $COP_{c,\min}$ is a minimum reference value, depending on $f_{\min}$ reference value, given by Eq. (6.68). The current COP_c is expressed by Eq. (6.69). The relative heat source input savings we look for are given by Eq. (6.70) via Eqs. (6.71) and (6.72). In Fig. 6.18a, b we plotted Eqs. (6.68–6.72). The $q_{G,n,rel}$ function is proportional to $q_{D,rel}$ and $COP_{c,rel}$. The $q_{D,rel}$ factor is direct

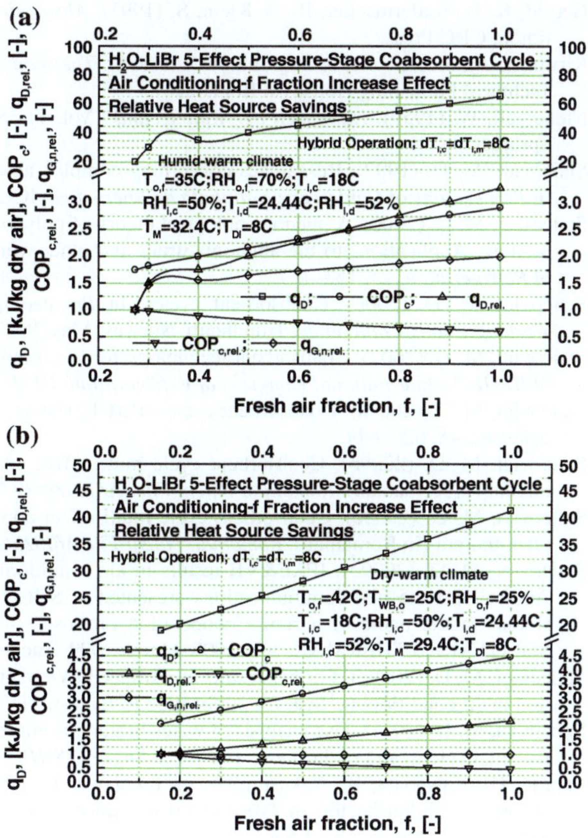

Fig. 6.18 Relative heat source savings of H_2O–LiBr 5-effect hybrid PSMECCC-ACS versus *f* fraction, for Table 6.7 data, humid–warm climate (**a**) and dry-warm climate (**b**)

proportional to *f* in both condensing and coabsorbent chiller cases, as Fig. 6.18 show, while $COP_{c,rel}$ factor is the one which differentiates these two chiller cases. Indeed, in case of a condensing chiller, $COP_{c,rel}$ is constant with respect to f increase, $COP_{c,rel} = const. = 1.0$, so $q_{G,n,rel}$ and $q_{D,rel}$ have the same trend, of increasing with f increase. On the contrary, in case of PSMECCC, the $COP_{c,rel}$ factor is much smaller, $0.5 \leq COP_{c,rel} \leq 1.0$, decreasing with f increase, as both Fig. 6.18 show. Consequently, the $COP_{c,rel}$ factor alters Eq. (6.70) right member value and succeeds to counterbalance $q_{G,n,rel}$ heat input increase with the f increase, up to 60 %, in case of the humid–warm climate, Fig. 6.18a, and totally (>100 %), in case of the dry-warm climate, Fig. 6.18b.

References

Alefeld, G. (1982). Regeln für den Entwurf von mehrstufigen Absorbermaschinen. *Brennstoff-Wärme-Kraft, 43*(2), 64–73.

Harris, N. C., Miller, C. E., & Thomas, I. E. (1985). *Solar energy systems design*. New York: Wiley.

Herold, K. E., Radermacher, R., & Klein, S. (1995). *Absorption chillers and heat pumps*. Boca Raton: CRC Press.

Kirilin, V. A., Sicev, V. V., & Seindlin, A. E. (1985). *Thermodynamics*. Bucharest: Scientific and Technical Publishing House (in Romanian).

Niebergall, W. (1959). *Handbuch der Kältetechnik* (Vol. 7)., Sorptions-Kältemaschinen Berlin: Springer.

Staicovici, M. D. (1992). Researches concerning complex absorption machines (in Romanian). In *Proceedings of Conference Nationala de Energetica*, ICEMENERG, Neptun (pp 89–95).

Staicovici, M. D. (2000). A phenomenological theory of polycomponent interactions in non-ideal mixture. Applications to the ammonia/water and other working pairs. *International Journal of Refrigeration, 23*, 153–167.

Staicovici, M. D. (2006). Coabsorbent cycles. In *Proceedings of Gustav Lorentzen Natural Working Fluids Conference*, Trondheim, Norway, May 28–31, IIF/IIR: 219–222.

Staicovici, M. D. (2007). Coabsorbent technology for heat pumping applications. In *Proceedings of IIRIIF 22 International Congress of Refrigeration, E1-2-1039*, Beijing.

Staicovici, M. D. (2009a). Coabsorbent cycles. Part 1: Theory. *International Journal of Thermal Sciences*, *48*, 626–644.

Staicovici, M. D. (2009b). Coabsorbent cycle multi-effect cooling method and applying plant thereof (in Romanian). *Romanian Patent Files A/00089/29.01.2009.*

Staicovici, M. D. (2009c). Coabsorbent cycle multi-effect cooling hybrid method and applying plant thereof (in Romanian). *Romanian Patent File A/00256/23.03.2009.*

Staicovici, M. D. (2009d). IPAE&VE results in air conditioning with coabsorbent technology, Author's private free communications presented to SOLITEM (Germany) and THERMAX (India) in the name of a future cooperation promise which never has been respected by SOLITEM and THERMAX. July 2008–March 2009, Aachen, Germany.

Staicovici, M. D. (2009e). Further results in advanced (*N*-effect) coabsorbent cycles and application in air conditioning. Sent for publication to *Energy, the International Journal.*

Staicovici, M. D. (2011). Use analysis of water-lithium bromide multi-effect pressure-stage coabsorbent cycle in air conditioning. In *Proceedings of ISHPC 2011 Conference*.

Yabase, H., Kawabata, K., Yakushiji, F., & Takigawa, T. (2005). Development of triple-effect absorption chiller-heater. In *Proceedings of International Sorption Heat pump Conference*, June 22–24, Denver, CO, USA.

Ziegler, F., Kahn, R., & Alefeld, G. (1993). Multi-effect absorption chillers. *International Journal of Refrigeration, 16*(5), 301–311.

Chapter 7
Coabsorbent Cycles Exergy Evaluation

In order to optimize an energy-process, the 1st principle has as main goal to reduce the primary energy consumption, expressed in our case by its effectiveness maximization (see Eqs. 3.53–3.56 of the COP_c, COP_h, COP_w, etc.). The 1st principle recommendations for improving an energy-process relay mainly on the design which affords the internal heat recovery [liquid–liquid (latent-gax and specific), vapor–liquid (latent and specific)]. On the other side, the 2nd principle recommends the heat exchange processes to take place with reduced temperature pinch, in order to reduce the exergy loss and improve the energy-process *COP*. Diminishing the consumption of the primary energy is a necessary condition for saving it, but not enough, its quality, which should matter as well, is not accounted for by the 1st principle. Moreover, the energy can be transformed, only, but not consumed, because it is conservating itself, and from this point of view the 1st principle effectiveness definition is ambiguous and its reliance on energy as a basic unit of measure is not adequate (Kreith and Kreider 1978). There are a few applications for which the 1st principle *COP* is an adequate tool for measuring the energy-process effectiveness, such as those supplied by unique or free sources, or by sources identified as matching a certain process task, etc. For the rest, however, implying the rank of different nature and quality energy-processes and supplying sources with respect to a given task, the 1st principle is no more effective. In this case, bearing in mind the previous assertions and having in view especially the environment protection problems the human activity is confronted with lately, and not only, a designer has the responsability to use more refined and effective investigation methods in order to minimize the primary energy consumption. One of the most important to this extent is the exergy method (Chartier et al. 1975; Alefeld 1988a, b; Baehr 1988; Bejan 1996), based on the 2nd principle application. We remember that work is a unique form of energy, with the highest quality (lowest entropy), and remark that, for this reason, it could be used as an universal exchange currency, to measure the quality of all other lower quality forms of energy, if appropriate energy-to-work conversion means were used, furnished by thermodynamics and the 2nd principle. Following this idea, the exergy method evaluates in a first step the potential of a real energy-process

M.-D. Staicovici, *Coabsorbent and Thermal Recovery Compression Heat Pumping Technologies*, Heat and Mass Transfer, DOI: 10.1007/978-3-642-54684-6_7,

(cycle) to transform its every energy input and output in mechanical work, named availability, or exergy. The reference temperature for this tranformation is the sink source temperature (here considered $T_{M,cooling}$ or simpler noted $T_{M,c}$). Out of the maximum availability received from the heat and/or work supplying sources, the process consumes an amount, in order to perform the task. We remark that, unlike energy, the exergy can be consumed. In a second step, the remaining (minimum) availability is assessed and compared to the maximum availability, calculating an exergy (2nd principle) efficiency as the ratio of the sum of all retrievable (useful) exergy outputs through the sum of all exergy inputs (see Eq. 7.31). Actually, the exergy efficiency enables to rank different processes, checking the approach of this figure of merit of the reversible (ideal) process efficiency, which equals unit. Using practical terms, the better ranked process is characterized by a better source-task match (or smaller heat exchange temperature pinch), which means a process with a smaller availability consumption, comparatively. In this way, the goal of an energy-process optimization is moved from the 1st principle aim, of minimizing the primary energy consumption, to that based on the 2nd principle and exergy method recommendations, of minimizing the availability consumption (Kreith and Kreider 1978). This assertion shows that the availability and primary energy consumptions are directly connected, because availability is an extensive property, proportional to the mass of the primary energy (Kreith and Kreider 1978). Another explanation of this direct link is proposed by the particular case of the coabsorbent heat pumping. Indeed, the increase of the 1st principle effectiveness, Eqs. (3.53–3.56), will decrease correspondingly the specific primary consumption. However, this increase is not possible but only when simultaneously the 2nd principle exergy efficiency of the cooling and heating processes at hand (see next equations) increases as well, in such a way that when Eqs. (3.53–3.56) approach and take the ideal values, given to Chap. 5, the exergy efficiency approaches and equals unit (see Appendix 1 of this chapter). In this way we have shown again, in a more analytical manner this time, that availability consumption and the primary energy consumption move in the same direction and that the former consumption is decisive for the latter.

The systems we are dealing with lay at the base of the cooling and heating applications. These are closed cycles, involving in their interactions with sources heat currents, only. In this case, unlike the open cycles, which, besides heat, involve mass exchange with sources, as well, the exergy balance becomes simpler, accounting for the availability of these heat currents, only (see Chap. 6 equation). The method is applied to coabsorbent and mechanical vapor compression heat pumping cycles, and is extended to the tri- and cogeneration as well. When a process heat source is cogenerated by a power station, the algorithm is completed in order to account for the exergy loss generated by substituting the station condensation operation with that in counterpressure. Continuing this idea, the exergy balance is extended on boundaries comprising not only the processes at hand, but their different supplying sources of energy as well (the free and the low entropy heat sources, solar, photovoltaic, wind and the hydro energies), in order to have a better measure of a process availability consumption for a given task, hence a

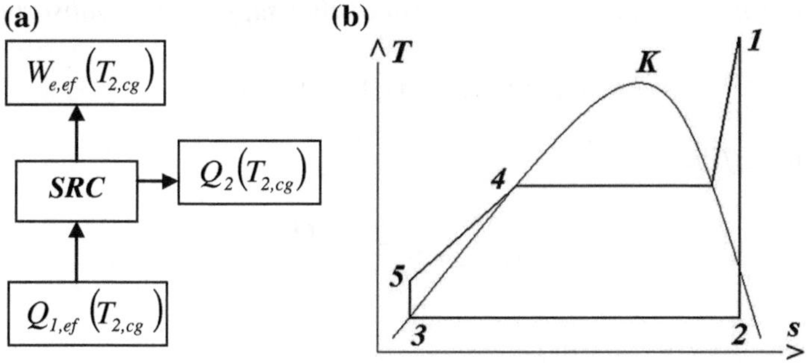

Fig. 7.1 Schematic of the thermal power plant working in cogeneration of power and heat (**a**) and its Rankine cycle with vapor superheating (**b**)

better rank of different nature processes from this point of view. Out of the sources mentioned above, some of them consume exergy for being produced (e.g. the fossil fuels, hydro). However, keeping track of exergy consumption history is not useful, as long as such sources are supplying simultaneously a large category of energy-processes on middle and long term, including those at hand as well, and for this reason it was skipped from the exergy balance boundary.

7.1 Simple Algorithm of the Heat Pumping Supplied in Cogeneration

Finding the most effective operation conditions of a thermal power station producing simultaneously electrical power and heat (cogeneration) was a permanent concern of many energy involved authors (e.g. Gerz 1982; Gasparovic 1982; Alefeld 1988c). A (SRC) thermal power plant working in counterpressure, $T_{2,cg} > T_{2,cond}$, has a high quality heat input, $Q_{1,ef}(T_{2,cg})$, and can deliver heat in cogeneration of electrical power, $W_{e,ef}(T_{2,cg})$, and heat, $Q_2(T_{2,cg})$, Fig. 7.1a, b, sacrificing (consuming) an amount of electrical power, calculated with respect to the condensing operation and equal approximately to the difference between $W_{e,ef}(T_{2,cond})$ and $W_{e,ef}(T_{2,cg})$. From this item point of view, a SRC is a heat pump, as known. The simple approach to find the effectiveness of this heat producing process is given next for a SRC with vapor superheating, shown in the T-s diagram in Fig. 7.1b:

1. *SRC boiler theoretical heat input*:

$$Q_1\left(T_{2,cond(cg,cbs)}\right) = h_1 - h_5 \tag{7.1}$$

2. *SRC boiler effective heat input*:

$$Q_{1,ef}\left(T_{2,cond(cg,cbs)}\right) = \frac{Q_1\left(T_{2,cond(cg,cbs)}\right)}{\eta_{boiler}\eta_{vt}} \tag{7.2}$$

3. *Condensing heat, or cogenerated heat or that supplying a coabsorbent plant*:

$$Q_2\left(T_{2,cond(cg,cbs)}\right) = h_2 - h_3 \tag{7.3}$$

4. *Thermal efficiency*:

$$\eta_{thermal} = \frac{Q_1 - Q_2}{Q_1} \tag{7.4}$$

5. *Turbine effective efficiency*:

$$\eta_{ef,turbine} = \eta_{thermal}\eta_{gen}\eta_{mech}\eta_{0,i} \tag{7.5}$$

6. *Electrical power effective efficiency*:

$$\eta_{ef,SRC} = \eta_{ef,turbine}\eta_{boiler}\eta_{vt} \tag{7.6}$$

7. *Electrical effective power*:

$$W_{e,ef}\left(T_{2,cond(cg,cbs)}\right) = \eta_{ef,SRC}Q_{1,ef}\left(T_{2,cond(cg,cbs)}\right) \tag{7.7}$$

Equation (7.7) is not taking into account the auxiliary electrical power consumption, which modern power plants of confine it to small values, $W_{e,aux} \leq (1-2)\,\%$. The model considered the vapor superheating temperature as $T_1 = 550\,°\mathrm{C} = const.$ The symbols in Eqs. (7.1–7.7) are supposed to have constant values, $\eta_{0,i} = 0.85 = const.$, $\eta_{gen} = 0.98 = const.$, $\eta_{mech} = 0.97 = const.$, $\eta_{boiler} = 0.91 = const.$, and $\eta_{vt} = 0.99 = const.$ (Kirilin et al. 1985), holding for the turbine isentropic, electrical generator, mechanical, boiler and vapor tubes efficiencies, respectively. The SRC heat pump effect was assessed in relation to Eq. (7.8), using literature available steam data:

$$COP_{hp,cg} = \left[\frac{Q_2\left(T_{2,cg}\right)}{W_{e,ef}\left(T_{2,cond}\right)\frac{Q_1\left(T_{2,cg}\right)}{Q_1\left(T_{2,cond}\right)} - W_{e,ef}\left(T_{2,cg}\right)}\right]_{Q_1\left(T_{2,cond}\right)} \tag{7.8}$$

The results of the $COP_{hp,cg}$ evaluation, Eq. (7.8), for one condensing temperature $T_{2,cond} = 39\,°\mathrm{C}$ and $Q_2\left(T_{2,cg}\right)$, are given in Fig. 7.2, versus the cogeneration temperature $T_{2,cg}$. The SRC effectiveness exceeds much those of today heat pumps, $COP_{hp} \leq 4-6 \ll COP_{hp,cg}$, therefore primary energy saving reasons require that this should be used in cogeneration mainly, to district heating, while the large heat pumps extracting heat from depleted sources should be used just in a complementary way (Radcenco et al. 1985).

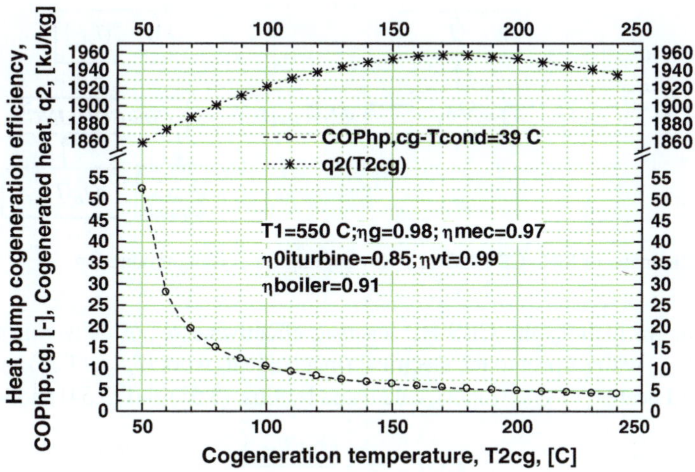

Fig. 7.2 SRC heat pump working mode assessment

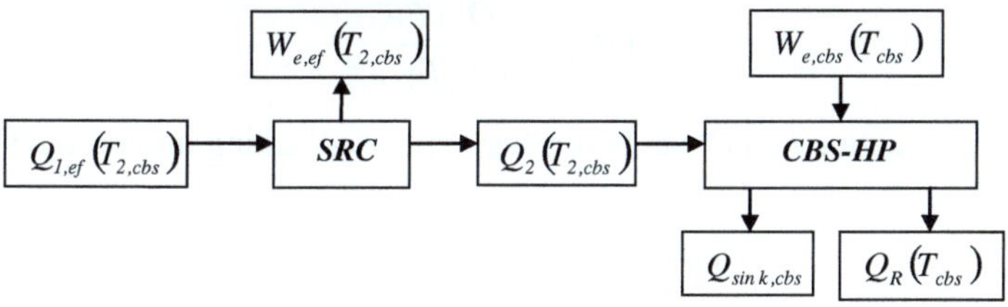

Fig. 7.3 Schematic of SRC-CBS-HP coupling for heat production

7.1.1 Steam Rankine Cycle-Coabsorbent Heat Pump Link

The coabsorbent heat pumps have the potential to achieve higher effectiveness comparatively only, and this will be shown later in this chapter. On this purpose, next we shall analyze the link between SRC and a coabsorbent heat pump, CBS-HP, Fig. 7.3, with respect to its electrical consumption, delivered by SRC. Here, the coabsorbent heat pump, supplied by SRC with the cogenerated heat $Q_2(T_{2,cbs})$ and the power $W_{e,cbs}(T_{cbs})$, provides the useful heat $Q_R(T_{cbs})$.

The effectiveness of $Q_2(T_{2,cbs})$ cogeneration is given by Eq. (7.9), calculated for $Q_2(T_{2,cg}) = Q_R(T_{cbs})$ and $T_{2,cg} = T_{cbs}$, similarly to Eq. (7.8) and Fig. 7.2.

$$
\begin{aligned}
COP_{hp(c),cg,cbs} &= \left[\frac{Q_2(T_{2,cbs})}{\Delta W_{e,ef}}\right]_{Q_1(T_{2,cond})} \\
&= \left[\frac{Q_2(T_{2,cbs})}{W_{e,ef}(T_{2,cond})\frac{Q_1(T_{2,cbs})}{Q_1(T_{2,cond})} - W_{e,ef}(T_{2,cbs})}\right]_{Q_1(T_{2,cond})}
\end{aligned}
\tag{7.9}
$$

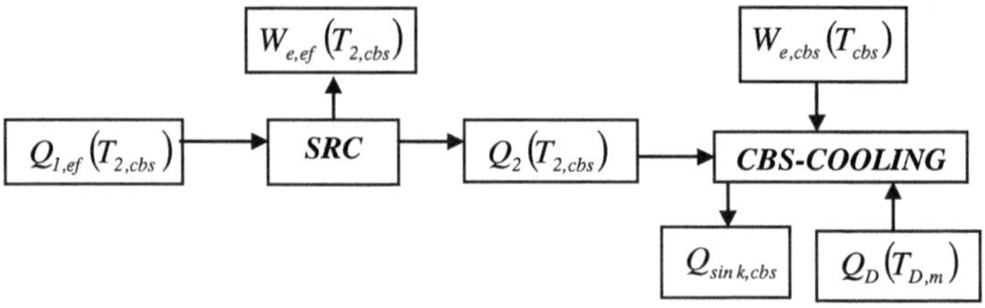

Fig. 7.4 Schematic of SRC-CBS-COOLING coupling for cooling production

Further on, we identify the CBS-HP intrinsic effectivenesses of delivering the useful heat $Q_R(T_{cbs})$ with respect to the electrical (mechanical) work $W_{e,cbs}(T_{cbs})$ and with the cogenerated heat $Q_2\left(T_{2,cbs}\right)$ inputs, calculated by Eqs. (3.55) and (3.54) and written as:

$$COP_w = \frac{Q_R(T_{cbs})}{W_{e,cbs}(T_{cbs})} \tag{7.10}$$

and

$$COP_h = \frac{Q_R(T_{cbs})}{Q_2\left(T_{2,cbs}\right)} \tag{7.11}$$

respectively. Equation (7.11) can be further expressed, taking into account Eq. (7.9), as follows:

$$COP_h = \frac{Q_R(T_{cbs})}{\Delta W_{e,ef} COP_{hp(c),cg,cbs}} \tag{7.12}$$

The CBS-HP delivers the useful heat $Q_R(T_{cbs})$, Fig. 7.3, consuming the electrical power $W_{e,cbs} + \Delta W_{e,ef}$, wherefrom, with Eqs. (7.10) and (7.12), its effectiveness we are looking for results in:

$$COP_{hp,cbs} = \frac{Q_R(T_{cbs})}{W_{e,cbs} + \Delta W_{e,ef}} = \left(COP_w^{-1} + COP_h^{-1} COP_{hp(c),cg,cbs}^{-1}\right)^{-1} \tag{7.13}$$

7.1.2 Steam Rankine Cycle-Coabsorbent Cooling Cycle Link

Next, we shall analyze the link between SRC and a coabsorbent cooling cycle, CBS-COOLING, Fig. 7.4, with respect to its electrical consumption. In this case, the coabsorbent cooling cycle, supplied by SRC with the cogenerated heat $Q_2\left(T_{2,cbs}\right)$ and the power $W_{e,cbs}(T_{cbs})$, extracts the useful heat $Q_D\left(T_{D,m}\right)$. The effectiveness of $Q_2\left(T_{2,cbs}\right)$ cogeneration is given by Eq. (7.9), as well.

Further on, we remember the CBS-COOLING intrinsic effectivenesses of extracting the useful heat $Q_D\left(T_{D,m}\right)$, with respect to the electrical work $W_{e,cbs}(T_{cbs})$ and with the cogenerated heat $Q_2\left(T_{2,cbs}\right)$ inputs, by:

$$COP_w = \frac{Q_D\left(T_{D,m}\right)}{W_{e,cbs}(T_{cbs})} \tag{7.14}$$

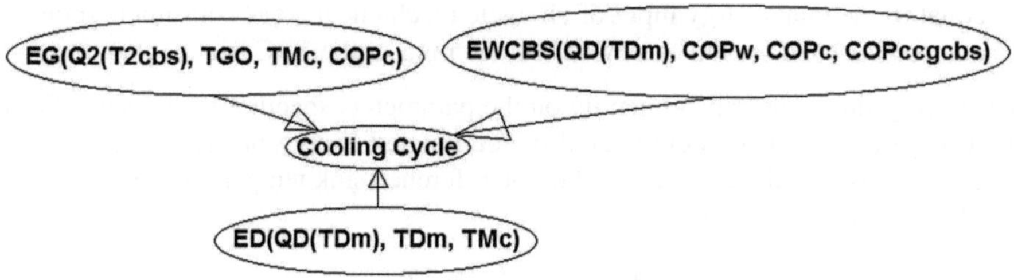

Fig. 7.5 Balance of the most important exergy items of a coabsorbent cooling cycle

and

$$COP_c = \frac{Q_D(T_{D,m})}{Q_2(T_{2,cbs})} \tag{7.15}$$

rewritting Eqs. (3.55) and (3.53), respectively. Equation (7.15) can be further expressed, taking into account Eq. (7.9), as:

$$COP_c = \frac{Q_D(T_{D,m})}{\Delta W_{e,ef} COP_{hp(c),cg,cbs}} \tag{7.16}$$

The CBS-COOLING extracts the useful heat $Q_D(T_{D,m})$, Fig. 7.4, consuming the electrical power $W_{e,cbs} + \Delta W_{e,ef}$, wherefrom, with Eqs. (7.14) and (7.16), its effectiveness we are looking for results in:

$$COP_{c,cbs} = \frac{Q_D(T_{D,m})}{W_{e,cbs} + \Delta W_{e,ef}} = \left(COP_w^{-1} + COP_c^{-1} COP_{hp(c),cg,cbs}^{-1}\right)^{-1} \tag{7.17}$$

calculated for $T_{2,cg} = T_{cbs}$.

7.2 Exergy Efficiency Algorithm of Coabsorbent Cooling Cycles

The more general exergy balance of a coabsorbent cooling cycle, linked to a cogeneration SRC which is providing its necessary electrical power and heat, is depicted in Fig. 7.5. In this figure the legend is:

1. $E_G(Q_2(T_{2,cbs}), T_{GO}, T_{M,c}, COP_c)$ holds for the total exergy input of all cycle generation stages;
2. $E_D(Q_D(T_{D,m}), T_{D,m}, T_{M,c})$ holds for the useful exergy output of the cycle desorber; and
3. $E_{W,SRC}(Q_D(T_{D,m}), COP_w, COP_c, COP_{c,cg,cbs})$ accounts for the SRC exergy input which enables the cooling cycle to be supplied with an amount of exergy

equal to the total exergy input of all cycle mechanical work consumers (pumps, compressor) and for the exergy destroy coming of the SRC heat cogeneration.

The exergy functions depend mainly on the parameters specified in brackets. From the exergy balance it was eliminated the cumulated exergy of the resorption heat output, as this is null, being rejected to the reference sink temperature, $T_{M,c}$.

The E_G exergy is calculated by:

$$E_G = Q_2\left(T_{2,cbs}\right)\theta_{w,G} \tag{7.18}$$

where $Q_2\left(T_{2,cbs}\right)$ and $\theta_{w,G}$ hold for the cycle cumulated generation heat and the heat-to-work Carnot efficiency factor, respectively. This factor is given here by:

$$\theta_{w,G} = 1 - \frac{T_{M,c}}{T_{GO}} = 1 - \frac{T_{M,c}}{T_{M,h}} \tag{7.19}$$

The cooling cycle exergy input with respect to the mechanical (electrical) work is calculated using Eq. (7.17):

$$E_W = W_{e,cbs} + \Delta W_{e,ef} = Q_D\left(T_{D,m}\right)\left(COP_w^{-1} + COP_c^{-1}COP_{hp(c),cg,cbs}^{-1}\right) \tag{7.20}$$

SRC supplies the cooling cycle with the E_W exergy, consuming an amount of exergy input, noted by $E_{W,SRC}$, Fig. 7.5. In order to find it, we bear in mind that SRC consumed an amount of heat equal to $\left(E_W\eta_{el,SRC}^{-1}\right)$ kJ for each unit of work (exergy) E_W produced for cycle supply, where η_{el} is the SRC electrical power efficiency. Hence, the SRC exergy input, $E_{W,SRC}$, which we are looking for, can be calculated with Eq. (7.20), as:

$$\begin{aligned} E_{W,SRC} &= \left(E_W\eta_{el,SRC}^{-1}\right)\theta_{w,SRC} \\ &= Q_D\left(T_{D,m}\right)\left(COP_w^{-1} + COP_c^{-1}COP_{hp(c),cg,cbs}^{-1}\right)\theta_{w,SRC}\eta_{el,SRC}^{-1} \end{aligned} \tag{7.21}$$

where $\theta_{w,SRC}$ is the Carnot factor of the SRC, Fig. 7.1b, given by:

$$\theta_{w,SRC} = 1 - \frac{T_{2,cond}}{T_1} > \eta_{el,SRC} \tag{7.22}$$

In Eq. (7.22), T_1 is replaced by the fuel burning temperature, $T_{burning}$, if $T_{burning} \ll T_1$. From Eq. (7.21) it results also that the SRC exergy efficiency is:

$$\eta_{E,SRC} = \frac{E_W}{E_{W,SRC}} = \frac{\eta_{el,SRC}}{\theta_{w,SRC}} \tag{7.23}$$

known also as the relative to Carnot (electrical) SRC power efficiency. Actual values of $\eta_{E,SRC}$ are for example ~0.4, 0.45–0.5, 0.45–0.65 and 0.65–0.7, holding true for the Organic Rankine Cycle (ORC), Steam Rankine Cycle (SRC), Gas Turbine Prime Movers (GTPM) and Combined Power Cycle (CPC), respectively.

Finally, the E_D exergy output is assessed as a positive amount. When the cooling cycle produces the temperature $T_{D,m} = (T_{DI} + T_{DO})/2$ below its sink temperature $T_{M,c}$, that is $T_{D,m} < T_{M,c}$, it is generating a sink source for an imaginary direct (power) ideal cycle. This could operate between the old sink temperature, $T_{M,c}$, which now is becoming the heat temperature for the direct cycle and the new sink temperature, $T_{D,m}$. The direct cycle is supplied by an amount of thermal energy equal to $(Q_D(T_{D,m}) + w_D)$ of temperature level equal to $T_{M,c}$, in order to produce a mechanical work amount w_D and to reject to the sink an amount of heat equal to $Q_D(T_{D,m})$. The maximum work output is calculated by:

$$w_D = \left(Q_D\left(T_{D,m}\right) + w_D\right)\theta_{w,D} \tag{7.24}$$

where $\theta_{w,D}$ is the direct ideal cycle heat-to-work Carnot efficiency (Kirilin 1985):

$$\theta_{w,D} = 1 - \frac{T_{D,m}}{T_{M,c}} \tag{7.25}$$

Combining Eqs. (7.24) and (7.25), the direct ideal cycle work output results in:

$$w_D = Q_D\left(T_{D,m}\right)\left(\frac{T_{M,c}}{T_{D,m}} - 1\right) \tag{7.26}$$

Bearing in mind that we deal with reversible processes, the imagined direct ideal cycle is the reverse cycle of an ideal cycle which extracts an amount of heat $Q_D(T_{D,m})$ from a cooled source having the temperature $T_{D,m}$ and delivers to the sink source of temperature $T_{M,c}$ an amount of thermal energy equal to $(Q_D(T_{D,m}) + w_c)$ with the help of a work input equal to w_c. This reverse cycle is nothing else but an ideal compression cooling cycle, which the ideal compression work consumption of amounts to:

$$|w_c| = Q_D\left(T_{D,m}\right)\theta_c^{-1} \tag{7.27}$$

In Eq. (7.27), θ_c is the known Carnot effectiveness of the ideal compression cooling cycle (e.g. Kirilin 1985), which here has the expression:

$$\theta_c = \left(\frac{T_{M,c}}{T_{D,m}} - 1\right)^{-1} \tag{7.28}$$

Of course, the two ideal cycles, direct and reverse, have equal works in absolute value:

$$w_D = |w_c| \tag{7.29}$$

This statement can be verified taking into account Eq. (7.26) on one side and Eqs. (7.27) and (7.28), on the other side. Returning now to the E_D exergy assessment, this results from the above considerations as being equal to:

$$E_D = Q_D\left(T_{D,m}\right)\theta_c^{-1} \tag{7.30}$$

The cooling cycles exergy comparison is done calculating for each of them an exergy efficiency, defined by (Chartier et al. 1975):

$$\eta_E = \frac{Retrievable\_(useful)\_exergy}{Input\_exergy} \leq 1 \tag{7.31}$$

Equation (7.31) has the following particular form in our case (see Fig. 7.5):

$$\eta_{E,c} = \frac{E_D}{E_G + E_{W,SRC}} \tag{7.32}$$

Equation (7.32) rewrites further introducing the exergy expressions given by Eqs. (7.18, 7.21, 7.23) and (7.30), as:

$$\eta_{E,c} = \frac{Q_D(T_{D,m})\theta_c^{-1}}{Q_2(T_{2,cbs})\theta_{w,G} + Q_D(T_{D,m})\eta_{E,SRC}^{-1}\left(COP_w^{-1} + COP_c^{-1}COP_{hp(c),cg,cbs}^{-1}\right)} \tag{7.33}$$

Equation (7.33) is rearranged, dividing it by $Q_D(T_{D,m})$ and taking into account Eq. (7.15), in order to obtain:

$$\eta_{E,c} = \frac{\theta_c^{-1}}{\theta_{w,G}COP_c^{-1} + \eta_{E,SRC}^{-1}\left(COP_w^{-1} + COP_c^{-1}COP_{hp(c),cg,cbs}^{-1}\right)} \tag{7.34}$$

Equation (7.34) fully expresses the exergy efficiency of the coabsorbent cooling cycles linked to a SRC. Depending on the supplying sources nature, this can take several particular forms, given next:

(i) *The exergy efficiency of the pure mechanical vapor compression cooling cycle.* The topic was the concern of previous works (e.g. McGovern and Harte 1995). Here, following a simplified way, the exergy efficiency of the mvc cooling cycle is obtained from Eq. (7.34), considering that the cycle generation exergy input vanishes $E_G = 0$ (see Fig. 7.5), therefore $Q_2(T_{2,cbs}) = 0$. Bearing this in mind, from Eq. (7.15) it results that $COP_c = \infty$, and further $COP_c^{-1} = 0$, wherefrom the exergy efficiency of the mvc cooling cycle has the following expression:

$$\eta_{E,c,mvc} = \eta_{E,SRC}\theta_c^{-1}COP_w \tag{7.35}$$

In Eq. (7.35), θ_c is given by Eq. (7.28), written for the mvc cycle, that is:

$$\theta_c = \left(\frac{T_C}{T_E} - 1\right)^{-1} \tag{7.36}$$

where indexes C and E hold for the evaporator and condenser.

(ii) *The coabsorbent cooling cycle is supplied by free heat sources.* In this case, SRC delivers to the cooling cycle electrical power, only, i.e. $W_{e,cbs}(T_{cbs}) > 0$. Consequently, the exergy destroy, caused by the heat cogeneration, is null, $\Delta W_{e,ef} = 0$ (see Fig. 7.4 and Eq. 7.9). Hence, the effectiveness of the $Q_2(T_{2,cbs})$ heat cogeneration becomes infinite, according to Eq. (7.9), $COP_{hp(c),cg,cbs} = \infty$, and because $COP_{hp(c),cg,cbs}^{-1} = 0$, Eq. (7.34) becomes:

$$\eta_{E,c,free} = \frac{\theta_c^{-1}}{\theta_{w,G}COP_c^{-1} + \eta_{E,SRC}^{-1}COP_w^{-1}} \tag{7.37}$$

The supply with free heat sources is one of the best operation mode of the coabsorbent cooling cycle, as Eq. (7.37) takes satisfactory values and the results will confirm this.

(iii) *The coabsorbent cooling cycle is supplied by low entropy heat sources.* Here are included those (low entropy) heat sources of high temperature (e.g. $T_{burning} = (800–1,200)\,°C$), created by burning fuels like those which are powering the ORC, SRC, GTPM, CPC, etc., thermal power plants (e.g. fossil fuel, biomass). Using such sources to supply quite low temperature generators (e.g. $T_G = (80–160)\,°C$) is a common practice in the absorption refrigeration. However, the task-source is clearly mismatched and the global exergy efficiency of the process at hand is penalized. Indeed, in such cases, the E_G exergy in Eq. (7.32) writes:

$$E_G = \eta_{G,fuel}^{-1}\theta_{w,fuel}q_G \tag{7.38}$$

where $\eta_{G,fuel}$ is the 1st principle heating effectiveness of the fuel fired generator (e.g. $\eta_{G,fuel} = 0.8 - 0.92$) and $\theta_{w,fuel}$ is the heat-to-work Carnot efficiency factor written for the case of fuel fired heating:

$$\theta_{w,fuel} = 1 - \frac{T_{M,c}}{T_{burning}} \tag{7.39}$$

Considering Eqs. (7.38) and (7.39), Eq. (7.37) rewrites:

$$\eta_{E,c,fuel} = \frac{\theta_c^{-1}}{\eta_{G,fuel}^{-1}\theta_{w,fuel}COP_c^{-1} + \eta_{E,SRC}^{-1}COP_w^{-1}} \tag{7.40}$$

The penalization we have referred to results comparing Eqs. (7.37) and (7.40) and showing that $\eta_{E,c,free} > \eta_{E,c,fuel}$. Indeed, from the obvious in equations $\theta_{w,fuel} > \theta_{w,G}$ and $\eta_{G,fuel} < 1$, it results that in Eq. $\eta_{G,fuel}^{-1}\theta_{w,fuel} > \theta_{w,G}$ holds true so much more. Hence, the denominator of Eq. (7.40) exceeds that of Eq. (7.37), wherefrom in Eq. $\eta_{E,c,free} > \eta_{E,c,fuel}$ holds true.

(iv) *The coabsorbent cooling cycle is supplied by solar energy.* In this case, the E_G exergy in Eq. (7.32) writes:

$$E_G = \tau_{atm}^{-1}\eta_{G,solar}^{-1}\theta_{w,solar}q_G \tag{7.41}$$

where τ_{atm} and $\eta_{G,solar}$ hold for the Earth atmosphere transmittance, (e.g. $\tau_{atm} = 0.15 - 0.70$), and the 1st principle instantaneous or long-term solar collecting effectiveness of the solar powered generator (e.g. $\eta_{G,solar} = 0.20 - 0.70$), (Kreith and Kreider 1978), respectively, and $\theta_{w,solar}$ is the heat-to-work Carnot efficiency factor written for the case of solar energy (heat) supply:

$$\theta_{w,solar} = 1 - \frac{T_{M,c}}{T_{sun}} \tag{7.42}$$

where $T_{sun} \cong 5,762$ K (Duffie and Beckman 1974; Kreith and Kreider 1978). Introducing Eq. (7.41) in Eq. (7.37), this rewrites:

$$\eta_{E,c,solar} = \frac{\theta_c^{-1}}{\tau_{atm}^{-1}\eta_{G,solar}^{-1}\theta_{w,solar}COP_c^{-1} + \eta_{E,SRC}^{-1}COP_w^{-1}} \tag{7.43}$$

The use of solar heating protects the environment, therefore is highly encouraged. However, it is characterized by a high source-task temperature mismatch, because the sun surface temperature is $T_{sun} \cong 5,762$ K, while a cycle generator operates mostly below 400 K. Hence, thermodynamically, this renders the exergy efficiency of Eq. (7.43) be small (see model results), and economically, the solar absorption cooling be marginal (Staicovici 1986, 1989).

(v) *The coabsorbent cooling cycle is supplied by clean work (electricity) produced by concentrated solar irradiance.* In this case, the solar irradiance is highly concentrated, $CR = 0.5 \times 10^2 - 3 \times 10^3$ and more, in order to obtain heat sources of elevated temperature, $T_{conc.} = (600-1,250)$ K. The concentration solar collectors can be build in different ways, chosen in function of the required temperature level. It can be usually of parabolic or linear Fresnel and of paraboloidal or Fresnel mirror (power tower) concepts, needing single-axis or two-axis sun tracking, respectively (Duffie and Beckman 1974; Kreith and Kreider 1978). These heat sources are supplying thermal power stations, of SRC or CPC type, for electricity production (Angelino 2008). The maximum (ideal) availability of the solar exergy has been previously evaluated (e.g. Press 1976; Parrott 1978; Izquierdo Millan et al. 1996). Here, we need the actual value of the solar exergy availability. In order to find it, suppose the sun is delivering an amount of radiative heat q. Its corresponding exergy, considered as an input data, equals:

$$E_{W,sun} = q\theta_{w,solar} \tag{7.44}$$

where $\theta_{w,solar}$ is the Carnot efficiency factor given by Eq. (7.42). Out of the radiative heat q, only a fraction reaches the receiver of the solar thermal power station:

$$q_{sr} = \eta_0 q \tag{7.45}$$

In Eq. (7.45), $\eta_0 < 1$ accounts for the overall optical losses, which the beam sun irradiance q is suffering until reaching the receiver. The optical losses depend on the Earth atmosphere transmittance (clearness index) and on the solar collector type and its physical and technical parameters, like mirror shape accuracy, mirror reflectance, sun tracking accuracy and transparent covers transmittance and reflectance (Duffie and Beckman 1974; Kreith and Kreider 1978). The q_{sr} irradiance is converted in electrical power w, which is the power station exergy output. With Eq. (7.45), the 1st principle efficiency, $\eta_{el,SRC}$, of this conversion process is given as known by:

$$\eta_{el,SRC} = \frac{w}{q_{sr}} = \frac{w}{\eta_0 q} \tag{7.46}$$

The exergy efficiency which we are looking for is computed with Eq. (7.47):

$$\eta_{E,SRC,solar\_conc.} = \frac{w}{E_{W,sun}} \tag{7.47}$$

or, with the help of Eqs. (7.44) and (7.46), it results:

$$\eta_{E,SRC,solar\_conc.} = \eta_0 \eta_{el,SRC} \theta_{w,solar}^{-1} \tag{7.48}$$

(vi) *The coabsorbent cooling cycle is supplied with clean work (electricity) produced by solar photovoltaic (PV) panels.* In case of the PV, the exergy efficiency is:

$$\eta_{E,PV} = \tau_{atm} \eta_{el,PV} \theta_{w,solar}^{-1} \tag{7.49}$$

where τ_{atm} has been defined by Eq. (7.41), $\eta_{el,PV}$ is the PV solar-to-electrical conversion efficiency and $\theta_{w,solar}$ is given by Eq. (7.42). The best actual PV solar panels available for consumers have an $\eta_{el,PV}$ efficiency of cca. 15.4–17.68 % (SRoeCo Solar 2012), depending mainly on the PV cell type and solar irradiance. When the sun light is concentrated onto the solar panel (CPV), the panel can achieve up to 27.0–31.0 % efficiency (Amonix & NREL 2011; Soitec 2013).

(vii) *The coabsorbent cooling cycle is supplied with clean work (electricity) produced by windmills.* In case of a windmill, the input and output exergies are of the same type, i.e. mechanical work, so the exergy efficiency is simply the ratio between the useful wind turbine rotation energy output done per unit of time P divided by the kinetic energy input per unit of time of the fluid passing through the area swept by the windmill rotor with the velocity V_∞ and the mass flow rate $\dot{m}$ (Kreith and Kreider 1978), or:

$$\eta_{E,windmill} = \frac{P}{\frac{1}{2} V_\infty^2 \dot{m}} \tag{7.50}$$

Equation (7.50) gives at the same time the windmill 1st principle efficiency, or, equally, that of the wind turbine, $\eta_{E,windmill} = \eta_{t,windmill}$. The windmill exergy efficiency depends mainly on the wind speed, propeller diameter, air density and wind turbine type, $\eta_{E,windmill} \cong 0.0 - 0.50$ (Kreith and Kreider 1978).

(viii) *The coabsorbent cooling cycle is supplied with clean work (electricity) produced by hydro power stations.* In case of a hydro power station, the input and output exergies are again mechanical work. Hence, similar to the item (vii), the exergy efficiency is the ratio between the turbine useful rotation energy P divided by the turbine energy input per unit of time, equal to the hydraulic dynamic head of the water mass flow rate passing through the turbine blades, $\rho g h \dot{V}$, or equal to the potential energy, $\dot{m} g h$, of the water mass flow rate,

$\dot{m} = \rho \dot{V}$, situated in amonte of the turbine and having the static head h with respect to the turbine height:

$$\eta_{E,hydro} = \frac{P}{\rho g h \dot{V}} \tag{7.51}$$

Equation (7.51) represents in fact the turbine dimensionless efficiency, $\eta_{t,hydro}$, obtained by measuring the power output P, water density ρ, static head h, and the volumetric flow rate $\dot{V}$ ($g = 9.81\,\mathrm{ms}^{-2}$ is the gravitational constant). The actual turbine efficiency is high, $\eta_{E,hydro} = \eta_{t,hydro} = 0.85$ (ASME PTC 18, IEC 60041). The conclusions formed to this item (viii) hold true for wave and tidal power plants supplying with clean electrical energy a coabsorbent or mvc cooling cycle as well. From the literature it results that this type of power plants is capable of exergy efficiencies equal to $\eta_{E,wave,tidal} = 0.25 - 0.85$, depending mainly of the type of sea converting equipment, period of the year and the distribution of the power frequency (Kreith and Kreider 1978).

In Table 7.1 we have listed the mvc and coabsorbent cooling cycle exergy efficiencies resulting from Eq. (7.34), considered without heat cogeneration and for all exergy efficiencies of the heat (*free*, low entropy *fuel* and the *solar*) and electricity (*SRC*, *SRC* with *solar concentration*, *PV*, *windmill* and the *hydro*) sources, involved in the (i–viii) items.

7.2.1 Exergy Efficiency Results of Coabsorbent and mvc Cooling Cycles

Exergy efficiency results have been calculated with the algorithm given in Tables 7.1 and 7.2 for cooling and heating applications. The computation didn't account for the auxiliary power consumption consumed (e.g. for pumping the sources fluids). The first results were obtained for mvc cooling, Fig. 7.6, the challenger and comparison term of all bare and hybrid coabsorbent cooling cycles at hand. In this figure, the exergy efficiency of mvc cooling cycle supplied by *SRC*, *SRC* with *solar concentration*, *PV*, *windmill* and the *hydro* electrical power sources (see Eqs. 7.35, 7.52, 7.56, 7.60 and 7.64, respectively), is plotted against the mvc COP_w cooling effectiveness, for two evaporator and condenser temperature study cases, $T_E = -50\,°\mathrm{C}$, $T_C = 32\,°\mathrm{C}$ (solid line) and $T_E = -25\,°\mathrm{C}$, $T_C = 35\,°\mathrm{C}$ (dash line), respectively. The lowest and higher COP_w values of each study case are the actual and improved ones, as proposed for the future (see Chap. 8), respectively. The same figure shows as well the exergy efficiency of the $NH_3 - H_2O$ nontruncated coabsorbent (NT) deep cooling cycle supplied in *SRC* cogeneration, marked by the left bottom dash line. From Fig. 7.6 it results that both cases have the same relative position with respect to different sources plotts, but of course, have different efficiency values. The highest and lowest efficiency values are due to the hydro—CPC and *SRC* solar concentration-PV electrical supply, respectively,

Table 7.1 Exergy efficiency of mvc and coabsorbent cooling cycle, operated by different heat and electrical sources

Eq. no.	Exergy efficiency
(7.35)	$\eta_{E,c,mvc,SRC} = \eta_{E,SRC}\theta_c^{-1}COP_w$
(7.37)	$\eta_{E,c,free,SRC} = \theta_c^{-1}\left(\theta_{w,G}COP_c^{-1} + \eta_{E,SRC}^{-1}COP_w^{-1}\right)^{-1}$
(7.40)	$\eta_{E,c,fuel,SRC} = \theta_c^{-1}\left(\eta_{G,fuel}^{-1}\theta_{w,fuel}COP_c^{-1} + \eta_{E,SRC}^{-1}COP_w^{-1}\right)^{-1}$
(7.43)	$\eta_{E,c,solar,SRC} = \theta_c^{-1}\left(\tau_{atm}^{-1}\eta_{G,solar}^{-1}\theta_{w,solar}COP_c^{-1} + \eta_{E,SRC}^{-1}COP_w^{-1}\right)^{-1}$
(7.52)	$\eta_{E,c,mvc,solar\_conc.} = \eta_{E,SRC,solar\_conc.}\theta_c^{-1}COP_w$
(7.53)	$\eta_{E,c,free,solar\_conc.} = \theta_c^{-1}\left(\theta_{w,G}COP_c^{-1} + \eta_{E,SRC,solar\_conc.}^{-1}COP_w^{-1}\right)^{-1}$
(7.54)	$\eta_{E,c,fuel,solar\_conc.} = \theta_c^{-1}\left(\eta_{G,fuel}^{-1}\theta_{w,fuel}COP_c^{-1} + \eta_{E,SRC,solar\_conc.}^{-1}COP_w^{-1}\right)^{-1}$
(7.55)	$\eta_{E,c,solar,solar\_conc.} = \theta_c^{-1}\left(\tau_{atm}^{-1}\eta_{G,solar}^{-1}\theta_{w,solar}COP_c^{-1} + \eta_{E,SRC,solar\_conc.}^{-1}COP_w^{-1}\right)^{-1}$
(7.56)	$\eta_{E,c,mvc,PV} = \eta_{E,PV}\theta_c^{-1}COP_w$
(7.57)	$\eta_{E,c,free,PV} = \theta_c^{-1}\left(\theta_{w,G}COP_c^{-1} + \eta_{E,PV}^{-1}COP_w^{-1}\right)^{-1}$
(7.58)	$\eta_{E,c,fuel,PV} = \theta_c^{-1}\left(\eta_{G,fuel}^{-1}\theta_{w,fuel}COP_c^{-1} + \eta_{E,PV}^{-1}COP_w^{-1}\right)^{-1}$
(7.59)	$\eta_{E,c,solar,PV} = \theta_c^{-1}\left(\tau_{atm}^{-1}\eta_{G,solar}^{-1}\theta_{w,solar}COP_c^{-1} + \eta_{E,PV}^{-1}COP_w^{-1}\right)^{-1}$
(7.60)	$\eta_{E,c,mvc,windmill} = \eta_{E,windmill}\theta_c^{-1}COP_w$
(7.61)	$\eta_{E,c,free,windmill} = \theta_c^{-1}\left(\theta_{w,G}COP_c^{-1} + \eta_{E,windmill}^{-1}COP_w^{-1}\right)^{-1}$
(7.62)	$\eta_{E,c,fuel,windmill} = \theta_c^{-1}\left(\eta_{G,fuel}^{-1}\theta_{w,fuel}COP_c^{-1} + \eta_{E,windmill}^{-1}COP_w^{-1}\right)^{-1}$
(7.63)	$\eta_{E,c,solar,windmill} = \theta_c^{-1}\left(\tau_{atm}^{-1}\eta_{G,solar}^{-1}\theta_{w,solar}COP_c^{-1} + \eta_{E,windmill}^{-1}COP_w^{-1}\right)^{-1}$
(7.64)	$\eta_{E,c,mvc,hydro} = \eta_{E,hydro}\theta_c^{-1}COP_w$
(7.65)	$\eta_{E,c,free,hydro} = \theta_c^{-1}\left(\theta_{w,G}COP_c^{-1} + \eta_{E,hydro}^{-1}COP_w^{-1}\right)^{-1}$
(7.66)	$\eta_{E,c,fuel,hydro} = \theta_c^{-1}\left(\eta_{G,fuel}^{-1}\theta_{w,fuel}COP_c^{-1} + \eta_{E,hydro}^{-1}COP_w^{-1}\right)^{-1}$
(7.67)	$\eta_{E,c,solar,hydro} = \theta_c^{-1}\left(\tau_{atm}^{-1}\eta_{G,solar}^{-1}\theta_{w,solar}COP_c^{-1} + \eta_{E,hydro}^{-1}COP_w^{-1}\right)^{-1}$

while *SRC* and windmill electrical supply is ranked in the low-to-middle position. Because all electrical suppliers we considered have a common source of energy, the sun, from Fig. 7.6 it becomes clear that the higher the degree of concentration of sun energy on Earth, the better is ranked the cooling with respect to the exergy efficiency. This assertion is valid for heating, as well, as it will be shown in the next section. Further, it is shown also that, thermodynamically, supplying the NT deep cooling in cogeneration is less effective than mvc supplied by the same *SRC* with electrical power ($\eta_{E,c,NT} = 0.1474 < 0.1959 = \eta_{E,c,mvc,SRC}$). However,

Table 7.2 Exergy efficiency of mvc and coabsorbent heating cycle, operated by different heat and electrical sources

Eq. no.	Exergy efficiency
(7.77)	$\eta_{E,c,mvc,SRC} = \eta_{E,SRC}\theta_{w,R}^{-1}COP_w$
(7.79)	$\eta_{E,c,free,SRC} = \theta_{w,R}^{-1}\left(\theta_{w,G}COP_h^{-1} + \eta_{E,SRC}^{-1}COP_w^{-1}\right)^{-1}$
(7.80)	$\eta_{E,c,fuel,SRC} = \theta_{w,R}^{-1}\left(\eta_{G,fuel}^{-1}\theta_{w,fuel}COP_h^{-1} + \eta_{E,SRC}^{-1}COP_w^{-1}\right)^{-1}$
(7.81)	$\eta_{E,c,solar,SRC} = \theta_{w,R}^{-1}\left(\tau_{atm}^{-1}\eta_{G,solar}^{-1}\theta_{w,solar}COP_h^{-1} + \eta_{E,SRC}^{-1}COP_w^{-1}\right)^{-1}$
(7.84)	$\eta_{E,c,mvc,solar\_conc.} = \eta_{E,SRC,solar\_conc.}\theta_{w,R}^{-1}COP_w$
(7.85)	$\eta_{E,c,free,solar\_conc.} = \theta_{w,R}^{-1}\left(\theta_{w,G}COP_h^{-1} + \eta_{E,SRC,solar\_conc.}^{-1}COP_w^{-1}\right)^{-1}$
(7.86)	$\eta_{E,c,fuel,solar\_conc.} = \theta_{w,R}^{-1}\left(\eta_{G,fuel}^{-1}\theta_{w,fuel}COP_h^{-1} + \eta_{E,SRC,solar\_conc.}^{-1}COP_w^{-1}\right)^{-1}$
(7.87)	$\eta_{E,c,solar,solar\_conc.} = \theta_{w,R}^{-1}\left(\tau_{atm}^{-1}\eta_{G,solar}^{-1}\theta_{w,solar}COP_h^{-1} + \eta_{E,SRC,solar\_conc.}^{-1}COP_w^{-1}\right)^{-1}$
(7.88)	$\eta_{E,c,mvc,PV} = \eta_{E,PV}\theta_{w,R}^{-1}COP_w$
(7.89)	$\eta_{E,c,free,PV} = \theta_{w,R}^{-1}\left(\theta_{w,G}COP_h^{-1} + \eta_{E,PV}^{-1}COP_w^{-1}\right)^{-1}$
(7.90)	$\eta_{E,c,fuel,PV} = \theta_{w,R}^{-1}\left(\eta_{G,fuel}^{-1}\theta_{w,fuel}COP_h^{-1} + \eta_{E,PV}^{-1}COP_w^{-1}\right)^{-1}$
(7.91)	$\eta_{E,c,solar,PV} = \theta_{w,R}^{-1}\left(\tau_{atm}^{-1}\eta_{G,solar}^{-1}\theta_{w,solar}COP_h^{-1} + \eta_{E,PV}^{-1}COP_w^{-1}\right)^{-1}$
(7.92)	$\eta_{E,c,mvc,windmill} = \eta_{E,windmill}\theta_{w,R}^{-1}COP_w$
(7.93)	$\eta_{E,c,free,windmill} = \theta_{w,R}^{-1}\left(\theta_{w,G}COP_h^{-1} + \eta_{E,windmill}^{-1}COP_w^{-1}\right)^{-1}$
(7.94)	$\eta_{E,c,fuel,windmill} = \theta_{w,R}^{-1}\left(\eta_{G,fuel}^{-1}\theta_{w,fuel}COP_h^{-1} + \eta_{E,windmill}^{-1}COP_w^{-1}\right)^{-1}$
(7.95)	$\eta_{E,c,solar,windmill} = \theta_{w,R}^{-1}\left(\tau_{atm}^{-1}\eta_{G,solar}^{-1}\theta_{w,solar}COP_h^{-1} + \eta_{E,windmill}^{-1}COP_w^{-1}\right)^{-1}$
(7.96)	$\eta_{E,c,mvc,hydro} = \eta_{E,hydro}\theta_{w,R}^{-1}COP_w$
(7.97)	$\eta_{E,c,free,hydro} = \theta_{w,R}^{-1}\left(\theta_{w,G}COP_h^{-1} + \eta_{E,hydro}^{-1}COP_w^{-1}\right)^{-1}$
(7.98)	$\eta_{E,c,fuel,hydro} = \theta_{w,R}^{-1}\left(\eta_{G,fuel}^{-1}\theta_{w,fuel}COP_h^{-1} + \eta_{E,hydro}^{-1}COP_w^{-1}\right)^{-1}$
(7.99)	$\eta_{E,c,solar,hydro} = \theta_{w,R}^{-1}\left(\tau_{atm}^{-1}\eta_{G,solar}^{-1}\theta_{w,solar}COP_h^{-1} + \eta_{E,hydro}^{-1}COP_w^{-1}\right)^{-1}$

despite this power reason, the former system was prefered many times to the latter in deep cooling applications during the last century, because absorption offered other advantages in operation, comparatively, such as lower upkeep, replacement of large compressors sucking the low pressure-high specific volume refrigerant by absorption and absorbent pumping, low noise and a better adaptation to higher cooling capacity. Normally, a national electrical grid is supplied with electrical power coming from power stations of different types, characterized by different $\eta_{E,i}$ figures of merit, $(i = 1, \ldots, n)$. In this case, η_E in Eqs. (7.35–7.67) should be

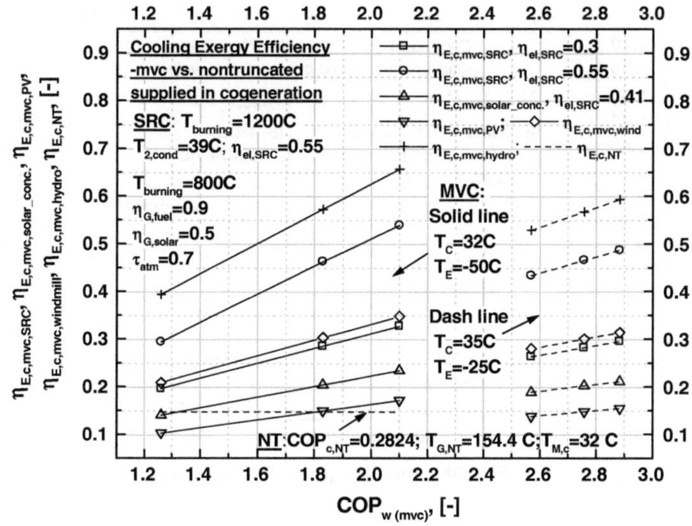

Fig. 7.6 Exergy efficiency of mvc cooling cycle plotted for two evaporator/condenser temperatures versus actual and future COP_w cooling effectiveness and six electrical sources as parameter

calculated as a weighted mean of the $\eta_{E,i}$ values. In introduction we emphasized the exergy method feature to enable the ranking of different nature energy-cycles and supplying sources performing a same task. Here, it is used to compare the mvc cooling results of Fig. 7.6 with those of Figs. 7.7 and 7.8, giving exergy efficiency results of the coabsorbent cooling cycles studied in Chaps. 3 and 4 for different heat supplying sources ($T_{M,c} = 32\,°C$). As we can remark from Figs. 7.7 and 7.8, the efficiency of coabsorbent cycles supplied by free heat sources is significantly higher than the actual mvc efficiency in both study cases. However, improving the mvc COP_w in the future is capable to challenge the coabsorbent technology, provided that mvc be supplied by high efficiency electrical stations, like hydro and CPC. Referring now just to the results of Figs. 7.7 and 7.8, first, the cogeneration heat supply penalizes much the efficiency of both types of coabsorbent cycles, as compared to the free heat supply, and from this point of view, it should be avoided, whenever possible. Second, the solar heat supply is ranked the last, with very low values, while the fuel heat supply increases to some extent the efficiency, comparatively, but is still low, less than 8 and 17 % in Figs. 7.7 and 7.8, respectively, and much below the efficiency of the cogeneration supply. In this respect, the cogeneration must be preferred always to fuel supply, from the power point of view. And last, but not the least, it is important to clarify an important aspect concerning the interpretation and credibility of the method at hand results. Indeed, the reader may ask himself why the exergy method is considered so powerful, because in fact it deals with an ideal concept, that of availability, which seemingly is far from the reality. The things are not at all like this, the method is deeply anchored in reality (see Appendix 2 of this chapter).

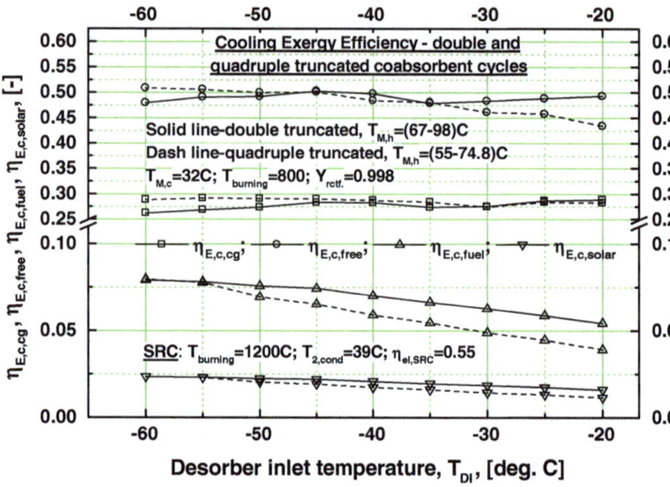

Fig. 7.7 Exergy efficiency of double and quadruple truncated coabsorbent cooling cycles versus desorber temperature and four heat sources as parameter

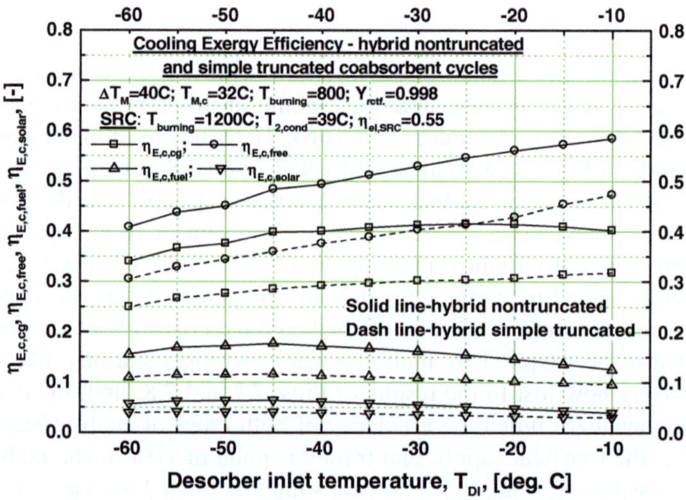

Fig. 7.8 Exergy efficiency of hybrid nontruncated and simple-truncated coabsorbent cooling cycles versus desorber temperature and four heat sources as parameter

7.3 Exergy Efficiency Algorithm of Coabsorbent Heating Cycles

The more general exergy balance of a coabsorbent heating cycle, linked to a cogeneration SRC which is providing its necessary electrical power and heat, is depicted in Fig. 7.9. In this figure the legend is:

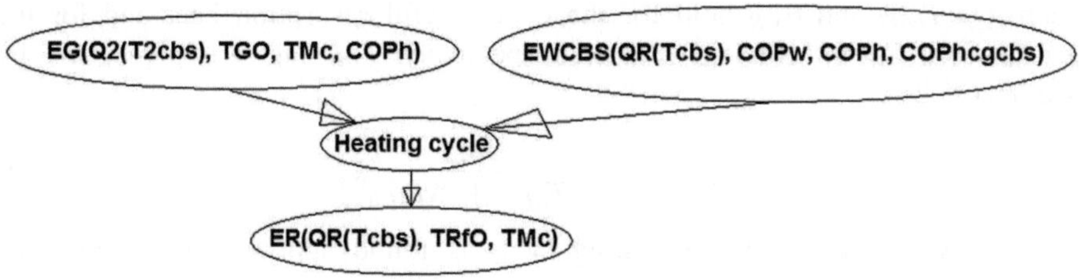

Fig. 7.9 Balance of the most important exergy items of a coabsorbent heating cycle

(i) $E_G(Q_2(T_{2,cbs}), T_{GO}, T_{M,c}, COP_h)$—see Sect. 7.2;
(ii) $E_R(Q_R(T_{cbs}), T_{Rf,O}, T_{M,c})$ holds for the useful exergy output of the cycle resorber; and
(iii) $E_{W,SRC}(Q_R(T_{cbs}), COP_w, COP_h, COP_{c,cg,cbs})$—see Sect. 7.2.

The E_G exergy is calculated in Sect. 7.2. The heating cycle exergy input with respect to the mechanical (electrical) work is calculated using Eq. (7.17), updated for the heating cycle:

$$E_W = W_{e,cbs} + \Delta W_{e,ef} = Q_R(T_{cbs})\left(COP_w^{-1} + COP_h^{-1} COP_{hp(c),cg,cbs}^{-1}\right) \quad (7.68)$$

The SRC exergy input, $E_{W,SRC}$, is calculated similarly to Sect. 7.2, with Eq. (7. 68), resulting in:

$$\begin{aligned} E_{W,SRC} &= \left(E_W \eta_{el,SRC}^{-1}\right)\theta_{w,SRC} \\ &= Q_R(T_{cbs})\left(COP_w^{-1} + COP_h^{-1} COP_{hp(c),cg,cbs}^{-1}\right)\theta_{w,SRC}\eta_{el,SRC}^{-1} \end{aligned} \quad (7.69)$$

In Eqs. (7.68) and (7.69), the $COP_{hp(c),cg,cbs}$ function is given by Eq. (7.9), and the COP_w and COP_h functions rewrite Eqs. (3.55) and (3.54), respectively. These are the CBS-HP intrinsic effectivenesses of delivering the useful heat $Q_R(T_{cbs})$, with respect to the electrical work $W_{e,cbs}(T_{cbs})$ and with the cogenerated heat $Q_2(T_{2,cbs})$ inputs, and are given by:

$$COP_w = \frac{Q_R(T_{cbs})}{W_{e,cbs}(T_{cbs})} \quad (7.70)$$

and

$$COP_h = \frac{Q_R(T_{cbs})}{Q_2(T_{2,cbs})} \quad (7.71)$$

respectively. Finally, the E_R exergy output is assessed by:

$$E_R = Q_R(T_{cbs})\theta_{w,R} \quad (7.72)$$

where $Q_R(T_{cbs})$ and $\theta_{w,R}$ hold for the cycle useful resorption heat and for the heat-to-work Carnot efficiency factor, respectively. This factor is given here by (see Sect. 5.2):

$$\theta_{w,R} = 1 - \frac{T_{M,c}}{T_{Rf,O} + \Delta T_{M,h}}, \tag{7.73}$$

The exergy efficiency of the heating cycle is found applying Eq. (7.31) (see Fig. 7.9):

$$\eta_{E,h} = \frac{E_R}{E_G + E_{W,SRC}} \tag{7.74}$$

Equation (7.74) rewrites further introducing the exergy expressions given by Eqs. (7.18, 7.72) and (7. 69), as:

$$\eta_{E,h} = \frac{Q_R(T_{cbs})\theta_{w,R}}{Q_2(T_{2,cbs})\theta_{w,G} + Q_R(T_{cbs})\eta_{E,SRC}^{-1}\left(COP_w^{-1} + COP_h^{-1}COP_{hp(c),cg,cbs}^{-1}\right)} \tag{7.75}$$

Equation (7.75) is rearranged, dividing it by $Q_R(T_{cbs})$ and taking into account Eq. 7.71), in order to obtain:

$$\eta_{E,h} = \frac{\theta_{w,R}}{\theta_{w,G}COP_h^{-1} + \eta_{E,SRC}^{-1}\left(COP_w^{-1} + COP_h^{-1}COP_{hp(c),cg,cbs}^{-1}\right)} \tag{7.76}$$

Equation (7.76) fully expresses the exergy efficiency of the coabsorbent heating cycles linked to a SRC. Similar to Sect. 7.2, depending on the operation conditions, this can take as well several particular forms, given next:

(i) *The exergy efficiency of the pure mechanical vapor compression heating (heat pump) cycle* is obtained from Eq. (7.76), proceeding similarly to Sect. 7.2. This time, for $E_G = 0$ and according to Eq. (7.71) $Q_2(T_{2,cbs}) = 0$, it results that $COP_h = \infty$, and further $COP_h^{-1} = 0$, wherefrom the exergy efficiency of the mvc heating (heat pump) cycle has the following expression:

$$\eta_{E,h,mvc} = \eta_{E,SRC}\theta_{w,R}COP_w \tag{7.77}$$

In Eq. (7.77) $\theta_{w,R}$ is given by Eq. (7.73), written for the mvc cycle, that is:

$$\theta_{w,R} = \theta_{w,C} = \left(1 - \frac{T_E}{T_C}\right) \tag{7.78}$$

where indexes C and E hold for the evaporator and condenser.

(ii) *The coabsorbent heating cycle is supplied by free heat sources*. This case has been emphasized by Sect. 7.2, as well, and the searched exergy efficiency is:

$$\eta_{E,h,free\_h.s} = \frac{\theta_{w,R}}{\theta_{w,G}COP_h^{-1} + \eta_{E,SRC}^{-1}COP_w^{-1}} \tag{7.79}$$

(iii) *The coabsorbent heating cycle is supplied by low entropy heat sources.* The case is similar to Sect. 7.2. The exergy efficiency results in:

$$\eta_{E,h,fuel} = \frac{\theta_{w,R}}{\eta_{G,fuel}^{-1}\theta_{w,fuel}COP_h^{-1} + \eta_{E,SRC}^{-1}COP_w^{-1}} \tag{7.80}$$

(iv) *The coabsorbent heating cycle is supplied by solar energy.* Again, the case is similar to Sect. 7.2 and the exergy efficiency results as:

$$\eta_{E,h,solar} = \frac{\theta_{w,R}}{\tau_{atm}^{-1}\eta_{G,solar}^{-1}\theta_{w,solar}COP_h^{-1} + \eta_{E,SRC}^{-1}COP_w^{-1}} \tag{7.81}$$

In our days, characterized by environmental problems due to energy production, an important particular case is represented by the solar heating using just solar collectors of flat-plate or low concentration types provided or not with selective surfaces and evacuated tubes (Duffie and Beckman 1974; Kreith and Kreider 1978). These can be utilized during the sunny days of a year directly for domestic warm water preparation and house heating, without involving a coabsorbent heating cycle. In this case, the collector own electrical consumption is zero (except that auxiliary for the intermediary heat exchange fluid pumping), therefore $COP_w = \infty$ and $COP_w^{-1} = 0$, while $COP_h = 1$, so Eq. (7.82) becomes:

$$\eta_{E,h,solar\_coll.} = \tau_{atm}\eta_{G,solar}\theta_{w,solar}^{-1}\theta_{w,R} \tag{7.82}$$

In Eq. (7.82), $\theta_{w,R}$ has the expression:

$$\theta_{w,R} = \left(1 - \frac{T_{M,c}}{T_{solar\_coll.}}\right) \tag{7.83}$$

where $T_{solar\_coll.}$ is the collector usefully heated fluid or mean collecting plate temperatures, etc. The expected $\eta_{E,h,solar\_coll.}$ values are small, below (6–8) %, for domestic applications.

The items (v–viii) of Sect. 7.2. hold true in case of the heating cycles at hand as well. Proceeding similarly to Sect. 7.2, in Table 7.2 we have listed the mvc and coabsorbent heating cycle exergy efficiencies resulting from Eq. (7.76) and involved in the (i–viii) items, considered without heat cogeneration and for all exergy efficiencies of the heat and electricity sources as in Table 7.1.

7.3.1 *Exergy Efficiency Results of Coabsorbent and Mechanical Vapor Compression Heating Cycles*

Similar to Sect. 7.2.1, the first results were obtained for the exergy efficiency of mvc heating cycle, Fig. 7.10, supplied again by *SRC*, *SRC* with *solar concentration*, *PV*, *windmill* and the *hydro* electrical power sources (see Eqs. 7.77, 7.84, 7.88, 7.92 and 7.96, respectively), and plotted against the mvc COP_w heating effectiveness, for

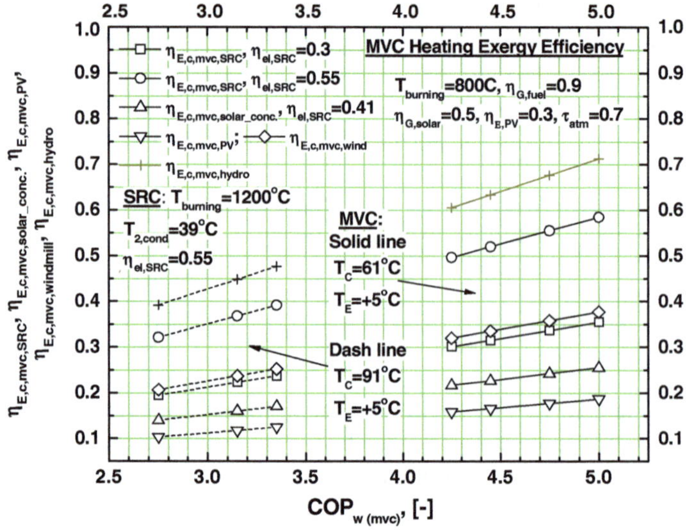

Fig. 7.10 Exergy efficiency of mvc heating cycle plotted for two evaporator/condenser temperatures versus actual and future COP_w heating effectiveness and six electrical sources as parameter

two evaporator and condenser temperature study cases, $T_E = +5\,°C$, $T_C = 61\,°C$ (solid line) and $T_E = +5\,°C$, $T_C = 91\,°C$ (dash line), respectively. The results of Fig. 7.10, shown similarly to Fig. 7.6 for the lowest (actual) and the increased (see Chap. 8 issues) COP_w values of each study case, prove that heating supplied by electrical sources with higher degree of concentration of sun energy is again better ranked with respect to the exergy efficiency, comparatively. Further results have been obtained for the double and quadruple truncated heating cycles, Fig. 7.11, and for the nontruncated and simple truncated hybrid heating cycles, Fig. 7.12a and b, respectively, Chaps. 3 and 4. Comparing the results of Figs. 7.10 and 7.11, it results that when supplied by free heat sources (SRC, $T_{M,h} = 37\,°C$), the truncated cycles could operate more effectively as the actual mvc heating supplied by high efficiency SRC power plants, at least to house heating and domestic hot water (dhw) preparation applications. However, the future COP_w improvements could change this situation in favor of the mvc heating cycle, especially when hydro and SRC are the electrical suppliers. Concerning the nontruncated and simple truncated hybrid heating cycles, Fig. 7.12a and b, the 2nd principle comparison with mvc heating, Fig. 7.10, done for a same internal heat temperature, $T_{M,h} = 5\,°C$, shows that coabsorbent cycles are better ranked, comparatively. However, despite the savings in electrical power which these cycles could bring, their practical application must overcome a few technical difficulties, like the appropriate operation with low and variable sink source temperatures.

The last part of this paragraph is devoted to a study case of the coabsorbent heat pumps, which, according to Sect. 7.1, have the potential to achieve higher effectiveness as compared to a SRC cogeneration. In Table 7.3 we give heating results of the NH_3–H_2O double and quadruple truncated cycles operation supplied by free

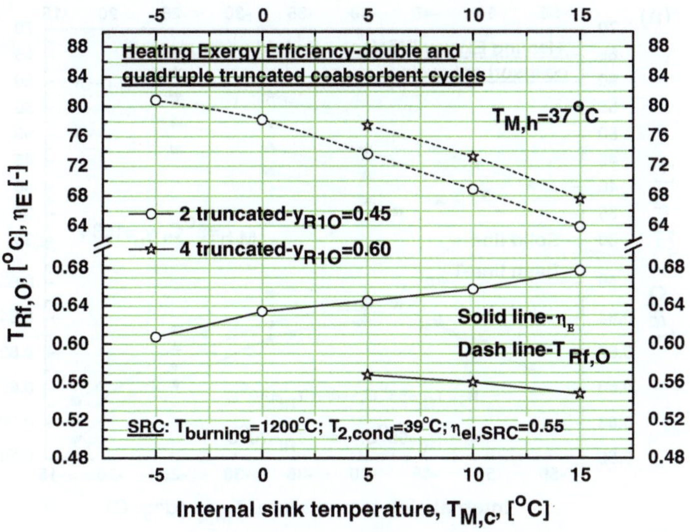

Fig. 7.11 Exergy efficiency of double and quadruple truncated coabsorbent heating cycles, supplied by SRC free heat source, $T_{M,h} = 37\,°C$, and plotted versus the internal sink source temperature, $T_{M,c}$

and low cogeneration heat sources, $T_{M,h} = 37\,°C$ and $T_{M,h} = 48\,°C$, respectively. The results, obtained considering gax recovery, but not accounting for the auxiliary electrical power consumption, prove our assertion. Indeed, during the cold and transition seasons, $T_{M,c} = (5-15)\,°C$, with free heat supply it is possible to produce heated fluids up to 63.84 and 77.42 °C (see $T_{Rf,O} = T_{cbs} = T_{2,cg}$ functions) with a relative to SRC heating effectiveness, $COP_{hp,cbs}/COP_{hp,cg}$, of 20.22 and 26.96, respectively, while in case of the cogeneration heat supply the heated fluids can reach up to 92.78 and 91.3 °C with the relative effectiveness of 1.43 and 1.18, respectively. The $COP_{hp,cbs}/COP_{hp,cg}$ figure of merit shows that coabsorbent technology could bring important primary energy savings in district heating applications, when supplied by sources coming of SRC, either free or cogenerated. In our case, the savings in electrical power are small, equal to $1/COP_{hp,cg} - 1/COP_{hp,cbs}$, but the environmental impact is high. The reader is encouraged to check the coabsorbent cycle feature, emphasized above, for the hybrid heating configuration, as well. In this case, $COP_{hp,cbs}$ could be similar to those of Table 7.3, because COP_w decreases, but COP_h increases, comparatively (see Eq. 7.13).

7.4 Cogeneration and Trigeneration Exergy Efficiency Algorithm of Coabsorbent Cooling and Heating Cycles

As already mentioned, thermodynamically, the main effect is decreased when a cycle is producing a second and so much more a triple effect at the same time, and the practice confirmed this behavior. However, when the cogeneration and

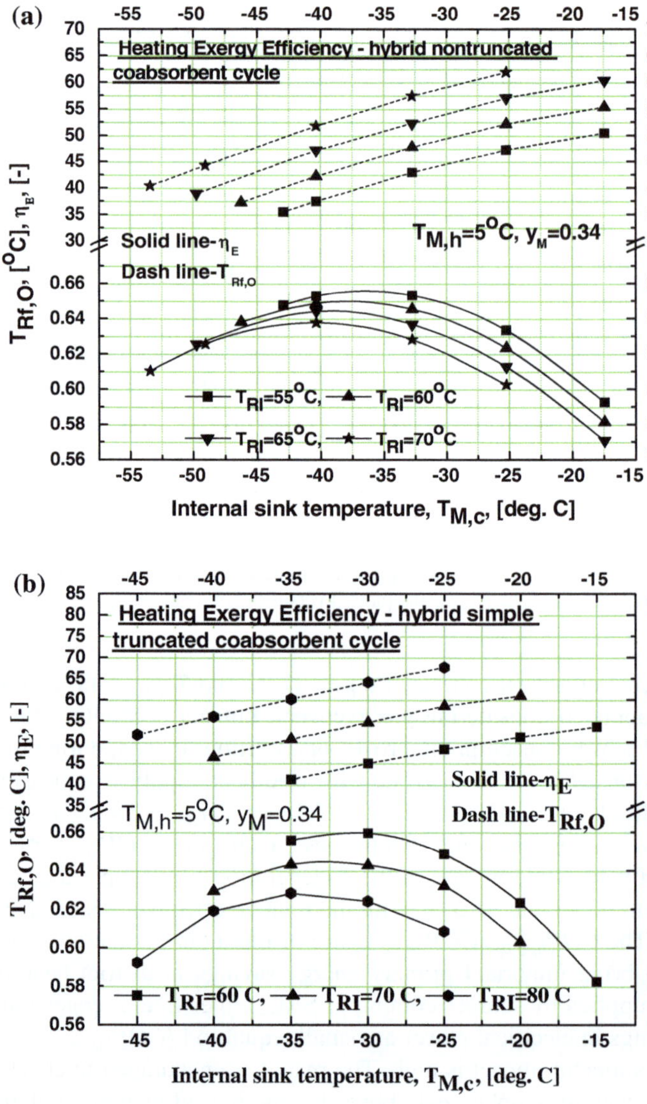

Fig. 7.12 Exergy efficiency of hybrid nontruncated (**a**) and simple-truncated (**b**) coabsorbent heating cycles, supplied by a free heat source, $T_{M,h} = 5\,°\text{C}$, and plotted versus internal sink source temperature, $T_{M,c}$

trigeneration are keeping the main (first) effect effectiveness with acceptable values, these energy solutions are still interesting in many heat pumping applications because of the equipment investment savings which these could bring (see Chap. 3 cogeneration and trigeneration). For the sake of completeness, next the trigeneration exergy efficiency will be given for cooling and heating cycles. However, in our approach the $Q_2(T_{2,cbs})$ heat supply of the SRC doesn't make too much sense

Table 7.3 Heating results of NH_3–H_2O coabsorbent truncated cycles operation supplied by free and low cogeneration heat sources versus SRC cogeneration

Functions	Coabsorbent heating cycles, Fig. 4.38a, b				Eq.
	Double truncated		Quadruple truncated		
$T_{M,c}$	5.0	5.0	5.0	15.0	–
$T_{M,h}$	37.0	48.0	37.0	48.0	–
T_{RI}	79.0	137.0	140.0	162.0	–
$T_{Rf,O} = T_{cbs} = T_{2,cg}$	63.84	92.78	77.42	91.3	–
$T_{2,cbs}$	39.0	50.0	39.0	50.0	–
$COP_{hp,cg,cbs}$	∞	52.45	∞	52.45	(7.9)
COP_w	444.9	340.9	417.9	278.9	(7.10)
COP_h	0.3248	0.3298	0.2798	0.2808	(7.12)
$COP_{hp,cbs}$	444.9	16.46	417.9	13.99	(7.13)
$COP_{hp,cg}$	22.0	11.50	15.0	11.90	(7.8)
$COP_{hp,cbs}/COP_{hp,cg}$	20.22	1.43	26.96	1.18	–
$y_{R1,O}$	0.45		0.60		–

(independent heat sources are more likely), so in the algorithm it is presented we shall skip it, but the SRC electrical power supply will be preserved. Suppose first, a cooling cycle produced additionally a heating effect exergy output equal to $E_R = q_R\theta_{w,R}$ and a mechanical (electrical) work, equal to $W_{output} = \eta_w q_G$, where the η_w work efficiency is defined, according to Eq. (3.138), by:

$$\eta_w = \frac{W_{output}}{q_G} \tag{7.100}$$

The trigeneration exergy efficiency is assessed by (see Eq. 7.31):

$$\eta_{E,c+h+w} = \frac{E_D + E_R + W_{output}}{E_G + E_{W,SRC}} \tag{7.101}$$

Taking into account Eq. (7.34) and completing it according to the new issues, the trigeneration exergy efficiency (cooling + heating + work) of a coabsorbent cooling cycle, is given by:

$$\eta_{E,c+h+w} = \frac{\theta_c^{-1} + \theta_{w,R}COP_c^{-1}COP_h + \eta_w COP_c^{-1}}{\theta_{w,G}COP_c^{-1} + \eta_{E,SRC}^{-1}COP_w^{-1}} \tag{7.102}$$

In case of the remote trigeneration, the denominator term $\eta_{E,SRC}^{-1}COP_w^{-1}$ is removed from Eq. (7.102), while W_{output} work represents the net work output, after all cycle work consumers (e.g. pumps, fans, compressors) have been subtracted, so Eq. (7.102) rewrites this time:

$$\eta_{E,c+h+w,remote} = \theta_{w,G}^{-1}\left(\theta_c^{-1}COP_c + \theta_{w,R}COP_h + \eta_w\right) \tag{7.103}$$

In practice, it is less likely that a heating cycle works in multi-generation mode. However, in this case, we add to the heating cycle a cooling effect exergy

$E_D = q_D\theta_c^{-1}$ and the work $W_{output} = \eta_w q_G$. Proceeding similarly, the trigeneration exergy efficiency (heating + cooling + work) of a coabsorbent heating cycle, is given by:

$$\eta_{E,h+c+w} = \frac{\theta_{w,R} + \theta_c^{-1} COP_c COP_h^{-1} + \eta_w COP_h^{-1}}{\theta_{w,G} COP_h^{-1} + \eta_{E,SRC}^{-1} COP_w^{-1}} \tag{7.104}$$

The cogeneration case can be derived easily from the trigeneration Eqs. (7.100–7.104). Also, all particular cases (i–iv) of Sects. 7.2 and 7.3 hold here true as well.

Appendix 1

Equation (7.34) is written as ideal (the boundary of the exergy balance is confined just to the cycle itself, without including the sources), therefore

$$\eta_{E,c} = \frac{\theta_c^{-1}}{\theta_{w,G} COP_c^{-1}} \tag{A7.1}$$

Considering in Eqs. (A7.1, 7.20, 7.28) and the ideal COPc, Eq. (5.6), it results

$$\eta_{E,c} = \left(\frac{T_{M,c}}{T_{D,m}} - 1\right)\left(1 - \frac{T_{M,c}}{T_{M,h}}\right)^{-1}\left(1 - \frac{T_{M,c}}{T_{M,h}}\right)\left(\frac{T_{M,c}}{T_{D,m}} - 1\right)^{-1} \equiv 1 \quad \text{q.e.d.} \tag{A7.2}$$

Appendix 2

A simple, yet not simplistic, explanation of the relationship between exergy efficiency and its economic impact is given next. A source is characterized by an energy flux (energy per unit of time and area), Φ_{source}, which is determining univocally its surface temperature, in such a way that the higher the energy flux, the higher the associated temperature (see the black body emissive power and peak temperature laws, Popa and Vintila 1977) and the higher the potential of the flux to be transformed in availability (mechanical work), E_{source},

$$E_{source} \sim \Phi_{source} \tag{A7.3}$$

Considering an end-user energy-process, according to the Lambert laws (Popa and Vintila 1977), the process receives a fraction of the source flux, only,

$$\Phi_{process} \leq \Phi_{source} \tag{A7.4}$$

characterized by the process incoming flux and temperature which are proportionally smaller. Bearing in mind the prior assertion, we conclude that:

$$E_{process} \sim \Phi_{process} \tag{A7.5}$$

and that the energy flux transfer has been done with an availability (exergy) destroy,

$$E_{process} \leq E_{source} \tag{A7.6}$$

Suppose now we have to design an energy-process, taking into account the 1st principle only, when choosing the supplying source. In this case, according to this principle requirement, the source energy and process energy needed to perform a certain task must be equal per unit of time, i.e.

$$\Phi_{source}A_{source} = \Phi_{process}\left(COP_{c,(h),(w)}\right)A_{process} \tag{A7.7}$$

where A holds for energy exchange area (m^2). Equation (A7.7) can be rearranged and taking into account Eqs. (A7.3) and (A7.5), it is obtained:

$$\frac{\Phi_{source}}{\Phi_{process}\left(COP_{c,(h),(w)}\right)} = \frac{A_{process}}{A_{source}} \sim \frac{E_{source}}{E_{process}} = \eta^{-1}_{E,source-process} \geq 1 \tag{A7.8}$$

Equation (A7.8) emphasizes the economical impact of the source-process exergy destroy, or of the exergy efficiency. Indeed the exergy destroy, expressed by $E_{process}/E_{source}$ ratio, which is nothing else but the source-process exergy efficiency, costs money, because the energy transfer area ratio $A_{process}/A_{source}$ increases correspondingly.

References

Alefeld, G. (1988a). Probleme mit der Exergie. *BWK* Bd. *40*, Nr. 3.

Alefeld, G. (1988b). Die Exergie und der II. Hauptsatz der Thermodynamik. *BWK* Bd. *40*, Nr.11.

Alefeld, G. (1988c). Zur Bewertung von Heizkraftwerken. *BWK* Bd. *40*, Nr. 6.

Amonix 7700 and National Renewable Energy Laboratory, 2011 data.

Angelino, G., & Invernizzi, C. (2008). *Binary conversion cycle for concentrating solar power technology, Solar Energy* (vol. 82/7, pp. 637–647).

Baehr, H. D. (1988). Probleme mit der Exergie? *BWK* Bd. *40*, Nr. 11.

Bejan A. (1996). *Advanced Technical Thermodynamics (in Romanian)*. Bucharest: Technical Publishing House.

Chartier, P., Gross, M., & Spiegler, K. S. (1975). *Applications de la thermody- namique du non-équilibre*. Paris: Hermann.

Duffie, J. A., & Beckman, W. A. (1974). *Solar energy thermal processes*. New York: Wiley.

Gasparovic, N. (1982). Die Wirkungsgrade der Kraft-Wärme-Kopplung – ein Vergleich. *BWK* Bd. *34*, Nr. 2.

Gerz, H.-B. (1982). Die Leistungszahl ε als thermodynamische Vergleichsgröße der Heizsysteme. *BWK* Bd. *34*, Nr. 2.

Izquierdo Millan, M., Hernandez, F., & Martin, E. (1996). Available solar exergy in an absorption cooling process. *Solar Energy, 56*(6), 505–511.

Kirilin, V. A., Sicev, V. V., & Seindlin, A. E. (1985). *Thermodynamics*. Bucharest: Scientific and Technical Publishing House. (540p, in Romanian).

Kreith, F., & Kreider, J. F. (1978). Principles of solar energy engineering. Washington, London: Hemisphere Publishing Corporation, New York: Mc. Graw-Hill Book Company.

McGovern, J. A., & Harte, S. (1995). An exergy method for compressor performance analysis. *International Journal of Refrigeration, 18*(6), 421–433.

Parrott, J. E. (1978). Theoretical upper limit to the conversion efficiency of solar energy. *Solar Energy, 21*, 227–229.

Popa, B., & Vintila, C. (1977). *Termotechnics and thermal machines*. Bucharest: The Didactic and Pedagogic Publishing House. (in Romanian).

Press, W. H. (1976). Theoretical maximum for energy from direct and diffuse sunlight. *Nature, 264*, 734–735.

Radcenco, V., et al. (1985). *Heat pump installations*. Bucharest: Technical Publishing House. (384p, in Romanian).

Soitec Company, Solar CPV (Concentrix) 2013 data.

SRoeCo Solar Company, 2012 data.

Staicovici, M. D. (1986). An autonomous solar ammonia-water refrigeration system. *Solar Energy, 36*, 115–124.

Staicovici, M. D. (1989). Preliminary feasibility study of solar cooling using multistage absorption cooling cycles. *Energetica (supliment), XIX, 2*, 38–51. (in Romanian).

Chapter 8
A Thermodynamic Approach of Mechanical Vapor Compression Refrigeration and Heating COP Increase

8.1 Introduction

The mechanical vapour compression (mvc) technology is laying down the basis of many important industrial, agricultural and household applications. Amongst these we could mention the heat pumping (cooling and/or heating production), gases compression, air liquefaction and separation and cryogenics (Stamatescu 1972; Stamatescu et al. 1982). Next we shall deal mainly with the heat pumping applications. Besides mvc, the absorption technology is used as well widely on this purpose (Niebergall 1959). The mvc and absorption channels have been always competing through the years in order to supply heat pumping practice with feasible solutions, as they have each advantages and disadvantages. On one side, the mvc needs for operation only two thermal sources (sink and heat) and a power source (electrical usually, or mechanical), satisfactorily available in almost all cases. Its main disadvantage is that it is suffering of a mismatch between the source it uses (electrical, of low entropy) and the task it accomplishes (heat pumping, of high entropy), and for this reason it is consuming an increased amount of primary energy (e.g. fossil fuel), with negative environmental consequences (see also Chap. 7 results). On the other hand, the absorption needs for operation three thermal sources (sink, low temperature heat and high temperature heat) and a very small pumping power consumption. In this case, the source-task match is better, in accordance with the 2nd principle recommendation, but it is however confronted with disadvantages as well, the high heat temperature source is not always available in applications and it has a higher investment cost and a longer starting period as compared to the mvc plants.

Lately, the coabsorbent technology has been introduced (Staicovici 2006, 2009a). This has the potential of improving absorption relative position with respect to its competitor, creating favorable pre-requisites for extending absorption channel efficient use in the centralized and distributed energy production (Chicco and Mancarella 2009; Staicovici 2009a). On the contrary, in last period the mvc focussed attention mainly on replacing the environmentally non-friendly working fluids with fluids or fluid-blends (Cox et al. 2009), having small impact with respect to global warming

M.-D. Staicovici, *Coabsorbent and Thermal Recovery Compression Heat Pumping Technologies*, Heat and Mass Transfer, DOI: 10.1007/978-3-642-54684-6_8,

effect and less for finding ways which could lead to a substantial improving of the compression effectiveness and of the COP of heat pumping units using it. For the reasons we showed here it is therefore the right time to think to a major improvement in the mvc COP as well. To this extent, in this chapter we shall present two methods for improving the compression process effectiveness in itself and the COP of refrigeration and heating cycles based on it.

For the sake of clarity, first a known introductory reminding part will follow. A gas compression process is performed with the help of a compressor device (Stamatescu et al. 1982; Kirilin et al. 1985). The sucked gas, having an initial low pressure and temperature, is compressed progressively in the compressor by a piston of various shapes, from inlet till outlet, with the help of a mechanical work input acting the piston, or a piston shaft. The discharge gas leaving the compressor has a high pressure (the useful effect we are looking for). Its magnitude with respect to the initial low pressure value is expressed numerically by the compression ratio. During compression, the gas has a natural tendency to warm itself, transforming an amount of the mechanical work input in heat. The generated heat is caused by an irreversible normal process of exergy destruction. As a consequence, the gas will increase its temperature. Depending on the way the generated heat is released during the compression process, the discharge gas temperature is equal to the initial one (total heat release, or no heat generated, therefore no exergy destruction takes place—ideal isothermic compression), it is reaching its maximum value (no heat release, maximum exergy destruction takes place—adiabatic compression), or it is taking values between the initial and the maximum ones (partial heat release, or partial exergy destruction it is happening—polytropic compression). Given the above interpretation of the relation between mechanical work (exergy) input and exergy destruction amount, it becomes clear that, for a same task, the isothermic compression needs a minimum mechanical work input, the adiabatic one needs the highest mechanical work input, and the polytropic one needs amounts of mechanical work input situated between the two extreme values mentioned.

Our goal is to use a specific mechanical compression work having values as low as possible, ideally equal to the isothermic work of each application. According to the present thinking, decreasing the specific compression work consumption is intrinsically related to the technical possibility to evacuate the generated heat during the compression process. Unfortunately, this heat release is critical in practice. Indeed, at least three causes hamper a good heat rejection during compression: (a) due to the high compressor speed, the gas staying time in the compressor is very short (less 50 ms); (b) the gas-cooling fluid heat exchange, via the compression chamber wall, is poor, and (c) the heat exchange area is very limited. For these reasons, the compression is only to a small extent polytropic, practically more often than not it can be considered adiabatic. The compression is therefore confronted in operation with the highest temperatures of the discharge gas and with the highest specific compression work. So far, limiting the discharge temperature of a compressor is a permanent concern of a designer, especially when the compressor is oil lubricated, in order to preserve the oil greasing quality as much as possible. Depending on the gas nature, this will confine correspondingly the compression ratio of an application as well.

A solution to decrease the discharge gas heating and still keeping the same compression ratio has been proposed, working with fluid blends (Cox et al. 2009). However, this will decrease the gas temperature, but the application COP is diminished as well. And last, but not the least, keeping the discharge temperature to nowadays accepted values forces the compression based refrigeration cycles to adopt a poor internal heat management (e.g. to renounce to the condensate subcooling by gas superheating, or to consume an extra-compression work in multi-stage compression in order to keep the sucked gas temperature at a low level). This compressor design policy is reflected in a low COP values achievement in refrigeration applications (e.g. industrial) and finally in a poor primary energy utilization.

8.2 Methods of Increasing the Refrigeration Effectiveness and Theirs Ideal Thermodynamic Limits

First, we shall deal with the refrigeration cycles. Later, our conclusions will be extended to heating cycles as well. Usually, the cycle cooling effectiveness is expressed by [see also Eq. (3.108)]:

$$COP_c = \frac{q_E}{|l_a|} \tag{8.1}$$

The COP_c in Eq. (8.1) can be increased applying two major methods. First, the discharge gas superheat of a polytropic or quasi-adiabatic mvc process is converted into mechanical (electrical) work, $\Delta|l_a|$, which will diminish correspondingly the compressor and implicitly the cycle work input (Staicovici 2011a). This method and the cycle applying it will be identified hereinafter as *thermal-to-work recovery compression* (TWRC). Equation (8.1) becomes in this case:

$$COP_{c,TWRC} = \frac{q_E}{|l_a| - \Delta|l_a|} > COP_c, \quad \Delta|l_a| > 0 \tag{8.2}$$

Second, the discharge gas superheat of the polytropic or quasi-adiabatic process is converted into a cooling useful effect, Δq_E, which will increase correspondingly the cycle cooling output (Staicovici 2011b, d). This second method and the cycle applying it will be identified hereinafter as *thermal-to-thermal recovery compression* (TTRC). Equation (8.1) becomes in this case:

$$COP_{c,TTRC} = \frac{q_E + \Delta q_E}{|l_a|} > COP_c, \quad \Delta q_E > 0 \tag{8.3}$$

8.2.1 TWRC Method

According to TWRC method, capitalizing the compressor discharge gas superheat and transforming it into mechanical or electrical power is done with the help of a recovering cycle. The recovering cycle can be supplied with heat directly by the

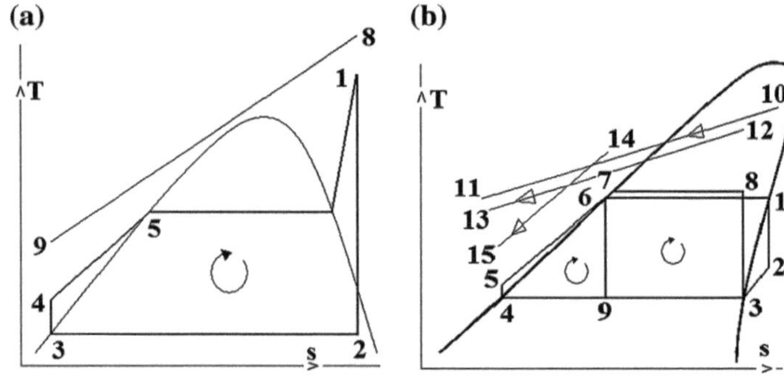

Fig. 8.1 Different power cycles. **a** Full Rankine cycle; **b** dry and wet ORC and TFC

superheated discharge gas, or by using intermediate heat transfer fluids, such as oils, with high normal boiling temperatures. The latter case leads obviously to a smaller global mvc cooling or heating effectiveness, but has the advantage of modifying little the classic mvc heat pumping cycle configuration and enabling to supply with heat the generator of either TWRC or TTRC recovering cycles at convenience. In this work, we have considered only the direct heat transfer with the recovery cycle. Going further with the recovery cycle analysis, this could be of various types, e.g. Rankine, Brayton, Stirling, Kalina, inverted Osenbrück, thermoelectric, Trilateral Flash (Kirilin et al. 1985; Smith 1993; Smith and da Silva 1994; Bryson 2007). The choice of the most adapted one to the particular task at hand must make the object of a careful consideration, including in the analysis all input data (application type, working fluid, safety requirements etc.). Except the thermoelectric conversion, once the right recovery cycle has been chosen, a second mandatory step is the selection of the most appropriate working fluid that covers it. The thermodynamic and the constructive criteria must be applied on this purpose. Bearing this in mind, probably ones of the best results are obtained if the recovering cycle were provided with the same working fluid as that utilized in the compression cycle, because it has convenient thermodynamic properties, verified by the refrigeration cycle, and it is simplifying the scheme. In this work, the mvc refrigeration and heating applications use the ammonia. According to the author analysis, the closed Rankine cycle seems to be the most recommended in our case, and has been considered in all applications throughout the present work. The working fluid for the Rankine cycle is ammonia too. Our choice is strengthened, presenting next a few technical proofs furnished by the comparison of Rankine cycle with the Trilateral Flash Cycle (TFC). Indeed, lately, considerable attention was paid to the TFC, thought a challenger of the Rankine cycle. The TFC is used in waste heat recovery power applications with two-phase screw expanders (Smith 1993, 1997; Smith and da Silva 1994; Bryson 2007). The TFC comes of a Rankine cycle with superheated vapor, Fig. 8.1a, T–s diagram, which reduced gradually its area, therefore its efficiency, passing through intermediary configurations, shown in Fig. 8.1b [(1–2–3–9–4–5–6–1)-dry Organic Rankine Cycle (ORC) with turbine and (8–3–9–4–5–7–8)-wet ORC with screw expander], until reached the smallest area configuration (6–9–4–5–6)-TFC. As a general remark, Fig. 8.1, the presumed higher cycle efficiency, the poorer the heat source-cycle match (see for comparison the heat sources 10–11, 12–13 and 14–15 relative position with respect to the cycles they supply). Bearing the above in mind, the interest in the TFC comes of its best heat source-cycle match out of all cycles in Fig. 8.1, as it enables the heat interacting fluids to keep a quasi-constant heat capacity $\dot{m}c_p$ and a minimum heat exchange temperature pinch, which means the least exergy loss. The most used TFC working fluids are the hydrocarbons, belonging to the isopentane group. A Rankine cycle-TFC comparison from the efficiency point of view follows. The ideal efficiencies of Rankine and TFC cycles write: $\eta_{c,C,RRC} = \left(1 - T_C/T_{CpO}\right)$ and $\eta_{c,C,TFC} = (1 - T_C/T_{CpO})/(1 + T_C/T_{CpO})$ (Bryson 2007, see also Appendix 1), respectively, and $\eta_{c,C,RRC} > \eta_{c,C,TFC}$. However, the TFC works close to reversible conditions, given the good matches of the heating transformation (see line 5–6, Fig. 8.1b) with the heat source, and of the condensation transformation with the sink source, respectively, therefore its actual efficiency may be close to the ideal one. On the contrary, the Rankine cycle is well matched to its condensation transformation with the sink source, only, because the heating transformation, with its sensible-latent-sensible heat supply needed characteristic is not well matched with the discharge gas sensible superheat delivery characteristic, therefore its actual efficiency may be quite far from the ideal one. Concluding up to this point, the higher efficiency advantage of the Rankine cycle over that of TFC has not been proved. However, despite the TFC favorable comments done so far, next, a few comments follow, analyzing critically and explain why we preferred the Rankine cycle to be the recovery cycle in our case, instead of TFC: (i) A high liquid phase temperature glide approaching the isopentane critical point (or similar) modifies the heat capacity constancy, reducing the TFC-heat source match; (ii) In our case, given that the ammonia critical point ($p_c = 113.3$ bar, $T_c = 132.3$ °C) has a favorable intermediary position between the sink and discharge gas temperatures, the heat source-NH_3 interaction can be well matched, conveniently establishing the two phase position, or even working with a transcritical NH_3 cycle, instead of the Rankine one. In this case, the high pressure operation resembles to that of the high pressure transcritical CO_2 heat pump cycle, well handled; (iii) The hydrocarbons used with the TFC are extremely flammable; (iv) Utilizing different working fluids in the refrigeration and recovery cycles, makes a TWRC to be manufactured difficultly in an unitary hermetic construction.

The calculations have been performed with a second choice, the water, as well. The work is including the ammonia results only, but comments about the water use are also given. Equally, the Rankine cycle can be of an open type, with water. It is simpler constructively, but less efficient.

Schematically, the method is presented by the example given in Fig. 8.2. Here, the dotted line indicates the Rankine Recovery Cycle (RRC), as everywhere throughout the figures to follow. The compressor (1) produces the discharge gas (4). It has the highest temperature and pressure and constitutes the heat source of the RRC generator (3), which recovers its specific heat within a large temperature glide (interval). The discharge gas (4) is cooled till a temperature close to the sink

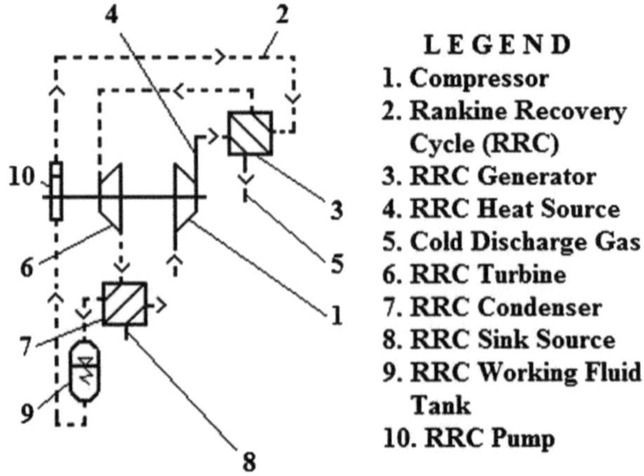

Fig. 8.2 One simple schematic of TWRC for improving the effectiveness of a mechanical vapor compression process. The *dotted line* indicates the Rankine recovery cycle (*RRC*)

value, becoming the cold discharge gas (5). At this stage, the cold discharge gas (5) is ready for a new compression process, serially connected to the first (1) and provided as well with a RRC, or it is obliged to cover further a refrigeration cycle, etc. The RRC working fluid (2) changes phase through the boiling in the RRC generator (3), it is generating vapor at a RRC high pressure, it is expanded in the RRC turbine (6), mounted preferrably on the same shaft as the compressor (1), it is producing a mechanical work amount which is being subtracted out of the compression work, it is further condensed at the RRC low pressure in the RRC condenser (7), provided with a RRC sink source (8), it is stored in a RRC working fluid tank (9), and it is taken away by the RRC pump (10), mounted preferrably on the same shaft as the compressor (1), and it is pumped from the low pressure to the RRC high pressure in the RRC generator (3), in order to close the RRC cycle. For a complete knowledge of the method, it is important to know its thermodynamic limit. This is given by the following:

Theorem 1 *The mechanical work of a polytropic or adiabatic theoretical compression process, provided with an ideal thermodynamic recovery and conversion means of its discharge gas superheat into work (TWRC), equals the isothermal work of the considered compression.*

Assumptions Next, we shall suppose the gas is suffering a theoretical mechanical compression process and is a perfect gas. Practically, this is not confining the generality of our results. First we deal with the adiabatic compression process. We operate with the following functions:

1. The discharge gas heat $q(t)$ is positively defined, $q(t) \geq 0$, as it supplies the recovery system; it is a function of real variable, the temperature t, $t \in I \subset R$, $I = [T_1, T_{2a}]$, $T_1 < T_{2a}$, $q(t) : I$, and has real values $q(t) \in \left[q(T_1), q(T_{2a})\right] \subset R$; it is continuous and strictly decreasing on I, $0 < q(T_1) > q(T_{2a}) = 0$, and the function $dq(t)$ enjoys a limited variation in the Stieltjes integral meaning within the I interval (Nicolescu 1977–1980);

2. The work $l(t)$, recovered through the ideal conversion of the gas heat $q(t)$, is positively defined, $l(t) \geq 0$ as it is the yield of the recovery system; it is defined and continuous on $t \in I$, $l(t) : I$, and has real values $l(t) \in [l_i, l_a] \subset R$; the work $l(t)$ is a strictly decreasing function on I, $0 < l(T_1) > l(T_{2a}) = 0$;
3. The Carnot efficiency is a real positive, continuous and strictly increasing function on $I, \eta_C(t) : I, \eta_C(t) \in [0, 1], 0 = \eta(T_1) < \eta(T_{2a}) > 0$.

Proof

The maximum work amount $dl(t)$ is obtainable converting with Carnot efficiency $\eta_C(t)$ the $dq(t)$ gas heat amount, delivered in the $t \in I$ point

$$dl(t) = \eta_C(t) dq(t) \tag{8.4}$$

where (Kirilin et al. 1985):

$$\eta_C(t) = 1 - \frac{T_1}{t} \tag{8.5}$$

We integrate Eq. (8.4) on the I interval:

$$\int_{T_{2a}}^{T_1} dl(t) = \int_{T_{2a}}^{T_1} \eta_C(t) dq(t) \tag{8.6}$$

The integration of the left member in Eq. (8.6) results in:

$$\int_{T_{2a}}^{T_1} dl(t) = l(T_1) - l(T_{2a}) \tag{8.7}$$

Equation (8.6) right member is a Stieltjes integral, fulfiling the existence conditions (Nicolescu 1977–1980). Given the assumptions, we apply the first mean formula of a Stieltjes integral (Nicolescu 1977–1980), and the right member of Eq. (8.6) becomes:

$$\int_{T_{2a}}^{T_1} \eta_C(t) dq(t) = \eta_C(\xi) \int_{T_{2a}}^{T_1} dq(t) = \eta_C(\xi) \left[q(T_1) - q(T_{2a}) \right] \tag{8.8}$$

In Eq. (8.8), $\eta_C(t)$ is calculated in a point $\xi \in I$. The $\eta_C(\xi)$ value is found applying the mean formula of a Riemann integral to function $\eta_C(t) : I$ (Nicolescu 1977–1980). Taking into account this and Eq. (8.5), it results:

$$\eta_C(\xi) = \frac{1}{T_1 - T_{2a}} \int_{T_{2a}}^{T_1} \eta_C(t) dt = 1 - \frac{ln \frac{T_{2a}}{T_1}}{\frac{T_{2a}}{T_1} - 1} \tag{8.9}$$

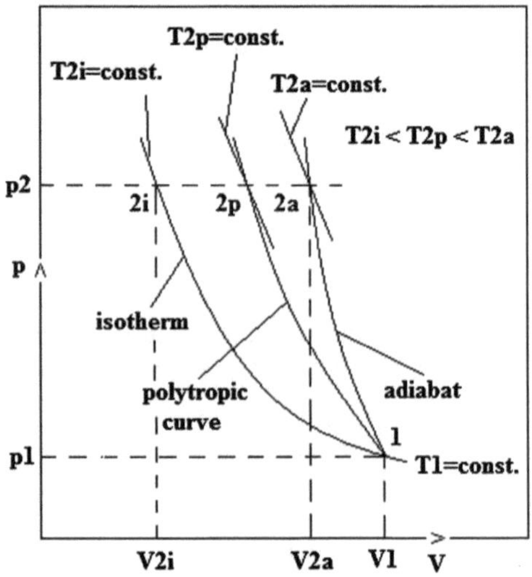

Fig. 8.3 Single-stage compression process plotted in the $p - V$ diagram, showing the isothermal, polytropic and *adiabatic lines* which characterize the gas possible evolutions during pressure increase from p_1 to p_2

For $\eta_C(\xi)$ of Eq. (8.9) an equivalent expression could be found, if the way shown in Appendix 1 of Chap. 1 had been followed (see Appendix 1 of this chapter). Continuing, we bear in mind that $q(T_{2a}) = 0$ and $l(T_{2a}) = 0$. Also, we note $q(T_1) = q_a$ and $l(T_1) = l_{a,r}$, and taking into account Eqs. (8.7) and (8.9), Eq. (8.6) becomes:

$$l_{a,r} = \left(1 - \frac{ln\frac{T_{2a}}{T_1}}{\frac{T_{2a}}{T_1} - 1}\right) q_a \tag{8.10}$$

Equation (8.10) is expressing the ideal recovery of the discharge gas heat q_a delivered in an adiabatic compression on the gliding temperature interval $[T_{2a}, T_1]$ and converted into the work $l_{a,r}$. In order to further evaluate Eq. (8.10), we have in view the single-stage compression process, plotted in the p − V diagram of Fig. 8.3. The compression works (isothermal, adiabatic, polytropic) and the discharge gas superheat are negative quantities, as input and output quantities, respectively, according to thermodynamics convention. Here we need their positive values. The isothermal work can be calculated by the following equation:

$$|l_i| = \left[\int_{V1}^{V_{2i}} pdV\right]_T \tag{8.11}$$

The gas evolution along the isothermal 1–2i, Fig. 8.3, is characterized by equation:

$$pV = p_1 V_1 = p_{2i} V_{2i} = const. \tag{8.12}$$

Using Eq. (8.12), and noting that $p_{2i} = p_{2a}$, Eq. (8.11) becomes (Kirilin et al. 1985):

$$|l_i| = (const.) \int_{V_1}^{V_{2i}} \frac{dV}{V} = p_1 V_1 \; ln \; \frac{p_{2i}}{p_1} = p_1 V_1 ln \frac{p_{2a}}{p_1} \tag{8.13}$$

The adiabatic work along the line 1–2a, Fig. 8.3, is given by (Kirilin et al. 1985):

$$|l_a| = \frac{k}{k-1} p_1 V_1 \left[\left(\frac{p_{2a}}{p_1} \right)^{\frac{k-1}{k}} - 1 \right] \tag{8.14}$$

The Poisson adiabatic equation, line 1–2a, Fig. 8.3, is given by (Kirilin et al. 1985):

$$pV^k = p_1 V_1^k = p_{2a} V_{2a}^k \tag{8.15}$$

The Clapeyron equation writes in points 1 and 2a for a mass of 1 kg, as follows, Eq. (1.10) (Kirilin et al. 1985):

$$p_1 V_1 = R' T_1 \tag{8.16}$$

and

$$p_{2a} V_{2a} = R' T_{2a} \tag{8.17}$$

respectively. Utilizing Eqs. (8.16) and (8.17) on one side, and Eq. (8.15), on the other side, the ratio (V_{2a}/V_1) can be eliminated, resulting in the known equation (Kirilin et al. 1985):

$$\frac{T_{2a}}{T_1} = \left(\frac{p_{2a}}{p_1} \right)^{\frac{k-1}{k}} \tag{8.18}$$

We apply Eq. (8.18) to Eq.(8.9). In the result, Eqs. (8.13) and (8.14) can be identified, and Eq. (8.9) rewrites as:

$$\eta_C(\xi) = \left(1 - \frac{ln \frac{T_{2a}}{T_1}}{\frac{T_{2a}}{T_1} - 1} \right) = 1 - \frac{|l_i|}{|l_a|} \tag{8.19}$$

The discharged gas superheat q_a can be calculated for a mass of 1 kg by:

$$|q_a| = c_p (T_{2a} - T_{2i}) \tag{8.20}$$

We note in Eq. (8.20) that $T_{2i} = T_1$. The specific heat at constant pressure is calculated by (Kirilin et al. 1985):

$$c_p = R' \frac{k}{k-1} \tag{8.21}$$

The Clapeyron equation is written in point 2i for a mass of 1 kg, Fig. 8.3, as:

$$p_{2i}V_{2i} = R'T_{2i} \tag{8.22}$$

We note that $p_{2i} = p_{2a}$, Fig. 8.3, and using Eqs. (8.12), (8.17), (8.21) and (8.22) in Eq. (8.20), we obtain:

$$|q_a| = \frac{k}{k-1}p_1V_1\left(\frac{V_{2a}}{V_{2i}} - 1\right) \tag{8.23}$$

From Eqs. (8.12) and (8.15), the volume ratio in Eq. (8.23) results as:

$$\frac{V_{2a}}{V_{2i}} = \left(\frac{p_{2a}}{p_1}\right)^{\frac{k-1}{k}} \tag{8.24}$$

We introduce Eq. (8.24) in Eq. (8.23), and then the result is compared to Eq. (8.14), finding that (Kirilin et al. 1985):

$$|q_a| = \frac{k}{k-1}p_1V_1\left(\left(\frac{p_{2a}}{p_1}\right)^{\frac{k-1}{k}} - 1\right) = |l_a| \tag{8.25}$$

Similar equations can be obtained in case of the polytropic compression as well. We remember that $|q_a|$ and q_a in Eqs. (8.25) and (8.10), respectively, are both positive quantities, for the reason already explained, and taking into consideration Eqs. (8.19) and (8.25), Eq. (8.10) becomes:

$$l_{a,r} = \left(1 - \frac{|l_i|}{|l_a|}\right)|l_a| = |l_a| - |l_i| \tag{8.26}$$

The compression work resulted from applying TWRC is simply the difference between the adiabatic work, $|l_a|$, and the recovered work given by Eq. (8.26), or:

$$\left|l_{a,TWRC}\right| = |l_a| - l_{a,r} = |l_i| \quad \text{q.e.d.} \tag{8.27}$$

The polytropic compression case is demonstrated similarly, replacing the index "*a*" by "*p*", and "*k*" by "*n*" in Eqs. (8.4)–(8.27), and hence the whole theorem is proved.

The theorem has both theoretical and practical importance enabling to make a rapid prediction about what it might be expected from TWRC application on each compression stage, useful in a preliminary design phase. Indeed, suppose we knew from the experience the practical efficiency relative to Carnot of converting the discharge gas heat into work, expressed by

$$[\eta(\xi)]_{practice} = \varepsilon\eta_C(\xi) \tag{8.28}$$

where $0 < \varepsilon < 1$. Taking into consideration Eqs. (8.26) and (8.28), Eq. (8.10) becomes:

$$l_{a,r} = \varepsilon\left(1 - \frac{ln\frac{T_{2a}}{T_1}}{\frac{T_{2a}}{T_1} - 1}\right)|l_a| \tag{8.29}$$

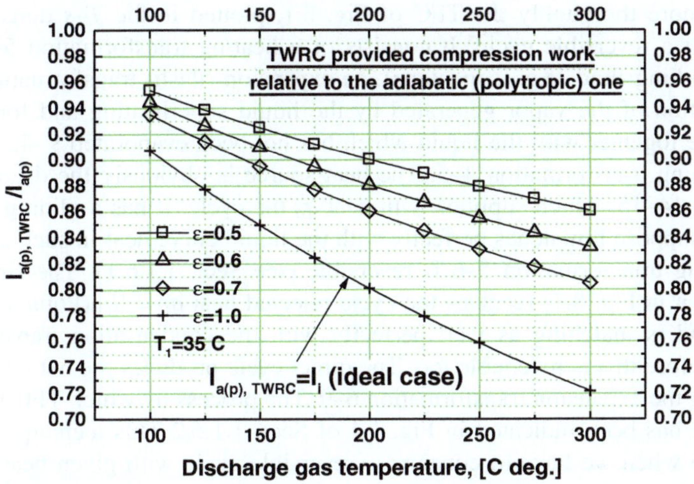

Fig. 8.4 Plot of Eq. (8.30): the compression work (adiabatic and polytropic) provided with TWRC, relative to same work but without recovery, against discharge temperature, having the relative to Carnot conversion efficiency ε as parameter

Further on, introducing Eq. (8.29) in Eq. (8.27), the following useful equation results:

$$\frac{|l_{a,TWRC}|}{|l_a|} = 1 - \varepsilon\left(1 - \frac{ln\frac{T_{2a}}{T_1}}{\frac{T_{2a}}{T_1} - 1}\right) \tag{8.30}$$

Next, two comments follow: (i) Eq. (8.30) allows one to predict the compression work of each TWRC provided stage, relative to adiabatic (polytropic) work, when the ε, T_{2a}, T_{2p} and T_1 parameters are known. The TWRC benefits of ε values which are by 20–30 % higher than those recorded by the Rankine steam electrical power stations (see Sect. 7.1). Equation (8.30) was plotted in Fig. 8.4 for several normal working parameters. The lines of constant ε have negative derivative with respect to the discharge gas temperature variable. This shows that, thermodynamically, the adiabatic operation at higher discharge gas temperatures is favouring the compression effectiveness improvement when applying TWRC, as compared to the polytropic operation. Moreover, the natural refrigerants having intrinsic properties to develop higher discharge gas temperatures for the same compression ratio [e.g. high "k" CH_4, CO_2, NH_3, air (Stamatescu 1972)] should be promoted so much more in applications, according to Eq. (8.30). Of course, translating this conclusion in practice should be carefully considered priorily, because the actual experience based on discharge gas temperature decrease (e.g. by refrigerant blends), opposes to our point of view, emphasized here. Our put in work order proposal requires the consideration of oil-free compressors or of those less sensitive to higher temperature operation, like screw or scroll type and the use of temperature-resistent materials (e.g. the high temperature greasing oils, ceramic); (ii) Let us analyze more thoroughly the TFC of Fig. 8.1, plotted in the *T–s* diagram, considering it a reversible cycle. It consists of a heating transformation 5–6 of the liquid working fluid (e.g. isopentane), an expansion (flash) transformation 6–9, a condensation of the vapor generated by the liquid superheating and forming the two-phase together with the liquid which has not been evaporated 9–4, and a liquid pumping transformation 4–5. The heat source is obviously the discharge gas superheat 14–15, which supplies with heat q_a the cycle. It has a gliding temperature and suppose it matches perfectly with the liquid curve heat in such a way that the heating transformation 5–6 is reversible (see Sect. 1.1.6.1). The sink source (not shown in Fig. 8.1) receives the cycle rejected heat $q_i = |l_i|$ (Popa and Vintila 1977), and is matching as well perfectly with the condensation transformation 9–4, therefore this is reversible too. The direct cycle produces the net output work $l_{a,r}$ during the expansion transformation 6–9. The process by which TFC was made reversible has been indicated in Fig. 1.4 of Sect. 1.1.6.2 This technique "works" each time when we have to conceive a reversible cycle with given heat and sink sources, like in our case. The method helps perfectly to find the ideal efficiency of a reversible cycle and particularly of the TFC, in a few steps. Indeed, first the elementary work produced by each TFC infinite small Carnot cycle (see also Fig. 1.4) is found with the help of Eq. (8.4). Then, the TFC work output $l_{a,r}$ is found totalizing the elementary work output, that is performing the integrals (8.8). The cycle efficiency, Eq. (8.9), results finally from its definition, that is applying Eq. (1.62). In our case, the efficiency can be further worked in order to obtain the known Eq. (A8.6) of the ideal TFC efficiency (see Appendix 1):

$$\eta_C(\xi) = 1 - \frac{ln\frac{T_{2a}}{T_1}}{\frac{T_{2a}}{T_1} - 1} \equiv \eta_{C,TFC} = \frac{T_{2a} - T_1}{T_{2a} + T_1} \qquad \text{(A8.6)}$$

which is identical to that given by Eq. (8.9). The proof that TFC is a reversible cycle in our ideal approach is included also in the Appendix 1.

8.2.2 TTRC Method

As TTRC recovering cycle we considered various types of absorption cooling cycles, which are capable to produce cooling when powered by the discharge gas heat source (e.g. H_2O–LiBr air conditioning, NH_3–H_2O for negative temperatures (Niebergall 1959). A TTRC method schematic is presented in Fig. 8.5, as an example. The dotted contour shows a classic single-stage absorption cooling cycle (ARC), described in detail e.g. in Niebergall (1959). The compressor (1) produces the high temperature and pressure discharge gas, which is the ARC generator (2) heat source. The discharge gas (3) is cooled till a temperature closer to the sink value and is ready for a new compression process, serially connected to the first (1) and provided as well with ARC, or covers further a refrigeration cycle, etc.

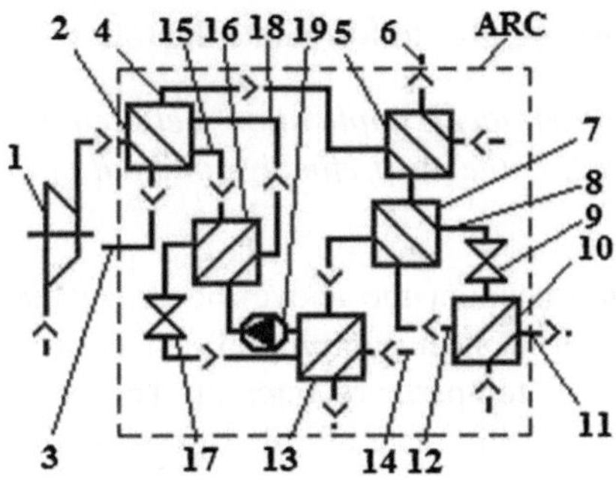

LEGEND
Dotted contour: Absorption Recovery Cycle
1. Compressor; 2. Generator; 3. Cold discharge gas; 4. Generated vapor; 5. Condenser; 6. Sink source; 7. Subcooler; 8. Subcooled refrigerant; 9. Expansion valve; 10. Evaporator; 11. Cooled source; 12. Cold refrigerant gas; 13. Absorber 14. Sink source; 15. Poor absorbent; 16. Heat exchanger for rich absorbent preheating; 17. Expansion valve; 18. Preheated rich absorbent; 19. Pump.

Fig. 8.5 One schematic of TTRC method. The *dotted contour* indicates a classic single-stage absorption (cooling) recovery cycle

The TTRC thermodynamic limit is shown by the following:

Theorem 2 *The cooling ideal (maximum) TTRC method effectiveness is higher than that of the TWRC method and has as upper limit the effectiveness of the mechanical vapor compression Carnot cooling cycle, only.*

Proof The ideal cooling effectiveness of TWRC and TTRC methods are given by Eqs. (8.31) and (8.32). They rewrite Eqs. (8.2) and (8.3), taking into account the theorem 1 conclusion and Eqs. (8.19) and (8.25):

$$COP_{c,TWRC,id} = \frac{q_E}{|l_i|} \tag{8.31}$$

and

$$COP_{c,TTRC,id} = \frac{q_E}{|l_a|} + \frac{\Delta q_{E,id}}{|l_a|} = \frac{q_E}{|l_i|}\frac{|l_i|}{|l_a|} + \frac{\theta_{c,C}\theta_{w,C}|q_a|}{|l_a|} = \frac{q_E}{|l_i|}(1-\eta_C(\xi)) + \theta_{c,C}\eta_C(\xi) \tag{8.32}$$

In Eq. (8.32), $\theta_{c,C} = (T_1/T_E - 1)^{-1}$ and $\theta_{w,C} = \eta_C(\xi)$ hold for the heat and work Carnot factors which product of expresses the Carnot effectiveness of ARC. Both theorem parts are proved because both in Eq. (8.33)

$$\theta_{c,C} > COP_{c,TTRC,id} = \frac{q_E}{|l_i|}(1-\eta_C(\xi)) + \theta_{c,C}\eta_C(\xi) > COP_{c,TWRC,id} = \frac{q_E}{|l_i|} \tag{8.33}$$

lead to the true statement (8.34), as $\theta_{c,C}$ is the Carnot cooling effectiveness of the mvc cycle, as well.

$$\frac{q_E}{|l_i|} < \theta_{c,C} = (T_1/T_E - 1)^{-1} \quad \text{q.e.d.} \tag{8.34}$$

Taking into account the result of Theorem 2, we are able to formulate the following:

Corollary 1 *The TTRC heat pumping method ideal application between two sources with temperatures T1 and T2, T1 < T2, is the most effective one out of all other existing heat pumping methods.*

Proof Indeed, the corollary is true taking into account the above theorem which states that the TTRC has as upper limit the Carnot heat pumping effectiveness of an ideal cycle operating within the T1 and T2 temperature interval. The corollary is useful in practice, as it helps a designer to eliminate all other possible time consuming solutions aiming at maximizing the COP of a certain heat pumping application.

8.3 Refrigeration Cycles Provided with TWRC

We analyze the single-stage (SSRC), two-stage (TSRC), three-stage (THSRC) and multi-stage (MSRC) refrigeration TWRC cycles (Staicovici 2009b, 2011a). In the most cases both compartments of these cycles (refrigeration + recovery) utilize a common working fluid (CWF). This option leads to a simpler construction, with a single common condenser cooled by the same sink source, and a common working fluid tank. In some special applications, the TWRC cycle can have two different condensers with two different sink source temperatures. It is the case of the heat pumps, or of the combined cooling + heating (cogeneration) for instance, which provide heat at a higher temperature level through the refrigeration condenser, but condense the working fluid in the second condenser, of the recovery compartment, cooled by a lower temperature sink source. Unlike known refrigeration cycles, when applying TWRC it is possible and recommended to practice the refrigerant condensate subcooling through suction gas superheating (CSTSGS). Also, in the TWRC–MSRC cycles the recovering turbines of each recovery cycle can be joined, forming a single equivalent turbine. For making the construction simpler and for avoiding excessive losses of useful work, all the recovering turbines (or the single equivalent one), the compressor and the recovery pump should be mounted on the same shaft. Many TWRC cycles can be elaborated, starting from these simple basic ideas. Hereinafter we shell present only the most important ones.

8.3.1 TWRC (SSRC, CWF)

This TWRC cycle type is plotted in the *T–s* diagram in Fig. 8.6. Schematically it is given in Fig. 8.7, which shows also Fig. 8.6 main equilibrium thermodynamic states. It is the simplest, constructively. The known compression refrigeration cycle, Fig. 8.6, is noted by a–b–c–d–e–f–a and the RRC by k–d–e–g–h–i–j–k. The refrigerant gas is compressed by an external power input (a–b) between p_E and

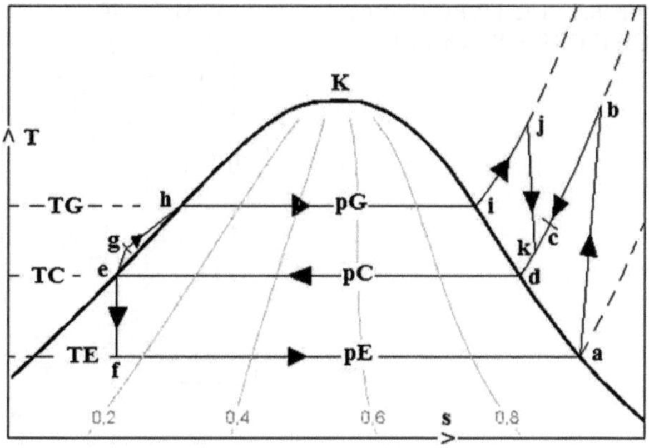

Fig. 8.6 The TWRC (SSRC, CWF) cycle represented in the T–s diagram

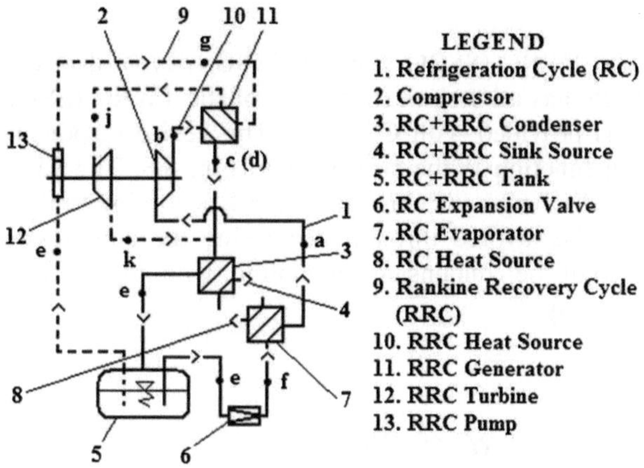

Fig. 8.7 The TWRC (SSRC, CWF) cycle schematic

p_C pressures ($p_C > p_E$), and reaches its highest discharge temperature in point b. Subsequently, it is cooled internally on the p_C isobar (b–c) in a recovering way, it is cooled optionally further externally (c–d), it is condensed externally (d–e) at T_C, it is expanded from p_C to p_E (e–f) and finally it is evaporated by external heat input (f–a) producing useful cooling at T_E and p_E, for closing the refrigeration cycle. In the recovery cycle, the working fluid is pumped by an external power input from p_C to p_G pressure (e–g) ($p_G > p_C$), it is suffering a (liquid (heating + boiling) + vapor superheating) process internally heated from T_g to T_j, $T_j < T_b$ at p_G (g–h–i–j), recovering the discharge gas sensible superheat (b–c), the resulting vapor is expanded from p_G to p_C, producing internal useful work (j–k) for decreasing the external compression power input, and finally it is externally condensed (k–d–e) at T_C and p_C, in order to close the recovery cycle. The TWRC (SSRC, CWF) model results are given in Sect. 8.4.

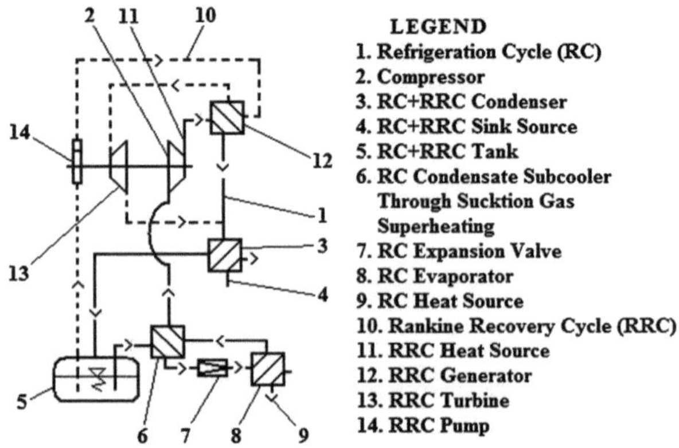

Fig. 8.8 The TWRC (SSRC, CWF, CSTSGS) schematic

8.3.2 *TWRC (SSRC, CWF, CSTSGS)*

The TWRC (SSRC, CWF, CSTSGS) cycle flow chart is given in Fig. 8.8. This cycle has all technical features of the TWRC (SSRC, CWF) cycle. Additionally, it is provided with CSTSGS. The device responsible for subcooling/superheating function is the subcooler 6, Fig. 8.8. The vapor superheating will increase the discharge gas temperature and, according to Eq. (8.30) and Fig. 8.4, the COP will be favored through decreasing the compression work. The cycle benefits also of the condensate subcooling, with positive effect on the refrigerant cooling capacity.

8.3.3 *TWRC (SSRC)*

The TWRC (SSRC) cycle flow chart is given in Fig. 8.9. The refrigeration cycle and the recovery one are completely separated here. This type of cycle can be used for heat pump or combined cooling + heating (cogeneration) applications. A TWRC cycle is unfavoured, to a small extent, in a heating application, as compared to a known conventional one. Indeed, its discharge gas sensible heat is converted in useful work, as we already have seen, and only its latent heat is delivered to the condenser heat sink for a useful heating effect. The heating COP of a TWRC cycle can compete with that of the conventional heat pump if its compression work were decreased enough by the TWRC process in order to compensate for the decrease of heat delivered to condenser. This is possible when the recovery sink source (see position 14, Fig. 8.9) decreases below a certain threshold value. The advantage of the TWRC (SSRC) cycle is more important in the cogeneration applications. The model results of these two working modes will be given further in Sect. 8.4.

Fig. 8.9 The TWRC (SSRC) schematic

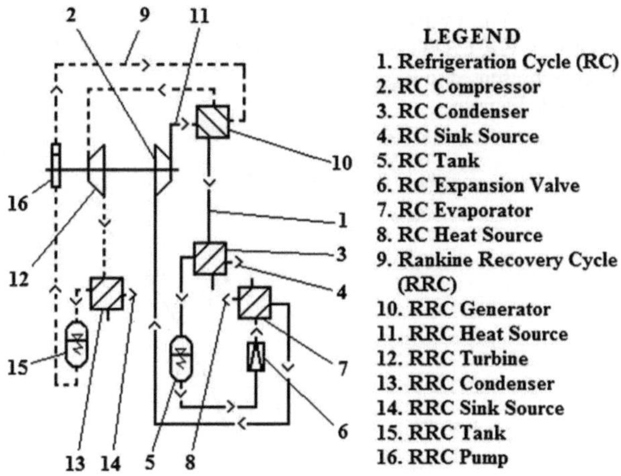

8.3.4 TWRC (TSRC, CWF, CSTSGS)

The advantage of the TWRC method is fully demonstrated by the applications with high temperature lift (more than 70 °C, for the sink at 30–35 °C). The existing mvc based equipment uses mainly the cascades of CO_2–NH_3 or the two-stage ammonia for producing industrial refrigeration (−40 to −50 °C) (Stamatescu 1972; Belozerov et al. 2007). Both cascade subcycles consume compression work, but only the first produces the useful cooling effect, the second one generates the cooling source for making possible the first subcycle operation, only. Thermodynamically, this means more exergy loss, therefore low effectiveness ($COP \sim 1.26$ at −50 °C with condensation at 32 °C (Belozerov et al. 2007). On the other side, a two stage compression needs stage intercooling below the sink source temperature. This is done at the expense of an additional compression work, in order to keep an acceptable discharge temperature in the last stage, which again penalizes the COP. Also, these two technical solutions avoid the benefic refrigerant condensate subcooling (CSTSGS). Taking into account the above considerations, we can remark that nowadays thinking is depriving the industrial cooling, and not only, of a better internal heat management and therefore of a much higher effectiveness. The TWRC method comes with a new thinking in order to eliminate the above mentioned drawbacks. Indeed, as it will be seen, the internal heat management is this time natural, capitalizing the intrinsic ways of the COP increase: (a) the cascade is avoided; (b) the two-stage compression is promoted in such a way that the stage intercooling is performed as well, but in a recovering mode, producing useful work through TWRC and decreasing the compression work input, and (c) the CSTSGS becomes here mandatory. The model results of the TWRC (TSRC, CWF, CSTSGS), promoting the (a)–(c) technical features, prove the potential of our proposal for COP improvement. They will be given further in Sect. 8.4.

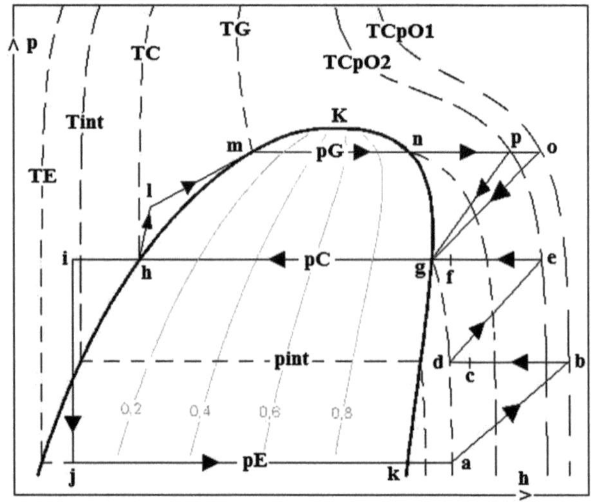

Fig. 8.10 The TWRC (TSRC, CWF, CSTSGS) cycle represented in the *p–hp–h* diagram

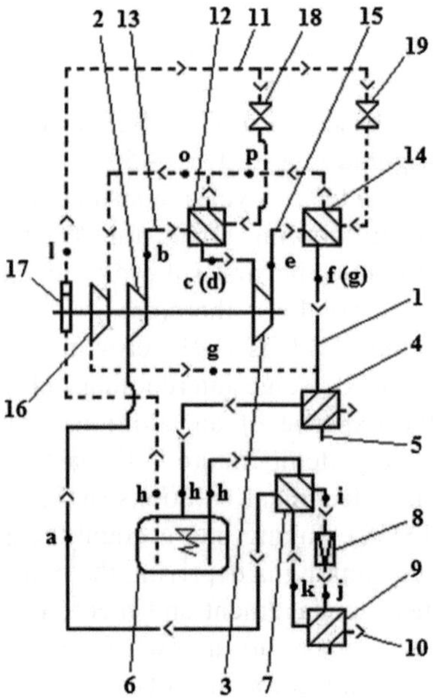

LEGEND

1. RC Refrigeration Cycle (RC)
2. RC First Stage Compressor
3. RC Second Stage Compressor
4. RC Condenser
5. RC Sink Source
6. RC Tank
7. RC Liquid Refrigerant Subcooler
8. RC Expansion Valve
9. RC Evaporator
10. RC Heat Source
11. Rankine Recovery Cycle (RRC)
12. RRC First Stage Generator
13. RRC First Stage Generator Heat Source
14. RRC Second Stage Generator
15. RRC Second Stage Generator Heat Source
16. RRC Turbine
17. RRC Pump
18., 19. Regulating Valves

Fig. 8.11 The TWRC (TSRC, CWF, CSTSGS) cycle schematic

This cycle is represented in the *p–h* diagram in Fig. 8.10. Schematically it is given in Fig. 8.11, which shows also Fig. 8.10 main equilibrium thermodynamic states. The two-stage compression Refrigeration Cycle (RC), Fig. 8.10, is noted by a–b–c–d–e–f–g–h–i–j–k–a. Each compression stage has its own Rankine Recovery Cycle (RRC). The discharge gas heat of the first compression stage is recovered by the RRC noted by h–l–m–n–o–g–h. Similarly, the discharge gas heat of the second compression stage is recovered by the RRC noted by h–l–m–n–p–g–h. The RC and both RRC cycles have a common condensing process. The refrigerant gas with superheated state ($\sim T_C, p_E$) is suffering the first compression between p_E and p_{int} pressures, ($p_{\text{int}} > p_E$), using external power input (a–b) and reaches its highest discharge temperature in point b $\left(T_{Cpo1}, p_{\text{int}}\right)$. Subsequently, it is cooled internally on the isobar p_{int} (b–c) in a recovering way, it is cooled optionally further externally (c–d), (T_C, p_{int}), is suffering the second compression between p_{int} and p_C ($p_C > p_{\text{int}}$), using external power input (d–e), and reaches its highest discharge temperature in point e $\left(T_{Cpo2}, p_C\right)$. Subsequently, it is cooled internally on the isobar p_C (e–f) in a recovering way, it is condensed externally (f–g–h) at (T_C, p_C), the condensate is subcooled in a recovering way at p_C, it is expanded from p_C to p_E (i–j), and finally it is evaporated by external heat input (j–k), producing useful cooling at (T_E, p_E) and the resulting gas is superheated in a recovering way subcooling the condensate (k–a), in order to close the RC. In the first compression stage recovery cycle h–l–m–n–o–g–h, the working fluid is pumped by an external power input from p_C to p_G pressure (h–l), ($p_G > p_C$), it is suffering a (liquid (heating + boiling) + vapor superheating) process internally heated from T_l to T_o, $T_o < T_{Cpo1}$ at p_G (l–m–n–o), recovering the first compression stage discharge gas sensible heat (b–c), the resulting vapor is expanded from p_G to p_C, producing internal useful work (o–g) for decreasing the external compression power input, and finally it is externally condensed (g–h) at T_C and p_C, in order to close this first recovery cycle. In the second compression stage recovery cycle h–l–m–n–p–g–h, the working fluid is pumped by an external power input from p_C to p_G pressure (h–l), ($p_G > p_C$), it is suffering a (liquid (heating + boiling) + vapor superheating) process internally heated from T_l to T_p, $T_p < T_{Cpo2}$ at p_G (l–m–n–p), recovering the second compression stage discharge gas sensible heat (e–f), the resulting vapor is expanded from p_G to p_C, producing internal useful work (p–g) for decreasing the external compression power input, and finally it is externally condensed (g–h) at T_C and p_C, in order to close the second recovery cycle. The TWRC (TSRC, CWF, CSTSGS) model results are given in Sect. 8.4.

8.3.5 TWRC (THSRC, CWF, CSTSGS) and TWRC (MSRC, CWF, CSTSGS)

The TWRC (THSRC, CWF, CSTSGS) and TWRC (MSRC, CWF, CSTSGS) cycles, have one, two, or more stages added to the TWRC (TSRC, CWF, CSTSGS) configuration. A general schematic is given in Fig. 8.12. The Legend of Fig. 8.11 can be adapted to hold true in Fig. 8.12 as well. A similar representation of the cycles at hand can be performed in the *p–h* diagram, as well. The TWRC (THSRC, CWF, CSTSGS) cycles have been modeled for temperature lifts of 77–107 °C (condensation at 32 °C) with ammonia. The model results are given further in Sect. 8.4.

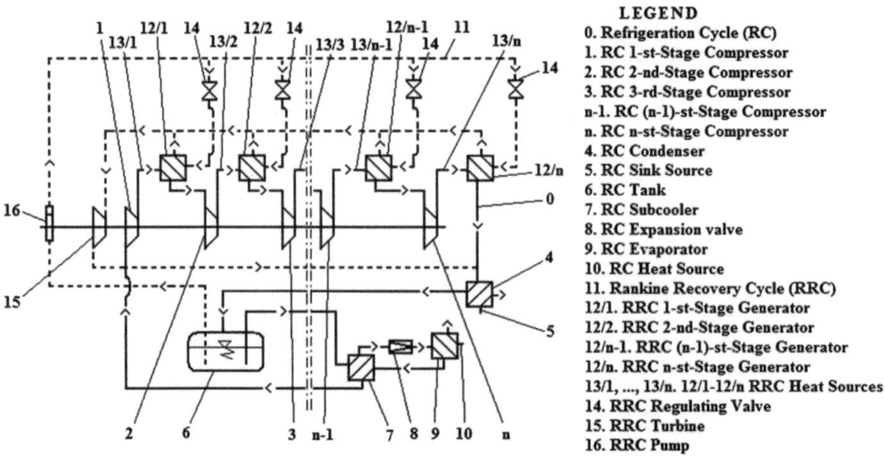

Fig. 8.12 TWRC (THSRC, CWF, CSTSGS) and TWRC (MSRC, CWF, CSTSGS) cycles schematic

8.3.6 Air Liquefaction and Separation Cycles Provided with TWRC

The specific compression work input is quite high in case of the air liquefaction and separation processes [(1.14–3.14) kWh/kg of liquid air (Stamatescu 1972)]. The TWRC could be applied to this type of applications, as well. In this case, unlike the heat pumping, the air cycle is open and the recovery one is closed. Besides this, the air critical point is low ($T_{K,air} = 132.6$ K; $p_{K,air} = 38.4$ bar), so, normally, the air cannot be used as an working fluid in a Rankine-TWRC cycle. For these reasons the two cycles must work with different nature fluids, therefore be completely separated one from the other. These remarks hold true in general, when air, or air-like fluids are compressed. In our case, the air is compressed to high pressures (100–200 bar), so a multi-stage compressor must be utilized. In Fig. 8.13, an air liquefaction schematic is shown. Here, the compression and the recovery sections are constructively of MSRC type, similar to those outlined in Fig. 8.12. With the help of Eq. (8.30) and Fig. 8.4, we have estimated the effect which the TWRC application might have in our case. Considering as input data the following values, $\varepsilon = 0.6$, $T_{2a} = 523.2$ K and $T_1 = 305.2$ K, it results that the potential in compression work savings is about 18 % per each stage, therefore per the whole air liquefaction unit.

The schematic given in Fig. 8.13 can be used for the achievement of very deep cooling (200–120 K) and criogenic temperature cooling ($T_E < 120$ K), as well, by simple changes. The air is replaced in this case by an appropriate working fluid, according to the task, the positions 5, 6 and 7 are replaced by an evaporator, and the circuit is closed, joining the connections 8 and 9.

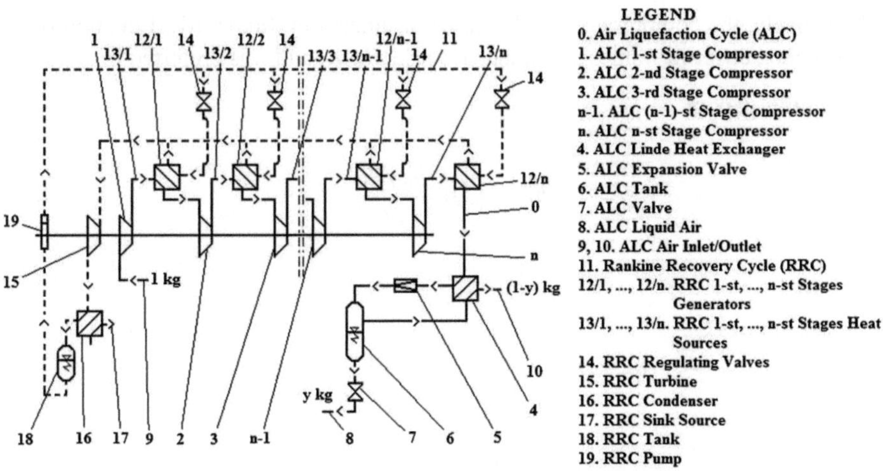

Fig. 8.13 A schematic of an air liquefaction unit with a single expansion, provided with TWRC

8.4 Results of Refrigeration Cycles Provided with TWRC

The heat pumping cycles presented in Sect. 8.3 have been modeled. The model main items are briefly given below. The extensive energy inputs or outputs, like heat and work, are calculated for 1 kg of working fluid. Figure 8.6 is referred to further for all notations.

The mass flow rate coefficient

The mass flow rate coefficient has been calculated by (Popa and Vintila 1977):

$$\eta_v = 1 - \varepsilon_0 \left[\left(\frac{p_C}{p_E} \right)^{\frac{1}{k}} - 1 \right] \; (\varepsilon_0 = 0.025) \tag{8.35}$$

The pumping, turbine and compression efficiencies

The values of the pumping and the turbine internal efficiencies, $\eta_p = 0.8$ and $\eta_t = 0.86$ have been taken from the literature (Kirilin et al. 1985). The stage compression efficiency was calculated by (Radcenco et al. 1985):

$$\eta_c = \frac{T_E}{T_C} \tag{8.36}$$

The adiabatic compression work

$$h_b = h_a + \frac{h_{b,s} - h_a}{\eta_c} \tag{8.37}$$

$$|l_a| = |h_b - h_a| \tag{8.38}$$

Alternatively, the compression work can be calculated by Eq. (8.14). Equation (8.38) gives higher, covering values as compared to Eq. (8.14) and was preferred in this work.

The pumping work

$$h_g = h_e + \frac{h_{g,s} - h_e}{\eta_p} \tag{8.39}$$

$$|l_p| = |h_g - h_e| \tag{8.40}$$

The turbine work

$$h_k = h_j - \frac{h_{k,s} - h_j}{\eta_t} \tag{8.41}$$

$$l_t = h_j - h_k \tag{8.42}$$

The evaporator heat

$$q_E = h_a - h_f \tag{8.43}$$

The condenser heat

$$q_C = h_d - h_e \tag{8.44}$$

The generator heat

$$q_G = h_j - h_g \tag{8.45}$$

The intermediary pressure in multi-stage compression

$$p_j = \sqrt[i]{p_E^{i-j} p_C^j} \tag{8.46}$$

where i and j hold for the stages number and each stage current number, respectively. In Appendix 1 it is shown that the intermediate pressures calculated in order to ensure a minimum work consumption, known from the compression classic theory, hold true in case of TWRC also.

The cooling COP

The cooling COP of the classic refrigeration cycle and of that provided with TWRC are expressed by:

$$COP_c = \frac{q_E}{|l_a|} \tag{8.47}$$

and

$$COP_{cr} = \frac{q_E}{|l_{a,TWRC}|}, \tag{8.48}$$

respectively.

The heating COP

Similarly, the heating COP of the classic refrigeration cycle and of that provided with TWRC are expressed by:

$$COP_h = \frac{q_C}{|l_a|} \tag{8.49}$$

and

$$COP_{hr} = \frac{q_C}{|l_{a,TWRC}|}, \tag{8.50}$$

respectively.

The cooling + heating (cogeneration) COP

The refrigeration + heating COP of the classic refrigeration cycle and of that provided with TWRC are expressed by:

$$COP_{cg} = \frac{q_C + q_E}{|l_a|} \tag{8.51}$$

and

$$COP_{cg,r} = \frac{q_C + q_E}{|l_{a,TWRC}|} \tag{8.52}$$

respectively.

The comparison with the isothermal compression

The comparison of a TWRC cycle with an isentropic compression refrigeration cycle can be made by the following ratio:

$$COP_{cr,rel,i} = \frac{COP_{cr}}{COP_{c,i}} \tag{8.53}$$

8.4.1 TWRC (SSRC, CWF), TWRC (SSRC, CWF, CSTSGS)

The model results, Table 8.1, cover the cooling range of $T_E = ((+5)-(-25))$ °C (1st column) with usual condensation temperatures ($T_C = 35-40$) °C (2nd column). The 3rd column gives the discharge gas temperatures $\left[T_{CpO} = (92.8-167.6)\ °C\right]$. The 5th and the 6th columns give TWRC refrigeration *COP* and that relative to the isothermal compression, COP_{cr} and $COP_{cr,rel,i}$, respectively. The last but one column gives the relative increase of the refrigeration *COP* when the TWRC is applied, as compared to the classic compression refrigeration cycle. The figures show here a relative improvement. The last column shows TWRC cycle generator pressure, p_G. A comparison between TWRC (SSRC, CWF) and TWRC (SSRC, CWF, CSTSGS) cycles is made in a single

Table. 8.1 The model results of TWRC (SSRC, CWF) cycle, Fig. 8.6, with ammonia in both RC and RRC cycles

T_E (K)	T_C (K)	T_{CpO} (K)	η_v (−)	COP_{cr} (−)	$COP_{cr,rel,i}$ (−)	COP_{cr}/COP_c (−)	p_G (bar)
278.2	313.2	366.0	0.9678	6.332	0.876	1.031	30
273.2	313.2	377.5	0.9595	5.267	0.839	1.037	34
268.2	308.2	375.9	0.9571	5.171	0.825	1.039	31
263.2	308.2	389.5	0.9466	4.374	0.791	1.049	36
258.2	308.2	404.7	0.9339	3.744	0.759	1.059	43
253.2	308.2	421.7	0.9184	3.234	0.727	1.070	52
248.2	308.2	440.8	0.8996	2.820	0.698	1.085	64
248.2[a]	308.2	546.2	0.8997	2.942	0.788	1.132	136

[a]Result for TWRC (SSRC, CWF, CSTSGS) cycle, Fig. 8.7

study case, given in the last two lines of Table 8.1. Both cycles have the same T_E and T_C values. As mentioned, the operation with CSTSGS benefits the COP_{cr}, but paying the tribute of an important increase of the T_{CpO} and p_G values. As a conclusion for this paragraph, the TWRC introduction has the potential to increase the single-stage compression refrigeration effectiveness. The feasibility seems to be reasonable in this case.

As we mentioned in Sect. 8.2.1, the cycle at hand has been run with water in its recovery cycle, as well. The model results show COP_{cr} which are smaller than those obtained for ammonia. Thermodynamically, in our case the discharge gas temperature is quite low for being an adequate heat source of such a working fluid in a Rankine cycle. Practically, given the low pressure of the steam until enough high temperatures, the resulting recovery turbine will result with high diameters, not acceptable. And last, but not the least, unlike ammonia, the water leakage in the recovery system makes the whole TWRC system to be manufactured difficultly in a hermetic construction. Concluding, the water use in the recovery cycle seems unfit, unpractical for heat pumping applications.

8.4.2 TWRC (SSRC)

The cogeneration of heat and cooling model results are given in Table 8.2. In this study case, it is supposed that TWRC cycle, Fig. 8.9, extracts heat from a low temperature heat source, $T_E = 285.2$ K (or lower), which could be for instance the underground water, ground, or air conditioning return. This heat is upgraded by the cycle in its condenser, which sink temperature of is increased to a higher temperature level, for satisfying the heating needs, $T_C = 323.2$ K (or higher). The sink source temperature of the recovery cycle, T_{Cr} (1st column), is considered for instance the ambient air, which during the summer-winter period can vary on a large domain (here considered $T_{Cr} = (323.2-223.2)$ K. The COP_{cr} increases constantly with respect to the classic refrigeration cycle (5th column), due to the recovery cycle contribution already discussed. It is interesting that, in case of TWRC (SSRC), this contribution can be so high (see 7th column), that the

Table 8.2 NH_3 TWRC (SSRC) cogeneration of heat and cooling model results, Fig. 8.9. Input data: $T_E = 285.2$ K; $T_C = 323.2$ K; $T_{CpO} = 428.0$ K; $COP_c = 5.206$; $COP_h = 6.206$

T_{Cr} (K)	COP_{cr} (−)	COP_{hr} (−)	$COP_{cg,r}$ (−)	COP_{cr}/COP_c (−)	COP_{hr}/COP_h (−)	$COP_{cr,rel,i}$ (−)	p_G (bar)
323.2	5.749	5.846	11.60	1.104	0.942	0.893	58.0
313.2	5.872	5.919	11.79	1.128	0.954	0.912	54.0
303.2	6.048	6.014	12.06	1.162	0.969	0.940	52.0
293.2	6.246	6.067	12.31	1.200	0.978	0.971	44.0
283.2	6.480	6.142	12.62	1.245	0.990	1.007	37.0
273.2	6.752	6.227	12.98	1.297	1.003	1.049	30.0
263.2	6.816	6.321	13.14	1.309	1.018	1.059	24.0
253.2	6.966	6.426	13.39	1.338	1.035	1.082	18.9
243.2	7.070	6.513	13.58	1.358	1.049	1.099	14.1
233.2	7.146	6.591	13.74	1.373	1.062	1.110	10.1
223.2	7.766	7.275	15.04	1.492	1.172	1.207	20.0
223.2	7.887	7.325	15.21	1.515	1.180	1.226	20.0

Table 8.3 NH_3 model results of TWRC (TSRC, CWF, CSTSGS) cycle, Fig. 8.11

T_E (K)	T_C (K)	T_{CpO1} (K)	T_{CpO2} (K)	η_{v1} (−)	η_{v2} (−)	COP_{cr} (−)	$COP_{cr,rel,i}$ (−)	p_G (bar)
253.2	313.2	411.1	404.9	0.970	0.971	3.235	0.865	25.0
248.2	313.2	426.6	416.8	0.967	0.965	2.948	0.866	36.0
243.2	308.2	424.9	415.6	0.965	0.963	2.918	0.868	48.0
238.2	308.2	443.4	429.1	0.959	0.957	2.636	0.855	55.0
233.2	308.2	465.0	443.8	0.954	0.950	2.407	0.845	60.0
228.2	305.2	475.2	450.5	0.949	0.944	2.302	0.839	66.0
223.2	305.2	503.7	467.5	0.942	0.935	2.147	0.836	78.0
223.2	305.2	514.9	467.5	0.943	0.935	2.198	0.841	79.0
218.2	305.2	538.3	486.2	0.935	0.924	2.045	0.840	94.0
213.2	303.2	565.3	499.6	0.928	0.914	1.996	0.844	101.0

total compression work consumption can be decreased well below the isothermal one (here for $T_{Cr} < 283.2$ K). Concerning the heating effect (see Sect. 8.3.3), the results confirm the existence of a threshold value (here $T_{Cr} \sim 273.2$ K). Above the threshold, COP_{hr} are relatively smaller as compared to COP_h (see 6th column) and become interesting energetically for T_{Cr} values below it, only.

8.4.3 TWRC (TSRC, CWF, CSTSGS)

The model results of TWRC (TSRC, CWF, CSTSGS) cycle, Fig. 8.11, are given in Table 8.3. The evaporator temperature covers the normal and industrial refrigeration, $T_E = ((-20)-(-60))$ °C (1st column), with condenser temperatures of $T_C = (30-40)$ °C (2nd column). The discharge gas temperatures of each compression stage are included in the 3rd and 4th columns. The refrigeration effectiveness, COP_{cr}, is shown in the 7th column. The figures indicate an important improvement

Table 8.4 NH_3 model results of the TWRC (THSRC, CWF, CSTSGS) cycle, Fig. 8.12

T_E (K)	T_C (K)	T_{CpO1} (K)	T_{CpO2} (K)	T_{CpO3} (K)	η_{v1} (–)	η_{v2} (–)	η_{v3} (–)	COP_{cr} (–)	$COP_{cr,rel,i}$ (–)	p_G (bar)
228.2	305.2	418.0	393.3	394.0	0.974	0.970	0.970	2.455	0.895	36.0
223.2	305.2	438.7	402.9	403.0	0.971	0.966	0.966	2.284	0.889	40.0
218.2	305.2	464.6	413.2	414.4	0.969	0.961	0.961	2.170	0.891	44.5
213.2	305.2	497.3	424.3	426.0	0.967	0.956	0.956	2.115	0.902	50.5
208.2	305.2	512.7	436.3	438.5	0.963	0.950	0.950	1.980	0.903	57.0
203.2	305.2	537.5	449.3	452.2	0.959	0.943	0.943	1.900	0.909	65.0
198.2	305.2	551.3	463.3	467.1	0.953	0.935	0.935	1.778	0.911	75.0

when TWRC is applied. This is obvious especially for the industrial refrigeration ($T_E < -40$ °C), where the rough COP increase is by 50–100 % as compared to the classic two-stage and cascade compression. Probably, the two-stage TWRC is the most attractive and worth to be translated in practice, for its relative low complexity and its great COP improvement. For this reason, the two-stage TWRC seems to be the most feasible for normal and industrial refrigeration production.

8.4.4 TWRC (THSRC, CWF, CSTSGS)

The model results of TWRC (THSRC, CWF, CSTSGS) cycle, Fig. 8.12, are given in Table 8.4. The evaporator temperature covers industrial refrigeration, $T_E = ((-40)-(-75))$ °C (1st column), with condenser temperature of $T_C = 32$ °C (2nd column). The discharge gas temperatures of each compression stage are included in the 3rd, 4th and 5th columns. The refrigeration effectiveness, COP_{cr}, is shown in the 9th column. The figures show an improvement as compared to the two-stage TWRC cycle, as we expected. Besides the two-stage, the three-stage TWRC cycle seems to be feasible as well in the industrial refrigeration.

As a general conclusion, the results of TWRC emphasize a reasonable feasibility of single-stage compression, while the three-stage and especially the two-stage compression are the most feasible, enjoying a great potential in refrigeration COP increase, and therefore in the primary energy savings and carbon footprint reduction (~50–100 % and more). Motivated by these encouraging issues, in the next section we shall go deeper with our analysis concerning theoretical and practical aspects of the TWRC application and feasibility.

8.5 Further Results Concerning TWRC Feasibility

The main problems related to TWRC successful application are at least two: (a) can a future compressor be adapted to operate satisfactorily with the high discharge gas temperatures resulting from applying the new concept?, and

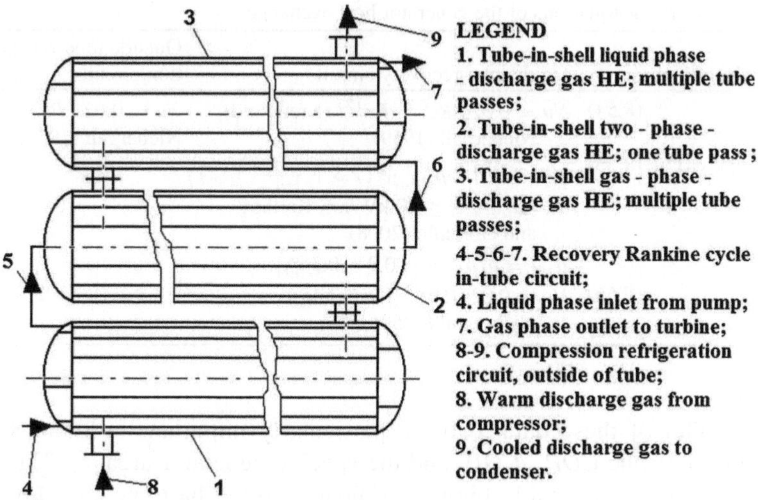

Fig. 8.14 TWRC generator schematic used in the model

(b) is it reliable and feasible to use the Rankine recovery cycle, which generator of could result with too high heat exchange surfaces, given the heat transfer with a gas (here discharge gas), considered in the literature to have low values? Referring to the first problem, it is basic for a compressor manufacturer. Its solving has favorable pre-requisites, using either oil free compressors, or high temperature lubricants etc., and will not be discussed here more. Concerning the second problem, we have performed a study case of a RRC generator, which operation of has been modeled and the results thereof will be presented next (Staicovici 2011d). As RRC generator we have considered the set of HE's depicted in Fig. 8.14, based on a tube-in-shell compact geometry. The heating fluid (discharge gas, refrigeration circuit) flows between tubes and shell and the fluid of the recovery circuit, changing the phase, flows in tubes. Despite the parallel arrangement of the three shells, both circuits, refrigeration and recovery, are serially connected each, in order to maximize fluids velocity. In this way, Re numbers characteristic to turbulent flow have been obtained, which maximize the heat exchange. Care has been paid also to minimize the pressure loss of the discharge gas, that is to minimize the additional consumption of compressor work. This is why we considered that the discharge gas must have a larger flow section, wherefrom must occupy the space between tubes and shell. The pressure loss in the recovery circuit was considered as being less important and can be compensated for, anyway, by the Rankine cycle pump, which is a much smaller power consumer, as compared to the compressor. The criteria equations governing the heat exchange in both circuits are included in Table 8.5. The model results for a single-stage TWRC, Figs. 8.6 and 8.8, are included in Table 8.6. The model results for a two-stage TWRC, Figs. 8.10 and 8.11, are included in

Table 8.5 Governing equations of the generator heat exchange

Recovery circuit phase		Inside tube-recovery circuit	Outside tube-refrigeration circuit	
Liquid ("l")	(8.54)	$Nu = 0.024\ Re^{0.24}\ Pr^{0.37}$ (Kraussold), Nieberball (1959)	$Nu = 0.027\ Pe^{0.78}$ Niebergall (1959)	(8.57)
Gas–liquid ("g–l")	(8.55)	$\frac{\alpha}{1.006} = \left(1.2\, p_{red}^{0.27} + \left(2.5 + \frac{1}{1-p_{red}}\right)p_{red}\right)$ $\alpha = \alpha_0 q^n;\ q \le 20\ \text{kW/m}^2$, Kotthoff and Gorenflo (2008) $p_{red} = p/p_c;\ n = 0.9 - 0.3p_{red}^{o.3}$		
Gas ("g")	(8.56)	$Nu = 0.027Pe^{0.78}$ (Kraussold), Niebergall (1959)		

Table 8.7. Out of these data, two are particularly important in Table 8.6 (see figures in bold): the COP_{cr}/COP_c and the specific generator area, S_t. They show the COP_{cr} can be reasonably high as compared to the baseline case, the lower T_E, the higher this increase. Also, the S_t area has reasonable values, because the global heat exchange coefficients are quite high for a gas phase implied in a heat transfer. Such figures indicate a relative short payback period of TWRC additional investment as compared to that of the baseline (2–3 years, according to the author estimate).

And last, but not the least, in our model we succeeded to keep the pressure losses of both HE sides with reasonable low values, especially to the refrigeration side, the most important in this respect as it was already mentioned priorly. Concerning the results of Table 8.7, they suggest the real benefit of introducing the TWRC. Same comments could be done here, similar to Table 8.6 as well. However, this time the COP_{cr} are higher by 50–100 %, as compared to the baseline case.

Two previous works enable to make a quantitative comparison with an already operating deep cooling equipment. Indeed, a CO_2/NH_3 cascade has to $T_E = -50$ °C and $T_C = 32$ °C a cooling effectiveness of $COP_c = 1.26$ (Belozerov et al. 2007; Prosper de Mulder 2004). According to our model, a TWRC two-stage cycle could have $COP_{cr} = 2.10$, for same input data (see Table 8.7), which is by 67 % higher. The two-stage TWRC investment payback is estimated to 1–3 years.

The final model results refer to the influence which RRC mass flow rate m_G has on the COP_{cr} and S_t. In Table 8.8 we give the output data of these functions for two study cases of the two-stage configuration, when m_G covers a range between the zero value and a maximum one. In Fig. 8.15 is plotted the $T_E = -30$ °C and $T_C = 35$ °C case. We remark that COP_{cr} increases slowly linearly (but importantly), while the S_t increases exponentially. In order to find an optimal cycle structure with respect to the effectiveness and investment value, a thermo-economical computation is necessary. It is worth mentioning that in case of $m_G = 0$, we find a TWRC structure without the discharge gas heat recovery at all. For instance, considering the two-stage cycle in Fig. 8.10, only the cycle plotted by the (a–b–c–d–e–f–g–h–i–j–k–a) contour works. Its

Table 8.6 Model results of a single-stage TWRC, Figs. 8.6 and 8.8

T_E, T_C, T_{CpO} (°C)	COP_c[a], COP_{cr} [b], $COP_{cr,rel,i}$, COP_{cr}/COP_c (−)	m_G (kg/kg) p_G p_{CpO} (bar), q_E (kJ/kg)	Generator of the recovery Rankine ammonia cycles								
			Recovery cycle			Refrigeration cycle					
			Phase	α (W/m²K)	Δp^f (bar)	α (W/m²K)	Δp^f (bar)	S (m²/kg)	S_t (m²/kg)	K (W/m²K)	$\Delta T_{m,lg}$ (°C)
−5	4.917	0.15	l	3,710	0.055	476.1	0.012	6.39	**27.1**	413.2	16.09
35	5.058	52	g–l	2,985	–	472.4	0.034	16.12		399.7	18.05
154.9	0.846	13.5									
	1.029	1,190	g	1,844	0.424	478.8	0.011	4.594		373.0	22.72
−10	4.122	0.19	l	4,525	0.181	474.6	0.026	13.72	**43.61**	420.5	11.44
35	4.321	66	g–l	3,911	–	474.0	0.05	22.74		414.0	13.48
178.6	0.826	13.5									
	1.048	1,196	g	2,339	0.824	485.3	0.018	7.15		394.0	23.23
−15	3.494	0.23	l	5,321	0.372	472.9	0.039	19.86	**52.55**	425.1	11.59
35	3.734	87	g–l	6,379	–	476.8	0.048	21.11		434.0	12.84
205.2	0.807	13.5									
	1.069	1,202	g	2,934	1.474	493.1	0.030	11.58		413.5	22.22
−20	3.037	0.29	l	3,657	0.258	473.9	0.061	31.72	**78.11**	410.9	9.228
35	3.292	106									
234.9	0.795	13.5	g–l	1,8086	–	475.0	0.043	19.15		452.4	9.976
	1.084	1,209	g	2,199	1.232[c]	496.9	0.073	27.24		397.3	20.02
−25	2.569	0.33	l	1,636	0.103	473.4	0.057	29.33	**86.04**	360.5	13.4
35	2.883[e]	116	g–l	–	–	475.6	0.023	10.07		470.0	13.75
268.3	0.772	13.5									
	1.122	1,217	g	1,345	0.754[d]	504.7	0.131	46.64		360.4	18.32

[a]+5 °C vapor superheating, Fig. 8.1, left
[b]Gas-gas inlet temp. pinch. = 8 °C
[c]Double tube passes
[d]Quadruple tube passes
[e]Gas-gas inlet temp. pinch. = 10 °C
[f]Line pressure losses, only

Table 8.7 Model results of a two-stage TWRC cycle, Figs. 8.10 and 8.11

T_E, T_C, (°C), p_{CpO}[a] (bar), q_E (kJ/kg)	COP_{cr}, $COP_{cr,rel,i}$ (−), p_{G1} (= p_{G2}) (bar)	T_{CpO} (°C), m_{G1} (= m_{G2}) (kg/kg)	Stage	Generator of the recovery Rankine ammonia cycles								
				Recovery cycle			Refrigeration cycle					
				Phase	α (W/m²K)	Δp (bar)[f]	α (W/m²K)	Δp (bar)[f]	S (m²/kg)[c]	S_t (m²/kg)	K (W/m²K)	$\Delta T_{m,\lg}$ (°C)
−30	**2.875**	151.7	1	l	3,617	0.041	335.8	0.014	4.316	**47.7**	302.6	24.14
35	0.855	0.12		g–l	2,762	–	352.0	0.048	12.26		307.4	25.69
13.5[a]												
1227	48			g	1,725	0.022	363.5	0.027	6.355		295.8	17.11
		142.4	2	l	3,617	0.042	381.5	0.004	4.485		339.2	20.72
		0.12		g–l	2,762	–	3,7817	0.016	14.37		327.6	20.57
				g	1,777	0.020	3,8117	0.007	5.911		309.4	15.47
−35	**2.594**	170.2	1	l	4,116	0.061	3,3416	0.019	4.897	**53.1**	304.7	27.95
35	0.8415	0.14		g–l	3,162	–	3,5514	0.052	11.1		314.4	30.15
13.5												
1240	55			g	1,973	0.034	3,6916	0.041	8.1		306.5	18.57
		155.9	2	l	4116	0.074	3,8015	0.006	5.982		342.3	20.36
		0.14		g–l	3,162	–	3,7912	0.018	15.37		333	20.57
				g	2,061	0.030	3,84.3	0.016	7.618		318.7	16.13
−40	**2.377**	191.9	1	l	333.7	0.116	4,836	0.029	6.592	**73.1**	307.4	28.41
35	0.8346	0.17		g–l	358.8	–	3,760	0.060	11.24		322.2	32.05
13.5	64											
1258				g	376.9	0,058	2,335	0.067	11		319.4	19.68
		170.6	2	l	380.4	0.203	4,836	0.012	11.5		346.6	14.45
		0.17		g–l	379.8	–	3,760	0.026	22.22		339.1	15.4
				g	387.4	0.052	2,486	0.014	10.56		329.6	16.23

(continued)

Table 8.7 (continued)

T_E, T_C, (°C), p_{CpO}[a] (bar), q_E (kJ/kg)	COP_{cr}, $COP_{cr,rel,i}$ (−), p_{G1} (= p_{G2}) (bar)	T_{CpO} (°C), m_{G1} (= m_{G2}) (kg/kg)	Stage	Generator of the recovery Rankine ammonia cycles								
				Recovery cycle			Refrigeration cycle					
				Phase	α (W/m^{2}K)	Δp (bar)[f]	α (W/m^{2}K)	Δp (bar)[f]	S (m^2/kg)[c]	S_t (m^2/kg)	K (W/m^{2}K)	$\Delta T_{m,lg}$ (°C)
−45	**2.194**	217.6	1	l	5526	0.170	333.5	0.037	7.196	**97.5**	309.7	32.63
35	0.8261	0.2		g–l	4,239	–	363.1	0.065	9.97		328.9	38.89
13.5	70			g	2,609	0.095	386	0.103	13.88		330.7	21.71
1282		186.6	2	l	5,526	0.497	380.5	0.022	21.03		349.7	9.886
		0.2		g-l	4,239	–	380.6	0.038	31.6		343.2	11.76
				g	2818	0.087	391	0.019	13.79		337.6	16.9
−50	**2.100**	230.5	1	l	3,916	0.076	232.9	0.018	9.701	**124**[d]	217.4	39.93
32												
12.38	0.8175	0.206		g–l	5,043	–	256.3	0.027	11.23		240.9	44.03
	78			g	1,896	0.039	273	0.059	21.86		235.8	22.13
1305		194.4	2	l	3,916	0.203	262.2	0.009	25.73		242.7	13.48
		0.206		g–l	5,043	–	264.8	0.013	33.24		248.4	14.42
				g	2,080	0.036	273.4	0.010	22.44		238.8	16.5
−55	**1.960**	265.1	1	l	4,322	0.081	234	0.018	8.289	**119**[d]	219.5	52.36
32												
12.38	0.8049	0.233		g–l	5,043	–	261.5	0.026	9.002		245.6	60.93
1348	78			g	1980	0.059	282.4	0.081	24.1		244.1	26.29
		213.0	2	l	4,322	0.24	262.2	0.008	24.43		244.2	19.65
		0.233		g–l	5,043	–	265.6	0.011	27.51		249.2	15.97
				g	2,169	0.055	276.9	0.012	25.84		242.6	18.4

(continued)

Table 8.7 (continued)

T_E, T_C, (°C), p_{CpO}[a] (bar), q_E (kJ/kg)	COP_{cr}, $COP_{cr,rel,i}$ (−), p_{G1} (= p_{G2}) (bar)	T_{CpO} (°C), m_{G1} (= m_{G2}) (kg/kg)	Stage	Generator of the recovery Rankine ammonia cycles								
				Recovery cycle			Refrigeration cycle					
				Phase	α (W/m^{2}K)	Δp (bar)[f]	α (W/m^{2}K)	Δp (bar)[f]	S (m^2/kg)[c]	S_t (m^2/kg)	K (W/m^{2}K)	$\Delta T_{m,\mathrm{lg}}$ (°C)
−60	**1.854**	265.3	1	l	2,886	0.039	135.4	0.008	13.54	**138**[e]	128.5	60.06
30	0.7842	0.22		g–l	8473	–	152.2	0.008	11.18		148.4	59.95
11.67[b]	95			g	1,430	0.022	162.9	0.037	44.52		145.2	24.11
1,391		253.6	2	l	2,886	0.037	152.3	0.001	12.94		143.6	56.22
		0.22		g–l	8,473	–	157.5	0.001	11.99		153.4	54.05
				g	1,460	0.021	165.5	0.004	44.11		147.6	22.63

a$p_{\mathrm{int}} = (p_E p_C)^{1/2}$

b$p_{\mathrm{int}} = (p_E p_C)^{1/2} - 0.3$

cShell-and-tube HE, one tube pass, $z = 31$

[d]Idem (c), $z = 37$

[e]Idem (c), $z = 43$

[f]Line pressure losses, only

Table 8.8 The m_G influence on S_t and COP_{cr} functions

$m_{G1} = m_{G2}$ (kg/kg s)	0.0	0.05	0.06	0.07	0.08	0.09	0.10	0.11	0.12	0.13	2-stage $T_E = -30$ °C $T_C = 35$ °C
S_t (m^2)	0.0	12.33	15.21	18.42	22.12	26.43	31.67	38.35	47.70	63.60	
COP_{cr}, (−)	2.643	2.735	2.754	2.774	2.794	2.814	2.834	2.854	2.875	2.896	
$m_{G1} = m_{G2}$ (kg/kg s)	0.0	0.08	0.10	0.12	0.14	0.16	0.18	0.19	0.20	0.206	2-stage $T_E = -50$ °C $T_C = 32$ °C
S_t (m^2)	0.0	22.87	29.61	37.39	46.61	57.96	72.84	82.52	94.82	124.2	
COP_{cr} (−)	1.800	1.905	1.933	1.962	1.993	2.024	2.056	2.073	2.089	2.100	

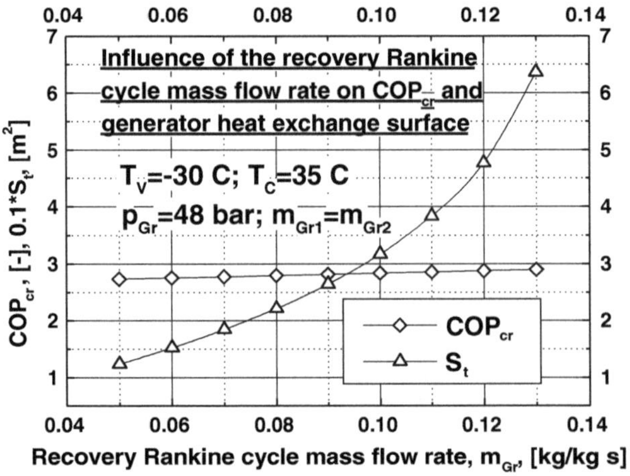

Fig. 8.15 m_G influence on S_t and COP_{cr} functions

effectiveness [$COP_{cr}(m_G = 0) = 1.80$, Table 8.8], it still much higher than that of the CO_2/NH_3, because a better internal heat recovery was applied (CSTSGS).

8.6 Refrigeration Cycles Provided with TTRC

The models of both TWRC and TTRC methods affording the assessment of mvc cooling and heating effectiveness will be given in detail in the next Sect. 8.7. Here, just a few TTRC refrigeration results are presented, obtained for a deep cooling 2-stage NH_3, with $T_E = -50$ °C, $T_C = 32$ °C (Staicovici 2011d). This cycle provides the discharge gas heat at $T_{2a} = 250$ °C for the coupling with: (a) an air conditioning H_2O–LiBr cycle, $T_E = +7$ °C, $T_C = 32$ °C, and (b) a deep cooling coabsorbent truncated NH_3–H_2O cycle, $T_{DI} = -50$ °C, $T_{M,c} = 32$ °C, Fig. 4.4. The effectiveness can be calculated simpler by: $COP_{c,TTRC} \cong COP_{cr}(m_G = 0) + (k_1 + k_2)COP_{c,ARC}$, $k_i \leq 1$. For the study case (a): $COP_{cr}(m_G = 0) = 1.688$ and $(k_1 + k_2)COP_{c,ARC} \sim 0.84 * 0.73 = 0.613$, so $COP_{c,TTRC} \cong 2.301$. For the study case (b): $COP_{cr}(m_G = 0) = 1.688$ and $(k_1 + k_2)COP_{c,ARC} \sim 0.775 * 0.2274 = 0.176$, so $COP_{c,TTRC} \cong 1.864$. We remark that in the study case (b), $1.864 = COP_{c,TTRC} < COP_{c,TWRC} = 2.10$ (see Table 8.7), although the Theorem 2 allows that $COP_{c,TTRC} > COP_{c,TWRC}$. The explanation is given by the higher source-task mismatch of the TTRC as compared to TWRC. The Carnot COP is 2.72, showing there is further room to find the most effective ARC cycle. The TTRC joins the ARC and the compression refrigeration units without changing much their standard construction and it is more flexible, therefore it is more feasible as compared to the TWRC.

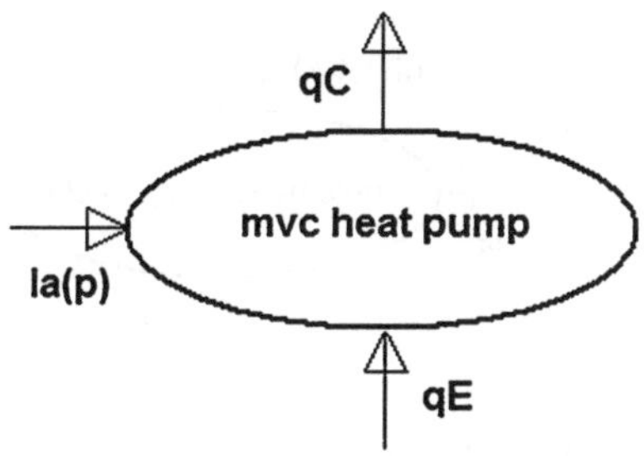

Fig. 8.16 A standard heat pump schematic

8.7 TWRC and TTRC Heat Pumping Theory and Recent Results

As we mentioned in Sect. 8.2, the TWRC and TTRC apply to both refrigeration and heating applications. The reason to introduce *TWRC* and *TTRC* in mvc refrigeration is easy to explain. Indeed, instead to loose the discharge gas superheat exergy in a low temperature condenser, or in an intercooler, it is wise to partially recover and transform it in useful exergy, either mechanical work, or of cooling. On the contrary, the reason to introduce *TWRC* and *TTRC* in case of a mvc heat pump is not that obvious, because one might think that the discharge gas superheat is anyway added to the condenser latent heat for increasing the heat pump effect. Our research has shown that even in this case the *TWRC* and *TTRC* introduction benefits the mvc heat pump effectiveness. Qualitatively, the 2nd thermodynamics principle proves this issue by the better match of the discharge gas source with the *RRC* and *ARC* generators than with the heat pump condenser. Indeed, the *RRC* and *ARC* generators operate to higher temperatures than the heat pump condenser, so the discharge gas looses less exergy in its heat exchange with *RRC*(*ARC*) generator than with the heat pump condenser, where the *TWRC* and *TTRC* use benefit comes of. The present section extends the research at hand to the cooling and combined cooling and heating, providing the theory of *TWRC* and *TTRC*. Further, these are modeled with this theory and results thereof are given.

8.7.1 TWRC and TTRC Heat Pumping Theory

A standard heat pump is depicted in Fig. 8.16. The heat pump extracts from heat sink in evaporator the lower temperature heat, $q_E > 0$, has an adiabatic or polytropic mechanical work input, $l_a < 0$, $l_p < 0$, and rejects to condenser the heat, $q_C < 0$, of the required higher temperature level. The adiabatic and polytropic cases are similar, so next only the first will be analysed. The first principle of thermodynamics writes simply:

$$|q_C| = q_E + |l_a| \tag{8.58}$$

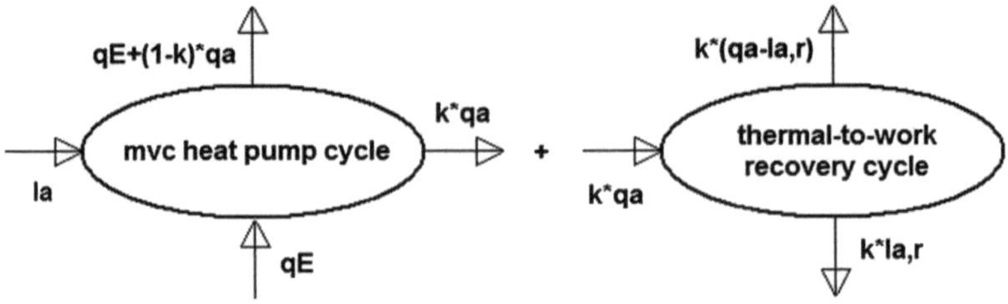

Fig. 8.17 TWRC cycle schematic

As a known result, l_a equals numerically the discharge gas superheat, $q_a < 0$ (e.g. Popa and Vintila 1977; Kirilin et al. 1985; Staicovici 2011a):

$$|l_a| = |q_a| \tag{8.59}$$

The standard heat pump COP is defined by:

$$COP_h = \frac{|q_C|}{|l_a|} = \frac{q_E + |l_a|}{|l_a|} = \frac{q_E}{|l_a|} + 1 = COP_c + 1 \tag{8.60}$$

Equation (8.60) shows the relationship existing between COP's of the two standard heat pumping cooling and heating cycles, i.e. COP_c and COP_h, respectively, referred to hereinafter. The TWRC and TTRC effectivenesses are characterized by the following basic:

Lemma 1 *Providing a standard mechanical vapor compression single-stage heat pumping cycle with TWRC or TTRC preserves the relationship between its heating and cooling COP's and increases theirs values, comparatively.*

Proof Let us consider first a TWRC cycle, schematically shown in Fig. 8.17. In this case, the heat pump condenser (see Fig. 8.17 left side picture) receives only the $(1-k)q_a$, $0 < k < 1$, discharge gas superheat fraction out of the q_a total. The rest of kq_a is used by the thermal-to-work recovery cycle (see Fig. 8.17 right side picture) in order to produce the work $kl_{a,r} > 0$ and to deliver to the same condenser an additional amount of heat $k\left(q_a - l_{a,r}\right)$. According to TWRC method, the $kl_{a,r}$ work diminishes the l_a input. We rewrite Eq. (8.60) taking into account Fig. 8.17, and with Eq. (8.59) we obtain successively:

$$\begin{aligned} COP_{h,TWRC} &= \frac{q_E + (1-k)|q_a| + k\left(|q_a| - l_{a,r}\right)}{|l_a| - kl_{a,r}} = \frac{q_E + |q_a| - kl_{a,r}}{|l_a| - kl_{a,r}} \\ &= \frac{q_E + |l_a| - kl_{a,r}}{|l_a| - kl_{a,r}} = \frac{q_E}{|l_a| - kl_{a,r}} + 1 \end{aligned} \tag{8.61}$$

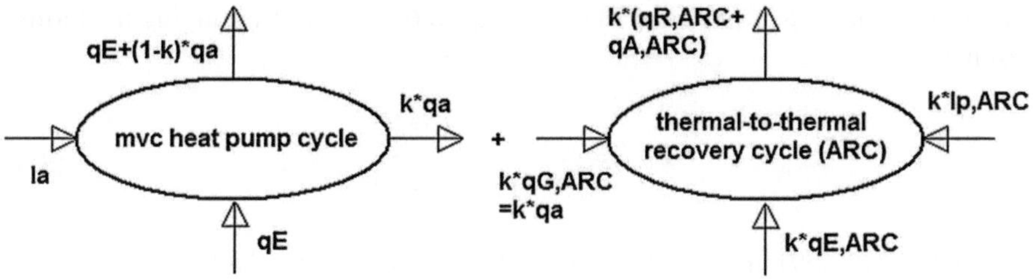

Fig. 8.18 TTRC cycle schematic

In Eq. (8.61), the first term of the right hand last equality is identified as the effectiveness of a mvc cooling cycle provided with TWRC (Staicovici 2011a, b, c, d):

$$\frac{q_E}{|l_a| - kl_{a,r}} = COP_{c,TWRC} > COP_c = \frac{q_E}{|l_a|} \tag{8.62}$$

Taking into account Eq. (8.62), Eq. (8.61) rewrites:

$$COP_{h,TWRC} = COP_{c,TWRC} + 1 \tag{8.63}$$

and the comparison with Eq. (8.60) leads to the proof of the first lemma part for TWRC. The second lemma part is proved for TWRC introducing in Eq. (8.62) result in Eq. (8.63):

$$COP_{h,TWRC} = COP_{c,TWRC} + 1 > COP_c + 1 = COP_h \tag{8.64}$$

Second, we consider the TTRC heat pumping cycle, schematically shown in Fig. 8.18. As recovery cycle we have chosen an absorption cycle, noted by ARC (see Fig. 8.18 right side picture), probably the most indicated in our case, as it belongs to the very common, mature and successful heat pumping absorption technology powered by a heat source. The ARC unit must be of cooling type only. This might contradict some readers intuition, as the ARC is used here to produce mainly heating and therefore it should be a cycle of heating type. Proving that only a cooling cycle matches to our task in this case, is left to the reader. Referring to Fig. 8.18 notations, the ARC unit receives to its generator the high (discharge gas) temperature heat input $kq_{G,ARC} = kq_a > 0$, extracts from the heat sink the $kq_{E,ARC} > 0$ low temperature heat, consumes the pumping work $kl_{p,ARC} < 0$ and delivers the useful heat $k(q_{R,ARC} + q_{A,ARC}) < 0$, rejected by the cycle resorber(s) and absorber and having the required temperature level, equal theoretically to that of the mvc heat pump condenser. From Fig. 8.18, the TTRC cycle effectiveness results as following:

$$COP_{h,TTRC} = \frac{q_E + (1-k)|q_a| + k\left(\left|q_{R,ARC}\right| + \left|q_{A,ARC}\right|\right)}{|l_a| + k\left|l_{p,ARC}\right|} \tag{8.65}$$

We bear in mind the heat balance of the ARC cycle:

$$k\left(\left|q_{R,ARC}\right| + \left|q_{A,ARC}\right|\right) = k\left(q_{E,ARC} + q_{G,ARC} + \left|l_{p,ARC}\right|\right) \tag{8.66}$$

and introducing Eqs. (8.59) and (8.66) in Eq. (8.65), after re-arranging the terms it results:

$$COP_{h,TTRC} = \frac{\frac{q_E}{|l_a|} + k\frac{q_{E,ARC}}{|l_a|}}{1 + k\frac{|l_{p,ARC}|}{|l_a|}} + 1 \tag{8.67}$$

We apply Eq. (8.59) in Eq. (8.67) and its different ratios are identified as:

$$\frac{q_E}{|l_a|} = \frac{q_E}{|q_a|} = COP_c \tag{8.68}$$

$$\frac{q_{E,ARC}}{|l_a|} = \frac{q_{E,ARC}}{|q_a|} = COP_{c,ARC} \tag{8.69}$$

$$\frac{|l_{p,ARC}|}{|l_a|} = \frac{\frac{q_{E,ARC}}{|q_a|}}{\frac{q_{E,ARC}}{|l_{p,ARC}|}} = \frac{COP_{c,ARC}}{COP_{w,ARC}} \tag{8.70}$$

Introducing Eqs. (8.68)–(8.70) in Eq. (8.67), this becomes:

$$COP_{h,TTRC} = \frac{COP_c + kCOP_{c,ARC}}{1 + k\frac{COP_{c,ARC}}{COP_{w,ARC}}} + 1 \tag{8.71}$$

The cooling effectiveness of TTRC writes, Fig. 8.18:

$$COP_{c,TTRC} = \frac{q_E + kq_{E,ARC}}{|l_a| + k|l_{p,ARC}|} \tag{8.72}$$

After re-arranging its terms and taking into account Eqs. (8.68)–(8.70), Eq. (8.72) becomes as:

$$COP_{c,TTRC} = \frac{COP_c + kCOP_{c,ARC}}{1 + k\frac{COP_{c,ARC}}{COP_{w,ARC}}} \tag{8.73}$$

Equation (8.73) is nothing else but the first term of Eq. (8.71), therefore the following result holds true:

$$COP_{h,TTRC} = COP_{c,TTRC} + 1 \tag{8.74}$$

This proves the lemma first part in case of TTRC. In order to prove the lemma second part of TTRC case, we remark that:

$$\frac{COP_{c,ARC}}{COP_{w,ARC}} << 1 \tag{8.75}$$

and make the approximation:

$$\frac{1}{1+\frac{COP_{c,ARC}}{COP_{w,ARC}}} \cong 1-\frac{COP_{c,ARC}}{COP_{w,ARC}} \tag{8.76}$$

Introducing Eq. (8.76) in Eq. (8.73) and arranging the terms, it results:

$$COP_{c,TTRC} = COP_c + kCOP_{c,ARC}\left(1-\frac{COP_c}{COP_{w,ARC}}\right) - k^2\frac{COP_{c,ARC}^2}{COP_{w,ARC}} \tag{8.77}$$

Equation (8.77) last term is of second order, $k^2\frac{COP_{c,ARC}^2}{COP_{w,ARC}} \cong 0.001$, and can be neglected as compared to the rest of terms and the bracket is positive, as $\frac{COP_c}{COP_{w,ARC}} << 1$, so Eq. (8.77) can be written:

$$COP_{c,TTRC} \cong COP_c + kCOP_{c,ARC}\left(1-\frac{COP_c}{COP_{w,ARC}}\right) > COP_c \tag{8.78}$$

Finally, we introduce this result in Eq. (8.74) for obtaining:

$$COP_{h,TTRC} = COP_{c,TTRC} + 1 > COP_c + 1 = COP_h \quad \text{q.e.d.} \tag{8.79}$$

and to prove in this way the whole lemma. The lemma results can be extended by the following:

Corollary 2 *A n-stage standard mechanical vapor compression heat pumping cycle, $n \in N$, provided with a TWRC or TTRC method application to each stage preserves the relationship between its heating and cooling COP's and increases theirs values, comparatively.*

The corollary can be easily obtained considering the n-stage case of the TWRC and TTRC cycles given schematically in Figs. 8.17 and 8.18, respectively. Referring first to the TWRC, Eqs. (8.63) and (8.64) hold true for the n-stage case as well and the cooling COP is given by the following equation:

$$COP_{c,TWRC,n} = \frac{q_E}{\sum_{i=1}^{n}\left(\left|l_{a,i}\right| - k_i l_{a,r,i}\right)} \tag{8.80}$$

where $0 < k_i < 1$. Referring now to the TTRC, Eqs. (8.74) and (8.79) hold true for the n-stage case as well and the cooling COP is given by the following equation:

$$COP_{c,TTRC,n} = \frac{COP_c + \sum_{i=1}^{n} k_i r_i COP_{c,ARC,i}}{1+\sum_{i=1}^{n} k_i r_i \frac{COP_{c,ARC,i}}{COP_{w,ARC,i}}} \tag{8.81}$$

Equation (8.81) is valid generally, when each stage uses a different ARC unit, with different operating parameters, that is $COP_{c(w),ARC,i} \neq COP_{c(w),ARC,j}$, $i \neq j$, $i, j \in N$. In Eq. (8.81), the r_i ratio is the fraction of the adiabatic work of the of the i-th stage, out of the total compression work input, and is given by:

$$r_i = \frac{|l_{a,i}|}{\sum_{j=1}^{n} |l_{a,j}|}, \quad i = 1, \ldots, n \tag{8.82}$$

The r_i ratios, $i = 1, \ldots, n$, enjoy the following obvious property:

$$\sum_{i=1}^{n} r_i = 1 \tag{8.83}$$

If $COP_{c(w),ARC,i} = COP_{c(w),ARC,j} = COP_{c(w),ARC}$ for all $i \neq j$, $i,j \in N$, Eq. (8.81) becomes:

$$COP_{c,TTRC,n} = \frac{COP_c + COP_{c,ARC} \sum_{i=1}^{n} k_i r_i}{1 + \frac{COP_{c,ARC}}{COP_{w,ARC}} \sum_{i=1}^{n} k_i r_i} \tag{8.84}$$

Finally, if $T_{CpO,i} = T_{CpO} = const.$ for all $i \in N$, then $k_i = k = const.$ for all $i \in N$, and taking into account Eqs. (8.83) and (8.84) becomes Eq. (8.73), valid for $n = 1$.

Corollary 3 *A single-stage standard mechanical vapor compression heat pumping cycle provided with the TTRC method applied serially m times, $m \in N$, preserves the relationship between its heating and cooling COP's and increases theirs values, comparatively.*

The corollary can be easily obtained considering the TTRC cycle provided with m ARC units serially connected on a single stage, given schematically in Fig. 8.18. The cooling effectiveness is obtained from Eq. (8.73):

$$COP_{c,TTRC} = \frac{COP_c + \sum_{i=0}^{m-1} k_i COP_{c,ARC,i}}{1 + \sum_{i=0}^{m-1} k_i \frac{COP_{c,ARC,i}}{COP_{w,ARC,i}}}, \quad i = 1, \ldots, m \tag{8.85}$$

In Eq. (8.85), k_i represents the fraction of heat delivered to the ith ARC unit, out of the total available discharge gas superheat, equal to $h(T_{CpO}) - h(T_C)$:

$$k_i = \frac{h(T_{GI,i}) - h(T_{GI,i+1})}{h(T_{CpO}) - h(T_C)}, \quad i = 1, \ldots, m \tag{8.86}$$

In Eq. (8.86), and $h(T_{GI,1}) = h(T_{CpO})$, $h(T_{GI,m}) > h(T_C)$, $h(T_{GI,m+1}) = h(T_C)$ and

$$\frac{h(T_{GI,m}) - h(T_C)}{h(T_{CpO}) - h(T_C)} + \sum_{i=1}^{m} k_i = 1, \tag{8.87}$$

or

$$\sum_{i=1}^{m} k_i = \frac{h(T_{CpO}) - h(T_{GI,m})}{h(T_{CpO}) - h(T_C)} < 1 \tag{8.88}$$

Combining the Corollaries 2 and 3, it is obtained the most general case, of a n-stage compression provided with m ARC units serially connected on each stage:

$$COP_{c,TTRC,nm} = \frac{COP_c + \sum_{i=1}^{n} r_i \sum_{j=1}^{m} k_{ij} COP_{c,ARC,ij}}{1 + \sum_{i=1}^{n} r_i \sum_{j=1}^{m} k_{ij} \frac{COP_{c,ARC,ij}}{COP_{w,ARC,ij}}} \tag{8.89}$$

$$k_{ij} = \frac{h(T_{GI,i,j}) - h(T_{GI,i,j+1})}{h(T_{CpO,i}) - h(T_{sat}(p_i))}, \quad i = 1, \ldots, n, j = 1, \ldots, m \tag{8.90}$$

where $h(T_{GI,i,1}) = h(T_{CpO,i})$, $h(T_{GI,i,m}) > h(T_{sat}(p_i))$, $h(T_{GI,i,m+1}) = h(T_{sat}(p_i))$, $T_{sat}(p_n) = T_C$ and

$$\sum_{j=1}^{m} k_{ij} = \frac{h(T_{CpO,i}) - h(T_{GI,i,m})}{h(T_{CpO,i}) - h(T_{sat}(p_i))} < 1 \tag{8.91}$$

In Sect. 8.2.2, the basic Theorem 2 showing the thermodynamic cooling COP limits of TTRC with respect to the ideal mvc and TWRC has been proved. The conclusions of this theorem can be extended to the heating cycles, as well, by a theorem which is proved next.

Theorem 3 *The heat pump ideal (maximum) TTRC method effectiveness is higher than that of the TWRC method and has as upper limit the effectiveness of the mechanical vapor compression Carnot heat pump cycle, only.*

Proof Let us consider Eqs. (8.63) and (8.74) of the lemma proved above and let us express the cooling COP's with respect to the heating COP's. Theirs ideal values become Eqs. (8.92) and (8.93), respectively:

$$COP_{c,TWRC,id} = COP_{h,TWRC,id} - 1 \tag{8.92}$$

$$COP_{c,TTRC,id} = COP_{h,TTRC,id} - 1 \tag{8.93}$$

We proceed similarly with Eq. (8.60), consider its ideal value form and express the cooling COP with respect to the heating one, in order to obtain:

$$COP_{c,id} = COP_{h,id} - 1 \tag{8.94}$$

We bear in mind the result of Theorem 2 and rewrite it as:

$$COP_{c,id} > COP_{c,TTRC,id} > COP_{c,TWRC,id} \tag{8.95}$$

and use it here, introducing Eqs. (8.92)–(8.94) in Eq. (8.95):

$$COP_{h,id} - 1 > COP_{h,TTRC,id} - 1 > COP_{h,TWRC.id} - 1 \tag{8.96}$$

The theorem is now proved because the following result holds true from Eq. (8.96):

$$COP_{h,id} > COP_{h,TTRC,id} > COP_{h,TWRC.id} \quad \text{q.e.d.} \tag{8.97}$$

Another fundamental theorem regarding the TWRC and TTRC methods application is given next:

Theorem 4 *A TWRC heat pumping cycle COP is favoured by the increase of its discharge gas superheating temperature, while a TTRC one is favored, on the contrary, by the decrease of this temperature value.*

Proof Indeed, let us consider first the TWRC cycle. Using the already known Eq. (8.30), we can write:

$$\frac{COP_c}{COP_{c,TWRC}} = \frac{|l_{a,TWRC}|}{|l_a|} = 1 - \varepsilon\left(1 - \frac{ln\frac{T_{2a}}{T_1}}{\frac{T_{2a}}{T_1} - 1}\right) \tag{8.98}$$

In order to prove the first part of the theorem it is enough to show that the members of Eq. (8.98) decrease when the internal heat source temperature T_{2a} increases for constant values of the relative to Carnot efficiency, $\varepsilon = const.$, and of the internal sink source temperature, $T_1 = const.$ Equally, we shall show this next, working with a new variable:

$$y = \frac{T_{2a}}{T_1},\ y \in (1, \infty),\ T_1 = const. \tag{8.99}$$

Mathematically, we have to prove that the derivative of Eq. (8.98) with respect to the y variable is negative.

$$\frac{d\left(\frac{|l_{a,TWRC}|}{|l_a|}\right)}{dy} = \frac{d}{dy}\left(1 - \varepsilon\left(1 - \frac{ln(y)}{y-1}\right)\right) = \varepsilon\frac{1 - \frac{1}{y} - ln(y)}{(y-1)^2} < 0 \tag{8.100}$$

or,

$$\frac{1}{y} + ln(y) > 1 \tag{8.101}$$

The y variable is written in the following form:

$$y = \frac{1+x}{1-x} \tag{8.102}$$

wherefrom,

$$x = \frac{y-1}{y+1} \tag{8.103}$$

Bearing in mind that $y \in (1, \infty)$, Eq. (8.99), from Eq. (8.103) it results that:

$$x \in (0, 1) \tag{8.104}$$

Using Eq. (8.102) in Eq. (8.101), this becomes:

$$ln\frac{1+x}{1-x} - \frac{2x}{1+x} > 0 \tag{8.105}$$

The two members of the left side in Eq. (8.105) can be rewritten using series of functions expressed by series of powers (Gaina et al. 1966), as:

$$ln\frac{1+x}{1-x} = 2x\sum_{n=0}^{\infty}\frac{x^{2n}}{2n+1}, \quad n \in N,\ x \in (-1, 1) \tag{8.106}$$

and

$$\frac{2x}{1+x} = 2x\sum_{n=0}^{\infty}(-1)^n x^n, \quad n \in N,\ x \in (0, \infty) \tag{8.107}$$

Introducing Eqs. (8.106) and (8.107) in Eq. (8.105), it results:

$$\sum_{n=1}^{\infty} x^n\left(1 - \frac{n+1}{n+2}x\right) > 0 \tag{8.108}$$

In order to prove the first theorem part it is enough to show that the series of functions (8.108) has a strictly positive general term:

$$x^n\left(1 - \frac{n+1}{n+2}x\right) > 0 \tag{8.109}$$

Indeed, the Eq. (8.109) leads to the true result (8.110):

$$1 > \frac{n+1}{n+2}x < 1 \tag{8.110}$$

because $\frac{n+1}{n+2} < 1$ and $x < 1$, according to Eq. (8.104), so the first theorem part is proved. Regarding a TTRC cycle, all equations given above in the lemma and the corollaries show the $COP_{c,TTRC}$ and $COP_{h,TTRC}$ direct proportionality with COP_c. Bearing in mind that COP_c increases with T_{2a} decrease, the theorem second part is also proved, hence the whole theorem is proved, q.e.d. (Note: A simpler proof of this theorem first part is outlined in Appendix 1 of this chapter).

8.7.2 TWRC and TTRC Heat Pumping Recent Results

The modeled *TWRC* single- and two-stage heat pumping cycles are shown in Figs. 8.6, 8.8, 8.10, and 8.11, respectively. The model of *TTRC* considered as ARC units belonging to coabsorbent technology (Staicovici 2009a), namely the cooling nontruncated and double truncated, plotted in Fig. 3.2, 4.4, and 4.12, respectively. The modeled single- and two-stage heat pumping *TTRC*are shown in Figs. 8.19a, b, respectively (see also Table 4.1). In case of the two-stage, both *TWRC* and *TTRC* are provided mandatorily with CSTSGS. Also, in the two-stage *TTRC*case, each of its stages can be provided with one or several serially connected *ARC*, as mentioned to Corollaries 2 and 3, but it is possible to join the each stage $ARC's$ units in a single equivalent one, too. A special attention has been paid to the $ARC's$ units using the gax, operating to high temperatures, with NH_3–H_2O. The results of a study case are presented in this respect in Figs. 8.20, 8.21, 8.22, showing an excellent resorber-generator thermal match.

The model of *TWRC* and *TTRC* cooling and heating cycles is a particular form of the theory of Sect. 8.7.1, and the results thereof are given in (Staicovici 2012). The NH_3 *TWRC* single- and double-stage cooling results are plotted versus evaporator and desorber inlet temperatures in Figs. 8.23a, b, respectively [see also Staicovici (2011d)]. The $ARC's$ have been run by $NH_3 - H_2O$ provided with vapor rectification and gax. The model results show that, essentially, the *TWRC* and *TTRC* single-stage refrigeration COP_c is better by 6–12 % than a standard configuration cycle (dash line), for condenser temperature $T_C = 35$ °C and evaporator temperature $T_E = ((-5)-(-25))$ °C. In case of the *TWRC* and *TTRC* two-stage refrigeration, the COP_c is roughly better by 7–18 % than a two-stage cycle, plotted in Figs. 8.10, 8.11, but by 50–80 % better than the present-day cycles, for $T_C = (35-30)$ °C and $T_E = ((-30)-(-60))$ °C (see e.g. the CO_2–NH_3 cascade, with $T_C = 32$ °C, $T_E = -50$ °C and $COP_c = 1.26$). Figure 8.23b shows also that *TWRC* performs better than *TTRC* for all cooling temperatures, contrary to theirs ideal limits, which indicate the opposite. It is therefore enough room for improving the *ARC* COP_c and its match with the discharge gas heat source in the future. However, even with the actual *ARC* technical limits, its comparative advantages brought in operation such, no additional noise, no moving parts and flexibility, makes the *TTRC* be a serious challenger of *TWRC*.

The heating applications results of the cycles given in Figs. 8.6, 8.8 and Figs. 8.10, 8.11, have been obtained for a cogeneration operation of heating, $T_c = (51-91)$ °C, and cooling, $T_E = ((-10)-(+10))$ °C. The single-stage heating COP_h are plotted for *TWRC* and *TTRC* in Fig. 8.24a, b, respectively, for $T_c = 51$ °C and $T_c = 61$ °C, while those of cogeneration, $COP_{cogen} = COP_c + COP_h$, are given in Fig. 8.25. From these figures it results that, as foreseen, both *TWRC* and *TTRC* methods applied to the standard heat pump configuration increase its effectiveness, here by ~2–5 %, but perform almost equally when compared one another.

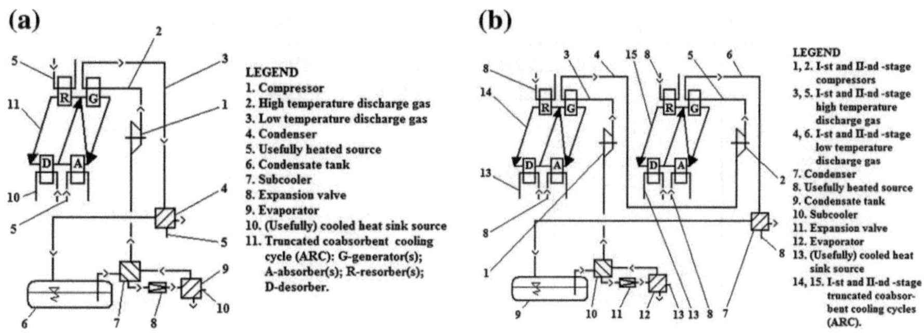

Fig. 8.19 *TTRC* cooling and/or heating cycles: **a** single-stage chart; **b** two-stage chart

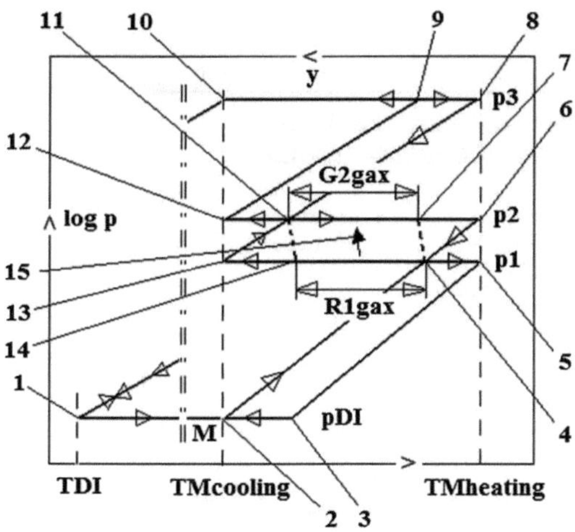

Pt.	p, [bar]	T, [°C]	y, [-]
1	4.952	5.0	0.9562
2	4.952	85.0	0.2607
3	4.952	96.45	0.2101
4	14.81	130.35	0.2607
5	14.81	142.0	0.2101
6	19.01	142.0	0.2607
7	19.01	126.35	0.3166
8	44.05	142.0	0.4708
9	44.05	129.75	0.5349
10	44.05	85.0	0.9562
11	19.01	96.25	0.4708
12	19.01	85.0	0.5349
13	14.81	85.0	0.4708
14	14.81	100.25	0.3993
15	$R_{1,gax}$ - $G_{2,gax}$ gax heat exchange		

Fig. 8.20 Full gax process study case of the double truncated cooling ARC typical of high temperature TTRC heating process and the joined data table

The two-stage heating results are plotted for *TWRC* and *TTRC* in Figs. 8.26a, b, respectively, for $T_c = (61-91)$ °C, while those of cogeneration are given in Fig. 8.27. Again, it results that both *TWRC* and *TTRC* methods applied to the standard heat pump configuration increase its effectiveness, but to a higher extent in this case, i.e. by ~5–30 %, as Figs. 8.26a, b show. Also, from Figs. 8.26 it results that this time *TWRC* performs slightly better than *TTRC*, therefore its COP_{cogen} is higher, comparatively, as Fig. 8.27 indicates. However, the *TTRC*

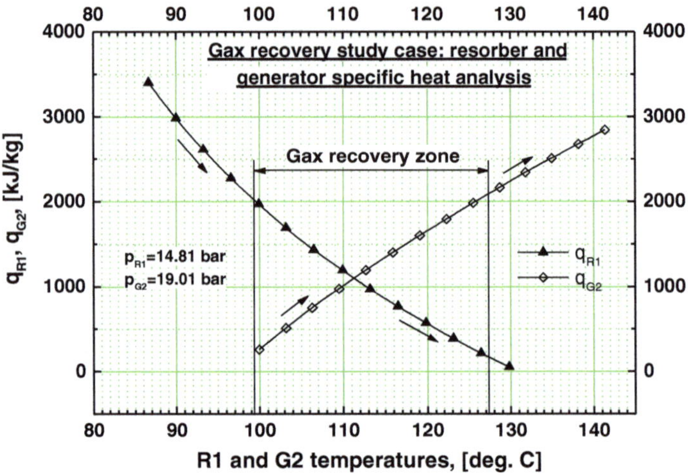

Fig. 8.21 q-plot high temperature gax recovery, Fig. 8.20 study case

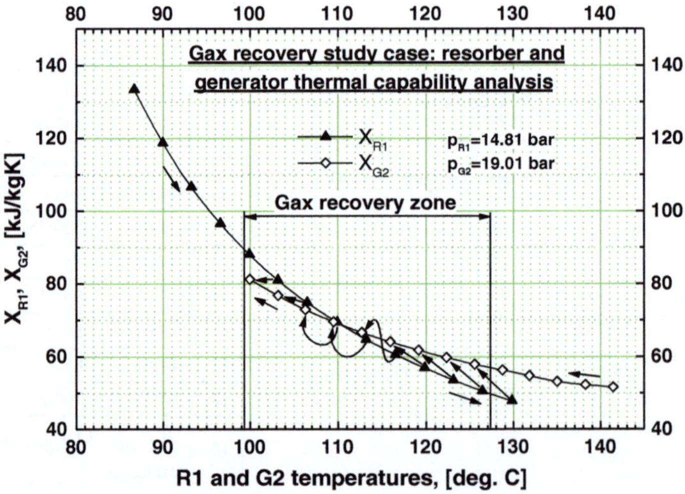

Fig. 8.22 X-plot high temperature gax recovery, Fig. 8.20 study case

effectiveness can be improved further. For the sake of completeness, in Figs. 8.28 and 8.29 are plotted the theoretical discharge gas temperatures of the single-stage and double-stage TWRC and TTRC heating and cooling, respectively, versus the evaporator and desorber temperatures.

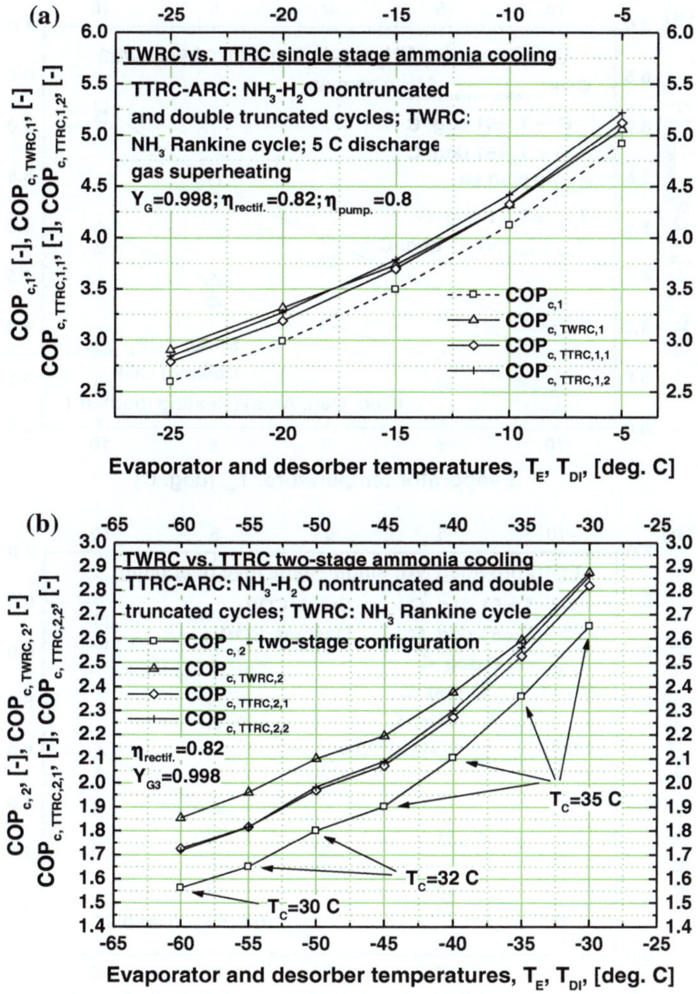

Fig. 8.23 **a** *TWRC* and *TTRC* ammonia single-stage cooling model results, **b** *TWRC* and *TTRC* ammonia two-stage cooling model results

Appendix 1

(i) First, we shall derive an equivalent expression of $\eta_C(\xi)$, Eq. (8.9), following the way covered in Appendix 1. In our case, Eqs. (A1.1) and (A1.2) become:

$$ln\frac{T_{2a}}{T_1} = ln\frac{1+x}{1-x} = f(x) \tag{A8.1}$$

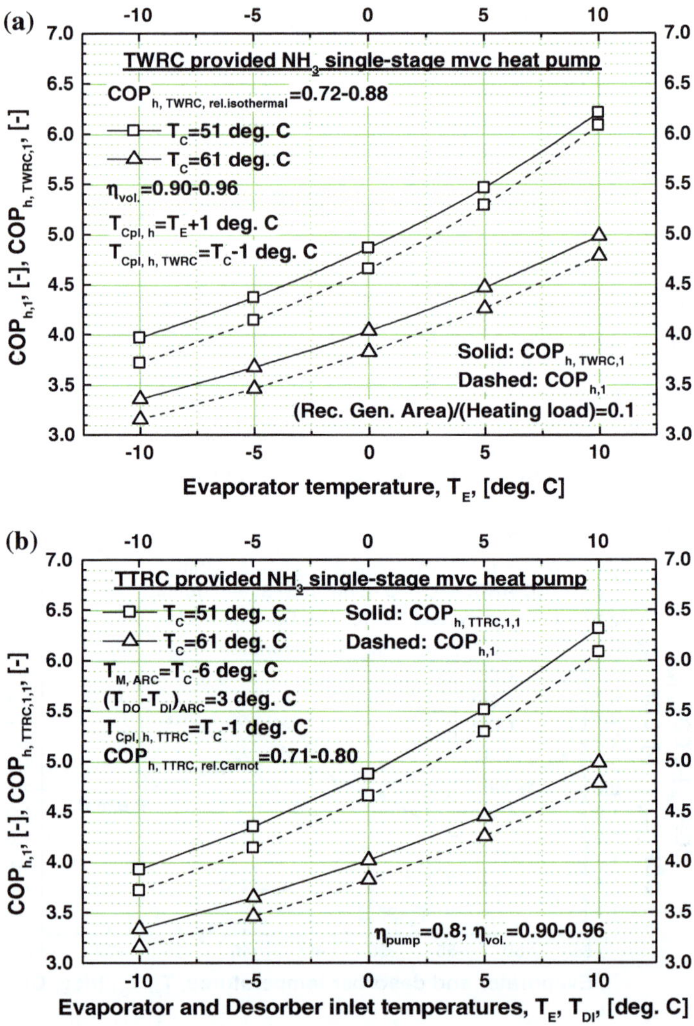

Fig. 8.24 Single-stage cogeneration model results: **a** *TWRC*; **b** *TTRC*

and

$$x = \frac{T_{2a} - T_1}{T_{2a} + T_1}, \quad x \in [0, 1), \; x \ll 1. \tag{A8.2}$$

respectively. Bearing in mind Eq. (A1.7), Eq. (A1.8) becomes in this case:

$$ln\frac{T_{2a}}{T_1} = 2\frac{T_{2a} - T_1}{T_{2a} + T_1} \tag{A8.3}$$

and Eq. (8.9) is rewritten in an equivalent form as follows:

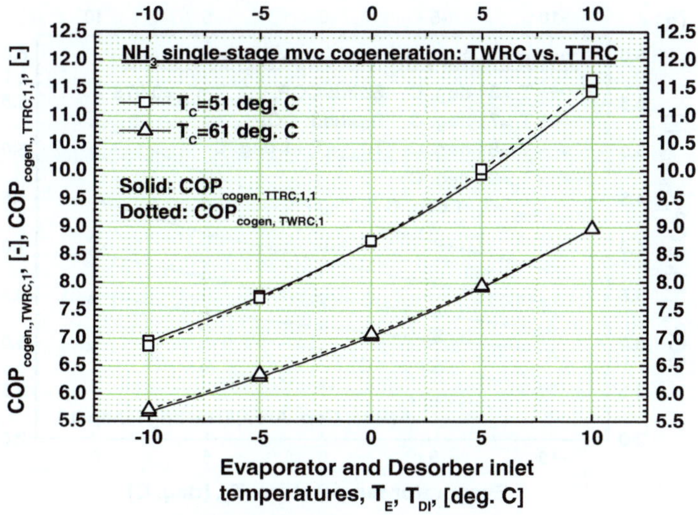

Fig. 8.25 Single-stage *TWRC* and *TTRC* cogeneration model results

or

$$\eta_{C,TFC} = 1 - \frac{T_1}{\frac{T_{2a}+T_1}{2}} \tag{A8.5}$$

Equation (A8.5) can be arranged also as:

$$\eta_{C,TFC} = \frac{T_{2a} - T_1}{T_{2a} + T_1} \tag{A8.6}$$

which is nothing else but the ideal thermal efficiency of the motor (direct) cycle named Trilateral Flash Cycle (TFC) (see Sect. 8.2.1).

(ii) Second, let us consider the ideal TFC, which recovers from the heat source the discharge gas superheat q_a, rejects to the sink source the amount of heat $-|q_i| = -|l_i|$ (Popa and Vintila 1977, p. 373, Table 15.3) and produces the work $l_{a,r} = |l_a| - |l_i|$, Eq. (8.26). Combining Eq. (8.10), written as:

$$l_{a,r} = \eta_{C,TFC} q_a \tag{A8.9}$$

with Eqs. (A8.5) and (8.26), it results:

$$\left(1 - \frac{T_1}{\frac{T_{2a}+T_1}{2}}\right) q_a = |l_a| - |l_i| \tag{A8.10}$$

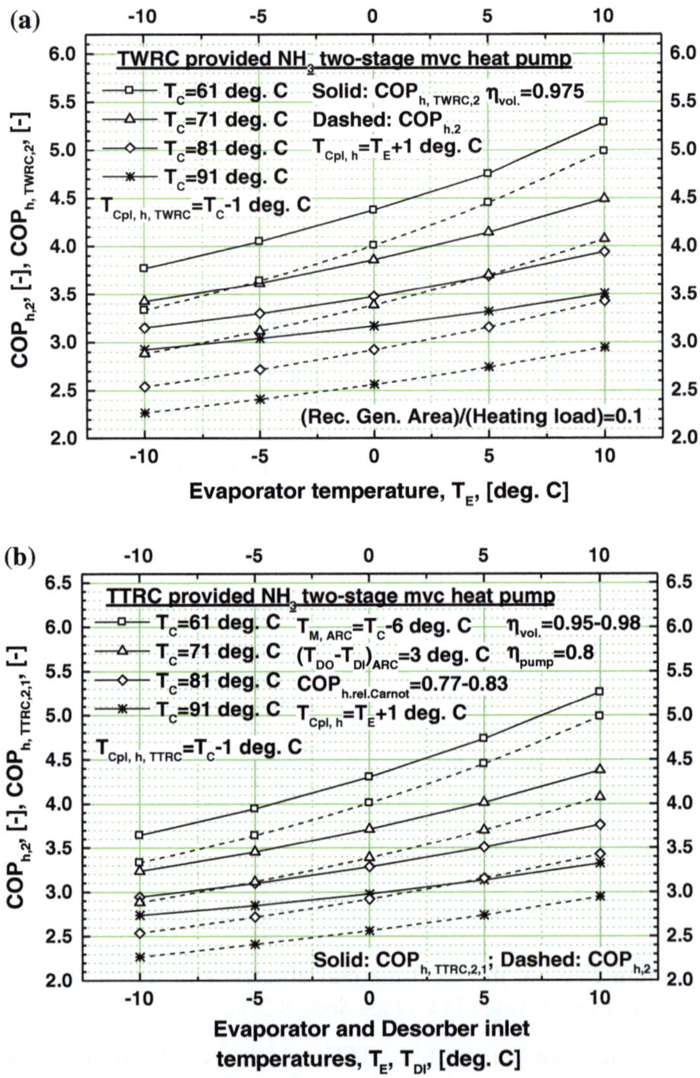

Fig. 8.26 Two-stage heating model results: **a** *TWRC*; **b** *TTRC*

Taking into account that $|l_a| = |q_a|$, Eq. (8.25), and that $-|q_i| = -|l_i|$ (see above), Eq. (A8.10) becomes:

$$\frac{q_a}{\frac{T_1 + T_{2a}}{2}} - \frac{|q_i|}{T_1} = 0 \tag{A8.11}$$

which proves that TFC entropy variation is null when is completely covered, $\Delta s = s_h - s_s = 0$, therefore the considered TFC is reversible [see Eqs. (1.86) and

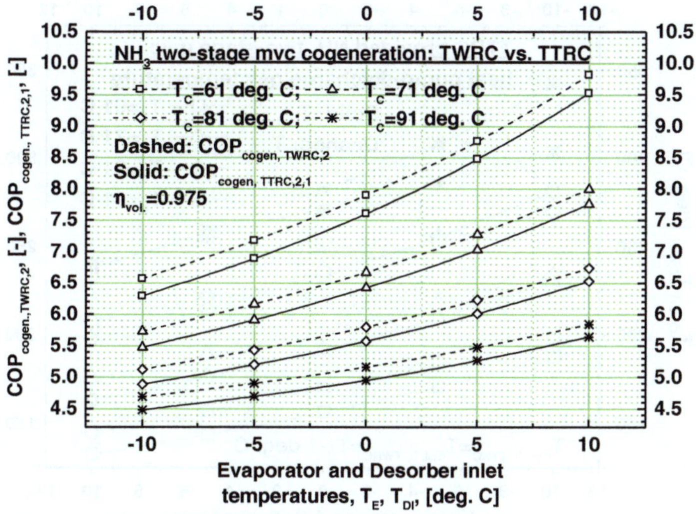

Fig. 8.27 Two-stage *TWRC* and *TTRC* cogeneration model results

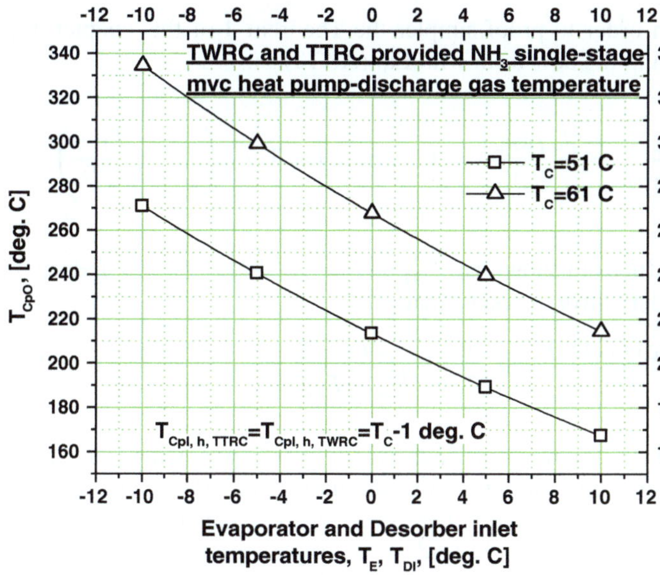

Fig. 8.28 TWRC and TTRC single-stage discharge gas temperatures

(1.88) as well]. From Eq. (A8.11) it results also that the heat source delivers in an equivalent way the heat q_a to the arithmetical mean temperature $\frac{T_1+T_{2a}}{2}$, as we expected.

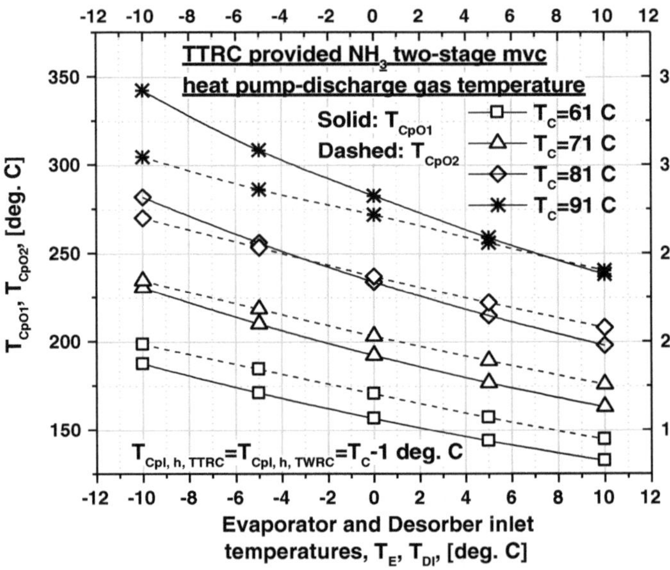

Fig. 8.29 TWRC and TTRC two-stage discharge gas temperatures

(iii) Third, in order to proof simpler the Theorem 4 of this chapter, Eq. (8.98) is rewritten in the new form with Eq. (A8.12) as:

$$\frac{COP_c}{COP_{c,TWRC}} = \frac{\left|l_{a,TWRC}\right|}{|l_a|} = 1 - \varepsilon\eta_C(\xi) = 1 - \varepsilon\left(1 - \frac{T_1}{\frac{T_{2a}+T_1}{2}}\right) = f(T_1, T_{2a}) \tag{A8.12}$$

and it suffices to show that $\frac{\partial f(T_1,T_{2a})}{\partial T_{2a}} = -\varepsilon\frac{\partial\eta_C(T_1,T_{2a})}{\partial T_{2a}} < 0$. Indeed, performing the last derivative, it results that:

$$-\varepsilon\frac{\partial\eta_C(T_1,T_{2a})}{\partial T_{2a}} = \varepsilon\frac{\partial}{\partial T_{2a}}\left(\frac{2T_1}{T_{2a}+T_1}\right) = -\varepsilon\frac{2T_1}{(T_{2a}+T_1)^2} < 0 \text{ q.e.d.} \tag{A8.13}$$

(iv) The classic compression saves input work, practicing staged compression with interstage cooling of the discharge gas (Popa and Vintila 1977). In Fig. 8.30 we plotted a compression process benefiting of two stages, with intermediary cooling, instead of a single one. The intermediary pressure p_x, $p_1 < p_x < p_2$ is established according to the known relation:

$$p_x = \sqrt{p_1 p_2}, \tag{A8.14}$$

which ensures the minimum work consumption in two-stage compression (Popa and Vintila 1977). The single-stage compression is the adiabatic transformation, represented in Fig. 8.30 by the adiabat $T_1 - T_{2a}$. In case of the

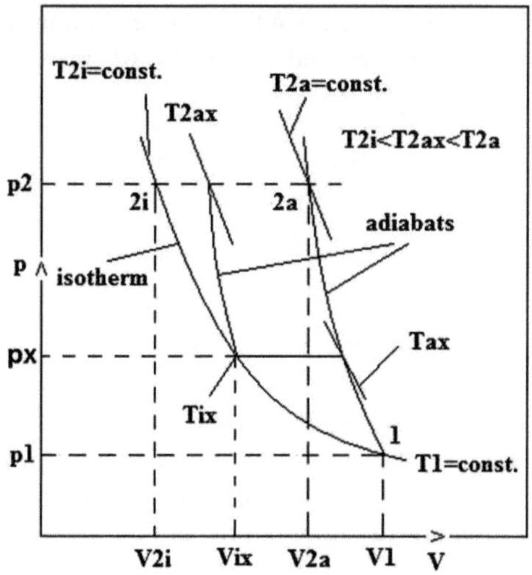

Fig. 8.30 Single-stage versus two-stage compression

two-stage compression, the first and second stage compressions are the adiabatic transformations $T_1 - T_{ax}$ and $T_{ix} - T_{2ax}$, respectively. The work savings is that corresponding to the $T_{ax} - T_{ix} - T_{2ax} - T_{2a} - T_{ax}$ area comprised between the two adiabats $T_{ax} - T_{T_{2a}}$ and $T_{ix} - T_{2ax}$ and the two isobar p_x and p_2 in the $p - V$ diagram. The relationship (A8.14) is valid for the compression practiced so far, without TWRC. The next lemma will prove that Eq. (A8.14) remains valid even in case of TWRC.

Lemma 2 *A two-stage adiabatic mechanical vapor compression process, provided with TWRC and operating between the initial and final pressures* p_1 *and* p_2*, respectively, has the same value of the intermediary pressure ensuring the minimum work consumption as that of the compression process without TWRC, that is* $p_x = \sqrt{p_1 p_2}$, $p_1 < p_x < p_2$.

Proof From considering Eqs. (8.27)–(8.30), the work of the single-stage compression provided with TWRC results in:

$$\left|l_{a,TWRC}\right| = |l_a|\left[1 - \varepsilon\left(1 - \frac{\ln\frac{T_{2a}}{T_1}}{\frac{T_{2a}}{T_1} - 1}\right)\right] \tag{A8.15}$$

where $|l_a|$ is given by Eq. (8.14):

$$|l_a| = \frac{k}{k-1} p_1 V_1 \left[\left(\frac{p_{2a}}{p_1}\right)^{\frac{k-1}{k}} - 1\right] \tag{A8.16}$$

Combining the two Eqs. (A8.15) and (A8.16), it is obtained:

$$\left|l_{a,TWRC}\right| = \frac{k}{k-1} p_1 V_1 \left[\left(\frac{p_{2a}}{p_1}\right)^{\frac{k-1}{k}} - 1\right]\left[1 - \varepsilon\left(1 - \frac{\ln\frac{T_{2a}}{T_1}}{\frac{T_{2a}}{T_1} - 1}\right)\right] \tag{A8.17}$$

The single-stage TWRC provided compression work of Eq. (A8.17) is written in case of the first and second TWRC provided stages works as follows:

$$\left|l_{a,TWRC,I}\right| = \frac{k}{k-1}p_1V_1\left[\left(\frac{p_x}{p_1}\right)^{\frac{k-1}{k}}-1\right]\left[1-\varepsilon\left(1-\frac{ln\frac{T_{ax}}{T_1}}{\frac{T_{ax}}{T_1}-1}\right)\right] \quad (A8.18)$$

and

$$\left|l_{a,TWRC,II}\right| = \frac{k}{k-1}p_1V_1\left[\left(\frac{p_{2a}}{p_x}\right)^{\frac{k-1}{k}}-1\right]\left[1-\varepsilon\left(1-\frac{ln\frac{T_{ax}}{T_{ix}}}{\frac{T_{ax}}{T_{ix}}-1}\right)\right] \quad (A8.19)$$

respectively. In Eq. (A8.19), the product p_1V_1 was used instead of p_xV_{ix} because the points (T_1,p_1) and (T_{ix},p_x) belong to the same isothermal curve $T_1 = T_{ix} = T_{2i} = const.$, and along of which we have $p_1V_1 = p_xV_{ix}$, Fig. 8.30. The ratios of temperatures in Eqs. (A8.18) and (A8.19) are written with the help of Eq. (8.18), valid along of an adiabat:

$$\frac{T_{ax}}{T_1} = \left(\frac{p_x}{p_1}\right)^{\frac{k-1}{k}} \quad (A8.20)$$

and

$$\frac{T_{2ax}}{T_{ix}} = \left(\frac{p_{2a}}{p_x}\right)^{\frac{k-1}{k}} \quad (A8.21)$$

respectively. Introducing Eqs. (A8.20) and (A8.21) in Eqs. (A8.18) and (A8.19), it is obtained;

$$\left|l_{a,TWRC,I}\right| = \frac{p_1V_1}{y}\left[\left(\frac{p_x}{p_1}\right)^y-1\right]\left[1-\varepsilon\left(1-\frac{yln\left(\frac{p_x}{p_1}\right)}{\left(\frac{p_x}{p_1}\right)^y-1}\right)\right] \quad (A8.22)$$

and

$$\left|l_{a,TWRC,II}\right| = \frac{p_1V_1}{y}\left[\left(\frac{p_{2a}}{p_x}\right)^y-1\right]\left[1-\varepsilon\left(1-\frac{yln\left(\frac{p_{2a}}{p_x}\right)}{\left(\frac{p_{2a}}{p_x}\right)^y-1}\right)\right] \quad (A8.23)$$

where $\frac{k-1}{k} = y$. Further, Eqs. (A8.22) and (A8.23) are added term by term, resulting:

$$\begin{aligned}\left|l_{a,TWRC,I+II}\right| &= \left|l_{a,TWRC,I}\right| + \left|l_{a,TWRC,II}\right| \\ &= \frac{p_1V_1}{y}\left\{(1-\varepsilon)\left[\left(\frac{p_x}{p_1}\right)^y+\left(\frac{p_{2a}}{p_x}\right)^y\right]+\varepsilon yln\frac{p_{2a}}{p_1}\right\} \quad (A8.23)\end{aligned}$$

The searched value of p_x is found solving equation below:

$$\frac{d}{dp_x}\left|l_{a,TWRC,I+II}\right| = 0 \tag{A8.24}$$

or,

$$\frac{d}{dp_x}\left|l_{a,TWRC,I+II}\right| = \frac{d}{dp_x}\left[\left(\frac{p_x}{p_1}\right)^y + \left(\frac{p_{2a}}{p_x}\right)^y\right] = 0 \tag{A8.25}$$

Performing the derivation in Eq. (A8.25), it results:

$$y\left(\frac{p_x}{p_1}\right)^{y-1}\frac{1}{p_1} - y\left(\frac{p_{2a}}{p_x}\right)^{y-1}\frac{p_{2a}}{p_x^2} = 0 \tag{A8.26}$$

wherefrom the lemma is proved, ($p_{2a} = p_2$):

$$p_x = \sqrt{p_1 p_2}, \text{ q.e.d.} \tag{A8.27}$$

Bearing in mind that both mechanical vapor compression processes, without and with TTRC method application, have the same work input [see Eq. (8.3)], and taking into account the lemma 2 result, it is concluded that:

Corollary 4 *Both TWRC and TTRC methods preserve the same relationship of the intermediary pressure calculus, in two-stage mechanical vapor compression,* Eq. (A8.27), *in order to minimize the work input.*

References

Belozerov, G., Mednikova, N., Pytchenko, V., & Serova, E. (2007). Cascade type refrigeration systems working on CO_2/NH_3 for technological processes of products freezing and storage. In *Proceedings of Ohrid 2007 IIR/IIF Conference*, April 19–21, Macedonia.

Bryson, M. J. (2007). The conversion of low grade heat into electricity using the thermosyphon rankine engine and trilateral flash cycle. PhD. thesis, School of Aerospace, Mechanical and Manufacturing Engineering Science, Engineering and Technology Portfolio RMIT University, August 2007.

Chicco, G., & Mancarella, P. (2009). Distributed multi-generation: A comprehensive view. *Renewable and Sustainable Energy Reviews, 13*, 535–551.

Cox, N., Mazur, V., & Colbourne, D. (2009). The development of azeotropic ammonia refrigerant blends for industrial process applications. In *Proceedings of Ohrid 2009 IIR/IIF Conference*, May 6–8, Macedonia.

Gaina, S., Campu, E., & Bucur, G. (1966). *Collection of problems of differential and integral calculus* (Vol. 2). Bucharest (in Romanian): Technical Publishing House.

Kirilin, V. A., Sicev, V. V., & Seindlin, A. E. (1985). *Thermodynamics*. Bucharest: Scientific and Technical Publishing House, 540p (in Romanian).

Kotthoff, S., & Gorenflo, D. (2008). Pool boiling heat transfer to hydrocarbons and ammonia: A state of the art review. *International Journal of Refrigeration, 31*(4), 573–602.

Popa, B., & Vintila, C. (1977). *Thermotechnics and thermal machines*. Bucharest: The Didactic and Pedagogic Publishing House. (in Romanian).

de Mulder, P. (2004). Vertical freezers with CO_2, Doncaster, *Star Refrigeration*, Glasgow, U.K.
Radcenco, V., Florescu, A., Duicu, T., et al. (1985). *Heat Pump Installations*. Bucharest: The Technical Publishing House. (in Romanian).
Nicolescu, M. (1977–1980). *Mathematical Analysis Vol. I, II*. Bucharest: The Didactic and Pedagogic Publishing House.
Niebergall, W. (1959). *Handbuch der Kältetechnik, Band 7, Sorptions—Kältemaschinen*. Berlin: Springer.
Smith, I. K. (1993). Development of the trilateral flash cycle system part 1: Fundamental consideration. *Proceedings of the Institution of Mechanical Engineers, Part A: Journal of Power and Energy, 207*(3), 179–194.
Smith, I. K., & da Silva, R. P. M. (1994). Development of the trilateral flash cycle system part 2: Increasing power output with working fluid mixtures. *Proceedings of the Institution of Mechanical Engineers, Part A: Journal of Power and Energy, 208*, 135–144.
Smith, I. K. (1997). Energy recovery from two-phase expansion by means of twin screw machines. In *Private Conference*.
Staicovici, M. D. (2006) Coabsorbent cycles. In *Proceedings of Gustav Lorentzen Conference on Natural Refrigerants* (pp. 219–222), Trondheim, Norway, May 28–31.
Staicovici, M. D. (2009a). Coabsorbent cycles. Part one: Theory. *International Journal of Thermal Sciences, 48*, 626–644.
Staicovici, M. D. (2009b). IPAE & VE offer prospectus I–IV—proposal for cooperation, Sent to the GRASSO GEA (Holland), AXIMA Refrigeration (France), BITZER (Germany), DORIN (Italy), EMBRACO (Brasil), SABROE (Danemark) and HOWDEN (UK) compressor manufacturers on 08.10.2009, Stuttgart, Germany.
Staicovici, M. D. (2011a). A method of improving the effectiveness of a mechanical vapor compression process and of its applications in refrigeration. *International Journal of Heat Mass Transfer, 54*(2011), 1752–1762. doi: 10.1016/j.ijheatmasstransfer.2011.01.016.
Staicovici, M. D. (2011b). A thermodynamic approach of mechanical vapor compression refrigeration COP increase. Part I: Methods and ideal limits. In *Proceedings of the 4th IIR Conference: Ammonia Refrigeration Technology*, Ohrid.
Staicovici, M. D. (2011c). A thermodynamic approach of mechanical vapor compression refrigeration COP increase. Part II: Thermal to work recovery method applications. In *Proceedings of the 4th IIR Conference: Ammonia Refrigeration Technology*, Ohrid.
Staicovici, M. D. (2011d). Climbing the thermodynamic effectiveness summits of mechanical vapor compression cooling through discharge gas heat conversion: a theoretical approach. In *Proceedings of the ICR 2011*, Prague.
Staicovici, M. D. (2012). Thermal-to-work and thermal-to-thermal recovery compression heat pumping. In *Proceedings of the Gustav Lorenzen Conference on Natural Refrigerants*, Delft, The Netherlands.
Stamatescu, C. (1972). *The cooling technique* (Vol. 1). Bucharest: The Technical Publishing House. (in Romanian).
Stamatescu, C., Peculea, M., Radcenco, V., Porneala, S., & Barbu, H. (1982). *Technical cryogenics*. Bucharest: The Technical Publishing House.

Chapter 9
A Non-equilibrium Phenomenological Two-Point Theory of Mass and Heat Transfer in Physical and Chemical Interactions

In this chapter, absorption and related phenomena will be analyzed with the help of a non-equilibrium phenomenological theory of mass and heat transfer, elaborated and published in past works by the author (Staicovici 1998a, b, 2000a, b, c, 2002).

9.1 Application to NH_3–H_2O and Other Working Systems

The NH_3–H_2O system is the most common working combination in the thermal absorption technology. Although there is quite a comprehensive theoretical and practical experience in the use of such combination, the gas–liquid interaction in this medium is, most surprisingly, still dealt with empirically. The analogy with heat transfer is commonly used for the calculation of interface mass transfer,

$$N = k\Delta \tag{9.1}$$

In Eq. (9.1), N is the mass flow transferred as a result of the action of the motive force Δ. The *motive force is equal to the deviation from the equilibrium conditions.* The proportionality factor k is the mass transfer coefficient. According to the well known concept of the "two films" (Lewis and Whitman 1924), N is proportional to the difference of the concentrations of the gas or liquid phases and to the global mass transfer coefficients of the phases:

$$N = K_g(y - y^*) = K_l(x - x^*) \tag{9.2}$$

In Eq. (9.2), x^* and y^* are the concentrations of the liquid and gas phases in equilibrium with the gas phase of concentration y and the liquid phase of concentration x, respectively. The coefficients K_g and K_l are expressed as functions of the global resistance to mass transfer, which have additive properties to the partial resistance of the phases, similar to the heat transfer. More often than not it is considered that the mass transfer is controlled by the phase with the greatest partial resistance. In order to decrease the global resistance, the dispersion of the process controlling a phase is usually used. That is why, according to the classic interpretation, it is implicitly admitted that *the* NH_3–H_2O *gas–liquid interaction is a surface phenomenon.*

M.-D. Staicovici, *Coabsorbent and Thermal Recovery Compression Heat Pumping Technologies*, Heat and Mass Transfer, DOI: 10.1007/978-3-642-54684-6_9,

The literature referring to mass transfer has mainly dealt with the dependence of transfer coefficients on molecular diffusivity. This approach includes "the two films" concept referred to above, as well as the theory of penetration (Higbie 1935), and renewal (Danckwerts 1954). The specialized literature has paid less attention to the relation between the absorption rate and the hydrodynamic conditions. So far it is experimentally proved that in certain given conditions there is a Reynolds analogy in case of the mass transfer

$$Sh = a_m \mathrm{Re}^{b_m} Sc^{c_m} \tag{9.3}$$

and

$$Nu = a_h \mathrm{Re}^{b_h} Pr^{c_h} \tag{9.4}$$

in the case of heat transfer, in two-phase gas–liquid flow in pipes (Hobler 1966; Ciborowski and Richlicki 1971). Equation (9.3) does not hold, for instance, when shear stresses exist at the gas–liquid interface, but successfully applies for the gas phase controlled mass transfer in wetted-wall columns.

Several authors have studied the heat and mass transfer in the NH_3–H_2O system (e.g. Ciborowski and Richlicki 1971; Anderson et al. 1964; Heuss et al. 1965; Chadzidakis 1971; Hera 1987; Chiriac et al. 1995; Mihaila et al. 1995). Most of them express their results by means of correlations similar to Eqs. (9.3) and (9.4). A detailed study was conducted by Keizer (1982) for two-phase NH_3–H_2O flow of the slug type in vertical pipe absorbers. For the purpose of evaluating mass transfer, first he made use of prediction correlations of partial coefficient k_l, which are due to past works by others (e.g. Banerjee et al. 1969, 1970; Gregory and Scott 1969; Jepsen 1970; Jagota et al. 1973; Kulic and Rhodes 1974; Kasturi and Stepanek 1974). Comparing his experimental data with the ones predicted and finding that they sensibly differ, sometimes even by an order of magnitude, Keizer places under a question mark the applicability of the correlations of the authors referred to above to the prediction of mass transfer in the case of the NH_3/H_2O system and comes to the conclusion that it cannot be precisely determined whose partial resistance, respectively that of the liquid phase or that of the gas phase, controls the NH_3–H_2O absorption process. Finally his correlations are satisfactory by adapting Eq. (9.3).

In an attempt to find a more accurate method of prediction of NH_3–H_2O gas–liquid interactions, the theory at hand will take into consideration the non-equilibrium thermodynamics (de Groot and Mazur 1964). Although it seems to be a natural direction of investigation, it has not been used in the analysis of the NH_3–H_2O system so far.

Herein below a simple experiment made by the author will be described, an experiment which strengthen the conviction that the method of approaching the interactions which form the main object of this study should be reviewed. The experiment consists in injecting a single gaseous ammonia bubble into the NH_3–H_2O small molar fractions solution at normal pressure and ambient temperature. Based upon the numerous observations recorded, it has been found out that, according to the gas phase feeding rate, the evolution of the bubble may be characterized as follows: (i) at low and moderate feeding rates (10^{-8}–10^{-7} kg/s), the dynamics of the bubble takes place in two distinct stages: (a) *growth*, when the bubble progressively expands its volume up to a maximum value, concurrently with gas absorption, followed by (b) *collapse,* when the bubble is rapidly absorbed, reducing its volume up to zero, although it continues to be fed with gas through the injection nozzle; throughout its evolution the bubble does not detach itself from the nozzle and has an almost spherical shape; in the case of a continuous gas feeding, the phenomenon becomes quasi-periodical; the frequency of complete evolution grows up with the increase of the flow rate; the duration of the growth stage is much longer than the one of the collapse stage; (ii) at high flow rates, the collapse does no longer occur, but a phenomenon of oscillation of the bubble volume between two values, a maximum and a minimum one, is noticed; the bubble is no longer spherical but elongates in the direction of the nozzle axis and has an irregular contour. The study of the bubble in its complete *growth-collapse* evolution gives rise to two questions which are difficult to answer by means of the current theory of interface mass transfer and absorption as a surface phenomenon, synthesized by Eqs. (9.1) and (9.2) (*the problem of the ammonia bubble absorption*, Staicovici 1998a): (i) if during the *growth* phase, as the bubble increases its volume, the molar fraction x of the interface approaches the saturation point so that the difference in the concentration of phases ($x^* - x$) and implicitly the current absorbed decrease (see Eq. 9.2), then how could be explained the occurrence of the *collapse* phase, which, quite on the contrary, assumes a continuous increase of the current absorbed in order to exceed the relatively constant value of the feeding flow rate? (ii) if absorption is considered to be a surface phenomenon, then, yet again, how could *collapse* be explained, a process during which the current absorbed, proportional to the gas–liquid contact area, decreases by the radius second power, instead of increasing as noted in the first question above? As it will be further demonstrated, these questions may be accurately answered by applying a non-equilibrium phenomenological theory, outlined in this chapter, to the interface of the subject interactions. The theory is not contradictory with respect to the classic equilibrium phenomenological approach, but it completes it, highlighting new important features of the mass and heat transport near an equilibrium state, like the ideal point approaching effect, analyzed further in this work.

9.1.1 A Non-equilibrium Phenomenological Approach of the Coupled Mass and Heat Transfer in Physical Mono-, Bi- and Particular Polycomponent Gas–liquid Interactions

In the general case, the interacting polycomponent, biphase evolution phenomena, of gas–liquid, solid–liquid, solid–gas or gas–gas, liquid–liquid type, may be considered. For the sake of clarity, here we shall use the notation for the binary physical interactions of gas–liquid type. The liquid phase may additionally contain other dissolved non-volatile components. The state parameters characterizing the interface, i.e. temperature, pressure and molar fraction, marked as T_g, p_g and y for gas and as T_l, p_l and x for liquid, are time functions. At interface, the surface tension forces, given by Laplace equation., or in a new approach by equality (9.5) (see Chap. 10, Staicovici 1998b)

$$div\vec{\sigma} = \Delta p \tag{9.5}$$

where $\vec{\sigma}$ is a normal positive vector to the surface and equal to the surface tension of the liquid, are generated. The two-phase system is considered to be open. It evolves towards the state of equilibrium through a finite sequence of stationary states. It is equally assumed that the total energy of the system is approximately equal to its internal energy, that is its macroscopic kinetic energy, the external fields of any nature which act on it as well as its internal tensions are negligible. Under these circumstances, when the mechanical work exchanged by the system with the exterior is negligible, the heat exchanged between the system and environment and between its various areas is sensibly equal to the total enthalpy exchange. Here a coupled mass and heat transfer process means a process involving the existence of (a) a phase (nature) change of components (reactants), (b) an interaction heat and (c) simultaneous mass and heat currents influencing each other. The gas–liquid interaction involves the simultaneous existence of coupled mass and heat currents "*j*", noted by indices "*i*" and respectively "*q*", of gas and liquid, marked by indices "*g*" and, respectively, "*l*", which get in and out of the system, marked as prime ($'$) and respectively secondary ($''$). The coupled currents are vectors, but here they will be treated as scalars, without loss of much information. Additionally, because energy is carried by the mass currents, they are considered the coupling currents and those of energy (heat etc.) are the coupled ones. Finally, the classic notion of "equilibrium" point (state) will be referred to as ideal point (state).

The entropy source, $\dot{S}$, shows the contribution of the currents to the global entropy yield. In a stationary regime this is written (see Eq. (1.145), Chartier et al. 1975):

$$\dot{S} = -\frac{\dot{E}}{T_0} = -\left(\sum j'_i s'_i + \sum j''_i s''_i\right) - \left(\sum \frac{j'_q}{T'} + \sum \frac{j''_q}{T''}\right) \tag{9.6}$$

where $\dot{E}$ is the exergy dissipation velocity and T_0 is the temperature of the infinite reservoir. Dimensionally j'_i and j'_q are expressed in mol/s or kg/s and in W, respectively. Equation (9.6) is particularized to two opposed elementary processes which are permanently encountered in the heat absorption technology: (i) the gas interacts with the liquid, resulting in a liquid with new properties (absorption); (ii) the liquid decomposes in liquid and gas with new properties (generation). Taking the above mentioned remarks into account, as well as the common conventional signs in thermodynamics, Eq. (9.6) can be rewritten for case (i) as

$$\dot{S} = -\left(j'_{i,g} \cdot s'_g + j'_{i,l} \cdot s'_l - j''_{i,l} \cdot s''_l\right) - \left[j'_{q,g} \cdot \frac{1}{T'_g} + j'_{q,l} \cdot \frac{1}{T'_l} - \left(j''_{q,l} + j_q\right) \cdot \frac{1}{T''_l}\right] \tag{9.7}$$

where j_q is a heat current, considered as an excess element resulting out of the deviation from the ideal state of the mixture. The following currents balances are considered in the system process heat:

$$j'_{q,g} + j'_{q,l} = j''_{q,l} + j_q \tag{9.8}$$

total mass:

$$j'_{i,g} + j'_{i,l} = j''_{i,l} \tag{9.9}$$

mass for one of the species:

$$\mathrm{y}\mathrm{j}'_{\mathrm{i,g}} + \mathrm{x}\mathrm{j}'_{\mathrm{i,l}} = (\mathrm{x} + \delta\mathrm{x}) \cdot \mathrm{j}''_{\mathrm{i,l}} \tag{9.10}$$

In Eq. (9.10) δx is the finite variation of molar fraction, suffered by the liquid as a result of its interaction with the gas phase. The heat flows are extensive elements. Taking into account the previous observation concerning the calculation of the heat exchange by means of total enthalpy, they can be written as:

$$\begin{aligned} j'_{q,g} &= j'_{i,g} \cdot h'_g \\ j'_{q,l} &= j'_{i,l} \cdot h'_l \\ j''_{q,l} &= j''_{i,l} \cdot h''_l \end{aligned} \tag{9.11}$$

Taking into consideration Eqs. (9.9) and (9.10), the first bracket of Eq. (9.7) can be written as:

$$-\left(j'_{i,g} \cdot s'_g + j'_{i,l} \cdot s'_l - j''_{i,l} \cdot s''_l\right) = -j'_{i,g}\left[\left(s'_g - s'_l\right) - (y - x)\frac{s''_l - s'_l}{\delta x}\right] \tag{9.12}$$

It is noted that the first bracket of Eq. (9.7) represents the contribution of the mass currents to the entropy source only, therefore the ratio in the right side of Eq. (9.12) becomes:

$$\frac{s''_l - s'_l}{\delta x} \cong \left(\frac{\partial s_l}{\partial x}\right)_{p,T} \tag{9.13}$$

The temperature T''_l is related to the finite variation of liquid temperature due to the deviation from the ideal state $\delta T = T''_l - T'_l$ by:

$$\frac{1}{T''_l} = \frac{1}{T'_l\left(1+\frac{\delta T}{T'_l}\right)} = \frac{1}{T'_l}\left[1 - \frac{\delta T}{T'_l} + \left(\frac{\delta T}{T'_l}\right)^2 - \cdots\right] \tag{9.14}$$

Following simple calculations by means of Eqs. (9.14) and (9.8)–(9.11) the second bracket of Eq. (9.7) may be written as:

$$\begin{aligned} &-\left[j'_{q,g}\frac{1}{T'_g} + j''_{q,l}\frac{1}{T'_l} - \left(j''_{q,l} + j_q\right)\frac{1}{T''_l}\right] \\ &= -j'_{i,g}\left\{\left(\frac{1}{T'_g} - \frac{1}{T'_l}\right)h'_g + \frac{1}{T'_l}\left(h'_g - h'_l + \frac{h'_l}{j_r}\right)\frac{\delta T}{T'_l\left(1+\frac{\delta T}{T'_l}\right)}\right\} \end{aligned} \tag{9.15}$$

where

$$j_r = \frac{j'_{i,g}}{j''_{i,l}} \tag{9.16}$$

is the reduced mass current of gas (mol. mol^{-1}), involved in interaction. The ratio $\frac{\delta T}{T'_l}$ is calculated by means of the liquid specific heat equation at constant pressure:

$$c_{p,l} = \left(\frac{\partial h_l}{\partial T}\right)_{p,x} \tag{9.17}$$

resulting in:

$$\frac{\delta T}{T'_l} = \frac{\delta h'_l}{T'_l \cdot c_{p,l}} = A j_r \tag{9.18}$$

where the non-dimensional factor A in Eq. (9.18), herein called "reduced excess heat" is obtained with Eqs. (9.9), (9.10) and (9.16) and has the expression:

$$A = (y - x)\left(\frac{\partial h'_l}{\partial x}\right)_{p,T} \cdot \frac{1}{T'_l \cdot c_{p,l}} \tag{9.19}$$

Equations (9.12)–(9.19) are introduced in Eq. (9.7) and the entropy source is written as:

$$\dot{S} = j'_{i,g}(-X) \tag{9.20}$$

where

$$-X = -\left\{\left[\left(s'_g - s'_l\right) - (y - x)\left(\frac{\partial s'_1}{\partial x}\right)_{p,T}\right]\right.$$
$$\left. + h'_g\left(\frac{1}{T'_g} - \frac{1}{T'_l}\right) + \frac{A}{T'_l(1 + Aj_r)}\left[\left(h'_g - h'_l\right) \cdot j_r + h'_l\right]\right\} \tag{9.21}$$

Equation (9.20) points out the proportionality of the entropy source with the coupling gas mass current $j'_{i,g}$ and with factor $(-X)$, identified as a thermodynamic force and named here *non-equilibrium (natural) thermodynamic force*. This force differs from the forces governing the coupled heat and mass transfer in continuous media, originating from Fourier's and Fick's laws, respectively and established with equilibrium thermodynamics tools. It only depends on the entry values of the system, a reason for which the marking primary (′) is hereinafter given up to. Because it depends only to the interface *p*, *T*, *x* and *y* state parameters, the force is a way independent function, therefore a state function. According to Eqs. (9.8)–(9.11), the coupled heat current j_q has the expression

$$j_q = j'_{i,g} \cdot C \tag{9.22}$$

where C is the local thermal capacity (availability) of the binary mixture (Staicovici 1995), obtained from Eqs. (9.8) to (9.11):

$$C = \left(h'_g - h'_l\right) - (y - x)\left(\frac{\partial h'_1}{\partial x}\right)_{p,T} \tag{9.23}$$

This theory is confined to considering that the currents involved are linear functions of the thermodynamical forces. Later on, this consideration will be justified. Hence, the phenomenological equations are reduced to the simplest linear expression:

$$j'_{i,g} = L_{11}(-X) \tag{9.24}$$

where L_{11} is the strictly positive phenomenological coefficient of proportionality, having the dimension of mol^2 KJ^{-1} s^{-1}. In a first approximation L_{11} is considered to be constant, independent of *X*. With Eq. (9.24), the entropy source [Eq. (9.20)], acquires the definite positive form [see also Eq. (1.145)]

$$\dot{S} = L_{11}X^2 \tag{9.25}$$

for all interface *p*, *T*, *x* and *y* values, except the ideal points, given by [Eqs. (1.202)–(1.205)]:

$$p_l = p_g = p_e$$
$$T_l = T_g = T_e$$
$$\mu_{1(2),l} = \mu_{1(2),g}$$

where it is not defined, according to the non-equilibrium phenomenological approach outlined in this theory. It is useful to further proceed to the non-dimensioning of equations referred to above. In order to do this, the following reduced elements are considered

$$L_{11,r} = \frac{L_{11}}{\left(\frac{j''_{i,l}}{R}\right)}; X_r = \frac{X}{R}; \dot{S}_r = \frac{\dot{S}}{R j''_{i,l}}$$
$$h_r = \frac{h}{T_b R}; c_{p,r} = c_p/R; s_r = s/R \tag{9.26}$$
$$p_r = \frac{p}{p_b}; T_r = \frac{T}{T_b}$$

In Eq. (9.26), R, p_b and T_b are the gas constant, a reduction pressure and an absolute reduction temperature, respectively. As to the above mentioned, the previously deduced equations can be re-written in a reduced form (index "r") as follows

$$\dot{S}_r = j_r(-X_r) = L_{11,r} \cdot X_r^2 \tag{9.27}$$

$$j_r = L_{11,r}(-X_r) \tag{9.28}$$

$$X_r = \left[(s_{g,r} - s_{l,r}) - (y - x) \left(\frac{\partial \mathrm{s}_{\mathrm{l,r}}}{\partial \mathrm{x}} \right)_{p_r T_r} \right] + h_{g,r} \left(\frac{1}{T_{g,r}} - \frac{1}{T_{l,r}} \right)$$
$$+ \frac{A_r}{T_{l,r}(1 + A_r \cdot j_r)} \left[(h_{g,r} - h_{l,r}) \cdot j_r + h_{l,r} \right] \tag{9.29}$$

$$A_r = \left(\frac{\partial \mathrm{h}_{\mathrm{l,r}}}{\partial \mathrm{x}} \right)_{p_r T_r} \frac{y - x}{T_{l,r} \cdot c_{p,l,r}} \tag{9.30}$$

$$j_{q,r} = j_r \cdot C_r \tag{9.31}$$

where

$$C_r = (h_{g,r} - h_{l,r}) - (y - x) \left(\frac{\partial \mathrm{h}_{\mathrm{l,r}}}{\partial \mathrm{x}} \right)_{p_r T_r} \tag{9.32}$$

The partial reduced thermodynamical force of a purely mass nature, given by the first square bracket of the right side of Eq. (9.29) is marked as $\mathrm{X}_{\mathrm{i,r}}$:

$$X_{i,r} = (s_{g,r} - s_{l,r}) - (y - x) \left(\frac{\partial \mathrm{s}_{\mathrm{l,r}}}{\partial \mathrm{x}} \right)_{p_r T_r} \tag{9.33}$$

We have to remark here the similitude of Eqs. (9.32) and (9.33). Taking into account that $L_{11,r}$ must be in a first approximation independent of X_r, from

Eqs. (9.28) to (9.29) the reduced natural thermodynamical force is expressed in an implicit form as (Staicovici 1998a):

$$-X_r = -\frac{X_{i,r} + h_{g,r}\left(\frac{1}{T_{g,r}} - \frac{1}{T_{l,r}}\right) + \frac{A_r h_{l,r}}{T_{l,r}\left[1 + A_r L_{11,r}(-X_r)\right]}}{1 + \frac{A_r(h_{g,r} - h_{l,r})}{T_{l,r}\left[1 + A_r L_{11,r}(-X_r)\right]} M\left(p_{r,e}, x_e\right)} \tag{9.34}$$

where $M(p_{r,e}, x_e)$ is a new function discussed later in this chapter. It may be derived solving Eq. (9.34) with respect to $-X_r$, by means of a simple iterative method (Dodescu 1979), or as a second degree algebraic equation (see Appendix of this chapter). The set of ideal points separates the strictly positive forces ($-X_r > 0$), characterized by absorption currents, $j_r > 0$, from the strictly negative ones ($-X_r < 0$), a fact which gives rise to generating currents, $j_r < 0$. The force is useful in approaching interaction mass and heat currents of mono-, bi- and particular polycomponent gas–liquid interactions with non-ideal mixture, coupling it with Eqs. (9.28) and (9.31) and the classic hydro-gaso-dynamics. For instance, the two last types of interactions are encountered to the binary systems NH_3–H_2O, H_2O–LiBr, NH_3–$LiNO_3$, NH_3–NaSCN, CH_3NH_2–H_2O etc., or to the ternary ones, such as NH_3–H_2O–LiBr, CH_3NH_2–H_2O–LiBr etc.

The case of pure component interactions is important and it is worth paying a little bit more attention to. Indeed, Eq. (9.34) holds here too, simply considering that x and y take close values to 0, or 1. However, the reduced excess heat A_r and the term $(y - x)\left(\frac{\partial s_{l,r}}{\partial x}\right)_{p_r,T_r}$ in Eq. (9.33) are vanishing when x and y approach 0 or 1, so that Eq. (9.34) simplifies in case of pure components, namely:

$$-X_{r,p} = -\left(s_{g,r} - s_{l,r}\right) - h_{g,r}\left(\frac{1}{T_{g,r}} - \frac{1}{T_{l,r}}\right) \tag{9.35}$$

In Eq. (9.35), the two terms of the right member have different signs, but $-X_{r,p}$ is strictly negative and holds for vaporization:

$$\begin{aligned} -X_{r,p,v} &= -X_{r,p} \\ j_{r,v} &= L_{11,r}\left(-X_{r,p,v}\right) \end{aligned} \tag{9.36}$$

In case of condensation, the currents are changing the sign, therefore:

$$\begin{aligned} -X_{r,p,c} &= -\left(-X_{r,p}\right) \\ j_{r,c} &= L_{11,r}\left(-X_{r,p,c}\right) \end{aligned} \tag{9.37}$$

When the system approaches an ideal point, the second term in the right side of Eq. (9.35) vanishes and the force takes maximum absolute values, according to a postulate to follow. The fourth paragraph includes case studies for binary and pure component interactions.

The phenomenological coefficient may be theoretically estimated by making a simple imaginary experiment. Let us consider an isolated gaseous mixture bubble, of volume $V_g(t)$, at pressure $p_g(t)$ and temperature T_g, which interacts with the liquid phase surrounding it. For convenience, it is assumed that the bubble has an isotherm adiabatic evolution in the liquid. According to the first principle of thermodynamics, the variation of its internal energy within the time unit is equal to the power exchanged between the gas and its environment, namely

$$\dot{U}_g = p_g \dot{V}_g$$

or,

$$-\mathrm{j_r}\rho_l \left|\dot{V}_g\right| u_g = p_g \dot{V}_g \tag{9.38}$$

where $u_g = c_{v,g}T_g$ is the gas specific internal energy. Upon absorption $\dot{\mathrm{V}}_\mathrm{g} < 0$, and upon generation $\dot{\mathrm{V}}_\mathrm{g} > 0$ and Eq. (9.38) is simplified to the expression:

$$j_r = \pm \frac{p_g}{\rho_1 c_{v,g} T_g} \tag{9.39}$$

Equation (9.39) shows that positive reduced currents are obtained in absorption, and negative ones in generation processes, as previously mentioned herein. This equation is valid when the homobaricity condition in the bubble is met (see further explanation in this chapter). Physically, the above condition is met where the pressure from the wall is sensibly equal to the pressure in the rest of the gas volume and has a known value. In this case the mass currents have low values and anticipating the further expressed conclusions, are far from the state of equilibrium. The determination of the phenomenological coefficient is not affected by the position of the system relative to the point of equilibrium in respect to which it has been assumed that $L_{11,r}$ is independent of $\mathrm{X_r}$. $L_{11,r}$ is calculated by means of Eq. (9.28):

$$L_{11,r} = \frac{j_r}{-X_r} \tag{9.40}$$

where $\mathrm{j_r}$ and $\mathrm{X_r}$ are determined through an iterative process from Eqs. (9.39) and respectively (9.34) (Staicovici 1998a). A similar result is obtained if, instead of the energy equation, the mass balance in the bubble is written,

$$-j_r \rho_1 \left|\dot{\mathrm{V}}\right| = \rho_\mathrm{g} \dot{\mathrm{V}} \tag{9.41}$$

resulting in:

$$j_r = \pm \frac{\rho_\mathrm{g}}{\rho_1} \tag{9.42}$$

For an ideal gas, Eq. (9.43) holds true:

$$\rho_\mathrm{g} = \frac{p_g}{RT_g} \tag{9.43}$$

wherefrom, the reduced mass flow acquires the form:

$$j_r = \pm \frac{p_g}{\rho_l R T_g} \tag{9.44}$$

which is equally valid and useful in homobaricity conditions.

9.1.2 A Non-equilibrium Phenomenological Approach of the Coupled Mass and Heat Transfer in Chemical Interactions

As soon as a chemical reaction was identified, much of a chemical problem involves coupled mass and heat transfer. The entropy source of a chemical interaction is expressed in a classic way by means of the chemical affinity, A, (Chartier et al. 1975; Vilcu and Dobrescu 1982). The linear form of equations is confined to cases having the ratio (A/RT) much less than unity. For our non-equilibrium approach, let us consider the general case of an independent stoichiometric chemical interaction:

$$\nu_1' A_1' + \nu_2' A_2' + \cdots + \nu_{n'}' A_{n'}' \Leftrightarrow \nu_1'' A_1'' + \nu_2'' A_2'' + \cdots + \nu_{n''}'' A_{n''}'' \tag{9.45}$$

In the conventional Eq. (9.45) we shall consider the direct reaction, where the reactants, A_i', $i = 1,\ldots, n'$, marked as prime ($'$), are to the left and the produces of reaction, A_i'', $i = 1,\ldots,n''$, marked as secondary ($''$), are to the right. The symbols ν_i' and ν_i'' stand for the stoichiometric coefficients of the reactants and produces of reaction, respectively. The inverse reaction has inverted prime and secondary superior index. Equation (9.45) implies the following mass and heat balances:

$$\sum_{i=1}^{n'} j_i' = \sum_{i=1}^{n''} j_i'' \tag{9.46}$$

and

$$\sum_{i=1}^{n'} j_{i,q}' = j_q + \sum_{i=1}^{n''} j_{i,q}'' \tag{9.47}$$

respectively. The mass currents j_i' and j_i'' are given by

$$j_i' = \lambda \nu_i' M_{A_i'}, \quad i = 1,\ldots,n' \tag{9.48}$$

and

$$j_i'' = \lambda \nu_i'' M_{A_i''}, \quad i = 1,\ldots,n'', \tag{9.49}$$

respectively, where $\lambda \in (0, 1)$ is the degree of chemical reaction (interaction) progress, according to Zemansky (1968), De Donder (1936) and Murgulescu and

Valcu (1982). The heat flows, coupled with the mass ones, are extensive elements here too. Similarly to the physical interactions, they can be written as:

$$\begin{aligned} j_{i,q}{}' &= h_i{}' j_i{}', \quad i = 1, \ldots, n' \\ j_{i,q}{}'' &= h_i{}'' j_i{}'', \quad i = 1, \ldots, n'' \end{aligned} \tag{9.50}$$

Next, it is supposed that all reactants and produces of reaction have the same pressure, equal to that of equilibrium, p_e. Additionally, the reactants have the temperatures T_i', $i = 1, \ldots, n'$, which can have different values, in a general case, but the produces of reaction have the same pre-established (controlled) temperature, equal to that of equilibrium, T_e. The state parameters may be time dependent or not. For given external conditions, an independent chemical reaction is a fixed, unique combination, where each j_i' mass current of the reactants is dependent of the other $n' - 1$ only through the number $\lambda \nu_k'$, $k = 1, \ldots, n'$, $k \neq i$, of moles involved in reaction, and the same applies to mass currents of the produces of reaction. In this case, the interaction behaves like having a single degree of freedom, for example that of a coupling mass current of one of the reactants, coupled with its corresponding heat current. Let us suppose that this current is j_1'. Using the current j_1', we define the following reduced mass currents of reactants and of produces of reaction:

$$j_{i,r}' = \frac{j_i'}{j_1'}, \quad i = 1, \ldots, n' \tag{9.51}$$

$$j_{i,r}'' = \frac{j_i''}{j_1'}, \quad i = 1, \ldots, n'' \tag{9.52}$$

where from it results:

$$j_i' = j_{i,r}' j_1', \quad i = 1, \ldots, n' \tag{9.53}$$

$$j_i'' = j_{i,r}'' j_1', \quad i = 1, \ldots, n'' \tag{9.54}$$

The entropy source is expressed again with Eq. (9.6), considering the same hypothesis as in the previous section. Further, taking into account Eqs. (9.46), (9.47), (9.50), (9.53) and (9.54) the entropy source can be written as:

$$\dot{S} = j_1'(-X) \tag{9.55}$$

where $(-X)$ will be termed *non-equilibrium (natural) thermodynamical force of an independent chemical interaction*. It is a function of p_e, T_i', $i = 1, \ldots, n'$ and T_e, and is given by Staicovici (2000b):

$$-X = -\left\{ \sum_{i=1}^{n'} j_{i,r}' \left[s_i' + h_i' \left(\frac{1}{T_i'} - \frac{1}{T_e} \right) \right] - \sum_{i=1}^{n''} j_{i,r}'' s_i'' \right\} \tag{9.56}$$

The force exhibits a same important property as that of the physical interactions, namely it is a function of state. The entropy and enthalpy in Eq. (9.56) must be evaluated on the basis of an absolute zero reference state, if the right measure of the interaction is desired to be obtained. The mass currents and reaction heat current are calculated by:

$$j_1' = L_{11}(-X) \tag{9.57}$$

$$j_i' = L_{11} j_{i,r}'(-X), \quad i = 2, \ldots, n' \tag{9.58}$$

and

$$j_q = j_1' \left(\sum_{i=1}^{n'} h_i' j_{i,r}' - \sum_{i=1}^{n''} h_i'' j_{i,r}'' \right) \tag{9.59}$$

respectively. Next, we shall consider the chemical interaction (9.45) at equilibrium. Its constant of equilibrium is given by Murgulescu and Valcu (1982) and Kirilin et al. (1985):

$$K_e = \frac{\prod_{i=1}^{n''} \left[A_i''\right]^{\nu_i''}}{\prod_{i=1}^{n'} \left[A_i'\right]^{\nu_i'}} \tag{9.60}$$

In Eq. (9.60), the molar concentrations of the reactants and of the produces of reaction are calculated by Murgulescu and Valcu (1982):

$$\left[A_i'\right] = \frac{n_{0,i}' - \lambda_e \nu_i'}{\sum_{i=1}^{n'} n_{0,i}' - \lambda_e \Delta\nu} \tag{9.61}$$

and

$$\left[A_i''\right] = \frac{\lambda_e \nu_i''}{\sum_{i=1}^{n'} n_{0,i}' - \lambda_e \Delta\nu} \tag{9.62}$$

respectively. In Eqs. (9.61) and (9.62), $n_{0,1}'$, $i = 1, \ldots, n'$, λ_e and $\Delta\nu$ hold for the initial number of moles of the reactants, the value of the degree of chemical reaction progress at equilibrium and the difference $\sum_{i=1}^{n''} \nu_i'' - \sum_{i=1}^{n'} \nu_i'$, respectively. The constant of equilibrium K_e can be determined through various known methods (e.g. experimentally, or can be calculated with the help of the chemical affinity, A etc., depending on the chemical reaction type). From Eqs. (9.60) to (9.62), the degree of chemical reaction progress at equilibrium, λ_e, is assessed. With this value, the phenomenological coefficient of a chemical interaction, $L_{11,e}$, can be found, writing Eq. (9.24) for the case at hand at equilibrium:

$$\lambda_e \nu_1' M_{A_1'} = L_{11,e}\left(-X_{1,e}'\right) \tag{9.63}$$

wherefrom:

$$L_{11,e} = \frac{\lambda_e \nu_1' M_{A_1'}}{\left(-X_{1,e}'\right)} \tag{9.64}$$

In Eq. (9.64), the natural force is written at equilibrium as:

$$\left(-X_{1,e}'\right) = -\left(\sum_{i=1}^{n'} j_{i,r}' s_i' - \sum_{i=1}^{n''} j_{i,r}'' s_i''\right) \tag{9.65}$$

As mentioned, the phenomenological coefficient $L_{11,e}$ is thought to be essentially constant along the entire reaction, that is:

$$L_{11,e} \approx L_{11} \approx const. \tag{9.66}$$

where

$$L_{11} = \frac{\lambda \nu_1' M_{A_1'}}{\left(-X_1'\right)} \tag{9.67}$$

is the current interaction phenomenological coefficient and $\left(-X_1'\right)$ is given by Eq. (9.56). Considering Eqs. (9.64), (9.66) and (9.67) together, the current degree of chemical reaction progress λ acquires the form:

$$\lambda = \frac{\left(-X_1'\right)}{\left(-X_{1,e}'\right)} \lambda_e \tag{9.68}$$

The chemical literature asserts that $\lambda = 0$ unconditionally to the beginning of a chemical interaction (e.g. Murgulescu and Valcu 1982). Phenomenologically, Eq. (9.68) shows that this is not true, but in case of two special circumstances, only. Indeed, the right member of Eq. (9.68) approaches zero value when: (a) the reactants temperature approaches zero absolute temperature, $T_i' \to 0\,\mathrm{K}$, and entropies and enthalpies vanish, and (b) the direct reaction is very poor, that is $\lambda_e \to 0$, therefore $\lambda_e > \lambda \to 0$.

9.2 Non-equilibrium Phenomenological Theory Applications. Case Studies of NH_3–H_2O, NH_3, H_2O and Other Working Pairs Gas–Liquid Interactions: Case Study of a Chemical Interaction Force

The thermal factors involved in the calculation of the natural thermodynamic force [Eq. (9.34)] have been expressed analytically in case of the ammonia-water system (Staicovici 1998a), by means of state equation, found by Ziegler and Trepp (1984). From experience, the process at hand is nonlinear and the assumption of a linear set of phenomenological equations needs to be justified. Indeed, the currents are linear with respect only to the force. In fact, they are strong non-linear with respect to all interface variables through the thermal factors s_r, h_r, A_r and C_r intervening in the calculation of the thermodynamical force (Ziegler and Trepp 1984). The phenomenological coefficient has been estimated by means of Eqs. (9.40), (9.39) and (9.34), resulting in $L_{11,r} = \frac{1}{|R|} = \frac{1}{8314}$. Before starting the application, a special attention was paid to the consistency of the force with respect to the equilibrium thermodynamics, regarding the ideal points calculation of a gas–liquid system and particularly of the ammonia-water mixtures. As it will be further mentioned, the force is not defined in an ideal point and never cancels. Practically, such a point is determined by progressively restricting the interval of definition throughout which the function has finite values, of equal absolute value and contrary signs, up to the desired level of accuracy (here usually obtained up to the sixth decimal). In principle, the two thermodynamical approaches, the equilibrium and the non-equilibrium phenomenological ones, are equivalent, as far as the determination of the ideal point is concerned. However, although in both cases the same state equation for the binary mixture under study has been used, the non-equilibrium phenomenological calculus has lead to sensibly different results in some parts of the solubility field because Ziegler and Trepp equation has an arbitrarily chosen zero reference of state. First attempts to approach the results of the two methods failed when a non-linear dependence of the mass current with respect to the natural force was considered. This explains why we used just only a linear dependence of the mass current with respect to the force [Eq. (9.24)]. Best results were obtained amplifying the second term of the denominator in Eq. (9.34) by a match function $M(p_{r,e}, x_e)$, already mentioned in Sect. 9.1.1. Several values of the match function are given in Fig. 9.1 for five reduced ideal pressures and five ideal solution mass fractions. These values were well correlated by a four-order double polynomial expansion:

$$M\left(p_{r,e}, x_e\right) = \exp\left(\sum_{i,j=0}^{4} A_{i,j} p_{r,e}^i x_e^j\right) \tag{9.69}$$

The coefficients $A_{i,j}$ are given below:

$A_{00} = -1.55107 \times 10^0; A_{01} = 4.30114 \times 10^0; A_{02} = -2.50709 \times 10^1;$
$A_{03} = 5.14562 \times 10^1; A_{04} = 3.53064 \times 10^1;$
$A_{10} = 2.20458 \times 10^1; A_{11} = -3.43605 \times 10^2; A_{12} = 2.15534 \times 10^3;$
$A_{13} = -5.56704 \times 10^3; A_{14} = 4.87992 \times 10^3;$
$A_{20} = -5.72617 \times 10^1; A_{21} = 9.54441 \times 10^2; A_{22} = -6.00841 \times 10^3;$
$A_{23} = 1.54652 \times 10^4; A_{24} = -1.35339 \times 10^4;$
$A_{30} = 5.53949 \times 10^1; A_{31} = -9.34452 \times 10^2; A_{32} = 5.88969 \times 10^3;$
$A_{33} = -1.51448 \times 10^4; A_{34} = 1.32402 \times 10^4;$
$A_{40} = -1.74164 \times 10^1; A_{41} = 2.94977 \times 10^2; A_{42} = -1.86140 \times 10^3;$
$A_{43} = 4.78901 \times 10^3; A_{44} = -4.18852 \times 10^3;$

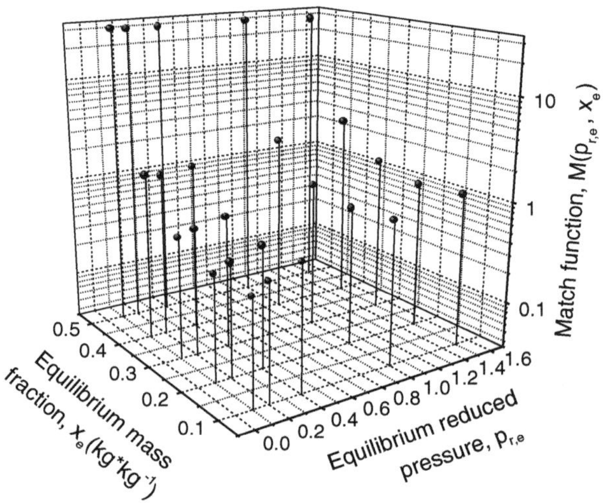

Fig. 9.1 Several values of the match function in Eq. (9.34) for ammonia/water gas–liquid mixtures ($p_b = 10$ bar; $T_b = 100$ K)

The relative differences (%) in the approximation of equilibrium mass fraction of the mixture solution with Eq. (9.34), are plotted in Fig. 9.2. The differences could come also from the different units used for x_e, namely mass fraction in the h–x diagram, respectively molar fraction in equation of state of Ziegler and Trepp, and the use of an unique computing value for the phenomenological coefficient. $M(p_{r,e}, x_e)$ compensates the arbitrarily chosen zero reference state of a system entropy and enthalpy. To this extent, considering the absolute zero temperature as zero reference state of an interacting system enthalpy and entropy can be a goal for the future non-equilibrium phenomenological approach applications, because in this case $M(p_{r,e}, x_e)$ is no more needed and the computation simplifies. However, in our case, before to use Eq. (9.34) in an evolution process assessment, we need the "nearest" ideal point which the system evolves to, in order to calculate the match function. Here it is helpful the Prigogine theorem, relative to the minimization of the entropy source in an irreversible stationary process, and the computation becomes an iterative one. However, practically with a good initial guess of the nearest ideal point, the use of this theorem would be not necessary and the calculus is correspondingly simplified (see further comments). For our application, Eq. (9.34) covers an usual domain for refrigeration purposes with the NH_3–H_2O system, namely of $1.0 \leq p \leq 15.0$ bar, $0.1 \leq x \leq 0.45$ and $0.5 \leq y \leq 0.998$.

Equations (9.27), (9.28), (9.31) and (9.34) have been represented for exemplification as partial functions of the reduced state parameters x, $t_{l,r}$, y and $t_{g,r}$ respectively in Figs. 9.3, 9.4, 9.5, 9.6 and 9.7. At interface, the gas and the liquid are in mechanical equilibrium, $p_l \cong p_g$, and no pressure partial functions exist in fact. For completeness, the natural thermodynamic force of pure components water and ammonia is given in two study cases in Fig. 9.8. The curves are different in shape,

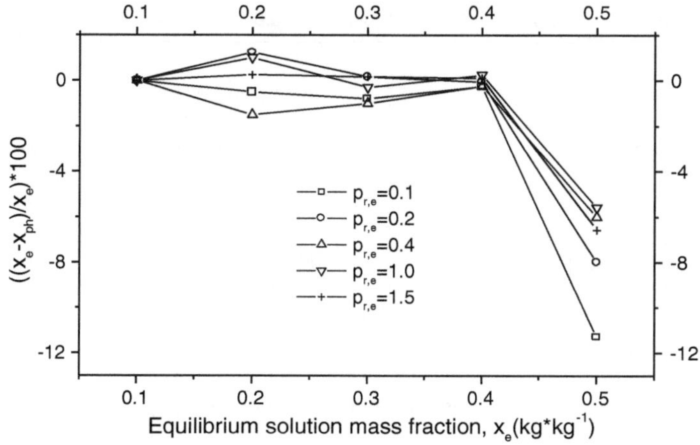

Fig. 9.2 The relative differences % in the approximation of equilibrium mass fraction of ammonia/water solutions with Eq. (9.34)

depending upon the variable used in the abscissa and the nature of the system. The partial functions which depend on the parameters of the same phase have the same absorption or generation monotonous variation. According to the postulate (see next section) and its consequence, the diagrams show the continuous increase of force, Figs. 9.3b, 9.5b, 9.6b and 9.7b, entropy source, Figs. 9.3a, 9.5a, 9.6a and 9.7a, and of the coupled mass and heat currents, Fig. 9.4a, b, respectively, approaching an ideal state. In case of currents, an *ideal point approaching (i.p.a) effect* is emphasized*: depending on how strong the currents coupling is, the approach to an ideal state determines an increase effect of the absolute value of the function by several percentages (in case of pure components), to several hundred times (for the binary system), as compared to the states which are far from the same ideal point.* In this way, referring to currents, the term of "far from equilibrium state" changes here its classical meaning and corresponds to its opposite, namely to the low interaction. The *i.p.a.* effect has been brought to light also in case of other working combinations, namely NH_3–$LiNO_3$ and NH_3–NaSCN, Figs. 9.9 and 9.10, respectively (see also further on the same for the H_2O–LiBr working pair). An *i.p.a.* effect behavior has been observed also calculating Eq. (9.56) for an independent chemical reaction. For the moment, we present a case study concerning the natural force of ammonia synthesis from nitrogen and hydrogen. The results are given in Fig. 9.11, where force [Eq. (9.56)] is plotted against several values of the reactants temperature $\left(T_1' = T_2'\right)$ and several pressures. The equilibrium temperature is equal to 600 K. The direct reaction, between hydrogen and nitrogen to form ammonia, has negative forces and currents and takes place until the temperature of reactants hydrogen and nitrogen is smaller than that of equilibrium. The inverse reaction, when ammonia decomposes to form hydrogen and nitrogen, has positive forces and currents and takes place as long as

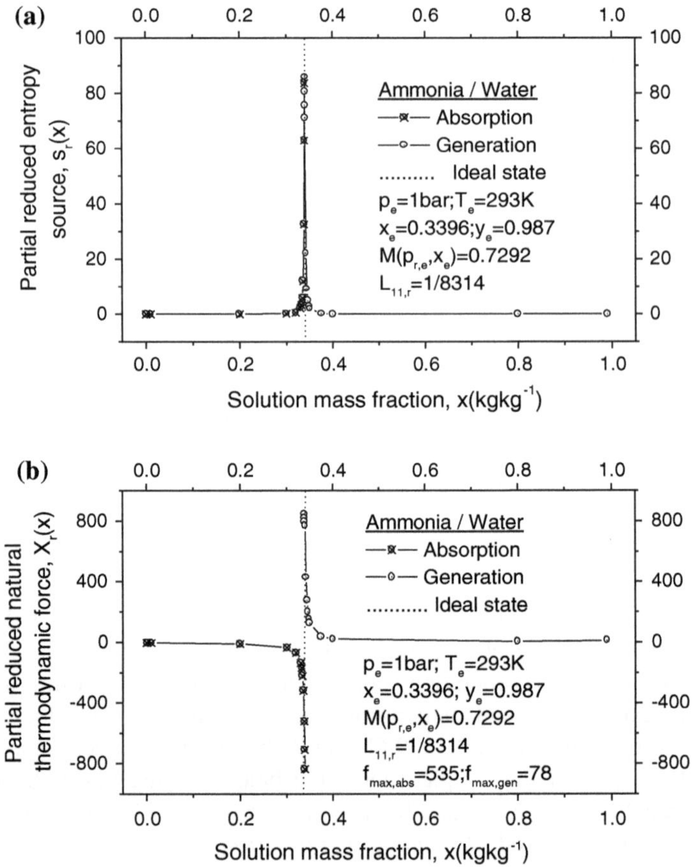

Fig. 9.3 The *i.p.a.* effect behaviour for the reduced entropy source (**a**) and natural thermodynamic force (**b**) as functions of solution mass fraction

the temperature of the reactant ammonia is higher than that of equilibrium. The mass and heat currents assessment is not included here.

The i.p.a. effect may be ranked, in case of a single interaction type, by a figure of merit, "f", equal to the ratio of two same type function values (e.g. absorption/condensation or generation/vaporization), corresponding to the states closer, respectively farther to the ideal state. Figures 9.3, 9.4, 9.5, 9.6, 9.7, 9.8, 9.9, 9.10 and 9.11 include also estimates of the maximal "f" values for the chosen examples. Besides a "f" evaluation, our phenomenological approach creates also the opportunity to compare the intensity of different interaction types, comparing the (reduced) values of natural forces calculated with Eqs. (9.34)–(9.37) and (9.56) for same interaction p, T, x and y state parameters. Obviously, the calculus must be performed in this case with entropy and enthalpy functions of state having a same zero reference state.

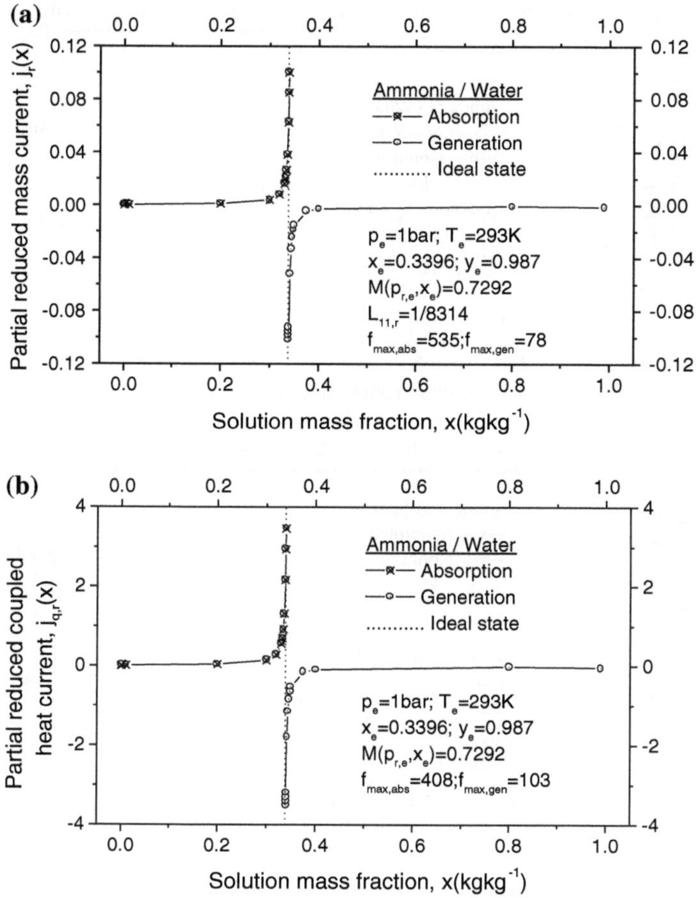

Fig. 9.4 The *i.p.a.* effect behaviour for the reduced mass (**a**) and heat current (**b**) as functions of solution mass fraction

The specialized literature does not contain systematic theoretical or experimental accounts of the non-equilibrium interaction in the NH_3–H_2O system, in order to compare them with the results obtained here. The only confirmation of such results is given by their compliance with a few experimental works which are in agreement with the graphs shown in Figs. 9.3, 9.4, 9.5, 9.6, 9.7, 9.8, 9.9 and 9.10, proving the increase of the gaseous mass flow driven as a two-phase binary system is approaching a state of saturation. To this effect, some mention should be made of the experimental work of (Hui and Thome 1985), who investigated the boiling of binary azeotropic ethanol-water and ethanol-benzene mixtures and ascertained the increase of the number of vaporization points, until it became impossible for such points to be individually distinguished, as the molar fraction of the mixtures approached the saturation value. Another work of NH_3–H_2O

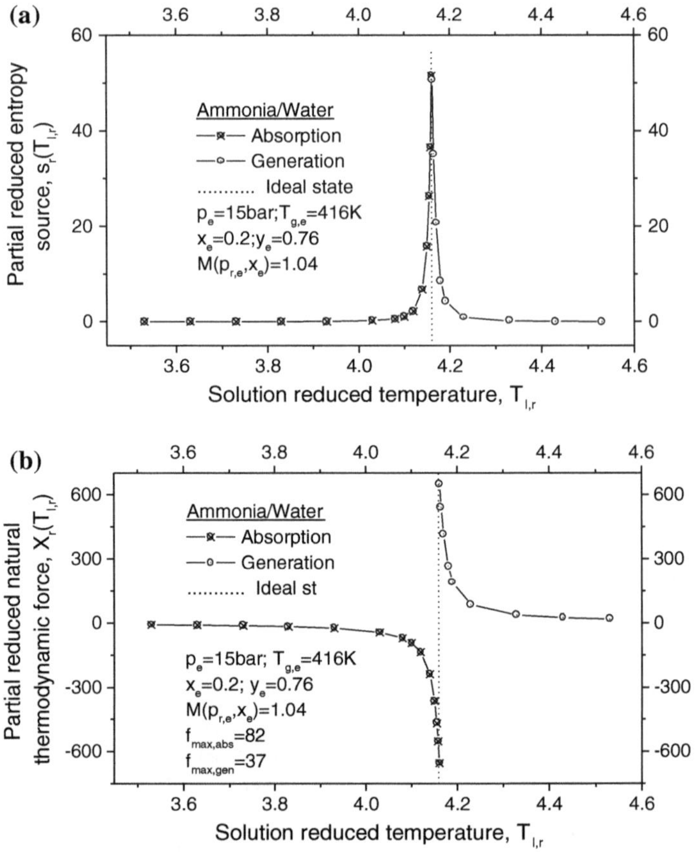

Fig. 9.5 The *i.p.a.* effect behaviour for the reduced entropy source (**a**) and natural thermodynamic force (**b**) as functions of solution temperature

falling film absorption in plate heat exchangers, done by Kang et al. (1998), has found experimental absorption data near equilibrium point which now the author can regard as a proof of the *i.p.a.* existence.

The importance of the currents coupling is worth to be mentioned. Indeed, let us consider the mathematical form of the chemical interactions given in Sect. 9.1.2 and particularize it for processes involving separate or simultaneous but non-coupled mass and heat transfer, characterized by the following features:

$$\begin{aligned} & n' = n'' \\ & j_i' = j_i''; \quad i = 1, \ldots, n' \\ & j_q = 0 \\ & \nu_i' = \nu_i''; \quad i = 1, \ldots, n' \end{aligned} \tag{9.70}$$

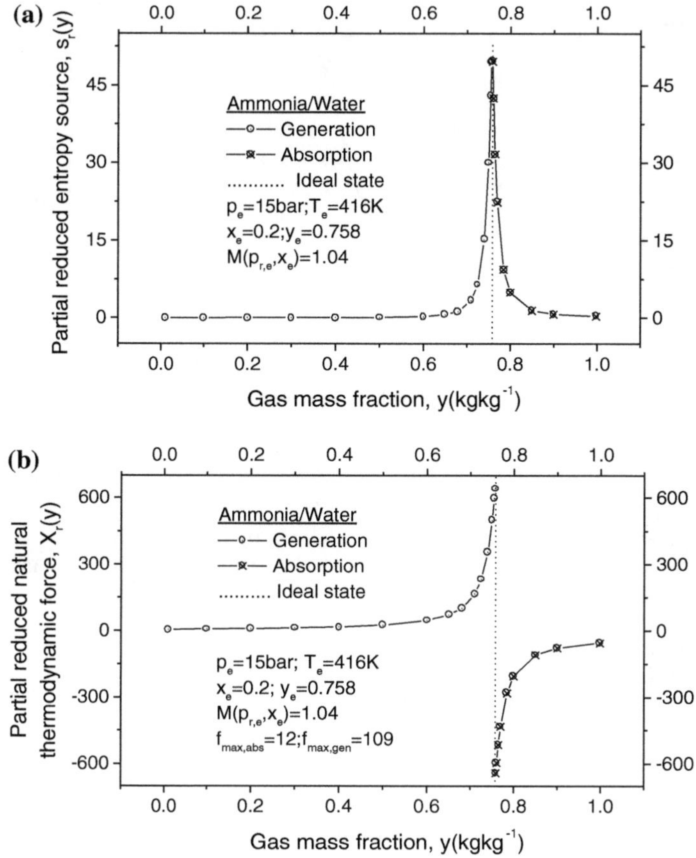

Fig. 9.6 The *i.p.a.* effect behaviour for the reduced entropy source (**a**) and natural thermodynamic force (**b**) as functions of gas mass fraction

In this case, *the force of a non-coupled mass and heat transfer process* is expressed with Eqs. (9.57) and (9.70) by:

$$-X = -\sum_{i=1}^{n'} j'_{i,r}\left[\left(s'_i - s''_i\right) + h'_i\left(\frac{1}{T'_i} - \frac{1}{T_e}\right)\right] \tag{9.71}$$

For a system referred to above, approaching an ideal point, we have $T'_i \to T_e$, and $s'_i \to s''_i$, $i = 1, \ldots, n'$, therefore $-X \to 0$, which means that the *i.p.a.* effect is rather a *coupling effect* .

At this point it is possible to give a satisfactory qualitative explanation to the problem of the ammonia bubble absorption, raised in the introduction of this chapter. Indeed, during the *growth* phase the absorbed mass flow j_r increases continuously as the state of the interface approaches an equilibrium point, according

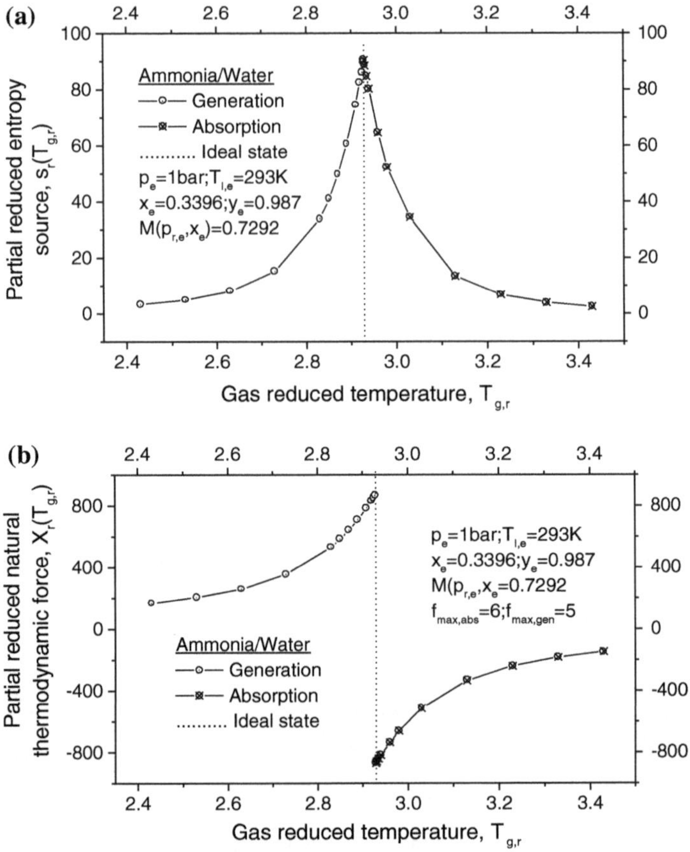

Fig. 9.7 The *i.p.a.* effect behaviour for the reduced entropy source (**a**) and natural thermodynamic force (**b**) as functions of gas reduced temperature

to the diagram shown in Fig. 9.3b plotted with p, T, x and y values close to the experimental conditions. At a certain moment, this increase exceeds the value of the feeding flow rate, causing the collapse of the bubble, and continues after the starting of the collapse at an even higher speed, up to the total absorption of the bubble. This further explains why the *collapse* duration is much shorter than the duration of the *growth.* Moreover, the non-equilibrium phenomenological qualitative explanation of the bubble collapse leads to the conclusion that: (i) the currents depend directly only to the interface variables and not to the interface magnitude, therefore *absorption in the* NH_3–H_2O *system should not be a surface process*, as it is usually considered and consequently (ii) the use of the mass transfer coefficients K_g and K_l as surface factors [Eq. (9.2)] is improper, at least in case of the NH_3–H_2O system and (iii) the estimation of the interface mass transfer by analogy with heat transfer is improper too, because the force in Eq. (9.2) tends to

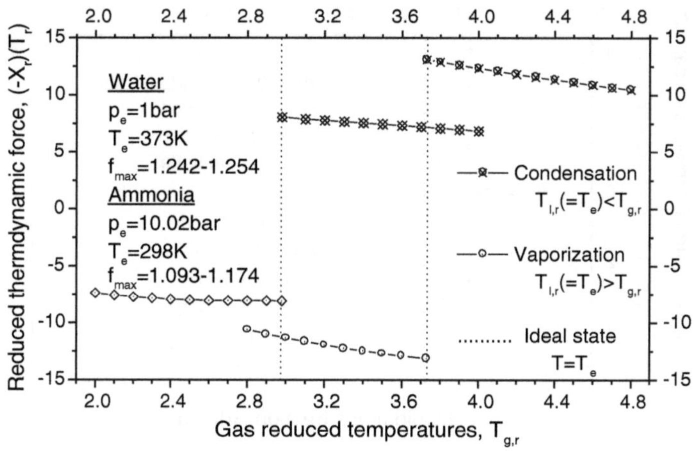

Fig. 9.8 The water and ammonia *i.p.a.* effect behaviour for the reduced natural thermodynamic force as function of liquid-gas reduced temperature

zero approaching an ideal point, which, from the non-equilibrium phenomenological point of view, is not true. A quantitative estimate of the evolution of the NH_3–H_2O gaseous mixture bubble in the NH_3–H_2O liquid absorber is given further in this chapter.

9.3 Non-equilibrium (Natural) and Equilibrium (Ideal) Thermodynamical Forces

Some readers may consider our results contradictory with respect to the classic approach of the mass and heat transfer. This is because the known thermodynamical forces governing coupled heat and mass transfer in continuous or discontinuous (interface) media, like $\nabla T/T^2$ and $\nabla \mu/T$ or ΔT and Δx, are vanishing in an ideal point, and the same applies to chemical forces, expressed with the help of chemical affinity (de Groot and Mazur 1964; Chartier et al. 1975; Vilcu and Dobrescu 1982), but the natural forces, given by Eqs. (9.34) and (9.56), are not. In order to distinguish between the two types of forces, the classic ones, involved in the coupled mass and heat transfer, will be referred to as *equilibrium (ideal) thermodynamical forces*. The difference between the natural and ideal forces comes from the different tools which they were obtained with, non-equilibrium and equilibrium, respectively. The natural forces describe the evolution of an interacting system, towards an ideal point, without to know if an equilibrium will be reached or no (this depends on the future external conditions of the interaction), while the ideal ones regard the same evolution supposing a priori that equilibrium already exists. The two approaches, natural and ideal, lead to same results and can be coupled in order to assess a mass

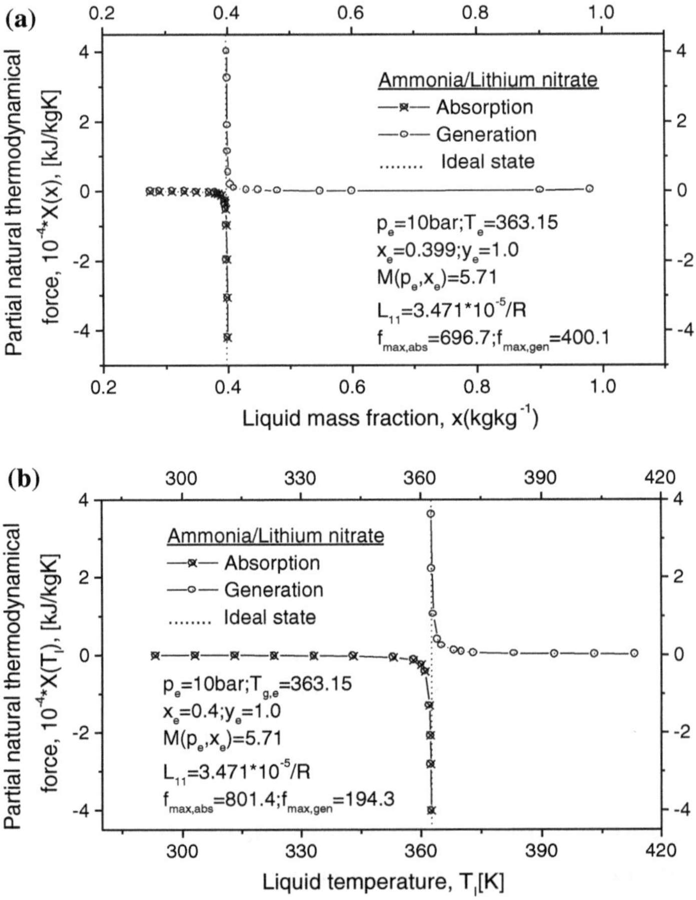

Fig. 9.9 The *i.p.a.* effect behaviour for partial natural thermodynamical forces of liquid mass fraction (**a**) and liquid temperature (**b**) in case of the NH_3–$LiNO_3$ system

and heat transfer process. A first successful example is shown further by the model of the ammonia bubble absorption in ammonia-water mixtures and its prove by experiment.

The non-equilibrium thermodynamical forces of physical and chemical interactions enjoy several important properties. First, as already mentioned, the force [Eq. (9.34)] applies also to physical interactions of solid–liquid, solid–gas or gas–gas, liquid–liquid type, provided that appropriate indexes be used (e.g. for a liquid–solid interaction with a solid final phase, the indices g and l change in l, respectively s). Unlike the ideal forces of Fourier and Fick type, which can not be applied to discontinuous media because ∇ is not defined in points of discontinuity (such those of an interface) the natural force is valid for both continuous or discontinuous media. Obviously, the currents assessment needs to determine

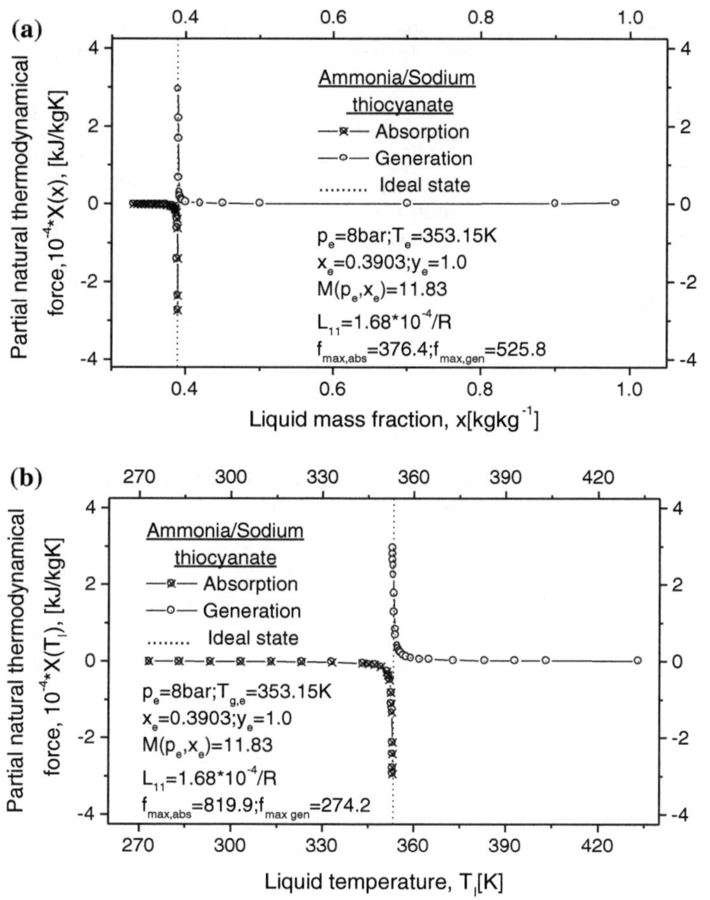

Fig. 9.10 The *i.p.a.* effect behaviour for partial natural thermodynamical forces of liquid mass fraction (**a**) and liquid temperature (**b**) in case of the NH_3–NaSCN system

analytically the right phenomenological coefficients, which often is a difficult task. The second property is related to the natural force behaviour close to an ideal point. This force being essentially new, the specialized literature does not point out this aspect and refers only to the total entropy increase to maximum values when the system approaches such a point, causing the source strictly positive definition. This is not enough to describe its behaviour near such a point. According to previous works (Chartier et al. 1975; Vilcu and Dobrescu 1982), the ideal forces are governing the coupled heat and mass transfer processes and the same is valid for the *non-equilibrium (natural) thermodynamical forces, which evaluate the natural tendency of the real systems making for equilibrium.* This essential feature shows that constant (zero included also) value forces or those with decreasing absolute value are not acceptable for systems approaching an ideal point. Hence, the only acceptable force behaviour remains the one expressed by the following:

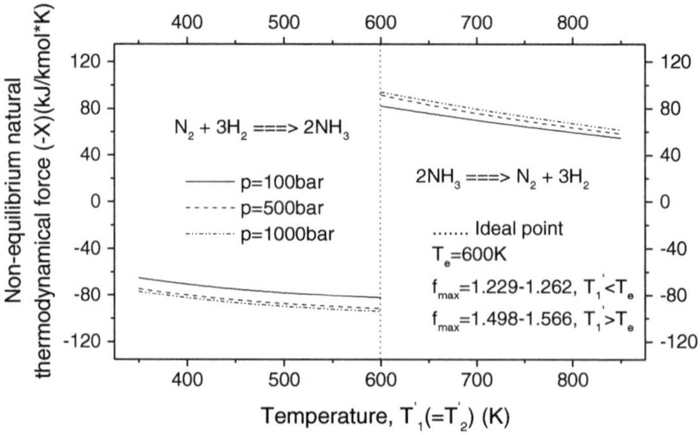

Fig. 9.11 The *i.p.a.* effect behaviour calculated in case of the natural chemical force for the ammonia synthesis, against the reactants same temperature

Postulate *The non-equilibrium (natural) thermodynamical force of a closed or open evolution system, involving coupled mass and heat transfer, continuously increases in absolute value to maximal finite values approaching an ideal point.*

The phenomenological factors, proportional to the force (entropy source, currents), must have the same force feature. The postulate and its consequence are verified by all study cases synthesized in Figs. 9.3, 9.4, 9.5, 9.6, 9.7, 9.8, 9.9, 9.10 and 9.11, including the physical interactions of pure components with ideal mixture, and binary with non-ideal mixture, as well as a chemical one. Third, we have the following:

Lemma *A natural thermodynamical force is an odd function at least in close proximity of an ideal point,* $z_e = z_e(p_{l,e}, T_{l,e}, x_e, p_{g,e}, T_{g,e}, y_e)$:

$$(-X)(z_e + \delta) = -(-X)(z_e - \delta).$$

This property results by integrating the force on a closed way, symmetrical with respect to the ideal point. Because $(-X)$ is a function of state, the integral is zero and making $\delta \to 0$, the lemma is demonstrated. Fourth, the force must not be defined in an ideal point, because, according to the lemma, it should take at the same time and with same probability two opposite values, corresponding to the elementary processes adjacent to an ideal point with opposite signs of the homologues currents, which is impossible, therefore we have the following

Corollary *A natural thermodynamical force is not defined in an ideal point.*

Here are included both the double ideal points [e.g. absorption (condensation) and generation (vaporization)] and the triple ones. This property extends to the currents and to the entropy source, not defined in an ideal point as well.

9.4 Modeling of the NH_3–H_2O Bubble Absorption, Analytical Study of Absorption and Experiments

For a more accurate evaluation of the NH_3–H_2O gas–liquid interactions, a combination of classic equilibrium and non-equilibrium phenomenological theories of coupled mass and heat transfer will be applied here in order to model absorption of the ammonia bubble. The resulting model is a non-empirical linear Phenomenological Hydro-Gaso-Dynamical (PhHGD) tool. By doing so, an attempt is made to find an appropriate local solution of the transport in the NH_3–H_2O medium, which might lead to its correct global evaluation, applicable in the calculation of the absorption installations.

9.4.1 Model of the Bubble Absorption Applying the PhHGD Tool

The solution of the problem at hand was faced from the very beginning with the following difficulties: (a) there are only a few papers in the specialized literature which deal with a thorough analysis of bubble absorption/generation in multi-component media; (b) although the process is in evolution, none of the selected papers approach it with non-equilibrium phenomenological tools, all the authors preferring to consider a priori that an interface thermodynamical equilibrium exists, and (c) so far, the PhHGD method results may be only compared with experiment. The papers falling into the category mentioned above are due to Merrill and Perez-Blanco (1997), who analyzed the bubble absorption in binary solutions, accompanied by combined mass and heat transfer and to Miyatake et al. (1997), who give a simple universal equation for the bubble growth in pure liquids and in solutions with non-volatile solute. Also, Sivagnanam et al. (1994) have developed correlations for the subcooled convection of binary mixtures during boiling, by using data for acetone-water, isopropanol-water and n-butanol-water, and Cooper and Stone (1981) have reported experiments where vapor bubbles are grown on a flat wall in a binary liquid which is initially stagnant and isothermal (hexane and octane in various proportions). Additionally, a lot of papers deal with the bubble dynamics in the context of boiling/condensation processes in pure liquids. For completion, some of them will be quoted hereinafter. Al-Hayes and Winterton (1981) have obtained experimental results for the mass transfer in air bubbles adhering to the wall of a pipe in which the fluid (water, water with surfactants and ethylene glycol) has a supersaturated history. Nigmatulin et al. (1981) examine a non-linear aspect of thermal and dynamic oscillating mass interaction, between a spherical vapor and gas bubble in a liquid, taking into account a non-uniform temperature in the bubble and the reciprocal diffusion in the gas–liquid mixture. Auracher and Maier (1983) find a correlation for experimental data of vapor bubble condensation in slightly subcooled pure liquids (R11, R113, and C_3H_8O) governed by a heat transfer mechanism. Klausner et al. (1993) determine the detachment time of vapor bubbles

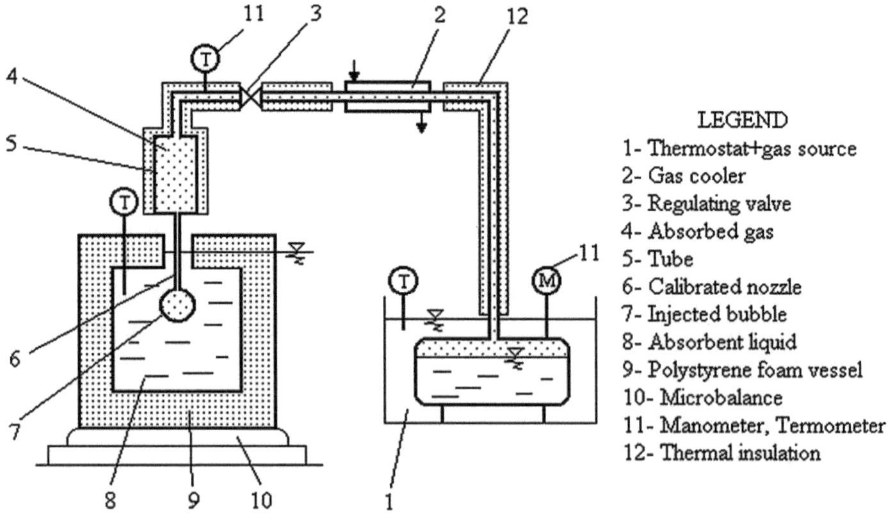

Fig. 9.12 Bubble generator scheme

in forced convection boiling. In a two part paper, Zeng et al. (1993a, b) suggest a new model of prediction for the detachment diameters of vapor bubbles in the saturated stagnant liquid and in the flowing liquid upon boiling. Brujan (1994) analyzes the question of the collapse of a spherical cavity bubble containing non-condensable gases and vapors in an infinite volume of non-newtonian liquid. Lee and Merte (1996a, b) have examined the growth of the bubble in uniformly superheated liquids and under microgravity conditions.

The bubble evolution in the NH_3–H_2O gas–liquid interaction has been modeled in relation to the bubble generator scheme, shown in Fig. 9.12. This scheme represents a portion of a vertical pipe absorber. The gaseous mixture, coming from a reservoir of liquid NH_3–H_2O at pressure 1.2–1.5 bar and temperature 5–40 °C, is expanded up to the value of the normal pressure and allowed to penetrate at flow rate $\dot{q}_t$ in a pressure-equalizing tube, which, in the case of vertical absorbers, has the role of a collector. The gas in the tube has pressure p_t, temperature T_g and volume V_t. From here, the gas passes through a calibrated nozzle and diameter d_0, and penetrates into the absorbing liquid mixture at the flow rate $\dot{m}$. The parameters of the gas injected in the liquid are p_g, T_g and y. At the gas–liquid interface and at infinity the liquid has the parameters p_l, T_l and x, respectively p_∞, T_∞, and x_∞.

9.4.1.1 Governing Equations

The model is based upon the following assumptions, which are generally those formulated by Rayleigh (1917) for the dynamics of a single bubble, namely a spherical symmetry flow and uniformity of the gas pressure in the bubble (homobaricity). When the bubble radius varies monotonously (growth or collapse), the homobaricity condition is written as $(\dot{R}/C)^2 \ll 1$, where $\dot{R}$ and C are the bubble radius velocity, and respectively the sound velocity in gas (Nigmatulin et al. 1981). During the bubble oscillations, homobaricity prevails if $\omega R \ll C$, where $\omega = 2\pi f$ is the annular frequency (f). Moreover, the following assumptions are considered for the liquid and gas medium: (a) the liquid gravitational forces are neglected; liquid is newtonian, viscous and incompressible; its physical and thermo-physical properties are constant in time (λ, ν, c_p, ρ_l, σ, D_x); (b) the gas has an ideal behaviour and has an isothermal adiabatic evolution, both in the tube and in the bubble; the flow in the tube assumes negligible charge losses; the gas–liquid interaction is an evolutional phenomenon, deemed to be an open system from the thermodynamical point of view; the molar fraction as well as the specific heat at a constant pressure are constant.

9.4.1.2 Liquid Phase Equations

The liquid motion is governed by the following equations (Landau and Lifschitz 1971):

global continuity

$$\frac{\partial \rho}{\partial t} + div(\rho \vec{v}) = 0 \tag{9.72}$$

continuity of a species

$$\rho \frac{dx}{dt} + div\vec{i} = 0 \tag{9.73}$$

laminar flow (Navier-Stokes)

$$\rho \frac{d\vec{v}}{dt} = -\nabla p + \frac{\partial \sigma'_{ik}}{\partial z_k} + \sum F_i \tag{9.74}$$

general coupled mass and heat transfer (thermodiffusion)

$$\rho\, T \frac{ds}{dt} = \sigma'_{ik} \frac{\partial v_i}{\partial z_k} - div\left(\vec{q} - \mu \vec{i}\right) - \vec{i} \nabla \mu \tag{9.75}$$

where (Staicovici 1998a):

$$\frac{ds}{dt} = c_p \frac{1}{T} \frac{dT}{dt} - \left(\frac{\partial \mu}{\partial T}\right)_{p,x} \frac{dx}{dt} \tag{9.76}$$

9.4.1.3 Interface Equations

At interface, the equations governing interaction are the following:

motion

$$p_{l,r=R} = p_g - \frac{2\sigma}{R} + \rho_g \left(v_g - v_{r=R}\right)\left(v_g - \dot{R}\right) \tag{9.77}$$

where v_g is the gas velocity, and $\dot{R} = \partial R/\partial t$;

gas mass balance

$$j_r \rho_l \dot{V} + \rho_g \left(v_g - \dot{R} \right) S = 0 \tag{9.78}$$

where S is the interface area, $\dot{V} = 4\pi R^2 \dot{R}$ is the time derivative of the bubble volume, and j_r is the reduced mass current involved in interaction (Staicovici 1998a, 2000c);

liquid mass balance

$$j_r \rho_l \left| \dot{V} \right| + | v_{r=R} | \rho_l S = \rho_l \left| \dot{V} \right| \tag{9.79}$$

mass balance for a species

$$j_r \rho_l \left| \dot{V} \right| y + \rho_l \left| \dot{V} \right| (1 - j_r) x + \rho_l D_x \nabla x S = \rho_l \left| \dot{V} \right| (x + \delta x) \tag{9.80}$$

where δx is the actual increase of the molar fraction x.

energy

$$\rho_l \left| \dot{V} \right| c_{p,l} \delta T_l = j_r \rho_l \left| \dot{V} \right| C + \lambda \nabla T_l S \tag{9.81}$$

where δT_l is the actual increase of the liquid temperature at interface, and C is the local thermal capacity (Staicovici 1995).

9.4.1.4 Gas Phase Equations

The gas phase is governed by the following equations:

1st principle of thermodynamics, for the gas in the tube

$$(\dot{q} - \dot{m}) h_g = \dot{p}_{g,t} V_t \tag{9.82}$$

where $\dot{q}$, $\dot{m}$ and V_t are the gas flow rate at which the tube is fed, the bubble and respectively the tube volume.

Bernoulli equation (stationary), applied to the flow in the tube (the flow rate through the orifice, Ionescu 1977)

$$\dot{m} = \alpha \Omega \sqrt{\rho_g \left| p_{g,t} - p_g \right|} \tag{9.83}$$

where α is the Coriolis coefficient of the kinetic energy, and Ω is the nozzle flow section.

mass balance in the bubble

$$\dot{m} - j_r \rho_l \left| \dot{V} \right| = \rho_g \dot{V} \tag{9.84}$$

9.4.1.5 Model Equations Initial and Boundary Conditions

The governing equations will be written herein below in a form which allows for the application of the known mathematical methods of numerical solution. To this effect, we shall firstly revert to the liquid phase equation. A special analysis of the mass, $\vec{i}$, and heat, $\vec{q} - \mu\vec{i}$, flows in Eq. (9.75), derived from the entropy source equation of thermodiffusion, showed that off-diagonal effects Soret and Dufour are small in case of our application (Staicovici 1998a). Hence, the currents may not be coupled, resulting in pure diffusion and pure conduction:

$$\vec{i} = -\rho\, \mathrm{D_x} \nabla x \tag{9.85}$$

$$\vec{q} - \mu\vec{\mathrm{i}} = -\lambda \nabla \mathrm{T} \tag{9.86}$$

respectively. The thermodiffusion Eq. (9.75) is further simplified, noting that terms $T(\partial\mu/\partial T)_{p,x}$ and $\vec{i}\nabla\mu$ may be neglected, as they are second-order terms, and the viscous tensor $\sigma'_{\mathrm{i,k}}$ is canceled in case of the spherical symmetry contraction ($i = k$) and of incompressible Newtonian liquid (Landau et Lifschitz 1971). Equally, the integration of Eq. (9.72), written on spherical coordinates, leads to

$$r^2\, \mathrm{v} = \mathrm{R}^2\dot{\mathrm{R}} = \mathrm{const} \tag{9.87}$$

where from

$$\mathrm{v} = \dot{\mathrm{R}}\left(\frac{\mathrm{R}}{\mathrm{r}}\right)^2 \tag{9.88}$$

At interface, from Eqs. (9.78) and (9.79) it results that

$$\rho_{\mathrm{g}}\left(\mathrm{v_g} - \dot{R}\right) = -j_r \rho_{\mathrm{l}} \dot{R} \tag{9.89}$$

and

$$\mathrm{v_{r=R}} = \dot{R}\left(\frac{R}{r}\right)^2 (1 - j_r) \tag{9.90}$$

respectively, so that the equation of motion is re-written as:

$$p_{l,r=R} = p_{\mathrm{g}} - \frac{2\sigma}{R} + \left(j_r \rho_{\mathrm{l}} \dot{R}\right)^2 \left(\frac{1}{\rho_{\mathrm{g}}} - \frac{1}{\rho_{\mathrm{l}}}\right) \tag{9.91}$$

Finally, the flow Eq. (9.74) written in spherical coordinates and without the viscous term (see the remark above) is integrated in relation to the radial coordinate r, between R and ∞. Using Eq. (9.89) it results that:

$$\left[\ddot{R}R + \dot{R}^2(1 + j_r)\right](1 - j_r) - \dot{R}R\frac{\partial \mathrm{j_r}}{\partial \mathrm{t}} = \frac{1}{\rho_{\mathrm{l}}}\left(p_{l,r=R} - p_\infty\right) \tag{9.92}$$

The second term of the left side of equation. is a second-order term and may be neglected. Considering the above, modeling the bubble absorption/generation leads to the solving of the following system of nonlinear ordinary differential equations and with partial derivatives (Staicovici 1998a, c)

$$\ddot{R} = \left[\left(p_{g,t} - \frac{\dot{m}^2}{\alpha^2 \Omega^2 \rho_g} - p_\infty - \frac{2\sigma}{R}\right)\frac{1}{\rho_l} + \dot{R}^2\left(j_r^2 \frac{\rho_l}{\rho_g} - 1\right)\right]\frac{1}{R(1 - j_r)} \tag{9.93}$$

$$\dot{p}_{g,t} = \left[\dot{q}_t - \dot{m}\right]\frac{h_g}{\mathrm{V_t}} \tag{9.94}$$

$$\frac{\partial \mathrm{x}}{\partial \mathrm{t}} + \dot{R}\left(\frac{R}{r}\right)^2 \frac{\partial \mathrm{x}}{\partial \mathrm{r}} = D_x\left(\frac{2}{r}\frac{\partial \mathrm{x}}{\partial \mathrm{r}} + \frac{\partial^2 \mathrm{x}}{\partial \mathrm{r}^2}\right) \tag{9.95}$$

$$\frac{\partial \mathrm{T_l}}{\partial \mathrm{t}} + \dot{R}\left(\frac{R}{r}\right)^2 \frac{\partial \mathrm{T_l}}{\partial \mathrm{r}} = D_T\left(\frac{2}{r}\frac{\partial \mathrm{T_l}}{\partial \mathrm{r}} + \frac{\partial^2 \mathrm{T_l}}{\partial \mathrm{r}^2}\right) \tag{9.96}$$

where

$$\dot{m} = \rho_g \dot{V} + j_r \rho_l \left|\dot{V}\right| \tag{9.97}$$

Equations (9.93)–(9.97) serve to determine the first four unknown factors, R, $\mathrm{p_{g,t}}$, x and respectively $\mathrm{T_l}$. In the next step the bubble radius velocity, $\dot{R}$, and the bubble radius acceleration, $\ddot{R}$, as well as the first time derivative of the gas pressure in the tube, $\dot{p}_{g,t}$, are determined. In order to determine the gas pressure in bubble, $\mathrm{p_g}$, and the liquid velocity, the equation

$$p_g = p_{g,t} \mp \frac{\dot{m}^2}{\alpha^2 \Omega^2 \rho_g} \tag{9.98}$$

and Eqs. (9.88) and (9.90), respectively, are used.

The initial and boundary conditions that ensure the uniqueness of the solution are as follows:

$$t = 0; r \geq R: \ \mathrm{R}(0) = \mathrm{R_0};\ \dot{\mathrm{R}}(0) = \mathrm{v_{r=R}}(0) = 0;\ \mathrm{T_l}(0, r) = T_\infty;\ \mathrm{x}(0, r) = x_\infty$$

$$\mathrm{p_{g,t}}(0) = p_\infty + \frac{2\sigma}{R_o} + \Delta p_{g,t_o};\ \mathrm{p_g}(0) = p_\infty + \frac{2\sigma}{R_o};\ \mathrm{p_l}(0, r) = p_\infty$$

and

$$t > 0;\ \mathrm{r} = \mathrm{R}: \ \delta\mathrm{x} = (\mathrm{y} - \mathrm{x})j_r + D_x\frac{\nabla x}{|\dot{\mathrm{R}}|};\ \delta\mathrm{T_l} = \frac{1}{c_{p,l}}\left(j_r C + \frac{\lambda \nabla \mathrm{T_l}}{\rho_l |\dot{\mathrm{R}}|}\right)$$

$$t \geq 0: \quad \mathrm{p_l}(t, \infty) = p_\infty;\ \mathrm{T_l}(t, \infty) = T_\infty;\ \mathrm{x}(t, \infty) = x_\infty$$

respectively.

9.4.1.6 Numerical Integration

The mathematical model presented in Sects. 9.3.1.1 and 9.3.1.2 has been numerically integrated by means of a Runge-Kutta-Gauss code of solving non-linear ordinary differential equation. systems applied to motion equations and by using an explicit scheme with finite differences for equations with partial derivatives. The forward time and the centered space have been used in the scheme with finite differences. The necessary thermodynamic factors (j_r, h_g, etc.) have been calculated at each time step by using the Ziegler and Trepp (1984) state equation. In order to provide greater accuracy of the simulation, the mass diffusivity coefficient, which is very important for the flow, has been adjusted at every time step according to equation

$$D_x = (3x + 2) \times 10^{-9} \tag{9.99}$$

(where the mass fraction x has a value calculated at the previous time step) without affecting the linearity of the equation with partial derivatives. Equation (9.99) is valid within the range $0.0 \leq x \leq 0.7$ and has been determined by linear interpolation of the results obtained by Kojima and Kashiwagi (1995).

9.4.1.7 Model Results of the Ammonia Bubble Absorption

The model solving needs firstly to find the nearest ideal point which the system evolves to, in order to calculate the match function $M(p_{r,e}, x_e)$ intervening in the natural force computation. This has been found applying a practical method (see Sect. 9.2), based on the remark that most systems have state parameters with quasi-constant values all along the evolution period, coming out of their connection with the infinite reservoirs or by fulfiling other engineering purposes etc. Depending on the system variance, these parameters partially or completely define the nearest ideal point. In our case, the ammonia/water system is biphase and bicomponent, so it suffice to have two such parameters, but the type of the parameters may change upon the specific application. For instance, for the absorption process of a closed system absorption refrigeration plant, these parameters are the rich solution final temperature, close to the sink source temperature and the absorbed gas molar fraction, but for the generation process, they change to the final poor solution temperature, close to the warm source temperature and the condensing pressure. In our case, of the bubble absorption, the near ideal point is defined by the quasi-constant absorption pressure ($p_l = p_g$) and gas molar fraction y.

The absorption takes place between gas and the subcooled absorbent, when forces and currents are positive. The set of entry data corresponds to the absorption of the ammonia bubble ($y = 0.992$) in water ($x_\infty = 10^{-4}$), at normal pressure ($p_\infty = 1.001$ bar) and environment temperature ($T_\infty = 293$ K). The solution of the motion equation is shown in Fig. 9.13. Figure 9.14 elucidates the problem of ammonia bubble absorption. Indeed, the reduced absorbed mass current j_r and the actual one

$$j'_{i,g} = j_r \rho_l |\dot{V}| \tag{9.100}$$

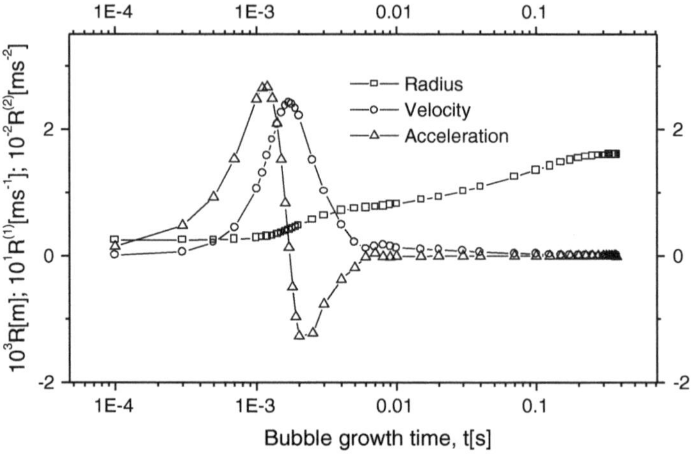

Fig. 9.13 Numerical solution of the dynamics of ammonia bubble absorption

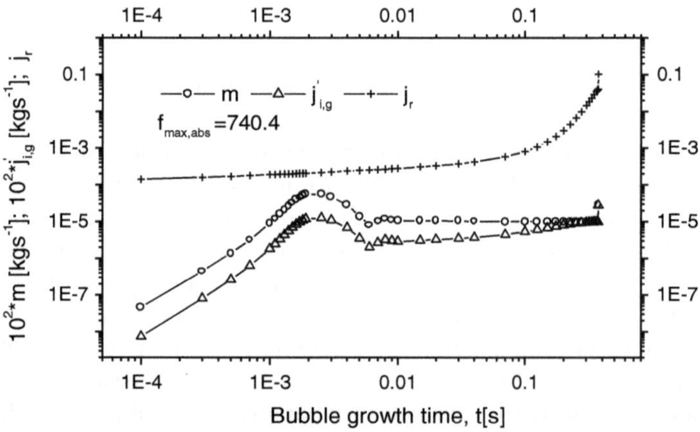

Fig. 9.14 Numerical phenomenological elucidation of the ammonia bubble absorption problem

continuously increase during bubble growth time, causing the bubble pressure decrease, Fig. 9.15. At a certain moment, the actual current equals the bubble mass feeding rate m and the collapse starts. The cumulated absorbed mass current

$$j'_{i,g,a}(t) = \int_o^t j'_{i,g}(z)dz \tag{9.101}$$

and the corresponding coupled heat current $j''_{q,l,a}$ are plotted against the bubble growth time in Fig. 9.16. The liquid mass fraction and temperature distributions are given in Figs. 9.17, and 9.18, respectively, against the same bubble growth time and the radial depth in liquid, measured from the interface. We note that the

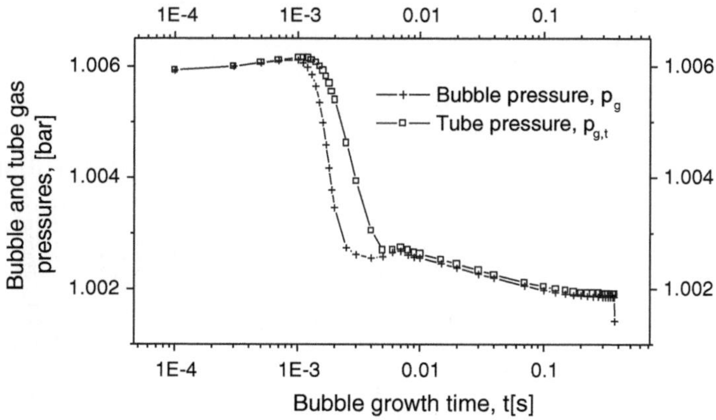

Fig. 9.15 Numerical solution of the tube and bubble pressure variation during ammonia bubble absorption

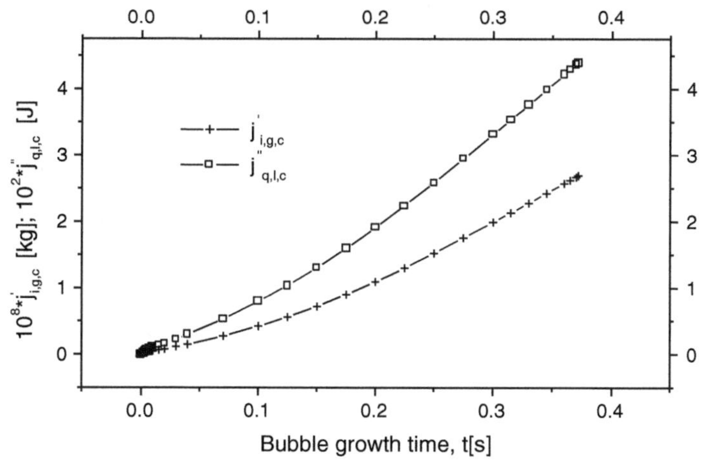

Fig. 9.16 Numerical solution of the cumulated absorbed mass and coupled heat currents during ammonia bubble absorption

liquid mass fraction and temperature almost linearly decrease with the radial depth during a complete evolution, and both ammonia and the thermal field diffuse over a distance of about 1.0×10^{-3} m.

The phenomenological modeling of the bubble absorption may be a valuable lesson for a future advanced thermal absorption technology. Next, we synthesize what is to be learned from it. First, the *absorption process in ammonia/water medium is a mass phenomenon and not a surface one*, as pointed out also by Eq. (9.100). Second, the absorption can be improved in an *intensive* way, seeking not the remoteness from the ideal point (see the classical point of view), but the

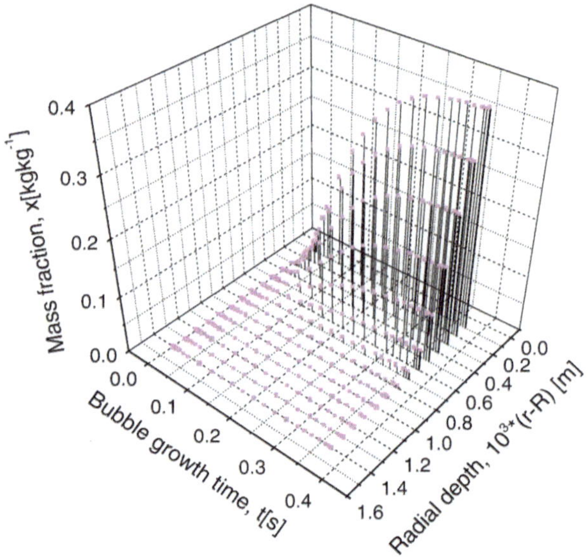

Fig. 9.17 Numerical solution of the liquid mass fraction evolution generated by the ammonia bubble absorption

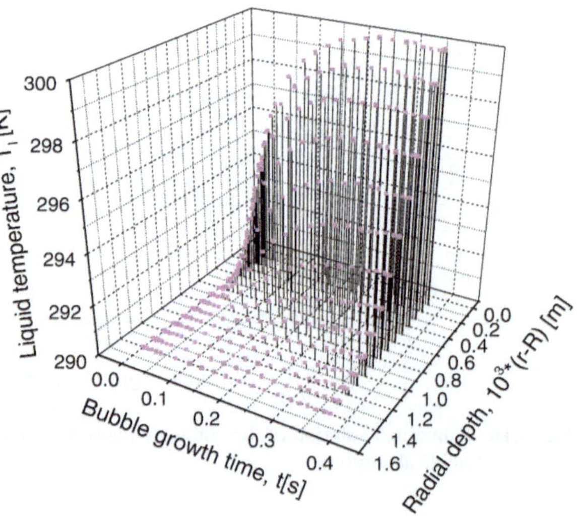

Fig. 9.18 Numerical solution of the liquid temperature evolution generated by the ammonia bubble absorption

nearness to it, when the currents naturally increase up to very high values without any additional technical improvements (see figure of merit $f = 740.7$ in Fig. 9.14). The tendency manifested in the construction of NH_3–H_2O absorbers of increasing the gas–liquid contact area through the dispersion of one of the phases in order to increase the absorption efficiency may not result in the anticipated effect. Quite on the contrary, by extending the contact area the more rapid evolution of the interface towards the equilibrium parameters is delayed (the mass fraction x of the liquid at interface decreases with the increase of the contact area, at the same gas feeding flow rate) and absorption decreases, as compared to the one achieved in a less dispersed phase. Moreover, the auxiliary pumping energy consumption is also increased. Thus, in the light of the above, the increase of the contact area is merely an *extensive* technical solution which helps to compensate the absorption decrease caused by an artificial remoteness from the ideal point, offering in exchange a more stable functioning of the absorbing device. In addition to the above, it is noted that quasi-spherical bubbles with a minimum contact area for the same volume stands for the greatest chances to achieve the best mass and heat transfer through i.p.a. effect appearance. This gives rise to the idea of constructing an NH_3–H_2O bubble flow absorber (Staicovici 1998a). It should be a flat plate and mounted horizontally. In comparison with a present NH_3–H_2O absorber with vertical pipes or with liquid phase dispersion, the bubble absorber offers the following advantages: (i) maximum absorption efficiency; (ii) minimum pressure loss on gas side; (iii) it is suited to a modern compact plate construction; (iv) minimum auxiliary energy consumption.

9.4.2 Analytical Study of NH_3–H_2O Absorption

The use of PhHGD code was extended to an analytical study of ammonia/water absorption (Staicovici 1998a). The modeled scheme in Fig. 9.12 is not provided with a cooling system, the absorption heat being naturally dissipated in the vessel with large heat capacity absorbent. Hence, from this point of view our results may differ to some extent to those which could be obtained on a better cooled absorber. However, they preserve the intrinsic important features of absorption, independent of the flow type, and constitute at the same time a first data base for the bubble absorber proposed in Sect. 9.3.1.4. The absorption is appreciated by the cumulated absorbed mass current $j'_{i,g,a}$ and a specific $j'_{i,g,a}$, which is defined by the integral mean where τ is

$$j'_{i,g,a,s} = \frac{1}{\tau}\int_{0}^{\tau} \mathrm{j}'_{\mathrm{i,g}}(t)dt = \frac{1}{\tau}j'_{i,g,a}(\tau) \qquad (9.102)$$

the bubble life time. The quantity expressed by Eq. (9.102) is a measure of absorption efficiency and is termed as *average indicator of absorption.* Additionally, the results include also information about the coupled heat currents $j''_{q,l,a}$ and $j''_{q,l,a,s}$, defined similarly to theirs mass homologues. The results presentation is confined here only to show the influence on the absorption of the initial liquid temperature, absorption pressure, gas temperature and gas mass fraction, plotted respectively in Figs. 9.19, 9.20, 9.21 and 9.22. Figure 9.19 recommends that the liquid initial temperature should not overcome to much the ideal point temperature, because the average absorption indicator decreases and its increase at higher temperatures is accompanied by important departures of liquid mass fraction x

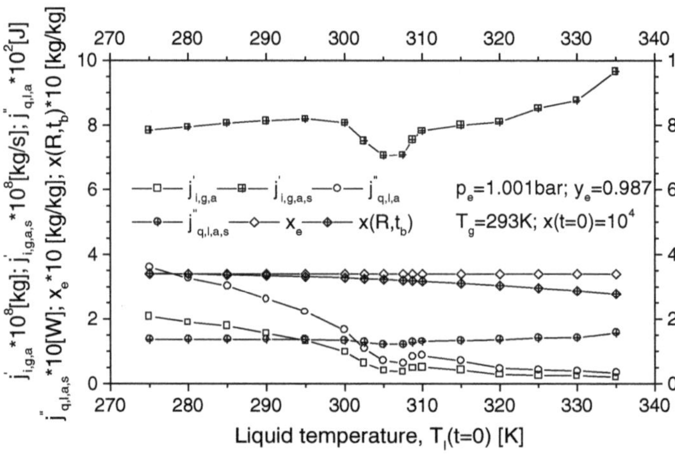

Fig. 9.19 Numerical study case of initial liquid temperature influence on mass and coupled heat currents in ammonia bubble absorption

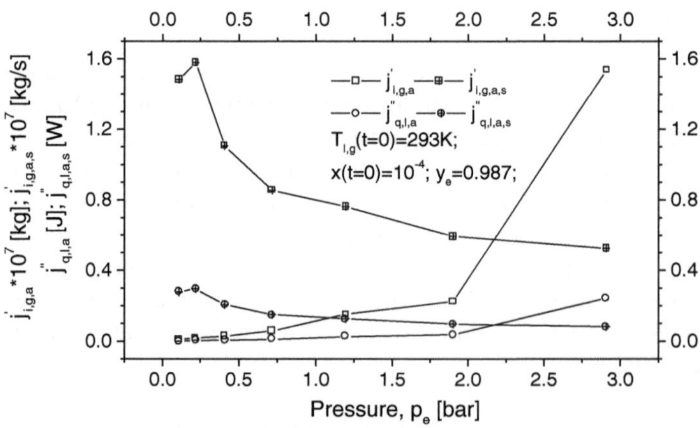

Fig. 9.20 Numerical study case of absorption pressure influence on mass and coupled heat currents

from the ideal value x_e. The absorption pressure shows considerable influence on absorption efficiency, the average indicator decreasing about three times when pressure increases from 0.2 to 2.9 bar, which is very surprising, Fig. 9.20. Working with as much as low absorption pressures is therefore desirable, but obviously this must also be correlated with the other design parameters of a plant. Figure 9.21 shows little influence of the gas temperature on absorption. However, low gas temperatures should be avoided and on this purpose gas superheating, a common practice with absorption refrigeration plants, has from the mass/heat transfer point

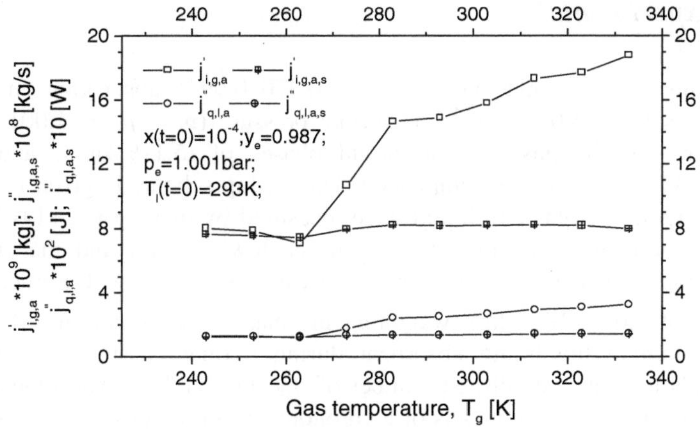

Fig. 9.21 Numerical study case of gas temperature influence on mass and coupled heat currents in ammonia bubble absorption

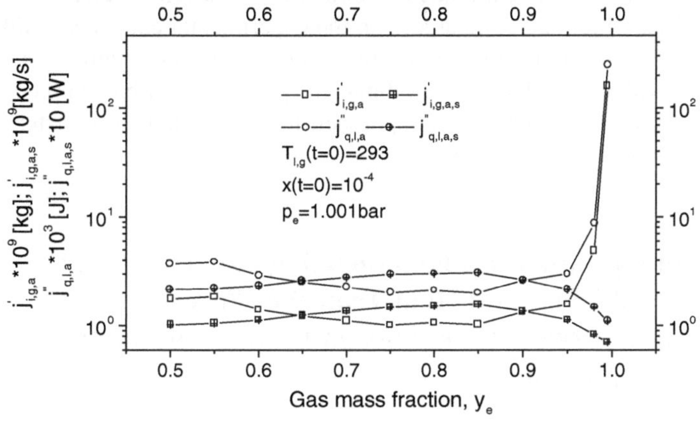

Fig. 9.22 Numerical study case of gas mass fraction influence on mass and coupled heat currents in ammonia bubble absorption

of view a second good reason. The last Fig. 9.22 shows a considerable influence which, this time, the gas mass fraction employed up on absorption efficiency. The cumulated absorbed mass current increases very much with y approaching unit. On the contrary, absorption efficiency decreases about twice from maximal values for same y variation, which is again very surprising. Bearing in mind the high energy consumption of the vapor rectification, also commonly used in absorption plants, this time there is an additional reason to avoid it, that of decreasing the mass and heat transfer in absorption processes. To this extent, on the opposite side, the resorption plants gain twice eliminating rectification and the coabsorbent technology, intrinsically related to the resorption cycle, is favored, besides a better thermal effectiveness, also by a better mass and heat transfer.

9.4.3 Experimental

Experiments of ammonia bubble (y = 0.970–0.991) absorption in distilled water ($x(t = 0) \leq 5.0 \times 10^{-4}$), at normal pressure ($p_l = p_g$ = 1.001 bar) and temperatures of the gaseous and liquid phases of (5.1–9.4) °C and respectively(1.2–17.0) °C, were accomplished. The testing scheme is given in Fig. 9.9. The temperatures of gas and liquid were measured by making use of 0.1 °C precision thermometers. The tube feeding mass flow rate, $\dot{q}_t$, and the cumulated absorbed mass current $j'_{1,g,a}$ were gravimetrically estimated, with a microbalance (precision of 1.0×10^{-7} kg), weighting the mass of ammonia absorbed in the vessel containing the liquid absorbent during a timed period of time which corresponds to a pre-established number of injected bubbles. The time measurements were carried out by means of a chronometer with a precision of 0.1 s. The gas mass fraction was assessed from readings of pressure and temperature of the ammonia/water solution contained in the reservoir and using current equilibrium data of ammonia/water combination. The mean experimental results are given in Table 9.1. It contains the cumulated absorbed mass currents, calculated by means of Eq. (9.101), which are compared with the experimental entry data. Although the experiments are in a reduced number, their good agreement, within ±15 % confidence limits with the model results, validate the proposed combined non-equilibrium and equilibrium phenomenological approach for mass and heat transfer local processes in ammonia/water mixtures.

9.5 A Non-equilibrium Phenomenological (Two-Point) Theory of Mass and Heat Transfer: Forces, System-Source Interactions and Thermodynamic Cycle Applications

In the classic theory, the heat and mass transfer are evaluated in conjuncture with an equilibrium state, where the equilibrium point, thought in a dynamical sense, plays a crucial role. To this extent, more often than not the same meaning is given to this point when a system interaction is taking place, e.g., with phase change in an isolated system, reaching equilibrium, or it occurs in an open system in the presence of a source (infinite reservoir) and finish attaining equilibrium with that, without changing the phase. Mathematically, an equilibrium point (marked by index "e") is expressed by equations of the following type [Eqs. (1.202)–(1.205)]:

$$\begin{aligned} p_l &= p_g = p_e \\ T_l &= T_g = T_e \\ \varphi_{1(2),l} &= \varphi_{1(2),g} \end{aligned} \qquad (9.103)$$

Table 9.1 Experimental versus model comparison for $j'_{i,g,a}$ in NH_3–H_2O bubble absorption tests

Mean experimental									Model	$j'_{i,g,a}$ relative error (%)
Run No.	T_g ± 0.05 (°C)	$T_l\,(t=0)$ ± 0.05 (°C)	$10^3 y \pm 0.5$ (–)	n	$\Delta t \pm 0.05$ (s)	$10^6 \Delta m \pm 5 \times 10^{-8}$ (kg)	$10^{10} q \left(= \frac{\Delta m}{\Delta t}\right)$(kg s^{-1})	$10^{10} j'_{i,g,a} \left(= \frac{\Delta m}{n}\right)$ (kg/bubble)	$10^{10} j'_{i,g,a}$ (kg/bubble)	
1	12.9	1.20	980	40	36.3	1.80	496 ± 14.5	450 ± 12.5	453 ± 11.2	±5.3
2	19.0	2.40	970	40	30.0	1.10	367 ± 17.3	275 ± 12.5	278 ± 13.5	+8.2 −10.4
3	18.5	3.20	988	40	20.5	1.78	867 ± 26.5	447 ± 12.5	431 ± 12.0	+9.5 −2.1
4	17.9	4.75	991	40	26.7	1.70	637 ± 19.9	425 ± 12.5	394 ± 14.2	+14.9 −0.8
5	17.5	6.50	987	40	24.2	1.43	593 ± 21.9	358 ± 12.5	331 ± 12.1	+15.9 −0.4
6	19.2	7.70	983	20	14.6	0.83	572 ± 36.2	417 ± 25.0	404 ± 23.2	+15.3 −8.9
7	19.4	9.80	991	20	8.4	0.85	995 ± 65.5	425 ± 25.0	436 ± 26.3	+9.1 −14.1
8	5.1	11.70	991	40	20.9	1.92	919 ± 26.1	450 ± 12.5	442 ± 11.0	+7.2 −3.6
9	18.9	13.9	989	40	15.2	1.37	903 ± 35.9	342 ± 12.5	321 ± 13.3	+14.9 −1.8
10	18.6	14.8	987	40	15.1	1.21	805 ± 35.8	303 ± 12.5	308 ± 12.1	+6.3 9.5
11	18.4	16.6	985	40	18.5	1.22	659 ± 28.8	304 ± 12.5	292 ± 14.3	+13.5 −5.3
12	18.3	17.0	987	40	16.5	1.25	760 ± 32.6	312 ± 12.5	329 ± 12.2	+2.1 −12.5

$p_l = p_g = 1.001$; $x(t=0) \leq 5.0 \times 10^{-4}$; $\sigma = 6.0 \times 10^{-2}$; $= 0.5$; $R_0 = d = 2.5 \times 10^{-4}$; $V_t = 4.0 \times 10^{-6}$; n number of bubbles

In Eq. (9.103), we have chosen, as example, the particular case of a two-phase gas/liquid and bi-component system. According to the classic assessment, heat and mass transfer are proportional to the system departure from the equilibrium point, which plays the role of a driving force. When this departure is null, the heat and mass currents become zero. As a corollary, the classic heat and mass transfer assessment bases on a "single-point" (equilibrium point) theory (SPT).

In Sects. 9.2 and 9.3, a good part of the non-equilibrium phenomenological theory of mass and heat transfer (NEPhTMHT) was presented. This section will set forth new theoretical and practical aspects of this theory. In the frame of NEPhTMHT, the mass currents are considered the coupling currents and those of heat, the coupled ones, therefore, when writing about the coupled processes at hand, we shall use the expression mass and heat transfer, rather than heat and mass transfer. As it will be further shown, unlike classic point of view, the NEPhTMHT is a "two-point" theory (TPT) (preferred abbreviation next). Accordingly, the forces governing the mass and heat transfer will have different properties as compared to the classic driving forces, depending on the type of point involved in interaction. Finally, a few words for a better understanding of the TPT role: while the two theories, SPT and TPT, are not contradictory but can reciprocally complete, the letter is capable to explain processes which are difficult to be explained by SPT, like the problem of the ammonia bubble absorption, or the heat transfer mechanism in heat pipes, outlined in this section.

9.5.1 Natural Forces of the Coupled and Non-coupled Mass and Heat Transfer

First we shall give a detailed explanation of the physical system, main principle and nomenclature of the TPT.

Basic hypothesis. The interactions at hand are considered to take place in a global system which has as constitutive parts one, two, or more systems and sources (infinite reservoirs). The systems can be mono-, bi-, or poly-component and can behave as isolated, adiabatic, closed, or open, when exchange amongst them and/or with sources mass, heat and mechanical work. The internal mechanical work annuls reciprocally among systems. The sources being of constant parameters, only systems evolution is searched. This theory is based upon the hypothesis according to which the total energy of the systems is considerably equal to their internal energy, that is their macroscopic kinetic energy, and the external fields of any nature that act upon them, as well as their internal tensions are negligible (simple global system at rest). Under the circumstances, the heat exchanged by the global system among its various constitutive parts is based upon the exchange of total enthalpy.

Coupled/non-coupled mass and heat transfer. A coupled mass and heat transfer is a process characterized by:

(a) the change of phase (nature) of the reacting systems;
(b) the presence of a reaction (latent) heat; and
(c) mass and heat currents that influence each other.

On the contrary, a non-coupled transfer does not involve reaction heat, and the systems do not modify their number and initial phase (nature) during interaction.

Mixed transfer of mass and heat. The transfer between systems and sources, simultaneously taking place by coupled and non-coupled mass and heat currents, will be referred to hereinafter as *mixed transfer.*

Interaction type. In the general case, n' constitutive parts A_i', $i = 1, \ldots, n'$, of the global system interact amongst them by coupled and/or non-coupled mass and heat transfer, giving rise to n'' constitutive parts A_i'', $i = 1, \ldots, n''$. Additionally, only the interactions through an intermediary system between two or more sources have sense.

Natural forces. The thermodynamic non-equilibrium (natural) forces govern the coupled/non-coupled mass and heat transfer. They differ from the classic thermodynamic ones (driving forces), here called equilibrium (ideal) forces (see Sect. 9.3). The natural forces are vector dimensions, that is two mass and heat transfer processes serially connected, occurring between states A and B, and B and C, respectively, have the same resulting natural force $\left(-\vec{X}\right)_{AB} + \left(-\vec{X}\right)_{BC}$ as that of mass and heat transfer process taking place from state A directly to state C, $\left(-\vec{X}\right)_{AC}$, or

$$\left(-\vec{X}\right)_{AB} + \left(-\vec{X}\right)_{BC} = \left(-\vec{X}\right)_{AC} \tag{9.104}$$

The same vector dimension exhibit also the mass and heat currents, proportional to the natural forces. Throughout this section, the transfer is deemed to be axial and the forces to be mainly scalar dimensions, without losing any useful information.

Basic points of interactions. The TPT makes distinction between two kinds of points towards which systems could "evolve". These points are defined by the state parameters (p, T, x, y,...) of the interacting constitutive parts. If such parameters define a source of constant parameters (p, T, x, y,...) = *const.*, then the respective point will be called *equilibrium (absolute) point.* Such point will be marked here with the index "ep". The terms "equilibrium" and "absolute" are understood herein in the static sense, i.e. which does not depend on the searched system and does not alter unless the parameters of the source alter. The natural forces are defined in an equilibrium point, where take the value zero. As specified in Sects. 9.2 and 9.3 and as it is going to be further shown, the points of a second type are the *ideal (relative) points.* Such points will be marked in this chapter with the index "ip". A point in this category is found in the interactions with coupled transfer. Depending on their nature, the systems can have different coupled transfers (e.g., condensation (absorption), vaporization (generation), solidification, melting, sublimation, de-sublimation). For this reason, the ideal points will be called *ideal points of condensation*, *vaporization*, etc. The ideal point is an equilibrium point in the classic dynamic (statistical) sense (Kirilin et al. 1985) where the static equilibrium cannot be reached, for which reason it has been called "ideal". The ideal point is defined by equations similar to those given in Eq. (9.103). A natural force is not defined in an ideal point (see Sect. 9.3). Depending on the interacting system, the ideal point has also been called "relative", in order to differentiate it even further from the absolute equilibrium point. The isolated systems with coupled transfer evolve towards an ideal point under the action of natural forces which comply with the postulate, lemma and corollary of Sect. 9.3. In a mixed transfer process, the ideal points also intervene, but their aspect is influenced by the interaction with the source. For this reason, in such points the natural force complies only partially with the lemma of Sect. 9.3, in the sense that the signs of the forces adjacent to the point are different, but their absolute value differs.

Entropy source, currents, natural forces and phenomenological coefficients. The stationary mass and heat transfer is studied within the frame of TPT by means of the entropy source, given by

$$\dot{S} = -\frac{\dot{E}}{T_0} = -\left(\sum j's' + \sum j''s''\right)_{mass} - \left(\sum \frac{j'_q}{T'} + \sum \frac{j''_q}{T''}\right)_{heat} \tag{9.105}$$

[see Eq. (9.6)], where j and j_q are mass and heat currents, respectively, marked as prime (′) and second (″), before and after interaction, respectively. The summation in Eq. (9.105) is done over all systems of the global system. Here sources were not included simply because the corresponding terms in the entropy source equation, which shows their contribution to the global increase of entropy, are annulled. Equation (9.105) can be also written as follows:

$$\dot{S} = j'_1(-X) \tag{9.106}$$

where j'_1 is one of the coupling mass currents (herein chosen the first), and $(-X)$ is the natural force. In its turn, the mass current is proportional to the force $(-X)$ and to a phenomenological coefficient, L_1, according to the equation below:

$$j'_1 = L_1(-X) \tag{9.107}$$

In its general form, according to the working hypothesis of the theory the force has the expression:

$$(-X) = -\left(\sum j'_{i,r}s'_i + \sum j''_{k,r}s''_k\right) - \left(\sum j'_{i,r}\frac{h'_i}{T'_i} + \sum j''_{k,r}\frac{h''_k + j_q}{T''_k}\right) \tag{9.108}$$

where $j'_{i,r} = j'_i/j'_1$ and $j''_{k,r} = j''_k/j'_1$, $i = 1,\ldots,n'$, $k = 1,\ldots,n''$, j_q is the reaction, latent or sensible heat resulting from interactions, and h' and h'' are the specific enthalpies of the interacting constitutive parts before and after interaction, respectively.

Natural forces of the coupled transfer. The coupled transfer is dealt with in detail in Sect. 9.2. It occurs in case of the physical or chemical interactions.

1. *Mono-, bi- and particularly poly-component systems with physical interaction in non-ideal mixture.*

$$(-X_{cpl}) = -\frac{X_i + h_g\left(\frac{1}{T_g} - \frac{1}{T_l}\right) + \frac{Ah_l}{T_l[1+AL_{cpl}(-X_{cpl})]}}{1 + \frac{A(h_g - h_l)}{T_l[1+AL_{cpl}(-X_{cpl})]}} \tag{9.109}$$

where

$$X_i = (s_g - s_l) - (y - x)\left(\frac{\partial s_l}{\partial x}\right)_{p,T} \tag{9.110}$$

and

$$A = \frac{(y - x)}{T_l \cdot c_{p,l}}\left(\frac{\partial h_l}{\partial x}\right)_{p,T} \tag{9.111}$$

Equation (9.109) was written for the interactions of gas–liquid generation/absorption type, but if adequate indices are used it is also valid for the other types of solid–liquid or solid–gas two-phase two-component physical interactions where the phase is changing. The sign of the force $(-X)$ was established so as to have a physical sense: in case of absorption the currents and forces are positive, therefore $(-X)_{cpl,abs} > 0$, and in case of generation they are negative, therefore $(-X)_{cpl,gen} < 0$. The typical aspect of a force described by Eq. (9.109) is shown in Fig. 9.23, for the NH_3–H_2O system with physical absorption/generation interaction. For the case study, the ideal point (state) is at $x_{ip} = 0.3396$ (marked by the dotted vertical line). The gas mixture absorption and generation take place for $x < x_{ip}$ and $x > x_{ip}$ liquid mass fractions, respectively. The partial forces $(-X)_{cpl,abs} > 0$ and $(-X)_{cpl,gen} < 0$, and the currents as well, respect the natural force postulate. The figure of merit, "f", ranking the *i.p.a.* effect, equals here the values $f_{\max,abs} = 535$ and $f_{\max,gen} = 78$.

2. *Interaction of pure component systems which change the phase.*

$$(-X)_{cpl,p} = -(s' - s'') - \max(h', h'')\left(\frac{1}{T'} - \frac{1}{T''}\right) \tag{9.112}$$

Equation (9.112) was obtained from Eq. (9.108), considering $n' = n'' = 1$ and the thermal balance to the level of the system for the direct and reverse interaction $h' = h'' \pm j_q$, where j_q is in this case the latent heat of phase change.

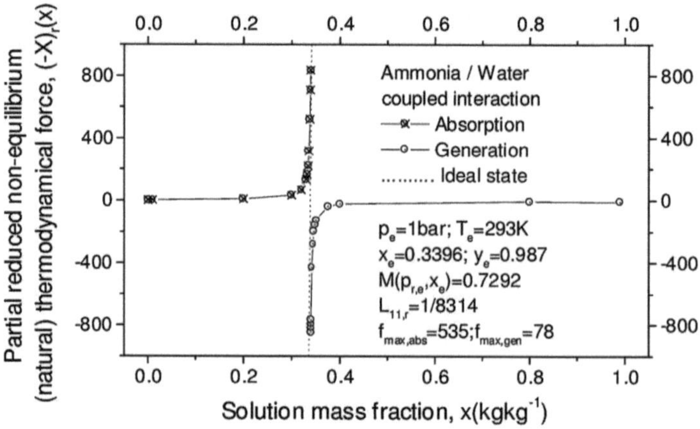

Fig. 9.23 Typical force behavior in case of a physical coupled interaction

Particularizing Eq. (9.112) for condensation or solidification processes, it results that:

$$(-X)_{cpl,c(s)} = -(-X)_{cpl,p} > 0 \tag{9.113}$$

while, for vaporization or melting, it results that:

$$(-X)_{cpl,v(m)} = (-X)_{cpl,p} < 0 \tag{9.114}$$

Typical behavior of force governing an interaction between pure component systems which change the phase is shown in Fig. 9.24 for nitrogen and oxygen vaporization/condensation. Again, the ideal points position is marked by dotted vertical lines and the forces respect the natural force postulate. However, while the interactions considered are once more of a physical type, the i.p.a. effect is this time much less pronounced, as compared to that of bi-component system interactions, Fig. 9.23.

3. *Independent chemical interaction*. The force of an independent chemical interaction is given in detail in Sect. 9.4. Its expression follows:

$$(-X)_{cpl,ch} = -\left\{\sum_{i=1}^{n'} j'_{i,r}\left[s'_i + \max\left(h'_i, h''_i\right)\left(\frac{1}{T'_i} - \frac{1}{T_e}\right)\right] - \sum_{i=1}^{n''} j''_{i,r}s''_i\right\} \tag{9.115}$$

As previously mentioned, the summation in Eq. (9.115) is over all reactants and produces of reaction. In case of the direct and reverse chemical reactions the next equations are valid:

$$(-X)_{cpl,ch,d} = (-X)_{cpl,ch} \tag{9.116}$$

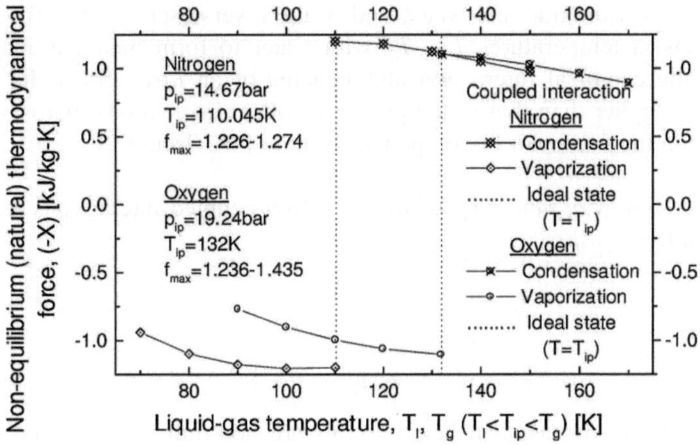

Fig. 9.24 Typical force behavior in case of a physical coupled gas–liquid interaction with pure systems

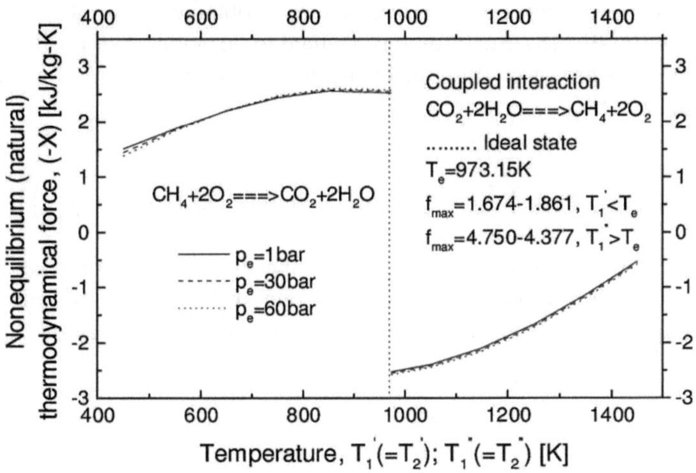

Fig. 9.25 Typical force behavior in case of an independent chemical interaction

and

$$(-X)_{cpl,ch,i} = -(-X)_{cpl,ch} \tag{9.117}$$

respectively. A typical natural force graph of an independent chemical interaction is shown in Fig. 9.25 (see also Fig. 9.11). The case study analyses the natural force behavior around an ideal point ($T_{ip} = 973.15$ K) for three constant reaction pressures (T_{ip} and pressures are chosen at random), when methane interacts in the direct sense at temperatures $T < T_{ip}$ with the oxygen to form carbon dioxide and water, and in the reverse sense the carbon dioxide interacts at temperatures $T > T_{ip}$ with water to form methane and oxygen. Here the chemical interaction also benefits of an *i.p.a.* effect. Its figure of merit is higher than that of the pure component interactions, but is still much less than the *f*-value of bi-component systems (see Sect. 9.2).

Natural forces of non-coupled transfer. Non-coupled interactions are characterized by:

$$\begin{aligned} & n' = n'' \\ & j_i' = j_i''; \quad i = 1, \ldots, n' \\ & j_q = 0 \\ & \nu_i' = \nu_i''; A_i' \equiv A_i''; \quad i = 1, \ldots, n' \end{aligned} \tag{9.118}$$

While involved mass and heat currents are non-null, as it can be remarked from Eqs. (9.117) in case of the non-coupled interactions we do not expect to have mass transfer, but only heat and/or mechanical work transfer.

1. *Heat exchange based interaction.* The natural force of an isobar heat exchange based non-coupled interaction is obtained from Eq. (9.115), by applying Eq. (9.118). Such force is deemed to be negative or positive, whether the processes are of cooling, or of heating, respectively, and is given by (see also Eq. 9.71):

$$(-X)_{ncpl} = -\sum_{i-1}^{n'} j_{i,r}' \left[\left(s_i' - s_i'' \right) + \max\left(h_i', h_i'' \right) \left(\frac{1}{T_i'} - \frac{1}{T_{ep}} \right) \right] \tag{9.119}$$

In case of interactions without sources, the unknown factor of the problem (the final equilibrium temperature T_{ep} or one of the currents) is calculated out of a simple thermal balance of the system in equilibrium given by Eq. (9.120):

$$\sum_{i=1}^{n'} j_{i,r}' c_{p,i} \left(T_{ep} - T_i' \right) = 0 \tag{9.120}$$

Referring to interactions with a source, Eq. (9.119) holds true again and, as previously mentioned, the summation is over all systems interacting with that source through non-coupled transfer. If T_s is the source constant temperature, then in Eq. (9.119) we must consider $T_{ep} = T_s$.

As previously mentioned, the natural forces have a vector additive property. By using this property, the natural force of a non-coupled isobar interaction may be expressed in a simpler form by means of Eq. (9.121) (see Appendix):

$$(-X)_{ncpl,s,p} = \sum_{i=1}^{n'} j_{i,r}' \left[\left(\frac{h}{T} \right)_i' - \left(\frac{h}{T} \right)_{ep,i} \right]_p \tag{9.121}$$

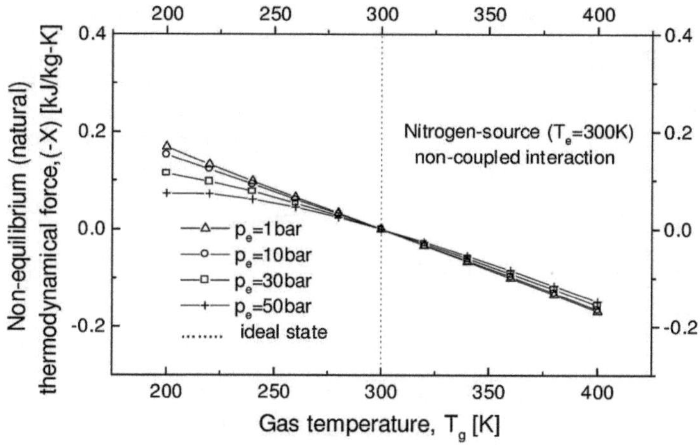

Fig. 9.26 Typical force behavior for a non-coupled interaction. Case study of nitrogen-source interaction

It can be easily demonstrated that Eq. (9.121) is also valid for $n' = 1$ in case of the forces of coupled isobaric interactions with pure component systems [Eq. (9.112)]. It is noted that the state function $(h/T) = s_{p,t}$ acquires a special significance in this context. For this reason, it will be referred to hereinafter as *isobar (mass and heat) transfer entropy*. The natural force of an isocore heat exchange based non-coupled interaction is obtained from Eq. (9.121), replacing enthalpy, h, by the internal energy, u:

$$(-X)_{ncpl,s,v} = \sum_{i=1}^{n'} j'_{i,r} \left[\left(\frac{u}{T} \right)'_i - \left(\frac{u}{T} \right)_{ep,i} \right]_v \tag{9.122}$$

In Eq. (9.122) the new thermodynamic function $(u/T) = s_{v,t}$ is termed *isocore (mass and heat) transfer entropy* (see Appendix).

The typical forces aspect of such non-coupled transfer is the one shown in Figs. 9.26 and 9.27. For the sake of simplicity, in Eq. (9.122) we considered $n' = 1$. The two case studies refer to isobar interactions of nitrogen and carbon dioxide with a constant temperature source, respectively. In this case the heat transfer is the classical conductive or convective one, proportional to the temperature difference of the systems in a reasonable broad vicinity of the equilibrium point. The forces have a monotonous variation in the cooling $T > T_{ep}$ and heating $T < T_{ep}$ processes and take value zero in the equilibrium points.

2. *Mechanical work (power) exchange based interaction*. Finally, the natural force of an adiabatic non-coupled interaction with a source will be established, that is the force whereby the systems of the global system exchange mechanical work (power) with a source. This force results from Eq. (9.106),

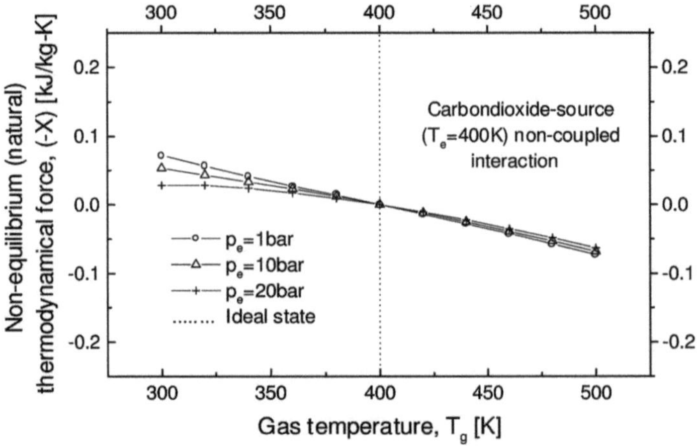

Fig. 9.27 Typical force behavior in case of a non-coupled interaction. Case study of carbondioxide-source interaction

assuming $j_q = 0$, given that the mechanical work is obtained by the consumption of the internal energy of the systems interacting with the source, having as parameters (p_{ep}, T_{ep}):

$$(-X)_{ncpl,w} = \sum_{i=1}^{n'-1} j'_{i,r}\left(e'_i - e''_i\right) \tag{9.123}$$

where

$$e = s + \frac{h}{T} \tag{9.124}$$

3. *Coupled/non-coupled transfer general characteristics*. Because, according to the postulate and its consequence (*i.p.a.* effect), the forces and currents continuously increase in absolute value to maximal finite values approaching an ideal point, the coupled transfer of mass and heat is a fast ("accelerated") transfer. Unlike these systems, the non-coupled interaction systems evolve towards an equilibrium point under the action of natural forces that tend to zero and have the value zero in this point. The resulting mass and heat currents preserve the characteristic of the force and do not benefit by the *i.p.a.* effect. That is why the non-coupled transfer is a slow ("decelerated") transfer.

9.5.2 System-Source Interactions

Given that the non-coupled transfer is slow, it is preferable to be avoided in power installations. The extents to which this can be achieved and the relation between the two types of transfer referred to above when a system and a source, as constitutive parts of the global system, interact one to the other, are presented hereinafter. To the beginning, the following first basic theorem will be demonstrated.

Theorem (Of the equilibrium (absolute) point)*A monocomponent or poly-component system and a source (infinite reservoir) of constant parameters may evolve and reach equilibrium together only by a non-coupled mass and heat transfer.*

Proof It will be first demonstrated by means of the reductio ad absurdum method that the system and the source cannot be in equilibrium by a coupled transfer. Basically, the sources are considered to preserve unaltered their physical state and parameters during any interaction with systems. For the sake of clarity only, here we shall consider that our source is an infinite heat reservoir of constant temperature T. It is assumed that the system is in equilibrium with the source by a coupled transfer. In this case, their interaction results in either a change of phase (nature) of the source, which is absurd, or it follows that the source, according to the corollary of Sect. 9.3, is not defined in $t = T$, but fluctuates at equilibrium between $T - \Delta T$ and $T + \Delta T$, where ΔT is a finite difference of temperature, which is again absurd, contradicting the hypothesis according to which T is constant and thus the first part of the theorem is demonstrated. A first conclusion which can be drawn out of the above is that at equilibrium the force and the currents of this interaction are annulled. In order to demonstrate that the system and the force not even cannot evolve together towards equilibrium by coupled transfer, it is sufficient to remind that, according to the postulate, the system would tend towards equilibrium with the source by forces and currents continuously increasing towards maximum finite absolute values, which are not annulled in the ideal point. This is absurd, as it contradicts the conclusion of the first part of the demonstration, and thus the theorem is demonstrated (q.e.d.).

The theorem is also valid in the event that the system evolving towards equilibrium with the source goes through various phases where coupled transfer also occurs. Let us assume, for instance, a gaseous system which is isobar cooled in its interaction with a sink source $T_s(= T_{ep})$ from temperature T_g to temperature $T_l(= T_{i.p} - \varepsilon)$, at which temperature it condenses, and the condensate is sub-cooled up to T_s, when reaches the equilibrium. The temperatures involved in this process are, as it can be noticed, in the relation $T_g > T_{ip} > T_l \geq T_s$. The coupled transfer of gas (cooling + condensation) and mixed [gas (cooling + condensation) + liquid sub-cooling] are described by the natural forces given by Eq. (9.112)

$$\begin{aligned} (-X)_{cpl,c} &= \left(s_g - s_l\right) + h_g\left(\frac{1}{T_g} - \frac{1}{T_l}\right) > 0, \\ T_l &= T_{ip} - \varepsilon \end{aligned} \tag{9.125}$$

and

$$(-X)_{mixed,c} = \left(s_g - s_{ep}\right) + h_g\left(\frac{1}{T_g} - \frac{1}{T_{ep}}\right) > 0 \tag{9.126}$$

respectively. If Eq. (9.126) is subtracted from Eq. (9.125), the result is a force of Eq. (9.119) type:

$$(-X)_{ncpl,c} = -\left(s_l - s_{ep}\right) - h_g\left(\frac{1}{T_l} - \frac{1}{T_{ep}}\right) < 0 \tag{9.127}$$

which is just the force of the non-coupled transfer process stipulated by the theorem. The non-coupled transfer is adjacent to the source and interposes between the source and the coupled process described by Eq. (9.112). This configuration of transfer with a source is general. It always appears whenever a mixed transfer process takes place. Developing the given example for the reversed process, of vaporization, the liquid warms isobar as a result of the interaction with a warm source $T_s(= T_{ep})$, from temperature T_l to temperature $T_g(= T_{i.p} + \varepsilon)$. At this temperature it vaporizes, and the resulting vapor is superheated up to T_s, when reaches the equilibrium. The temperatures involved in the process are in the relation $T_l < T_{ip} < T_g \leq T_s$. The homologous equations to Eqs. (9.125)–(9.127) are the following:

$$\begin{aligned} &(-X)_{cpl,v} = -\left(s_g - s_l\right) - h_g\left(\frac{1}{T_g} - \frac{1}{T_l}\right) = -(-X)_{cpl,c} < 0, \\ &T_g = T_{ip} + \varepsilon \end{aligned} \tag{9.128}$$

$$(-X)_{mixed,v} = -\left(s_{ep} - s_l\right) - h_g\left(\frac{1}{T_{ep}} - \frac{1}{T_l}\right) < 0 \tag{9.129}$$

and

$$(-X)_{ncpl,v} = -\left(s_g - s_{ep}\right) - h_g\left(\frac{1}{T_g} - \frac{1}{T_{ep}}\right) > 0 \tag{9.130}$$

respectively. From Eqs. (9.125)–(9.127) or (9.128)–(9.130) and in accordance with Eq. (9.104), it is noted that the force of the mixed transfer has additive properties as compared to the partial forces of the coupled, respectively, non-coupled transfer. Generalizing, the following vector equation is obtained:

$$\left(-\vec{X}\right)_{mixed} = \left(-\vec{X}\right)_{cpl} - \left(-\vec{X}\right)_{ncpl} \tag{9.131}$$

which is valid for all two-phase interactions with mixed transfer where the coupled transfer results from a physical or a chemical interaction [Eqs. (9.109)–(9.117)]. Coming back to the previous examples, for a certain chosen gas and preestablished

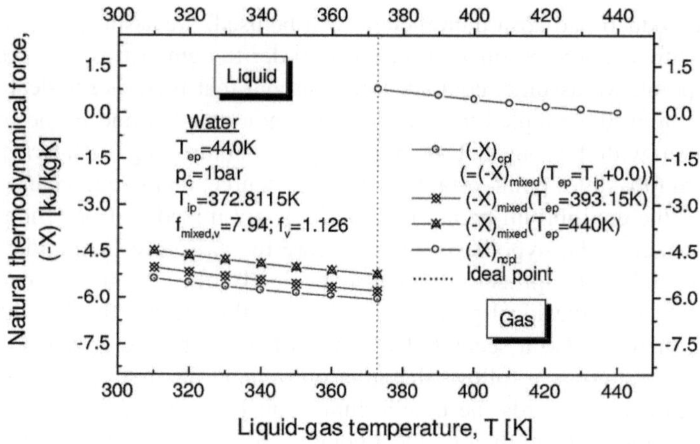

Fig. 9.28 Study case of natural forces of a mixed transfer in water [liquid (heating + vaporization) + gas super-heating]

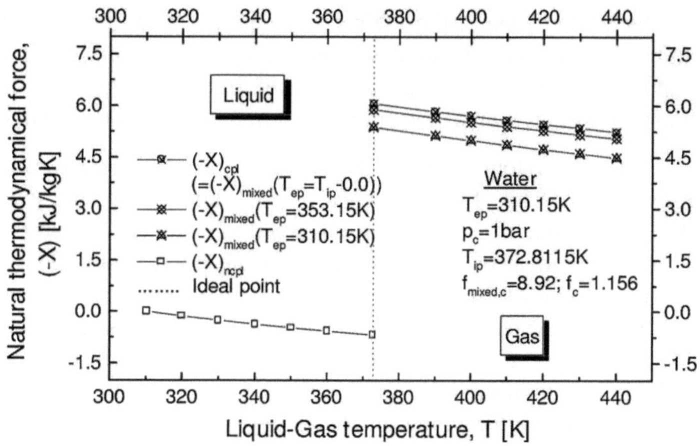

Fig. 9.29 Study case of natural forces of a mixed transfer in water [gas (cooling + condensation) + liquid sub-cooling]

temperatures T_g or T_l and T_s, the $\left(-\vec{X}\right)_{mixed}$ force has a determined value. A simple analysis of the two terms of the right member of Eq. (9.131) shows that forces have contrary signs (see also Figs. 9.28 and 9.29), but the non-coupled and coupled transfers do not have contrary effects. This observation precedes the making evident of a more general property of the mixed transfer. In order to identify such property, the following second basic theorem must be first demonstrated:

Theorem (Concerning the existence and uniqueness of the ideal (relative) point) *If a system interacts with a source (infinite reservoir) by mixed stationary mass and heat transfer, then the constitutive isobar processes, of coupled and non-coupled transfer, respectively, are always separated by an ideal (relative) point which is unique.*

Proof The reductio ad absurdum method will be used. We assume a system which interacts with a source by mixed transfer. In order to demonstrate the existence of the ideal point, we assume, contrary to all reason, that there is no ideal (relative) point separating the coupled transfer from the non-coupled one in the interaction of the system with the source. Then, the only point controlling the interaction is the equilibrium (absolute) point, established by the source, and in accordance with the theorem of the equilibrium point, the transfer is not mixed, but only non-coupled, which contradicts the hypothesis and therefore the existence of the ideal point is demonstrated. The reciprocal statement is evidently true. In order to demonstrate the uniqueness of the ideal point, it is assumed that beside the first ideal point identified, there is also a second. Then, according to the reciprocal of the existence of the ideal point, it follows that after outrunning the first ideal point, the system evolves again towards the equilibrium point by a mixed transfer, so that its natural force moves from the first ideal point to the second by having continuously increasing values in absolute value, according to the postulate of the natural force of the coupled transfer, set forth in Sect. 9.3. We will now imagine that the system evolves in the reverse direction. This leads to the conclusion that the coupled transfer process which takes place from the second ideal point to the first is characterized by natural forces continuously decreasing in absolute value, which is absurd, contradicting the postulate of the natural force referred to above. Thus the uniqueness of the ideal point, as well as the entire theorem, are demonstrated (q.e.d.).

It follows that a system with mixed transfer cannot contain two or more coupled transfers in series. Indeed, as we already know from the experimental physics, this means that if a system is subjected to a mixed transfer where the distance between the ideal point and the equilibrium point is sufficiently large so that the system is forced to undergo more than a change of phase, then it passes directly from the first phase to the last one, which corresponds to the equilibrium with the source, without going through the intermediary phase changes (e.g., the mixed transfer by sublimation or de-sublimation).

As a conclusion of the foregoing, it can be stated that the non-coupled transfer systems may reach equilibrium with sources by reversible transformations round the equilibrium point, to which, according to Eq. (9.105), the null velocities of exergy destruction $\left(\dot{E}\left(z_{ep}\right) = 0\right)$ correspond. On the contrary, the coupled transfer systems can never reach equilibrium with sources and that is why their transformations are irreversible, far from equilibrium, and reach maximum values of exergy destruction in the vicinity of the ideal points. In other words, equilibrium is not possible but only with the sources.

Now, taking into consideration the above theorem and the lemma demonstrated in Sect. 9.3, the following corollary could be formulated:

Corollary *If a system evolves towards the equilibrium dictated by a source through a mixed mass and heat transfer, then the natural forces of the partial coupled and non-coupled transfers composing it have always contrary signs.*

The above examples will be particularized to a case study of two mixed transfers in water at normal pressure: (i) (liquid heating + vaporization), Fig. 9.28, and (ii) (condensation + liquid cooling), Fig. 9.29. The natural forces describe them subject to the temperatures of the liquid and gas. In Fig. 9.28, the source equilibrium point is $T_{ep} = 440$ K. The system evolves towards this point through a coupled transfer of liquid heating and vaporization at $T_g = T_{ip}(= 372.8115 \text{ K}) + \varepsilon$ [see $(-X_{cpl})$], and non-coupled of gas superheating, whereby, in accordance with the equilibrium point theorem, it reaches equilibrium at $T_g = T_{ep} = 440$ K [see $(-X)_{ncpl}$]. Approaching T_{ep} and T_{ip}, the force of the entire process $(-X)_{mixed}$ increases in absolute value (see $(-X)_{mixed}$ at $T_{ep} = 393.15$ K and $T_{ep} = 440$ K). To the limit, when $T_{ep} = T_{ip} + \varepsilon$, we have the greatest absolute value of the mixed force $(-X)_{mixed} = (-X)_{cpl}$. It is noted that, in accordance with the ideal point theorem corollary and Eq. (9.114), the forces have contrary signs $(-X)_{cpl,v} < 0$ and $(-X)_{ncpl} > 0$, and the two transfers are separated by an ideal vaporization point with unequal branches, around which $\left|(-X)_{cpl}\right| \gg (-X)_{ncpl}$. Proceeding similarly as to the *i.p.a.* effect evaluation, in case of the mixed transfer a merit factor which emphasizes the difference between the forces adjacent to an ideal point could also be introduced,

$$f_{mixed} = \left| \frac{(-X)_{cpl}(z_{ip} \pm \varepsilon)}{(-X)_{ncpl}(z_{ip} \mp \varepsilon)} \right| \tag{9.132}$$

Here $f_{mixed,v} = 7.94$ and and for the *i.p.a.* effect, $f_v = 1.126$. Equation (9.132) prefigures the advantage of using the coupled transfer in transport phenomena, as compared to the noncoupled transfer (conductive or convective) [see the heat pipe (Stefan 1995; Bilegan 1982)].

In Fig. 9.29, the source has $T_{ep} = 310.15$ K. The system evolves towards this point by a coupled transfer of gas cooling and condensation at $T_l = T_{ip}(= 372.8115 \text{ K}) - \varepsilon$ [see $(-X_{cpl})$] and a non-coupled of liquid subcooling, up to equilibrium at $T_l = T_{ep}(= 310.15 \text{ K})$ [see $(-X)_{ncpl}$]. Approaching T_{ip} and T_{ep}, then again $(-X)_{mixed}$ increases, approaching $(-X)_{cpl}$ [see $(-X)_{mixed}$ at $T_{ep} = 310.15$ K and $T_{ep} = 353.15$ K]. To the limit, when $T_{ep} = T_{ip} - 0.0$, we have the greatest mixed force $(-X)_{mixed} = (-X)_{cpl}$. The forces signs are different here too $(-X)_{cpl,c} > 0$ and $(-X)_{ncpl} < 0$, and the two transfers are separated by an ideal vaporization point with unequal branches, where $\left|(-X)_{ncpl}\right| \gg (-X)_{cpl}$, so that $f_{mixed,c} = 8.92$ [Eq. (9.132)] and $f_c = 1.156$.

9.5.3 Phenomenological Coefficients of Mixed Transfer and the Theorem Concerning the Maximization Thereof

The mass and heat currents are calculated from the phenomenological point of view by means of equations (see Sect. 9.2):

$$\left(\vec{j_i}\right)_{phenom} = L\left(-\vec{X}\right) \tag{9.133}$$

and

$$\left(\vec{j}_q\right)_{phenom} = \left(\vec{j}_i\right)_{phenom} C = L\left(-\vec{X}\right)C \tag{9.134}$$

respectively, which are valid for the coupled, non-coupled or mixed transfer. In Eq. (9.134), C stands for the thermal capacity. In case of a bi-component system physically interacting with one source only, C is given by [see also Eq. (9.23)]:

$$C = \left(h'(z) - h''(z)\right) - (y - x)\left(\frac{\partial h'(z)}{\partial x}\right)_{p,T} \tag{9.135}$$

In case of a system interacting with different sources between two states the forces are modified accordingly and $C = h'(z)$.

According to the phenomenological equilibrium theory, the phenomenological forces and coefficients are deemed to be weakly dependent on each other. The relation between the natural force and the phenomenological coefficient is maintained here too. The phenomenological coefficient depends on the thermal properties and the dynamics of the interacting elements. The phenomenological coefficients involved in a mixed transfer will be established in this section. For the sake of clarity, the calculation will be particularized to the mixed transfer of [gas (cooling + condensation) + liquid sub-cooling], without diminishing the generality of the results. Indeed, Eqs. (9.133)–(9.135) are taken into account and the mass and heat currents of the transfer are written in accordance with Eq. (9.131), which leads to:

$$L_{mixed}\left(-\vec{X}\right)_{mixed} = L_{cpl}\left(-\vec{X}\right)_{cpl} - L_{ncpl}\left(-\vec{X}\right)_{ncpl} \tag{9.136}$$

and

$$L_{mixed}\left(-\vec{X}\right)_{mixed}\left(h_g - h_{ep}\right) = L_{cpl}\left(-\vec{X}\right)_{cpl}\left(h_g - h_l\right) - L_{ncpl}\left(-\vec{X}\right)_{ncpl}\left(h_l - h_{ep}\right) \tag{9.137}$$

respectively. Equations (9.136) and (9.137) are solved together with Eqs. (9.131) and (9.132), assuming that the forces $(-X)_{cpl}$ and $(-X)_{ncpl}$ are collinear and we obtain:

$$L_{ncpl} = -L_{cpl} f_{mixed}\, r_h < 0 \tag{9.138}$$

and

$$L_{mixed} = L_{cpl}\frac{f_{mixed}}{1 + f_{mixed}}(1 - r_h) > 0 \tag{9.139}$$

where

$$r_h = \frac{\min\left|h\left(z_{ep}\right) - h\left(z_{ip} \pm \varepsilon\right)\right|}{\max\left|h'(z) - h\left(z_{ip} \pm \varepsilon\right)\right|} > 0 \tag{9.140}$$

Equations (9.138) and (9.139) are valid for all interactions with mixed transfer. The phenomenological coefficient of the coupled transfer of a physical or chemical nature, L_{cpl} must be established prior to calculating L_{ncpl} and L_{mixed}. In the case of (ideal) (′)–liquid (″) (ideal) gas (′)–solid (″) and liquid (′)–solid (″) interactions, this coefficient has the expression:

$$L_{cpl} = \frac{j_r}{(-X)_{cpl}} j_i'' \tag{9.141}$$

where

$$j_r = \pm \frac{p'}{\rho'' c_v' T'} \tag{9.142}$$

or

$$j_r = \pm \frac{\rho'}{\rho''} \tag{9.143}$$

and $j_i'' = \rho'' \left| \dot{V}' \right|$ is established for phenomenological and dynamic reasons (see Sect. 9.4). The above results allow us to highlight the characteristics of the mixed transfer itself and in comparison with the non-coupled transfer. Indeed, let us to establish the ratio of the mass and heat currents of mixed and noncoupled transfer involved in the interaction with the source previously used:

$$r_i = \frac{j_{i,mixed}}{j_{i,ncpl}} = \frac{L_{mixed}(-X)_{mixed}}{L_{cpl}(-X)_{cpl}} \tag{9.144}$$

and

$$r_q = \frac{j_{q,mixed}}{j_{q,ncpl}} = \frac{L_{mixed}(-X)_{mixed}\left(h_g - h_{ep}\right)}{L_{cpl}(-X)_{cpl}\left(h_l - h_{ep}\right)} \tag{9.145}$$

respectively. If Eqs. (9.138)–(9.140) are taken into consideration, then Eqs. (9.144) and (9.145) are written as follows:

$$r_i = \frac{1 - r_h}{r_h} \tag{9.146}$$

and

$$r_q = \frac{1 - r_h^2}{r_h^2} \tag{9.147}$$

respectively. Given that:

$$\lim_{z_{ep} \to z_{ip}} r_h = 0 \tag{9.148}$$

it follows that the ratios r_i and r_q are maximized for $z_{ep} \to z_{ip}$. Figure 9.30 shows the effect of the initial temperature of the liquid and of the distance between the ideal point and the equilibrium point on r_q in case of the water vaporization at

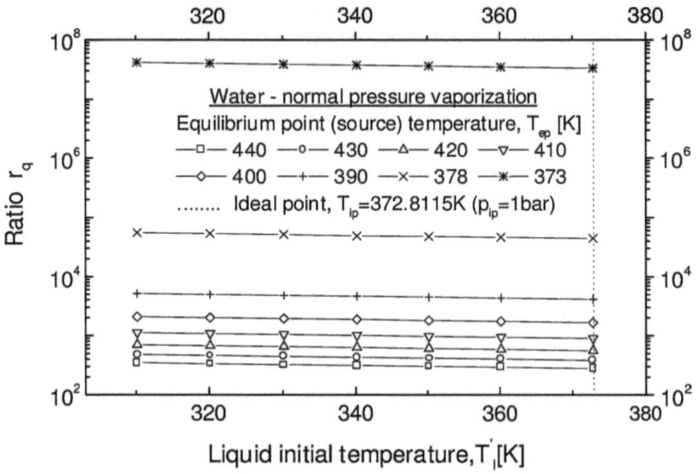

Fig. 9.30 The effect of the liquid initial temperature and of equilibrium point-to-ideal point distance on the ratio r_q in case of water at normal pressure

normal pressure. On the other hand, taking into account Eqs. (9.133), (9.134), (9.138), (9.141) and (9.148), it is noted that

$$\lim_{z_{ep} \to z_{ip}} j_{i,ncpl} = 0 \tag{9.149}$$

and

$$\lim_{z_{ep} \to z_{ip}} j_{q,ncpl} = 0 \tag{9.150}$$

so that, based upon the consequence of the natural force postulate:

$$\max\left|j_{i,mixed}\right| = \lim_{z_{ep} \to z_{ip}} j_{i,mixed} = j_{i,cpl}\left(z_{ip} \pm \varepsilon\right) \tag{9.151}$$

and

$$\max\left|j_{q,mixed}\right| = \lim_{z_{ep} \to z_{ip}} j_{q,mixed} = j_{q,cpl}\left(z_{ip} \pm \varepsilon\right) \tag{9.152}$$

Thus, the following theorem has been demonstrated:

Theorem (Concerning the maximization of the mixed transfer) *If a system interacts with a source by mixed transfer, the mass and heat currents are maximized in absolute value and in relation with the non-coupled currents when the equilibrium point tends towards the ideal point.*

Bearing in mind the problem raised at the beginning of Sect. 9.5, the theorem of the equilibrium point shows that, although it is slow, the non-coupled transfer is unavoidable in a power installation connected to sources. However, wherever possible, the theorem of the mixed transfer maximization indicates that this drawback could be diminished by conveniently replacing the non-coupled transfer with a mixed one. Efficient energetic plants already apply it, although this arises more from a valuable empirical experience than from a rigorous proof.

Equation (9.145) demonstrates that the i.p.a. effect is the main responsible for the increase of heat exchange with a source in case of the mixed transfer. Its highlighting seems to be the key of clarifying the coupled (mixed) mass and heat transfer phenomena which are difficult to be explained by the classic heat and mass transfer approach. This is the case of the refrigerating agent–absorbent systems interactions (e.g., ammonia-water, Sect. 9.2), or of the heat transfer mechanism in heat pipes, as it will be shown in the next work section, as a first application of the TPT.

9.5.4 Application of TPT to the Thermodynamic Cycles

9.5.4.1 Theoretical Aspects

The TPT may contribute to the analysis and optimization of the thermodynamic cycles. A cycle has a certain number of meaningful points $(p, T, x, y,\ldots)_1$, ..., $(p, T, x, y,\ldots)_n$ (degrees of freedom) characterizing the inlets/outlets and the evolution of a mono-, bi-, or poly-component/phase working fluid. The working fluid covers the cycle component devices where undergoes mass, heat, and work transfer processes. The processes are considered to be of non-coupled and/or mixed nature. Each device houses a global system where the working fluid materializes the multitude of systems interacting one to each other and with sources, connected to devices. While many times, not all, sources interact in fact indirectly with the working fluid through appropriate parts of the devices, here in our applications a direct contact fluid-source will be considered. Taking advantage of these statements, the following balances on the working fluid will be written in a cycle:

- property of conservation of the natural force $(-X)$, as a state function, Sect. 9.2

$$\oint d(-X) = 0 \tag{9.153}$$

- mass conservation law

$$\oint d(L(-X)) = 0 \tag{9.154}$$

- 1st principle of thermodynamics

$$\oint d(LC(-X)) = 0 \tag{9.155}$$

Equations (9.153)–(9.155) are valid for any type of natural forces presented in the second section of this chapter, which describe physical, chemical, work etc. interactions. If the cycle energy exchanged with the source is explicitly expressed in Eq. (9.155), the following expression is obtained

$$\int_{cycle} LC(-X) = \dot{W} \tag{9.156}$$

Equation (9.153) was applied in Sect. 9.3 to demonstrate the lemma for an isolated system. Here, if separate mixed and non-coupled transfers occur in the cycle, Eq. (9.153) is rewritten under the particular form as

$$\oint d\left[(-X)_{mixed} - (-X)_{ncpl}\right] = 0 \tag{9.157}$$

When, besides the non-coupled forces, the coupled natural forces also intervene, the cyclic evolution of the working fluid requires that

$$\oint d(-X)_{cpl} = 0 \tag{9.158}$$

Combining Eqs. (9.157) and (9.158) and bearing in mind Eq. (9.131), it follows that

$$\oint d(-X)_{ncpl} = 0 \tag{9.159}$$

Equations (9.153)–(9.155) can be useful in the first steps of a certain cycle design. Suppose a cycle to be conceived, thermodynamically speaking, Eqs. (9.155)–(9.156), already applied in their classic form, reduce the degree of freedom of the cycle. The phenomenological Eqs. (9.153)–(9.154) show that the degree of freedom of the cycle connected to sources must be further reduced, provided that no other additional constraints exist.

Next we shall apply the TPT to two simple study cases. As we will remark, at this stage of development this new theory will confine to reveal only some of its main qualities in the study of cycle basic thermodynamic properties and suggest which type of applications could be used in the future to.

9.5.4.2 Application to a Heat Pipe Type Cycle

The heat pipes are largely used today in energetic installations (Bilegan 1982). The literature presenting their detailed thermohydrodynamic modeling and experimental work is also very rich, where from some recent papers are cited (Zhu and Vafai 1999; Wang and Vafai 2000a, b). However, while the benefic role of the working fluid vapor as a rapid energy carrier between evaporator and condenser is well understood so far, not yet the evaporator/condenser heat transfers taking place with intrinsic high values are satisfactorily explained when the internal heat pipe heat transfer is discussed. Indeed, keeping our notations, the heat transfer is classically assessed in a mixed interaction of vaporization/condensation type by the empiric equation (Leontiev 1985; Hell 1982; Bejan 1996):

$$(j_q)_{classic} = \alpha S\left|\left(T_{ip} \pm \varepsilon\right) - T_{ep}\right| \tag{9.160}$$

where α and S are a coefficient of heat transfer and the transfer area, respectively. In principle, Eq. (9.160) may not be useful in the prediction of the transferred power from the heat pipe evaporator to condenser, simply because α is an empiric dimension. On the contrary, Eq. (9.160) is utilized to determine the α value in a second step through a current method which in the first step experimentally finds the transferred heat power and the working fluid temperature differences. Moreover, from the TPT point of view, Eq. (9.160) also lacks the physical support in order to assess a mixed transfer, because $(j_q)_{classic} \to 0$ when $T_{ep} \to T_{ip}$ and thus it contradicts the mixed transfer maximization theorem. Moreover, it can be noticed that, as expected, the distance max $\left|\left(T_{ip} \pm \varepsilon\right) - T_{ep}\right|$ is not factorable in the phenomenological Eqs. (9.126) and (9.129), describing the mixed transfer for pure components. In the frame of the TPT, the only transfer where this distance is factorable is the pure non-coupled one. Indeed, in this case $T_{ip} \pm \varepsilon$ becomes the temperature of the heated (cooled) fluid up to nearly T_{ep} without a change of phase and if this temperature is further written down as T_f, from Eqs. (9.119) to (9.133), for sufficiently small $T_{ep} - T_f$ distances it results that:

$$(j_q)_{phenom} = (\alpha S)_{phenom}\left(T_{ep} - T_f\right) \tag{9.161}$$

where

$$(\alpha S)_{phenom} = LC\left[\left(\frac{\partial s}{\partial T}\right)_p - \frac{\max\left(h', h''\right)}{T_f T_{ep}}\right] \tag{9.162}$$

Reverting to the problem at hand, given the remarks outlined above, in the followings we shall use the TPT to explain the real basic thermodynamic reason of the high heat transfer capability of a heat pipe type cycle. To do this, a simple mathematical apparatus will be used, applying Eqs. (9.153)–(9.157) to a case study of mass and heat transfer of a pure component, water, in a cyclic evolution between two sources, characterized by the equilibrium point of vaporization $z_{ep,v} = T_{ep,v}$ and of condensation $z_{ep,c} = T_{ep,c}$, respectively. The heat pipe cycle exchanges heat with the two sources by mixed transfer, it produces a small amount of mechanical work, used internally exclusively in order to overcome the pressure losses of the working fluid operating between evaporator and condenser, Δp_{loss}, and uses the capillary forces to overcome the evaporator—condenser pressure difference $\Delta p_{capillary} = p_{ep,v} - p_{ep,c}$. Under the circumstances, the forces are only mixed, so Eq. (9.157) is written as follows:

$$(-X)_{mixed,v} + (-X)_{mixed,c} = 0 \tag{9.163}$$

or, considering Eqs. (9.129) and (9.126),

$$\left\{-\left[s\left(T_{ep,v}\right)-s\left(T_{ip,c}-\varepsilon\right)\right]-h\left(T_{ip,v}+\varepsilon\right)\left(\frac{1}{T_{ep,v}}-\frac{1}{T_{ip,c}-\varepsilon}\right)\right\}$$
$$+\left\{\left[s\left(T_{ip,v}+\varepsilon\right)-s\left(T_{ep,c}\right)\right]+h\left(T_{ip,v}+\varepsilon\right)\left(\frac{1}{T_{ip,v}+\varepsilon}-\frac{1}{T_{ep,c}}\right)\right\}=0 \tag{9.164}$$

A second equation of the case study is obtained if Eqs. (9.163) and (9.154) were taken into consideration, resulting in:

$$L_{mixed,v}-L_{mixed,c}=0 \tag{9.165}$$

Further, accounting for Eqs. (9.157)–(9.159), the following equations hold true:

$$(-X)_{cpl,v}+(-X)_{cpl,c}=0 \tag{9.166}$$

and

$$(-X)_{ncpl,v}+(-X)_{ncpl,c}=0 \tag{9.167}$$

In our case, because of the mass conservation property, we have $\left(j_r j_i''\right)_v+\left(j_r j_i''\right)_c=0$ and further, with the help of Eqs. (9.141) and (9.166), the following result is obtained: $L_{cpl,v}-L_{cpl,c}=0$. Also, from Eqs. (9.166) and (9.167), it results that $f_{mixed,v}=f_{mixed,c}$. Introducing the last results in Eq. (9.165) this becomes:

$$(r_h)_v-(r_h)_c=0 \tag{9.168}$$

The final form of Eq. (9.168) is obtained, if here Eq. (9.140) were taken into account, resulting in,

$$\left[h\left(T_{ep,v}\right)-h\left(T_{ip,v}+\varepsilon\right)\right]_v-\left[h\left(T_{ip,c}-\varepsilon\right)-h\left(T_{ep,c}\right)\right]_c=0 \tag{9.169}$$

In Eqs. (9.164) and (9.169), $T_{ep,v}$ and $T_{ep,c}$ are input data, and $T_{ip,v}=T_{ip,v}(p_{ep,v})$ and $T_{ip,c}=T_{ip,c}(p_{ep,c})$ are the vaporization and condensation ideal points temperatures, respectively. The latters are calculated from the water equilibrium data, provided that the pressure values satisfying equation $\Delta p_{loss}=\Delta p_{capillary}$ were known. In our study case, it is not needed to know these pressures, it suffices to find the unknowns $T_{ip,v}+\varepsilon$ and $T_{ip,c}-\varepsilon$ from the implicit algebraic system made up by Eqs. (9.164) and (9.169). Before solving this system, it is noted that, generally, an always strictly positive distance between the equilibrium point and the ideal point characterizes a system connected to a source by mixed transfer

$$\left\|z_{ep}-z_{ip}\right\|>0 \tag{9.170}$$

for any $z_{ep}\in D$, with $(-X)_{mixed}$:D, because $z_{ip}\notin D$, according to the lemma of Sect. 9.3. Equally $(\exists)\varepsilon>0,\ \varepsilon=0.0$, so that:

$$\min\left\|z_{ep}-\left(z_{ip}\pm\varepsilon\right)\right\|>0 \tag{9.171}$$

From Eqs. (9.170) to (9.171), it results therefore that temperatures of the ideal and equilibrium points must fulfil in equations below:

$$T_{ep,c} < T_{ip,c} - \varepsilon \le T_{ip,v} + \varepsilon < T_{ep,v}$$

It can be noted from the above relationship that the isothermal transfer ($T_{ep,c} = T_{ep,v}$) is impossible. To this extent, the isothermal defect, equal to $T_{ep,v} - T_{ep,c}$ and considered so far a consequence of the imperfect heat pipe achievement in practice (Bilegan 1982), is in fact one of its intrinsic properties. It is difficult to analytically solve the system made up of Eqs. (9.164) and (9.169). In order to obtain the solution, a trial-and-error method and a commercial code suitable for the calculation of the thermodynamic functions of state and of algebraic solution (EES 1992), have been used. The results obtained as unique solutions with absolute errors below 3 % for various domains of water temperatures and various distances between equilibrium points, are shown in Table 9.2. Strictly depending on $T_{ep,v}$ and $T_{ep,c}$ and on the nature of the fluid, they reveal the basic heat pipe cycle property explaining its high heat transfer: while the distances between the equilibrium points are up to one order of magnitude higher, the distances $\min\|z_{ep} - (z_{ip} \pm \varepsilon)\|$ are very small, which according to the mixed transfer maximization theorem leads to a very good mixed transfer (see Table 9.2 $r_{q,c}(= r_{q,v})$ values). Our explanation remains valid if real heat pipes are considered, where the distances $\min\|z_{ep} - (z_{ip} \pm \varepsilon)\|$, resulting from the solving of the system will be altered, getting higher due to the intervention of additional non-coupled forces (e.g., those governing the transfer through walls and liquid working fluid + its transporting capillary structure) or forces of a different nature, such as Laplace, gravitational, etc., resulting in a reduction of the mass and heat transfer (lower $r_{q,c}(= r_{q,v})$ values). We can remark that the i.p.a. effect again lays down the basis of explaining a mixed transfer based process, which this time is the high heat transfer of the heat pipe. A previous work (Stefan 1995) thoroughly studying the evaporative heat transfer in capillary grooved heat pipes, is confirming this statement. Indeed, although the result passed not remarked by the author, it follows from this work that the vaporization heat current reaches its maximum at the minimum T_{ep} temperatures of the evaporator wall, that is the temperatures which are the closest to T_{ip}, in good agreement with the mixed transfer maximization theorem.

9.5.4.3 Role of the Natural Forces Conservation Equation Application to the CAN Theory

A second application emphasizes the importance of Eq. (9.153). Its introduction is beneficial as it completes the application of the second principle of thermodynamics in the thermodynamics of cycles. It is more general than the wellknown equation

$$\oint ds = 0 \tag{9.172}$$

Table 9.2 Phenomenological results of mixed mass and heat transfer in case of water suffering a heat pipe type cyclic evolution between two sources

$t_{ep,v} - t_{ep,c}$ (°C)	$t_{ep,v}$ (°C)	$t_{ep,c}$ (°C)	$t_{ip,v}$ (°C)	$t_{ip,c}$ (°C)	$t_{ep,v} - \begin{pmatrix} t_{ip,v} \\ -\varepsilon \end{pmatrix}$ (°C)	$t_{ep,c} - \begin{pmatrix} t_{ip,c} \\ -\varepsilon \end{pmatrix}$ (°C)	$10^{-8} r_{q,c}$ $(= r_{q,v})$	Absolute error, Eq. (9.153) (kJ kg^{-1} K^{-1})
1	50	49	49.87	49.10	0.06	0.03	3.63	0.030
	60	59	59.87	59.10	0.06	0.03	3.55	0.021
	80	79	79.87	79.10	0.06	0.03	3.38	0.012
	90	89	89.86	89.10	0.08	0.04	1.85	0.013
	100	99	99.85	99.10	0.09	0.04	1.80	0.011
5	85	80	84.86	80.10	0.08	0.04	1.90	0.013
	90	85	89.85	85.10	0.10	0.05	1.20	0.015
	95	90	94.96	90.10	0.08	0.04	1.86	0.011
	100	95	99.86	95.10	0.08	0.04	1.83	0.010
	155	150	154.85	150.10	0.11	0.06	0.67	0.005
	160	155	159.86	155.10	0.10	0.06	1.20	0.015
	205	200	204.87	200.10	0.09	0.06	0.52	0.002
	210	205	209.87	205.10	0.09	0.06	0.51	0.002
10	90	80	89.85	80.10	0.10	0.05	1,22	0.012
	100	90	99.84	90.10	0.12	0.06	0.83	0.012
	160	150	159.86	150.10	0.10	0.06	0.68	0.003
	210	200	209.87	200.10	0.09	0.06	0.52	0.002

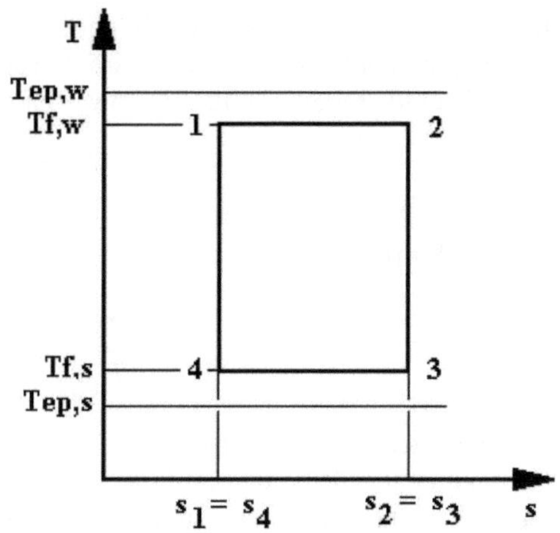

Fig. 9.31 The irreversible Carnot cycle connected to sources by non-coupled transfer and used in the phenomenological CAN theory assessment

which is valid for reversible cycles, only (Kirilin et al. 1985, see also Eqs. [1.88] and [1.89]). Equation (9.153) proves to be useful especially in the calculation of the irreversible cyclic transfer, where the Clausius integral (Kirilin et al. 1985, see also Eq. (1.97)),

$$\oint \frac{dq}{T} < 0 \tag{9.173}$$

cannot be practically used. Hereinafter Eq. (9.153) will be used for a brief phenomenological assessment of the Curson–Ahlborn–Novikov (CAN) theory (Bejan 1996; Ares de Parga et al. 1999; Ishida 1999; Gyftopoulos 1999; Sieniutycz 1998, 2001; Grazzini 1991). Originally, the CAN theory aimed at the power maximization of the Carnot cycle, subject to optimization of its finite-time heat transfer with sources. Later, the basic idea of this theory extended also to other cycles of work or heat pumping type (Bejan 1996). First of all, phenomenologically, only the pure non-coupled transfer with sources should be considered from the CAN's theory point of view in a cycle including both mixed and noncoupled transfer with sources, because, according to the results outlined in Sect. 9.5.3, in case of the mixed transfer, the cycle carnotization and finite-time heat transfer maximization occur simultaneously, when the working fluid temperatures approach those of the sources (see third theorem). Next, we assume the Carnot 1–2–3–4–1 cycle in Fig. 9.31, used in the CAN theory, represented by two isotherms 1–2 and 3–4 and two isentropics 2–3 and 4–1. The cycle receives the heat flow $\dot{q}_w$ from the warm source of temperature $T_{ep,w} = const.$, yields the heat flow $\dot{q}_s$ to the sink source of temperature $T_{ep,s} = const.$, and produces the mechanical power $\dot{w}$. Following the interaction with the warm and sink sources, the working fluid reaches the temperatures $T_{f,w} = const. \leq T_{ep,w}$ and $T_{f,s} = const. \geq T_{ep,s}$ by non-coupled transfer of the convective and radiative type. The CAN theory aims at the maximization of the power of cycle 1–2–3–4–1, which is considered endoreversible, by equation:

$$\frac{\dot{q}_w}{T_{f,w}} + \frac{\dot{q}_s}{T_{f,s}} = 0 \tag{9.174}$$

In Eq. (9.174), obtained by the integration and derivation of Eq. (9.172) in relation to time, the heat flows $\dot{q}_w$ and $\dot{q}_s$ are assessed by means of Eq. (9.160). The calculation is made together with the power balance equation:

$$\dot{q}_w + \dot{q}_s = \dot{w} \tag{9.175}$$

The cycle delivers the maximum power when $T_{f,s}$ and $T_{f,w}$ have the optimal ratio (Bejan 1996):

$$\frac{T_{f,s}}{T_{f,w}} = \sqrt{\frac{T_{ep,s}}{T_{ep,w}}} \tag{9.176}$$

In the phenomenological theory, Eq. (9.174) is replaced by the stationary Eq. (9.153). Hereinafter it will be demonstrated that it is in fact a Clausius integral, at least in the case of the cycle under study. Indeed, the application of Eq. (9.153) to the cycle 1–2–3–4–1 leads to the following balance of natural forces:

$$\begin{aligned}\left[(-X)_{ncpl,2} - (-X)_{ncpl,1}\right] + (-X)_{ncpl,w,2-3} + \left[(-X)_{ncpl,4} - (-X)_{ncpl,3}\right]\\ + (-X)_{ncpl,w,4-1} = 0\end{aligned} \tag{9.177}$$

where, according to Eqs. (9.119), (9.121) and (9.122),

$$\begin{aligned}
(-X)_{ncpl,1} &= -\left[\left(s_1 - s_{ep,w,1}\right) + h_{ep,w,1}\left(\frac{1}{T_1} - \frac{1}{T_{ep,w}}\right)\right]\\
(-X)_{ncpl,2} &= -\left[\left(s_2 - s_{ep,w,2}\right) + h_{ep,w,2}\left(\frac{1}{T_2} - \frac{1}{T_{ep,w}}\right)\right]\\
(-X)_{ncpl,w,2-3} &= -\left[\left(s_2 + \frac{h_2}{T_2}\right) - \left(s_3 + \frac{h_3}{T_3}\right)\right]\\
(-X)_{ncpl,3} &= -\left[\left(s_3 - s_{ep,s,3}\right) + h_{f,s,3}\left(\frac{1}{T_3} - \frac{1}{T_{ep,s}}\right)\right]\\
(-X)_{ncpl,4} &= -\left[\left(s_4 - s_{ep,s,4}\right) + h_{f,s,4}\left(\frac{1}{T_4} - \frac{1}{T_{ep,s}}\right)\right]\\
(-X)_{ncpl,w,4-1} &= -\left[\left(s_4 + \frac{h_4}{T_4}\right) - \left(s_1 + \frac{h_1}{T_1}\right)\right]
\end{aligned} \tag{9.178}$$

Account is taken of the fact that:

$$\begin{aligned}
T_1 &= T_2 = T_{f,w}\\
T_3 &= T_4 = T_{f,s}\\
s_2 &= s_3 = s_{ep,w,2} = s_{ep,s,3}\\
s_1 &= s_4 = s_{ep,w,1} = s_{ep,s,4}
\end{aligned} \tag{9.179}$$

and together with Eq. (9.178), Eq. (9.177) becomes:

$$\frac{h_1 - h_2}{T_{f,w}} + \frac{h_3 - h_4}{T_{f,s}} + \left(h_{ep,w,2} - h_{ep,w,1}\right)\left(\frac{1}{T_{f,w}} - \frac{1}{T_{ep,w}}\right) + \left(h_{f,s,4} - h_{f,s,3}\right)\left(\frac{1}{T_{f,s}} - \frac{1}{T_{ep,s}}\right) \quad (9.180)$$

The equation of the 1st principle of thermodynamics is written under the form

$$dq = dh - vdp \quad (9.181)$$

and is integrated between states 1 and 2 and then between states 3 and 4, obtaining:

$$q_w = (h_2 - h_1) - \int_1^2 vdp \quad (9.182)$$

and

$$q_S = (h_4 - h_3) - \int_3^4 vdp \quad (9.183)$$

Equations (9.182) and (9.183) are introduced in Eq. (9.180), obtaining:

$$\frac{q_w}{T_{f,w}} + \frac{q_s}{T_{f,s}} = d_r \leq 0 \quad (9.184)$$

where the quantity:

$$d_r = -\left(\frac{1}{T_{f,w}} \int_1^2 vdp + \frac{1}{T_{f,s}} \int_3^4 vdp\right) + \left(h_{ep,w,2} - h_{ep,w,1}\right)\left(\frac{1}{T_{f,w}} - \frac{1}{T_{ep,w}}\right) + \left(h_{f,s,4} - h_{f,s,3}\right)\left(\frac{1}{T_{f,s}} - \frac{1}{T_{ep,s}}\right) \leq 0 \quad (9.185)$$

is referred to herein as *reversibility defect* (of the Carnot cycle considered). It is easy to demonstrate that the reversibility defect is zero only if, simultaneously (i) the working fluid is an ideal gas, in which case the first term is annulled, and (ii) $T_{f,w} \to T_{ep,w}$ and $T_{f,s} \to T_{ep,s}$, in which case the last two terms are annulled. Obviously, in our case we have $d_r < 0$. Thus it has been demonstrated that Eq. (9.153) is the Clausius integral for the irreversible Carnot cycle 1–2–3–4–1, according to Eq. (9.184). It further follows that, from the phenomenological point of view, the endoreversibility is an unrealistic simplification, at least to the case at hand, and the CAN theory based thereon, appears to be simplistic, and insufficiently substantiated. This theory weak point is emphasized in a good part of past works by others (Ishida 1999; Gyftopoulos 1999), but is still neglected in many other advanced studies (Sieniutycz 1998; Sieniutycz 2001). Parallel studies were also accomplished to optimize the power output of irreversible cycles

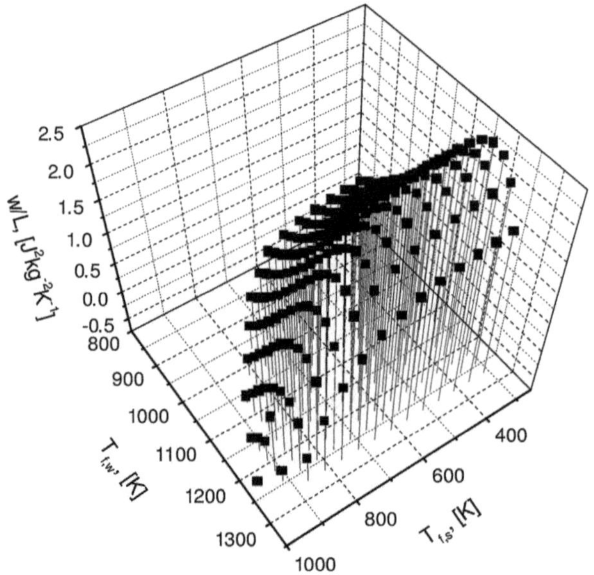

Fig. 9.32 Power diagram of the irreversible Carnot cycle against working fluid temperatures. Code inlet data: (1) Working fluid: CO_2 gas; (2) $T_{ep,w} = 1{,}275$ K; $T_{ep,s} = 310$ K; (3) $s_1 = s_4 = s(t = T_{ep,w}, p = 240\text{ bar}) = const.$; $s_2 = s_3 = s(t = T_{ep,w}; p = 60\text{ bar}) = const.$

for various scenarios of irreversibility [e.g., for the irreversible Carnot cycle (Grazzini 1991)]. The phenomenological approach of the problem at hand is hereinafter considered for a case study of the irreversible Carnot cycle shown in (Fig. 9.31), which operates with CO_2 gas. For this application the set of input data, mentioned in the explanatory notes of Fig. 9.32, were chosen at random. The power of the cycle was assessed by solving the algebraic system consisting of Eqs. (9.180) and (9.175), phenomenologically expressed by Eq. (9.186) below:

$$\begin{aligned}&\left[h_{f,w,2}(-X)_{ncpl,2} - h_{f,w,1}(-X)_{ncpl,1}\right]\\&\quad + \left[h_{f,s,4}(-X)_{ncpl,4} - h_{f,s,3}(-X)_{ncpl,3}\right] + \frac{\dot{w}}{L} = 0\end{aligned} \tag{9.186}$$

The phenomenological coefficients of pure non-coupled interactions are not available yet and further theoretical studies are necessary to obtain them. However, in Eq. (9.186), for the sake of simplicity, we considered them as constant ($L = const.$), without to change the main results of our presentation. The result of the calculation carried out by means of the same commercial code (EES 1992) is presented in the chart shown in Fig. 9.32, for various values of temperatures $T_{f,s}$ and $T_{f,w}$ ranging from $T_{ep,s}$ and $T_{ep,w}$. The surface $(\dot{w}/L)$ has a maximum value at the optimum temperatures $(T_{f,w})_{opt} \cong 1195$ K and $(T_{f,s})_{opt} \cong 315$ K, in agreement with the CAN theory, but the optimum values do not comply with Eq. (9.176). The optimum coordinates may be obtained directly, analytically, by searching for the extreme values of the $(\dot{w}/L)(T_{f,s}, T_{f,w})$ function. This approach is difficult and has been avoided even in the present case, which is one of the simplest. Although here the phenomenological methods and CAN differ from the value point of view, the fact that they are concordant from the point of view of the prediction of a maximum power existence is encouraging. For this reason it should be assumed that a coupling of these two theories could be beneficial in a more realistic preliminary calculation of the optimization of the power cycle, heat pump, etc. To this effect, however, it will be necessary to conduct more thorough studies in future, which studies are beyond the framework of this chapter.

Appendix 1

The natural thermodynamical force may be given also explicitly, solving the second degree algebraic equation in $-X_r$, obtained from Eq. (9.34)

$$d(-X_r)^2 + (1 - ad + c)(-X_r) - (a + b) = 0 \tag{A9.1}$$

where

$$\begin{aligned} a &= X_{i,r} + h_{g,r}\left(\frac{1}{T_{g,r}} - \frac{1}{T_{l,r}}\right) \\ b &= \frac{A_r h_{l,r}}{T_{l,r}} \\ c &= \frac{A_r\left(h_{g,r} - h_{l,r}\right)M\left(p_{r,e}, x_e\right)}{T_{l,r}} \\ d &= A_r L_{11,r} \end{aligned} \tag{A9.2}$$

Absorption processes occur when $1 - ad + c < 0$ and $d < 0$ and the force is given by

$$(-X_r) = \frac{|1 - ad + c| - \sqrt{(1 - ad + c)^2 + 4|d|(a + b)}}{2d} \tag{A9.3}$$

Generation processes take place when $1 - ad + c > 0$ and $d > 0$ and the force is expressed by

$$(-X_r) = \frac{-|1 - ad + c| + \sqrt{(1 - ad + c)^2 + 4|d|(a + b)}}{2d} \tag{A9.4}$$

or by Eq. (A9.3), when $1 - ad + c > 0$, but $d < 0$. From the above it can be remarked that the use of the explicit form of the force is no simpler than that of

Eq. (9.34), which the author always preferred in computations although the values obtained with it are comparatively smaller up to maximum 1.5 %.

Appendix 2

Using the vector additive property of the natural force, an isobaric interaction with a source [Eq. (9.15)] can be brought to a simpler form. In order to proof Eq. (9.16), it is enough to consider a system-source interaction. We shall start from an elementary interaction between $T' = T + dT$ and $T_{ep} = T$. With $n' = 1$, Eq. (9.15) leads to the elementary force

$$d(-X)_{ncpl,s,p} = \left[s(T + dT) - s(T) + h(T + dT) \left(\frac{1}{T + dT} - \frac{1}{T} \right)_p \right] \quad \text{(A9.5)}$$

or, developing in Taylor series and keeping the linear form terms only, we obtain:

$$d(-X)_{ncpl,s,p} = -\left(ds - \frac{h}{T^2} dT \right)_p \quad \text{(A9.6)}$$

We take into account the equation of the 2nd principle of thermodynamics for isobar processes in Eq. (A9.6)

$$ds = \left(\frac{dh}{T} \right)_p \quad \text{(A9.7)}$$

where from the elementary force rewrites:

$$d(-X)_{ncpl,s,p} = -d\left(\frac{h}{T} \right)_p \quad \text{(A9.8)}$$

The force between two states, initial T' and T_{ep}, is obtained integrating Eq. (A9.8)

$$(-X)_{ncpl,s,p} = -\int_{T_{ep}}^{T'} d\left(\frac{h}{T} \right)_p = -\left[\left(\frac{h}{T} \right)(T') - \left(\frac{h}{T} \right)(T_{ep}) \right] \quad \text{(A9.9)}$$

where the state thermodynamic function $\left(\frac{h}{T} \right)_p = s_{p,t}$ is termed here *(mass and heat) isobar transfer entropy*. In case of isocore processes, the enthalpy h in Eq. (A9.9) is replaced by the internal energy u, which leads to the new thermodynamic function $\left(\frac{u}{T} \right)_v = s_{v,t}$, termed here *(mass and heat) isocore transfer entropy.*

References

Al-Hayes, R. A. M., & Winterton, R. H. S. (1981). Bubble growth in flowing liquids. *International Journal of Heat and Mass Transfer, 24*, 213–221.

Anderson, J. D., Bollinger, R. E., & Lamb, D. E. (1964). Gas-phase controlled mass transfer in two-phase annular horizontal flow. *AIChEJ, 10*, 640–645.

Ares de Parga, G., Angulo-Brown, F., & Navarrete-Gonzales, T. D. (1999). A variational optimization of a finite-time thermal cycle with a nonlinear heat transfer law. *Energy, 24*(12), 997–1008.

Auracher, H. & Maier, M. (1983). Vapor bubble condensation in slightly subcooled pure liquids. In *16th International Congress of Refrigeration* (Vol. B1, pp. 269–274).

Banerjee, S., Scott, D. S., & Rhodes, E. (1969). Studies on cocurrent gas–liquid flow in helically coiled tubes I. *The Canadian Journal of Chemical Engineering, 47*, 445–453.

Banerjee, S., Scott, D. S., & Rhodes, E. (1970). Studies on cocurrent gas–liquid flow in helically coiled tubes II. *The Canadian Journal of Chemical Engineering, 48*, 542–551.

Bejan, A. (1996). *Advanced technical thermodynamics*. Bucharest: Technical Publishing House. (in Romanian).

Bilegan, I. C. (1982). The heat pipe (Ed.) *Tehnica Bucuresti* (in Romanian).

Brujan, E. A. (1994). Heat transfer in spherical bubble dynamics. In *Conference on Hydraulic Machinery and Hydrodynamics*, Timisoara (Vol. 2, pp. 103–110).

Chadzidakis, S. (1971). Wärmeübertragung am waagerechten mit Ammoniak-Wasser-Lösungen berieselten Rohr. Kältetechhnik-Klimatisierung, *23*(7), 202–208.

Chartier, P., Gross, M., & Spiegler, K. S. (1975). *Applications de la thermodynamique du non-équilibre*. Paris: Hermann.

Chiriac, F., Mihaila, C., Tarca, C., & Baltaretu, F. (1995, August 20–25). Heat and mass transfer in the boiling of ammonia/water solution. In *Proceedings of the 19th International Congress of Refrigeration*, The Hague (Vol. IIIa, pp. 67–71).

Ciborowski, J. W., & Richlicki, R. M. (1971). Mass transfer in two-phase gas–liquid wavy flow. *International Journal of Heat and Mass Transfer, 14*, 1261–1266.

Cooper, M. G., & Stone, C. R. (1981). Boiling of binary mixtures—study of individual bubbles. *International Journal of Heat and Mass Transfer, 24*(12), 1937–1950.

Danckwerts, P. V. (1951). Significance of liquid-film coefficients in gas absorption. *Industrial and Engineering Chemistry, 43*, 1460–1467.

de Donder, T. (1936). *L'Affinité*. Paris: Gauthier-Villars.

de Groot, S. R., & Mazur, P. (1964). *Non-equilibrium thermodynamics*. Moscow: Mir. (in Russian).

Dodescu, G. (1979). *Numerical methods in algebra*. Bucharest: Technical Publishing House. (in Romanian).

Grazzini, G. (1991). Work from irreversible heat engines. *Energy International Journal, 16*(4), 747–756.

Gregory, G. A., & Scott, D. S. (1969). In E. Rhodes & D. S. Scott. *Physical and chemical mass transfer in horizontal cocurrent gas–liquid slug flow* (pp. 633–665). New York: Plenum Press.

Gyftopoulos, E. P. (1999). Infinite time (reversible) versus finite time (irre-versible) thermodynamics: A misconceived distinction. *Energy, 24*(12), 997–1008.

Hell, F. (1982). Grundlagen der Wärmeübertragung. VDI-Verlag: Düsseldorf

Hera, D. (1987). Contributions to the study of heat and mass transfer in absorption processes in ammonia-water solutions in film flow in horizontal pipes. Ph. D. Thesis, Bucharest Technical Civil Engineering University (in Romanian).

Heuss, J. M., King, C. J., & Wilke, C. R. (1965). Gas–liquid mass transfer in cocurrent froth flow. *AIChEJ, 11*, 866–872.

Higbie, R. (1935). The rate of absorption of a pure gas into a still liquid during short periods of exposure. *Transaction on AIChE, 31*, 365–389.

Hobler, T. (1966). *Mass transfer and absorbers*. New York: Pergamon Press.
Hui, T. O., & Thome, J. R. (1985). Study of binary mixture boiling: Boiling site density and subcooled heat transfer. *International Journal of Heat and Mass Transfer, 28*(5), 919–928.
Ionescu, D. G. (1977). *Introduction to hydraulics*. Bucharest: Technical Publishing House. (in Romanian).
Ishida, M. (1999). The role and limitations of endoreversible thermodynamics. *Energy, 24*(12), 997–1008.
Jagota, A. K., Rhodes, E., & Scott, D. S. (1973). Mass transfer in upwards cocurrent gas–liquid annular flow. *The Chemical Engineering Journal, 5*, 23–31.
Jepsen, J. C. (1970). Mass transfer in two-phase flow in horizontal pipelines. *AIChE Journal, 16*, 705–711.
Kang, Y.T., Akisawa, A., & Kashiwagi, T. (1998, August 23–28). Experimental investigation of the NH_3–H_2O falling film absorption in plate heat exchangers. In *Proceedings of the 11th ITHC, Heat Transfer*, Kyongju, Korea (Vol. 6, pp. 84–89).
Kasturi, G., & Stepanek, J. B. (1974). Two-phase flow IV: Gas and liquid side mass transfer coefficients. *Chemical Engineering Science, 29*, 1849–1856.
Keizer, C. (1982). Absorption refrigeration machines. Ph.D. Thesis, Department of Mechanical Engineering, Delft University of Technology, Delft.
Kirilin, V. A., Sicev, V. V., & Seindlin, A. E. (1985). *Thermodynamics*. Bucharest: Scientific and Technical Publishing House. (in Romanian).
Klausner, I. F., Mei, R., & Zeng, L. Z. (1993). Vapor bubble departure in forced convection boiling. *International Journal of Heat and Mass Transfer, 36*(3), 651–662.
Klein, S. A., & Alvarado, F. L. (1992–2002). Engineering Equations Solver
Kojima, M., & Kashiwagi, T. (1995). Mass diffusivity measurements for ammonia-vapor absorption processes. In *19th International Congress of Refrigeration*, Hague (Vol. IVa, pp. 53–360).
Kulic, E., & Rhodes, E. (1974). Chemical mass transfer in concurrent gas–liquid slug flow in helical coils. *Canadian Journal of Chemical Engineering, 52*, 114–123.
Landau, L., & et Lifschitz, E. (1971). *Mécanique des fluides*. Moscow: Mir.
Lee, H. S., & Merte, H, Jr. (1996a). Spherical vapor bubble growth in uniformly superheated liquids. *International Journal of Heat and Mass Transfer, 39*(12), 2427–2447.
Lee, H. S., & Merte, H, Jr. (1996b). Hemispherical vapor bubble growth in microgravity: Experiments and model. *International Journal of Heat and Mass Transfer, 39*(12), 2449–2461.
Lewis, W. K., & Whitman, W. G. (1924). Principles of gas absorption. *Industrial and Engineering Chemistry, 16*, 1215.
Lewis, W. K., & Whitman, W. G. (1924). *Industrial and Engineering Chemistry, 16*, 1215.
Merrill, T. L., & Perez, B. H. (1997). Combined heat and mass transfer during bubble absorption in binary solutions. *International Journal of Heat and Mass Transfer, 40*(3), 589–603.
Mihaila, C. C., Chiriac, F. N., Sota, I. D., & Baltaretu, F. (1995 August 20–25) The modeling of a film absorber with ammonia/water solution. In *Proceedings of the 19th International Congress of Refrigeration*, The Hague (Vol. IIIa, pp. 154–160).
Miyatake, O., Tanaka, I., & Lior, N. (1997). A simple universal equation for bubble growth in pure liquids and binary solutions with a nonvolatile solute. *International Journal of Heat and Mass Transfer, 40*(7), 1557–1584.
Murgulescu, I. G., & Valcu, R. (1982). *Introduction in Physical Chemistry*. Bucharest: Publishing House of the Romanian Academy. (in Romanian).
Nigmatulin, R. I., Khabeev, N. S., & Nagiev, F. B. (1981). Dynamics, heat and mass transfer of vapour-gas bubbles in a liquid. *International Journal of Heat and Mass Transfer, 24*(6), 1033–1044.
Rayleigh, L. (1917). On the pressure developed in a liquid during the collapse of a spherical cavity. *Philosophical Magazine, 34*, 94–98.
Sieniutycz, S. (1998). Hamilton–Jacobi–Bellman analysis of irreversible thermal exergy. *IJHMT, 41*(1), 183–196.
Sieniutycz, S. (2001). Optimal work in sequential systems with complex heat exchange. *IJHMT, 44*(5), 897–918.

Sivagnanam, P., Balakrishnan, A. R., & Varma, Y. B. G. (1994). On the mechanism of subcooled flow boiling of binary mixtures. *International Journal of Heat and Mass Transfer, 37*(4), 681–689.

Staicovici, M. D. (1995). Polybranched regenerative GAX cooling cycles. *International Journal of Refrigeration, 18*, 318–329.

Staicovici, M. D. (1998a). Contributions to two-phase flows and heat transfer in solar refrigerating installations. Ph.D.Thesis, Bucharest Polytechnics University (in Romanian).

Staicovici, M. D. (1998b). Variational numerical and analytical approach of the liquid capillary rise effect. *Revue Roumaine des Sciences Techniques, série Mécanique Apliquée, 43*(1), 85–99.

Staicovici, M. D. (2000a). A phenemenological theory of polycomponent interactions in non-ideal mixtures. Application to NH_3/H_2O and other working pairs. *International Journal of Refrigeration, 23*(2), 153–167.

Staicovici, M. D. (2000b). A non-equilibrium phenomenological theory of the mass and heat transfer in physical and chemical interactions. Part I—Theory and its application to NH_3/H_2O and other working systems. *International Journal of Heat and Mass Transfer, 43*(22), 4153–4173.

Staicovici, M. D. (2000c). A non-equilibrium phenomenological theory of the mass and heat transfer in physical and chemical interactions. Part II- Modeling of the NH_3/H_2O bubble absorption, analytical study of absorption and experiments. *International Journal of Heat and Mass Transfer, 43*(22), 4175–4188.

Staicovici, M. D. (2002). A non-equilibrium phenomenological (two-point) theory of mass and heat transfer. Forces, system-source interactions and thermodynamic cycle application. *International Journal of Thermal Sciences, 41*(2002), 737–752.

Stefan, P. (1995). Evaporative heat transfer in capillary grooved heat pipes. In *Proceedings of the 19th International Congress of Refrigeration* (Vol. IIIa, pp. 720–727).

Vilcu, R., & Dobrescu, A. (1982). *Thermodynamics of irreversible processes*. Bucharest: Technical Publishing House. (in Romanian).

Wang, Y., Vafai K. (2000b). An experimental investigation of the thermal performance of an asymmetrical flat plate heat pipe. *IJHMT, 43*(15), 2657–2668.

Wang, Y., Vafai, K. (2000a). Transient characterization of flat plate heat pipes during startup and shutdown operations. *IJHMT, 43*(15), 2641–2656.

Zemansky, M. W. (1968). *Heat and thermodynamics*. New York: Mc Graw Hill Book Comp.

Zeng, L. Z., Klausner, J. F., Bernhard, D. M., & Mei, R. (1993a). A unified model for the prediction of bubble detachment diameters in boiling systems—II. Flow boiling. *International Journal of Heat and Mass Transfer, 36*(9), 2271–2279.

Zeng, L. Z., Klausner, J. F., & Mei, R. (1993b). A unified model for the prediction of bubble detachment diameters in boiling systems—I. Pool boiling. *International Journal of Heat and Mass Transfer, 36*(9), 2261–2270.

Zhu, N., Vafai, K. (1999). Analysis of cylindrical heat pipes incorporating the effects of liquid-vapor coupling and non-Darcian transport-a closed form solution. *IJHMT, 42*(18), 3405–3418.

Ziegler, B., & Trepp, C. (1984). Equation of state for ammonia-water mixtures. *International Journal of Refrigeration, 7*, 101–106.

Chapter 10
A New Wording of the Laplace Equation: Variational Numerical and Analytical Approach of the Liquid Capillary Rise Effect

10.1 Introduction

This chapter focuses on the computational aspects involved in the liquid capillary rise effect. It is well known that in the neighbourhood of continuous media interfaces the molecular interaction generates surface tension forces. When a liquid, in equilibrium with its vapor, interacts with the walls of the vessel containing it, the free surface of the liquid will curve near the wall. Due to the surface tension, here arise complementary (Laplace) forces of pressure. If the vessel has capillary dimensions (comparable to the curvature radii) these forces are governing, in gravity field, the shape of the interface and liquid rise, and further strongly influence the micro- and macro-scale operation of devices using capillary effects [e.g. heat pipes (Stephan 1995)].

First, we shall consider the Laplace equation and further the variational calculus will guide us in order to get a numerical and analytical approach of the problem at hand (Staicovici 1998b).

10.2 A New Wording of the Laplace Equation

So far, its derivation involved infinitesimal force and energy balances, and elementary differential geometry (Kirilin 1985; Leontiev 1985; Kikoine 1979; Landau 1971). In our attempt, for the sake of completeness, we shall adopt a particular framework of our problem, without loss of generality.

Let us consider a circular cylindrical capillary tube, of radius r_c, in a rectangular *xoy* coordinate system, Fig. 10.1. The *oy* axis is the capillary symmetry axis. The capillary is partially plunged in vertical position in a liquid of an infinite reservoir, of level $y = 0$. Depending on whether the liquid/solid forces are attractive or repulsive, the liquid wets or not the walls and the capillary rise will be positive

M.-D. Staicovici, *Coabsorbent and Thermal Recovery Compression Heat Pumping Technologies*, Heat and Mass Transfer, DOI: 10.1007/978-3-642-54684-6_10,

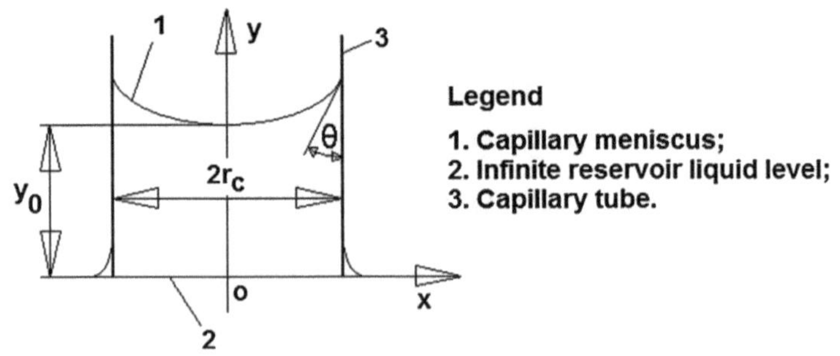

Fig. 10.1 Scheme of the capillary effect in a circular cylindric structure

(pointed towards the positive y values in Fig. 10.1), or negative. According to the most applications, suppose the liquid wets the capillary wall (contact angle is $0 < \theta < \pi/2$). It elevates in tube till $y(0) = y_0$, the capillary head. At equilibrium, the differential form of the 1st principle of thermodynamics reads:

$$dq = du + pdv + dl_\sigma + dl_c \tag{10.1}$$

In Eq. (10.1), $dl_\sigma < 0$ and $dl_c > 0$ stand for the elementary works of interface forces, received by the liquid interacting with the tube walls, and for the kinetic work done by the liquid, respectively. Measurements of various wick combinations showed that the maximum height the working fluid will attain may be slightly different for the same structure, according to whether the film was raising or falling. In Eq. (10.1) we neglected the secondary effect responsible for this behavior (e.g. friction losses). In the simplest case, liquid/solid conjugation conditions are adiabatic ($dq = 0$), isothermal ($du = c_v dT = 0$) and isochoric ($dv = 0$). For this type of interaction, met for example in liquid surface tension assessment, Eq. (10.1) becomes $dl_\sigma + dl_c = 0$. If l_c is approximated by the gravity potential energy, $l_c \cong l_p$, which the liquid gains during the process and integrate, we obtain:

$$l_{\sigma 1-2} + l_{p1-2} = const. \tag{10.2}$$

The isothermal and isochoric interactions suggest the use of the free energy thermodynamic potential, which at equilibrium has the form:

$$df = d(u - Ts) = -sdT - pdv + dl_\sigma + dl_c \tag{10.3}$$

For $T = const.$ and $v = const.$, the free energy has a minimum value related to Eq. (10.2) by:

$$f_{1-2} = l_{\sigma 1-2} + l_{p1-2} = l_{\sigma 1-2} + l_{c1-2} = const. = \min. \tag{10.4}$$

Other interactions may occur at quasi-constant temperature and constant pressure. This is the case of the heat pipe applications. Such interactions suggest the use of the free enthalpy potential, which at equilibrium is given by:

$$d\varphi = -sdT + vdp + dl_\sigma + dl_c \tag{10.5}$$

For $T = const.$ and $p = const.$, this time the free enthalpy has a minimum:

$$\varphi_{1-2} = dl_{\sigma 1-2} + dl_{p1-2} = const. = \min. \tag{10.6}$$

Equations of (10.4) and (10.6) type, further on referred to as energy equations, could be obtained, also, for other potentials appropriately describing the various solid/liquid conjugation conditions. However, commonly they all lead to the same extremum problems.

Apart from the energy equations, an optional tool in capillary investigations is the force equation, obtained from the balance of surface and gravity forces acting on the liquid at equilibrium:

$$\vec{F}_\sigma + \vec{F}_g = 0 \tag{10.7}$$

Let V be a cylindrical bounded volume of liquid in the capillary, whose boundary is a piecewise smooth orientable interface A, including the top, the bottom at $y = 0$ and the cylindrical surface. The interface work and the gravity potential energy have the expressions:

$$l_{\sigma 1-2} = -\int_A \vec{\sigma}\, dA \tag{10.8}$$

and

$$l_{p1-2} = \int_V \Delta p dV \tag{10.9}$$

respectively. The vector $\vec{\sigma}$ has a double aspect, referred to as a specific force, $\lfloor Nm^{-2} \rfloor$, or as a surface specific energy, $\lfloor Jm^{-2} \rfloor$. In Eq. (10.8), $\vec{\sigma}$ is considered as a specific interface free energy, pointed towards the interface normal. According to the Antonoff rule (Murgulescu 1979), $|\vec{\sigma}|$ equals the difference of the two media surface tensions. The vector $\vec{\sigma}$ would be different from zero on the interface, only (see position 1 in Fig. 10.1). Also, in Eq. (10.9), Δp stands for the bottom pressure difference after and before liquid rise, $\Delta p = p_2 - p_1$. The use of the Gauss divergence theorem for Eq. (10.8) yields:

$$-\int_A \vec{\sigma}\, dA = -\int_V div\vec{\sigma}\, dV \tag{10.10}$$

The introduction of Eqs. (10.8) and (10.10) in Eq. (10.2) gives:

$$\int_V (-div\vec{\sigma} + \Delta p) dV = const. \tag{10.11}$$

The total differentiation of Eq. (10.11) with respect to V settels the integrand to zero, whence the capillary pressure equation (Staicovici 1998b):

$$div\vec{\sigma} = \Delta p \tag{10.12}$$

Equation (10.12) is the new wording for the known Laplace equation:

$$\vec{\sigma}\left(\frac{1}{R_1}+\frac{1}{R_2}\right)=\Delta p \tag{10.13}$$

where R_1 and R_2 are the interface main curvature radii. Equations (10.13, 10.12) is more general, including as a particular case the Laplace equation. It states that the pressure forces of the surface tension, pointed along the surface gradient (towards its curvature centre) are equal to the liquid hydrostatic pressure at every point. This new pressure equation may be checked, for instance, for the spherical meniscus, where $div\vec{\sigma}=\left(1/R^2\right)\left(\partial\left(R^2\sigma\right)/\partial R\right)=2\sigma/R=\Delta p$, or, for the cylindrical one, where $div\vec{\sigma}=(1/R)(\partial(R\sigma)/\partial R)=\sigma/R=\Delta p$. The last equalities of each case are already known.

The pressure equation may be an alternative tool in capillary assessment. In order to simplify it, suppose that the second medium is air. Its pressure can be considered constant at every point of the liquid meniscus $(p_2)=const.$ and from Eqs. (10.12) or (10.13) it results (Landau 1971):

$$\left(\frac{1}{R_1}+\frac{1}{R_2}\right)-\frac{p_1}{\sigma}=const. \tag{10.14}$$

Mathematically, the energy, force and pressure equations are, obviously, equivalent. This will be checked in text at the right moment. They apply to both positive and negative capillary heads and to all structures. However, the symmetrical structures, like those of cylindrical or parallel walls type, simplify the computation. In the following, for the sake of clarity, we deal with these simpler structures.

10.3 Variational Numerical Approach

Let us consider the liquid rise in the capillary tube of Fig. 10.1. Its modelling means the determination of the bottom interface equation. In our case, this reduces to a single function $y=y(x)$, $x\in[0,r_c]$. The Eqs. (10.8) and (10.9) become:

$$l_{\sigma 1-2}=2\pi\sigma\int_0^{r_c}x\sqrt{1+y'^2}dx-\sigma\pi r_c^2 \tag{10.15}$$

and

$$l_{p1-2}=\frac{1}{2}2\pi\gamma_l\int_0^{r_c}xy^2dx \tag{10.16}$$

respectively. The introduction of Eqs. (10.15) and (10.16) in Eq. (10.4), leads to a variational problem of extremum for the functional:

$$F(y) = \int_0^{r_c} f(x, y, y') = \int_0^{r_c} x\left(l_c^2\sqrt{1 + y'^2} + y^2\right)dx = \min. \tag{10.17}$$

where

$$l_c = \sqrt{\frac{2\sigma}{\gamma_l}} \tag{10.18}$$

is the capillary length, dependent on the liquid nature, only. At extremum point, the functional argument $y(x)$ must satisfy the Euler equation (Silov 1989):

$$\left(\frac{\partial}{\partial y} - \frac{d}{dx}\frac{\partial}{\partial y'}\right)f = 0 \tag{10.19}$$

From Eqs. (10.17) and (10.19), we deduce that the searched extremal is a solution of the second-order nonlinear differential equation:

$$\frac{y}{l_c^2} = \frac{1}{2}\left(\frac{y''}{\left(1 + y'^2\right)^{3/2}} + \frac{y'}{x\sqrt{1 + y'^2}}\right) \tag{10.20}$$

satisfying some boundary conditions, e.g.:

$$y(0) = y_0; \quad y'(0) = 0; \quad y'(r_c) = \text{ctan}\theta \tag{10.21}$$

The meniscus shape is the result of superposition of two effects, as it can be seen from Eq. (10.20), that of the interface curving near a straight wall, described by the first term in the bracket, and that one due to the wall curving, described by the second term in bracket. Moreover, Eq. (10.20) is nothing else but Eq. (10.14), where the constant is zero. It describes the surface mean curvature with terms in the bracket equal to the surface main curvatures. The extremal cannot be expressed in terms of elementary functions, therefore we used a numerical approach. To this purpose, Eq. (10.20) was first solved for y'', obtaining:

$$y'' = \left(\frac{2}{l_c^2}y\sqrt{1 + y'^2} - \frac{y'}{x}\right)\left(1 + y'^2\right) \tag{10.22}$$

Secondly, we investigated the existence and uniqueness of the solution, of the Cauchy problem [Eq. (10.21)] for Eq. (10.22). We found that the Lipschitz condition from the theorem of existence and uniqueness is not fulfilled, therefore several solutions could exist. That means that apart from the first two boundary conditions, the third must be taken into account, as well, in order to single out a solution for the given parameters y_0, r_c and l_c. Sometimes it is suggested that a supplementary condition to an extremum problem in capillary would be:

$$\int dV = const. \tag{10.23}$$

expressing the constancy of the total volume of working fluid. We avoided it here, as being actually difficult to use. The numerical solution was obtained by the Runge-Kutta method. Prior to start the integration, we solved the limit at zero of the ratio (y'/x) which appears in Eq. (10.22). Using the Cauchy conditions (10.21) and assuming that Eq. (10.22) holds true at $x = 0$ too, we found:

$$\lim_{x\to 0} \frac{y'}{x} = \frac{y_0}{l_c^2} \tag{10.24}$$

The results of the numerical integration are given in Fig. 10.2a–d, where the capillary length is plotted against the capillary height, $l_c = l_c(y_0)$, in a double logarithmic scale. The families of curves depend on two parameters, r_c and θ and are almost straight lines in this plot, enabling easy extrapolation. It is possible to eliminate one out of the two parameters more difficult to be measured in practice, θ, or σ, if the solution $y(x)$, obtained from the numerical integration, is constrained to satisfy one of equations of the thermodynamic potentials which appropriately describe the liquid/solid capillary interaction. In case of the free energy potential, $y(x)$ must satisfy equation:

$$\pi \gamma_l \int_0^{r_c} x\left[l_c^2\left(\sqrt{1+y'^2}-1\right)+y^2\right]dx = f = h - pv - Ts \tag{10.25}$$

In Eq. (10.25) the free energy minimum is computed with available working fluid p, v, T data, in an iterative way. A similar equation can be obtained for the free enthalpy potential, as well.

As previously mentioned, the energy and force equations are mathematically, equivalent. This is simple to check in this case if one projects Eq. (10.7) on the oy axis, at some radius x. We have:

$$F_g(x) = -2\pi\gamma_l \int_0^x zy(z)dz \tag{10.26}$$

and

$$F_\sigma(x) = 2\pi x\sigma \frac{y'}{\sqrt{1+y'^2}} \tag{10.27}$$

where from, with Eqs. (10.7) and (10.18), it follows the integral-differential equation:

$$\int_0^x zy(z)dz = \frac{1}{2}xl_c^2\frac{y'}{\sqrt{1+y'^2}} \tag{10.28}$$

Its total differentiation with respect to x gives exactly Eq. (10.20). If Eq. (10.28) were considered at $x = r_c$, together with the $y'(r_c) = \text{ctan}\theta$ boundary condition of Eq. (10.21), then the refined version of the Jurin-Borelli law $y_0 = 2\sigma\cos\theta/\gamma_l r_c$

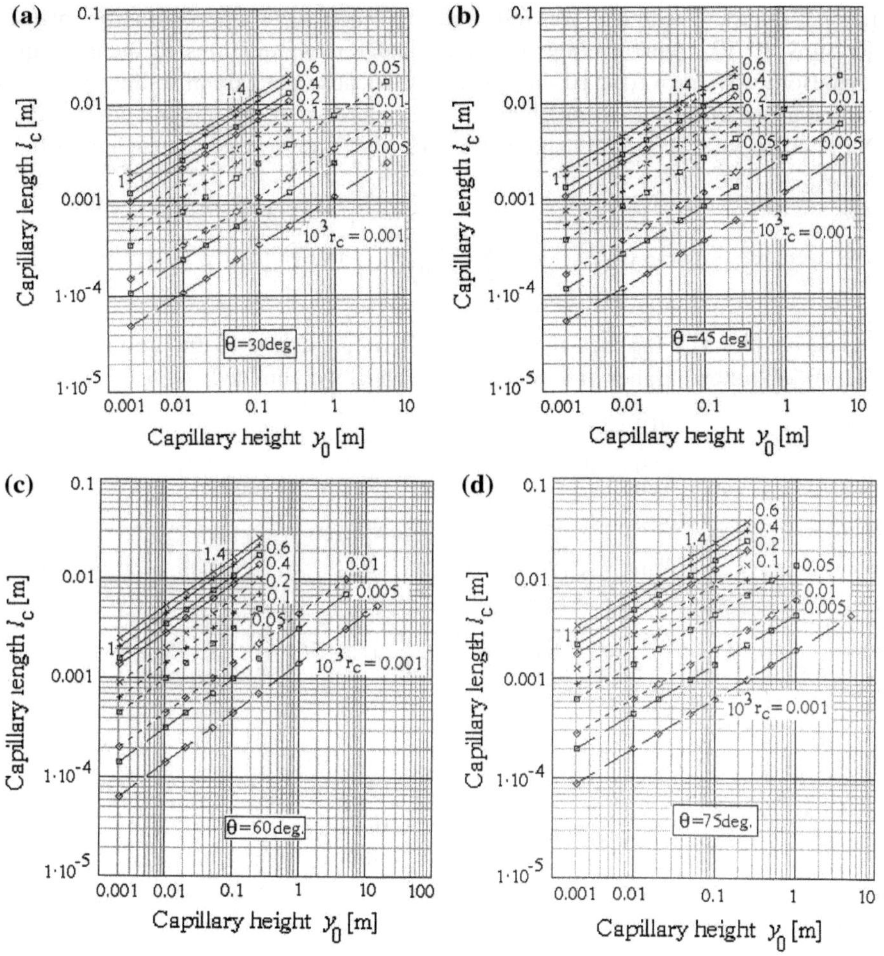

Fig. 10.2 Numerical approach of the cylindrical tube capillary rise, with contact angle $\theta = (30-75)°$ and capillary radius $r_c = \left(1.4 - 1.0 * 10^{-3}\right) * 10^{-3}$m as parameters. **a** $\theta = 30°$, **b** $\theta = 45°$, **c** $\theta = 60°$, **d** $\theta = 75°$

(an approximation of the capillary rise effect often used in practice), is obtained. The case of the capillary infinite parallel walls is briefly presented in the Appendix.

10.4 Analytical Approach

As already mentioned, we could not find a closed form solution of Eq. (10.20). However, besides its numerical integration, we also paid attention to report here an analytical approach. On this purpose, we investigated first a direct variational method (Ritz). Secondly, we followed the successive approximation method (Krasnov 1977), transforming the energy and forces equations in integral equations. Both ways proved to be unsatisfactory, because the gain in solution accuracy is done at the cost of a too much complexity increase of the integral computation. Our attempt was successful using the Taylor-Mac Laurin series expansion of function $y(x)$ in the vicinity of $x = 0$:

$$y(x) = y_0 + \sum_{i=1}^{(i)} y_{(0)}^{(i)} \frac{x^i}{i!}, \quad 0 \leq x \leq r_c \tag{10.29}$$

The derivatives $y_{(0)}^{(i)}$ may be calculated step by step, passing to the limit at $x \to 0$ in Eq. (10.22) and its higher order derivatives. Thus, we obtained:

$$\begin{aligned} y(0) &= y_0, \\ y_{(0)}^{(2)} &= \frac{y_0}{l_c^2}, \\ y_{(0)}^{(4)} &= \frac{3}{2}\frac{y_0}{l_c^2}\left[\frac{1}{l_c^2} + 2\left(\frac{y_0}{l_c^2}\right)^2\right], \\ y_{(0)}^{(6)} &= \frac{5}{6}\frac{y_0}{l_c^2}\left[\frac{3}{l_c^4} + \frac{63}{l_c^2}\left(\frac{y_0}{l_c^2}\right)^2 + 56\left(\frac{y_0}{l_c^2}\right)^4\right], \\ &\ldots \end{aligned} \tag{10.30}$$

and

$$y_{(0)}^{2n+1} = 0, \quad n = 0, 1, 2, \ldots \tag{10.31}$$

In case of the circular cylindrical tube capillary rise, the analytical expression of the liquid meniscus shape results now from Eqs. (10.29)–(10.31) in the form:

$$y(x) = y_0 + \frac{y_0}{l_c^2}\frac{x^2}{2!} + \frac{3}{2}\frac{y_0}{l_c^2}\left[\frac{1}{l_c^2} + 2\left(\frac{y_0}{l_c^2}\right)^2\right]\frac{x^4}{4!} + \frac{5}{6}\frac{y_0}{l_c^2}\left[\frac{3}{l_c^4} + \frac{63}{l_c^2}\left(\frac{y_0}{l_c^2}\right)^2 + 56\left(\frac{y_0}{l_c^2}\right)^4\right]\frac{x^6}{6!} + \cdots \tag{10.32}$$

Equation (10.32) must be supplemented with:

$$y'(r_c) = \text{ctan}\theta \tag{10.33}$$

where $y'(r_c)$ is the first derivative of Eq. (10.32) at $x = r_c$ [see Eq. (10.21)]. If necessary, higher-order terms can be added to the expansion (10.32) using formal differentiation codes, applied to (10.22). This series converges quite slowly. For example, in case of water, the terms decreasing rate is approximately 10 times. However, despite this, Eq. (10.32), completed with Eq. (10.33), has the same accuracy as the numerical solving used to obtain the results plotted in Fig. 10.2a–d, and additionally is much simpler mathematically and more rapid.

Frequently, not always, the liquid meniscus shape of the cylindrical tube is assimilated with a spherical one. How accurate is this approximation, can be appreciated by writing the circle equation describing the assimilated spherical interface in the *xoy* coordinate system, Fig. 10.1, and comparing it with the true surface given by Eq. (10.32). If r is the circle radius and y_0 the capillary head, we have:

$$y_{sph}(x) = y_0 + r - \sqrt{r^2 - x^2}, \quad |x| \le r \tag{10.34}$$

or, using the binomial series expansion:

$$y_{sph}(x) = y_0 + \frac{1}{2r}\frac{x^2}{1!} + \frac{1}{4r^3}\frac{x^4}{2!} + \frac{3}{8r^5}\frac{x^6}{3!} + \cdots \tag{10.35}$$

The identity of Eqs. (10.32) and (10.35) implies that:

$$r = \frac{l_c^2}{y_0} \tag{10.36}$$

for the first two terms, but this holds not true for those of higher-order. For this reason the assimilation with a spherical meniscus is not recommended normally.

Appendix

Let us consider two parallel infinitely-long walls symmetrical with respect to the *yoz* plane and separated by the capillary distance $2s$, Fig. 10.A.1 The assumptions made for the cylindrical tube hold here true, too. The meniscus would be obtained via Eqs. (10.8), (10.9) and (10.4), minimizing the functional (Staicovici 1998b):

$$F(y) = \int_0^s f(y, y')dx = \int_0^s \left(l_c^2\sqrt{1 + y'^2} + y^2\right)dx = \text{min}. \tag{10.A.1}$$

The Euler equation for integrand $f(y, y')$ has the form (Silov 1989):

$$f - y'f_{y'} = C = const. \tag{10.A.2}$$

and the extremals are found solving the first order nonlinear differential equation:

$$\frac{l_c^2}{\sqrt{1 + y'^2}} = C - y^2 \tag{10.A.3}$$

provided that:

$$y(0) = y_0; \quad y'(0) = 0; \quad y'(s) = \text{ctan}\theta \tag{10.A.4}$$

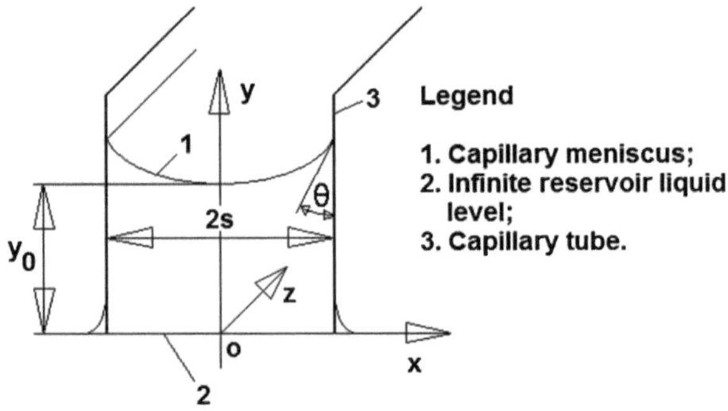

Fig. 10.A.1 Scheme of the capillary effect in an infinite parallel walls structure

The solution of Eq. (10.A.3) is expressed by the elliptic integral (Landau 1971):

$$x(y) = \int_{y_0}^{y} \frac{\left(A - \frac{t^2}{l_c^2}\right) dt}{\sqrt{1 - \left(A - \frac{t^2}{l_c^2}\right)^2}} \tag{10.A.5}$$

where

$$A = \frac{C}{l_c^2} = 1 + \frac{y_0^2}{l_c^2} \tag{10.A.6}$$

and $y \in \lfloor y_0, l_c\sqrt{A - \sin\theta} \rfloor$.

The projection of force, Eq. (10.7), on the *oy* axis gives the integral-differential equation:

$$\int_0^x y(t)dt = \frac{1}{2} l_c^2 \frac{y'}{\sqrt{1 + y'^2}} \tag{10.A.7}$$

Differentiating it totally with respect to x yields the second-order differential equation:

$$\frac{y}{l_c^2} = \frac{y''}{2\left(1 + y'^2\right)^{3/2}} \tag{10.A.8}$$

Equation (10.A.8) is exactly Eq. (10.14) for const. = 0. The right-hand member represents the mean curvature of the surface. Multiplying it by y' and integrating once, Eq. (10.A.3) is obtained, checking the equivalence of energy, force and pressure equations, in this case, also.

The application of the successive differentiation method, mentioned in Sect. 10.4., to Eq. (10.A.3) for Cauchy initial conditions (10.A.4), gives the following analytical expression of the liquid meniscus shape in case of the infinite parallel walls capillary:

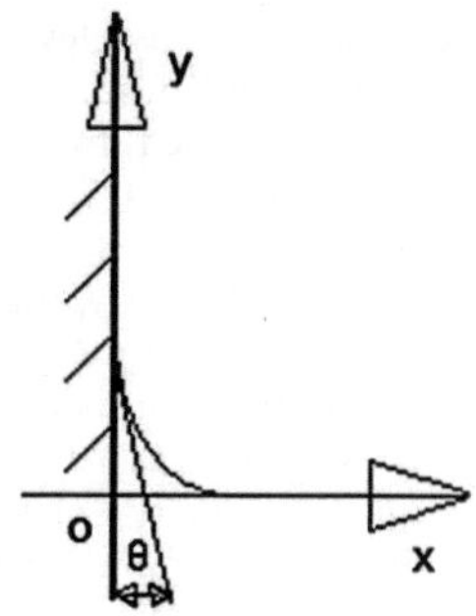

Fig. 10.A.2 Capillary effect near a straight wall in an infinite liquid reservoir

$$y(x) = y_0 + 2\frac{y_0}{l_c^2}\frac{x^2}{2!} + 4\frac{y_0}{l_c^2}\left[\frac{1}{l_c^2} + 6\left(\frac{y_0}{l_c^2}\right)^2\right]\frac{x^4}{4!} + 8\frac{y_0}{l_c^2}\left[\frac{1}{l_c^4} + \frac{66}{l_c^2}\left(\frac{y_0}{l_c^2}\right)^2 + 180\left(\frac{y_0}{l_c^2}\right)^4\right]\frac{x^6}{6!} + \cdots \tag{10.A.9}$$

where, $0 \le x \le s$. Equation (10.A.9) must be supplemented with:

$$y'(s) = c\tan\theta \tag{10.A.10}$$

Identifying the true meniscus shape given by Eq. (10.A.9) with an assimilated circular cylindrical surface, described by Eq. (10.35), implies that the curvature radius would be $r = l_c^2/(2y_0)$ for the two first terms, but this holds not true for the higher-order terms. For this reason, the assimilation with a circular cylindrical meniscus is not recommended.

Finally, the functional (10.A.1) describes, also, the conditions of equilibrium of the liquid contained in an infinite reservoir and curved near a straight wall, Fig. 10.A.2. Here, Eq. (10.A.3) must be solved together with the new boundary conditions (the origin of the *xoy* coordinate system is moved at the wall):

$$y'(0) = -c\tan\theta; \quad y(\infty) = y'(\infty) = 0 \tag{10.A.11}$$

This time A from Eq. (10.A.6) equals unit, $A = 1$, and Eq. (10.A.5) has the primitive:

$$x(y) = -\frac{l_c}{2\sqrt{2}} arg\tan\sqrt{1 - \frac{1}{2}\left(\frac{y}{l_c}\right)^2} + \sqrt{2l_c^2 - y^2} + x_0 \tag{10.A.12}$$

where the integration constant x_0 is found for $x = 0$ and $y = l_c\sqrt{1 - \sin\theta}$.

References

Kikoine, A., & Kikoine, I. (1979). *Physique moléculiare*. Moscow: Mir Publishing House.

Kirilin, V. A., Sicev, V. V., & Seindlin, A. E. (1985). *Thermodynamics* . Bucharest: Scientific and Encyclopedic Publishing House. (in Romanian).

Krasnov, M., Kisselev, A., & Makarenko, G. (1977). *Équations integrales*. Moscou: Mir Publishing House.

Landau, L. D., & Lifschitz, E. M. (1971). *Mécanique des fluides*. Moscow: Mir Publishing House.

Leontiev, A. (1985). *Théorie des echanges de masse et de chaleur*. Moscow: Mir Publishing House.

Murgulescu, I. G., & Segal, E. (1979). *Cinetic-moleculaire theory of matter, II.1. Introduction in physical chemistry*. Bucharest: Romanian Academy Publishing House. (in Romanian).

Silov, G. E. (1989). *Mathematical analysis* . Bucharest: Scientific and Encyclopedic Publishing House. (in Romanian).

Staicovici, M. D. (1998b). *Variational numerical and analytical approach of the liquid capillary rise effect*, Académie Roumaine, Revue Roumaine Des Sciences Techniques, série de Mecanique Appliquée, tome **43**, no. **1**, pp. 85–99, janvier-février, 1998.

Stephan, P. (1995). Evaporative heat transfer in capillary grooved heat pipes. In: *Proceedings of 19th International Congress of Refrigeration*, IIIa (pp. 720–727).

Chapter 11
Marangoni Convection Basic Mechanism Explanation, Pseudo-Marangoni Cells Model and Absorption–Desorption Mass and Heat Exchangers Model Application

11.1 Introduction

The Marangoni convection stimulation is a practical method used by the thermal absorption technology in the past decades in order that the absorption processes be significantly improved. The literature devoted to this effect is rich, including theoretical and experimental studies on various working fluid-absorbent-surfactant combinations (e.g. Rie and Kashiwagi 1991; Hihara and Saito 1994; Nordgren and Setterwall 1996; Möller and Knoche 1996; Kim et al. 1996; Hoffmann et al. 1996; Beutler et al. 1996; Kang et al. 1999; Ha 2002; Boyadjiev and Halatchev 1998; Velde et al. 2001; Ganzevles and van der Geld 1998; Yao et al. 1996; Li et al. 2002; Daiguji et al. 1997; Kumar et al. 2003; Bahloul et al. 2003). Besides this, attempts have been made also on explaining its nature. So far, a connection between the absorbent surface tension gradient and induced convection onset and stability is established, but its true mechanism is still not well known.

In previous papers (Staicovici and Isvoranu 2003, 2004; Isvoranu and Staicovici 2004), the basic mechanism of the Marangoni effect is explained, using the two-point theory (TPT) of mass and heat transfer (see Chap. 9) and a new wording of the Laplace equation (Staicovici 1998, see also Chap. 10). The results of this approach are applied to an ammonia-water absorption process in a Marangoni cell.

11.2 True Marangoni Effect Mechanism

The common way of inducing Marangoni convection is by obtaining interfacial tension gradients. In mono-component and multi-component media, this is achieved by creating surface temperature gradients ("thermo-capillarity") and both temperature and working fluid (dissolved surfactant) concentration gradients ("distill-capillarity"), respectively. Several models are proposed to explain distill-capillarity, out of which it can be mentioned that of the surfactant droplet (Kashiwagi 1988), or that of the

M.-D. Staicovici, *Coabsorbent and Thermal Recovery Compression Heat Pumping Technologies*, Heat and Mass Transfer, DOI: 10.1007/978-3-642-54684-6_11,

"salting (radical)-out" (Hozawa et al. 1991), depending on whether the surfactant exceeds the solubility limit in the absorbent or not. Supposing the surface tension depends on temperature T and working fluid concentration c, $\sigma = \sigma(T, c)$, than, the interfacial tension gradient along a direction x writes

$$\frac{\partial \sigma}{\partial x} = \frac{\partial \sigma}{\partial T}\frac{\partial T}{\partial x} + \frac{\partial \sigma}{\partial c}\frac{\partial c}{\partial x} \tag{11.1}$$

In case of bi-component and bi-phase interactions, the system is bi-variant and if absorption pressure were established, the temperature T and mass fraction c in Eq. (11.1) depend one another. Usually, authors assess distill-capillarity confining to simply evaluate the sign of Eq. (11.1), only. Additionally, in papers modeling the Marangoni convection (e.g. Rie and Kashiwagi 1991; Boyadjiev and Halatchev 1998; Bahloul et al. 2003), surface tension gradient equals the shear stress tensor τ when considering Navier-Stokes equations.

$$\tau_{z_i z_j} = \frac{\partial \sigma}{\partial z_i} \tag{11.2}$$

where z_i, z_j, i, j = 1, 2, 3 are local coordinates for each control volume node at the free surface. As a general remark, the basic factor that brings about the fluid motion in Marangoni convection is not mentioned in any topic published work. For this reason, next we shall give the simple explanation of its true mechanism.

First, in case of a liquid–gas interface, the liquidcgas interface pressure difference in a certain point of the fluid free surface, $\Delta p = p_l - p_g$, and surface tension are related by Laplace equation:

$$\Delta p = \sigma_f \left(\frac{1}{R_1} + \frac{1}{R_2}\right) \tag{11.3}$$

In Eq. (11.3), σ_f is the module of the surface tension vector $\vec{\sigma}_f$ considered as a specific force, (N/m), acting in the interface tangent plane. If the interface normal $\vec{n}$ were pointed in order to get out of the liquid, then the curvature radii are negative or positive with respect to $\vec{n}$ and from Eq. (11.3), Δp becomes positive or negative for convex and concave interfaces, respectively, as a classic result. Now, taking into account the other aspect of the surface tension, which equally might be considered as an interface specific mean free energy, (J/m^2), the new wording for the Laplace equation writes here (see Chap. 10):

$$div\vec{\sigma}_e = \Delta p \tag{11.4}$$

In Eq. (11.4), Δp has the same meaning and values as in Eq. (11.3) and the vector $\vec{\sigma}_e$ (J/m^2) has the same module as that of vector $\vec{\sigma}_f$ (N/m) in Eq. (11.3), $\sigma_e = \sigma_f = \sigma$. It is directed towards the interface normal (see comment above), or towards the opposite direction, in case of concave or convex surfaces, respectively, and acts in good agreement with the known principle of surface free energy minimization tendency through the interface area minimization. Equation (11.4) is more general than Eq. (11.3). In order to proof this statement, it is enough to apply the two equations to the particular case of an horizontal free surface of a Marangoni cell in a

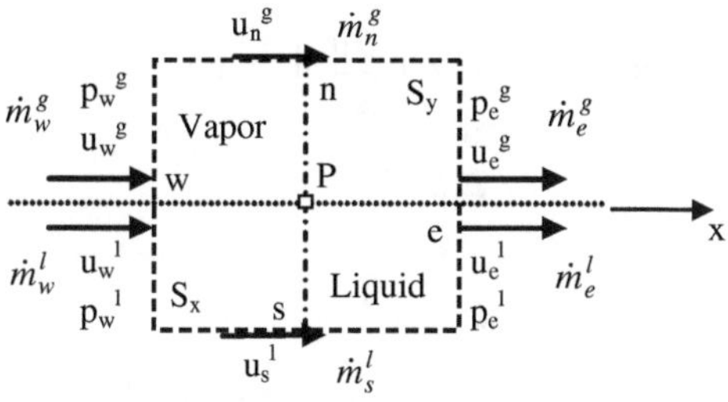

Fig. 11.1 Interface control volume

rectangular xy coordinate system, having the normal parallel to the coordinate y, and a non-constant tension gradient with respect to the coordinate, x, $\frac{\partial \sigma}{\partial x} \neq ct$ in point $(x, 0)$. The results would be different. Indeed, taking into account that

$$\frac{\partial \sigma_e}{\partial y} = \frac{\partial \sigma_f}{\partial x} \tag{11.5}$$

Eq. (11.4) emphasizes the generation of a non-zero fluid pressure gradient with respect to coordinate x, according to ($p_g = ct.$)

$$\frac{\partial p}{\partial x} = \frac{\partial}{\partial x}\left(\frac{\partial \sigma_e}{\partial y}\right) = \frac{\partial}{\partial x}\left(\frac{\partial \sigma_f}{\partial x}\right) = \frac{\partial^2 \sigma}{\partial x^2} \neq 0 \tag{11.6}$$

while Eq. (11.3) does not, because for R_1 and $R_2 \to \infty$, it results that $\frac{\partial p}{\partial x} = 0$ for any $\vec{\sigma}$ variation (Staicovici and Isvoranu 2003).

Moreover this result is in good agreement with the "le Chatelier" principle, the system acting in order to counteract the effect of the $\frac{\partial \sigma}{\partial x} \neq 0$ appearance at interface (Staicovici and Isvoranu 2004). Indeed, considering a finite interface control volume as in Fig. 11.1, Eq. (11.6) shows that when the surface free energy is decreasing in order that $\frac{\partial^2 \sigma}{\partial x^2} < 0$, or increasing in order that $\frac{\partial^2 \sigma}{\partial x^2} > 0$, then the pressure of the east-west control volume faces will be different with respect to coordinate x, expressed by $\nabla p = \frac{p_e^l - p_w^l}{x_e - x_w} < 0$ or $\nabla p = \frac{p_e^l - p_w^l}{x_e - x_w} > 0$ and a fluid movement will be generated towards positive or negative x coordinate, in order to increase or decrease its surface free energy, respectively, and preserve its initial specific minimum free energy (see Fig. 11.3). Further on, taking advantage of Eqs. (11.4) and (11.5), the fluid motion governed by interfacial tension gradients might be characterized by the following more general property:

$$grad(p) = grad(div\vec{\sigma}) \tag{11.7}$$

This mechanism—$grad(div\vec{\sigma}) \to grad(p) \to$ fluid movement—is shown schematically for the salting-out effect (Hozawa 1991), and for the Marangoni convection in binary drops in air (Ganzevles and van der Geld 1998), in Fig. 11.2a, b, respectively. In case of the ammonia-water system during absorption without additive, Fig. 11.2a, the crest region of free surface wave becomes ammonia richer, causing surface tension decrease, consequently $p_e - p_w < 0$, so fluid moves downward. In the wave trough, the reverse happens and fluid moves upward. The two opposite

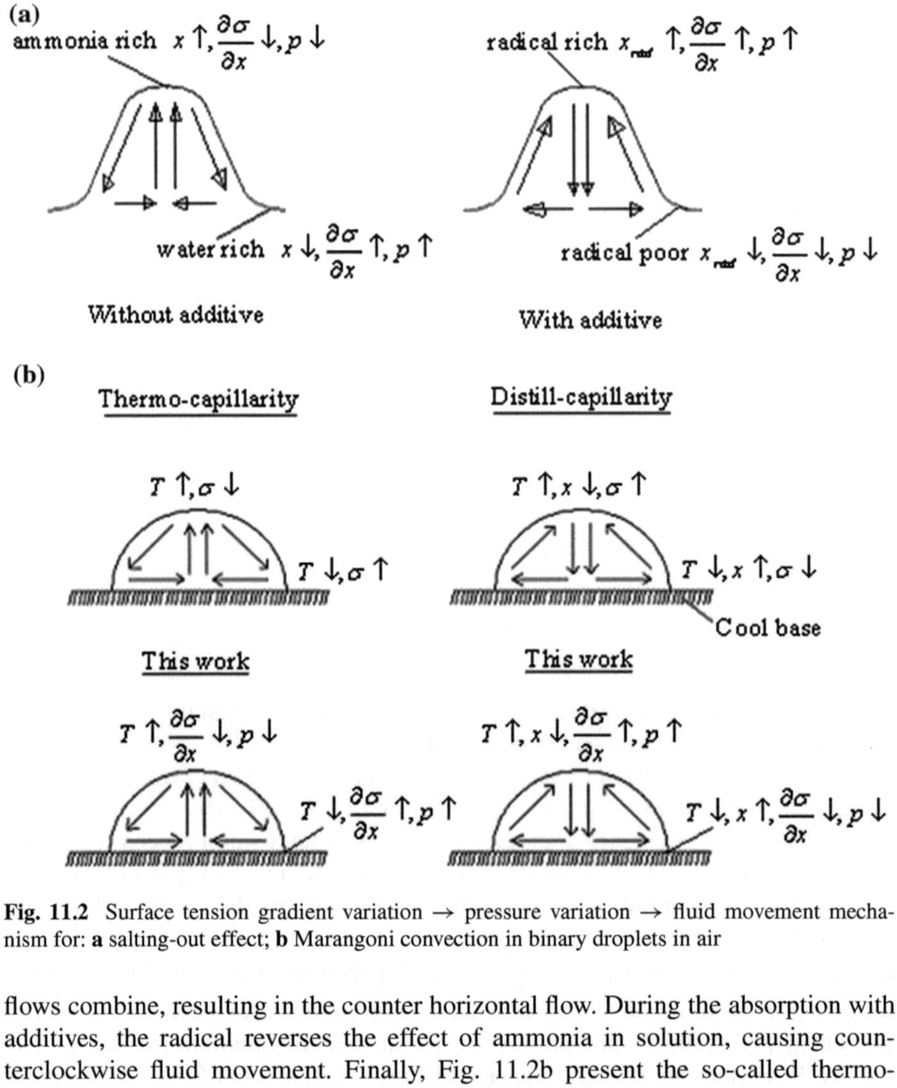

Fig. 11.2 Surface tension gradient variation → pressure variation → fluid movement mechanism for: **a** salting-out effect; **b** Marangoni convection in binary droplets in air

flows combine, resulting in the counter horizontal flow. During the absorption with additives, the radical reverses the effect of ammonia in solution, causing counterclockwise fluid movement. Finally, Fig. 11.2b present the so-called thermo-capillary and distill-capillary effects in mono-component and binary drops in air, condensed on a cool base. The upper figures give an incomplete explanation of the fluid movement (Ganzevles and van der Geld 1998). This work completes it. In case of the thermo-capillarity, Fig. 11.2a without additives applies. In case of the distill-capillarity, the lower-temperature evaporating component plays the role of additive, and Fig. 11.2a with additives applies.

In the next section we shall assess a Marangoni convection from a quantitative point of view, modeling an ammonia-water absorption process in a Marangoni cell. However, prior to do so, first we shall give a simpler mathematical formulation of the induced fluid movement, helpful in our complete understanding of the Marangoni convection and further modeling. We consider the cell in Fig. 11.3 during absorption for an ideal fluid and write equation of Bernoulli in its

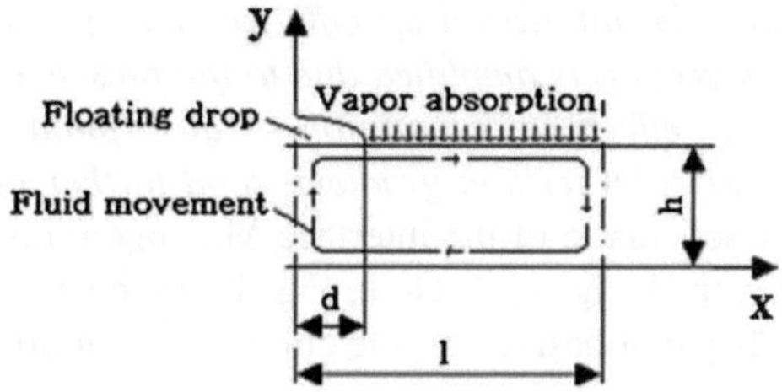

Fig. 11.3 Schematic of a Marangoni convection cell

differential form for the top fluid steady-state moving along the free surface and integrate it between two certain points 1 and x:

$$\int_{1}^{x} d\left(\frac{v^2}{2}\right) = -\int_{1}^{x} \frac{dp}{\rho} \tag{11.8}$$

For the sake of simplicity, we note that $v(1) \cong 0$ and $\rho \approx ct$. Taking into account Eq. (11.4) we obtain

$$v(x) = \pm\sqrt{\pm\frac{2}{\rho}\left[\frac{\partial\sigma}{\partial u}(x) - \frac{\partial\sigma}{\partial u}(1)\right]} \tag{11.9}$$

For a reduced form (index “r”), we put $v_r = v\frac{L}{\upsilon}$ (L-characteristic length, υ-cinematic viscosity) and Eq. (11.9) becomes:

$$v_r(x) = \pm\sqrt{\pm 2[Mg(x) - Mg(1)]} \tag{11.10}$$

Equation (11.10) emphasizes a reduced velocity dependence of a Marangoni number:

$$Mg = \frac{L^2}{\upsilon^2 \rho}\frac{\partial\sigma}{\partial z} \tag{11.11}$$

The common applications have $\left|\frac{\partial\sigma}{\partial z}\right| \approx 10^0 - 10^{-1}, L \approx 10^{-2}, \upsilon \approx 10^{-6}, \rho \approx 10^3$, so that $|Mg| \approx 10^5 - 10^4$. Equation (11.9) enables an analysis of the interfacial tension gradient in order to get a certain induced fluid velocity. To this extent, Eq. (11.1) was considered for $\sigma = \sigma(c(x))$ only. The analysis of several $\frac{\partial c}{\partial x}$ shapes (linear, polynomial and exponential) revealed that the exponential form got velocities with values met in current applications $\left(v \approx 10^{-2}\right)$, only. This simple approach suggests an *i.p.a.* behavior, showing that *i.p.a.* effects accompany Marangoni cells working (Staicovici 2000a, b, c; Staicovici and Isvoranu 2004). The modeling presented in the next section will show this.

Now, taking into account the ideas outlined above, we are able to formulate the Marangoni convection true explanation (Staicovici and Isvoranu 2003, 2004). Indeed, Eqs. (11.1), (11.4–11.6) and (11.9) can be considered the key of our problem: *the Marangoni convection is generated by non-constant interfacial tension gradients, which create parallel fluid non-zero pressure gradients, and consequently fluid motion, tending, according to the “le Chatelier” principle, to counteract in this way the interfacial specific free energy variation. In multi-component media, this convection is amplified due to the presence of adjacent surface zones with opposite i.p.a. effects, of generation ↔ absorption, which brings about high local non-zero interfacial tension gradients, and further high pressure gradients and fluid motion.* A schematic of the interface Marangoni convection basic mechanism explanation is given in Fig. 11.4. Here, Fig. 11.4a, b show the counter and parallel flow formation. They emphasize also the effect of $\frac{\partial\sigma}{\partial x}$ variation, mathematically expressed by:

$$\frac{\partial^2\sigma}{\partial x^2} < 0 \Rightarrow \frac{\partial u}{\partial x} > 0; \tag{11.12}$$

and

$$\frac{\partial^2\sigma}{\partial x^2} > 0 \Rightarrow \frac{\partial u}{\partial x} < 0.$$

The mechanism of flow deviation is explained next, in Fig. 11.4c, d. As they indicate, the main flow is deviated by an opposite, secondary flow, towards the bottom of the cell. Obviously, the stronger the secondary flow, the higher the main flow deviation is. The main flow deviation lays to the basis of the eddy formation.

11.3 Pseudo-Marangoni Ammonia–Water Cell Modeling

Induced convection in a pseudo-Marangoni cell, Fig. 11.3, was modeled with the ammonia-water system. The modeling uses known hydrodynamic, energy and species transport equations for interface and bulk fluid of the cell. The modeling used the TPT for absorption/generation processes at interface (Staicovici 2000a, b, c). Its first purpose is to analyze with known theoretical tools the Marangoni effect and compare it with the explanation outlined in paper second section. Pseudo-Marangoni cell comes from the fact that no additive effects were accounted for in the convection simulation, meaning that, when absorption starts, the fluid mixture surface located below the "drop" is free of gas absorption, has the same surface tension as that of a gas–liquid interface, and absorbing mixture equilibrium is described by its pure ammonia-water equation of state (Ziegler and Trepp 1984).

Physical model

The governing system of equations comprises Cartesian bi-dimensional equation of continuity, Navier-Stokes equations, energy and species transport equations:

$$\begin{aligned}
&\text{Continuity: } \frac{\partial\rho}{\partial t} + \rho\vec{\nabla}\cdot(\vec{V}) = 0\\
&\text{Navier-Stokes: } \rho\frac{\partial\vec{V}}{\partial t} + \rho\vec{V}\cdot\vec{\nabla}(\vec{V}) = -\vec{\nabla}p + \vec{\nabla}\cdot\vec{\Pi} + \rho g(\beta\Delta T + \gamma\Delta c) + \vec{F}_g\\
&\text{Energy: } \rho\frac{\partial T}{\partial t} + \rho\vec{V}\cdot\vec{\nabla}(T) = \frac{1}{c_p}\vec{\nabla}\cdot(\lambda\vec{\nabla}T)\\
&\text{Species: } \rho\frac{\partial c}{\partial t} + \rho\vec{V}\cdot\vec{\nabla}(c) = \vec{\nabla}\cdot(\rho D\vec{\nabla}c)
\end{aligned} \tag{11.13}$$

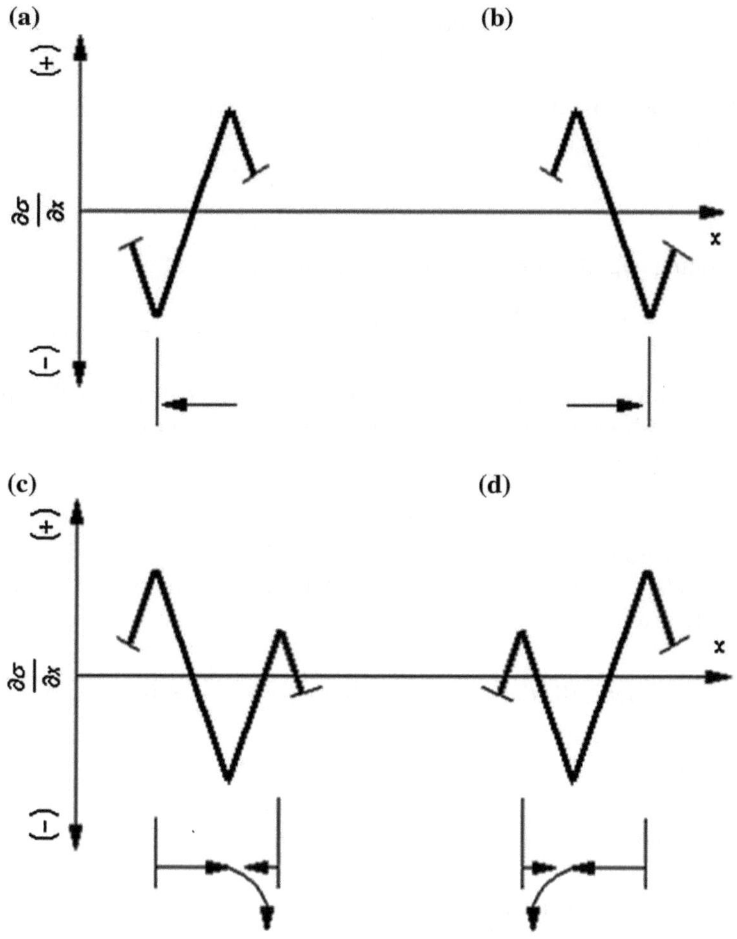

Fig. 11.4 Schematic of the interface Marangoni convection basic mechanism explanation. **a** Counter fluid flow formation $\frac{\partial^2\sigma}{\partial x^2} > 0 \Rightarrow \nabla_p > 0$, **b** Parallel fluid flow formation $\frac{\partial^2\sigma}{\partial x^2} < 0 \Rightarrow \nabla_p < 0$. **c** Parallel fluid flow deviation and clock wise eddy formation. **d** Counter fluid flow deviation and clock wise eddy formation

Numerical model

The system of four partial differential equations. was solved numerically, with the help of a CAFFA code (Computer Aided Fluid Flow Analysis), incorporating the Finite Volume Method. In order to solve the pressure, the SIMPLE algorithm was used. The calculus domain was discretized by an enough large number of cells in order that the computing accuracy be satifactory. We choosed a rectangular mesh of 96 × 96 volumes of control. The numerical method is segregated, non-staggered, semi-implicite in space and explicite (Euler) in time, based on a hybride scheme. It applies to the primitive variables velocity, temperature and concentration. Writing the transport equations in the general form

$$\frac{\partial \rho \varphi}{\partial t} + \frac{\partial J_{\varphi,k}}{\partial x_k} = \rho S_\varphi \tag{11.14}$$

where $J_{\varphi,k}$ is the convective and diffusive flux of the searched function φ along the direction x_k

$$J_{\varphi,k} = \rho u_k \varphi - \Gamma_\varphi \frac{\partial \varphi}{\partial x_k} \tag{11.15}$$

and integrating over the control volume ΔV, we obtain:

$$\left(\rho_P^n \varphi_P^n - \rho_P^{n-1} \varphi_P^{n-1}\right) \frac{\Delta V}{\Delta t} + J_e A_e - J_w A_w + J_n A_n - J_s A_s = \overline{\rho S_\varphi} \Delta V \tag{11.16}$$

The e, w, n and s indexes hold for the positions of the cell walls, while the uppercase localize the mesh knots. Putting flux $J_{\varphi,k}$ in a discretized form, we obtain a linear algebraic system as

$$a_P \varphi_P = a_E \varphi_E + a_W \varphi_W + a_N \varphi_N + a_S \varphi_S + b_P, \tag{11.17}$$

solved by an iterative semi-implicit procedure (SIP).

The boundary and initial conditions, which have made the subject of a thorough analysis, are given next:

$$\begin{aligned}
&1.\, y = h \text{ and } x \in (0, d):\; u = 0,\; v = 0,\; \frac{\partial T}{\partial y} = 0,\; \frac{\partial c}{\partial y} = 0;\\
&2.\, y = 0 \text{ and } x \in (0, l):\; u = 0,\; v = 0,\; T = 293\,K,\; \frac{\partial c}{\partial y} = 0\\
&3.\, x = 0 \text{ and } x = l:\; u = 0,\; \frac{\partial T}{\partial x} = 0,\; \frac{\partial c}{\partial x} = 0
\end{aligned} \tag{11.18}$$

The temperature T, the concentration c and the u velocity component can be found at interface, $y = h$ and $x \in (d, l)$, writing the transport equations for the interface control volume (Fig. 11.1) in the discrete form:

$$\begin{aligned}
U_P^{m+1} &= U_P^m\left(1 - \Delta \dot{m} \frac{\Delta t}{\rho_P^m \Delta V}\right) - \frac{\Delta t}{\rho_P^m \Delta V}\left(\dot{m}_e^l u_e^l - \dot{m}_w^l u_w^l + \dot{m}_n^g u_n^g - \dot{m}_s^l u_s^g\right)\\
&\quad - \left(p_e^l - p_w^l\right)\frac{S_y}{2}\frac{\Delta t}{\rho_P^m \Delta V} + \frac{\Delta t}{\rho_P^m \Delta V}\frac{S_y}{2}\left(\tau_{xx}^e - \tau_{xx}^w\right) - \frac{\Delta t}{\rho_P^n \Delta V} S_x \left(\frac{d\sigma}{dx}\right)_P\\
T_P^{m+1} &= T_P^m - \frac{C}{\left(c_p\right)_P^m}\dot{m}_n^g \frac{\Delta t}{\rho_P^m \Delta V} + \frac{\Delta t}{\rho_P^m \Delta V}\frac{S_y}{2}\left[\left(\frac{\lambda}{c_p}\frac{\partial T}{\partial x}\right)_e - \left(\frac{\lambda}{c_p}\frac{\partial T}{\partial x}\right)_w\right]\\
&\quad - \frac{\Delta t}{\rho_P^m \Delta V} S_x \left(\frac{\lambda}{c_p}\frac{\partial T}{\partial y}\right)_s\\
c_P^{m+1} &= c_P^m - \left(c_n^g - c_P^m\right)\dot{m}_n^g \frac{\Delta t}{\rho_P^m \Delta V} + \frac{\Delta t}{\rho_P^m \Delta V}\frac{S_y}{2}\left[\left(\rho D \frac{\partial c}{\partial x}\right)_e - \left(\rho D \frac{\partial c}{\partial x}\right)_w\right]\\
&\quad - \frac{\Delta t}{\rho_P^m \Delta V} S_x \left(\rho D \frac{\partial c}{\partial y}\right)_s
\end{aligned} \tag{11.19}$$

where C is the absorption heat and

$$d\dot{m} = -j_r[\max(0, \dot{m}_e) + \max(0, -\dot{m}_w) + \max(0, -\dot{m}_s)]$$

$$u_w^g = u_e^g = 0,\ \dot{m}_w^g = \dot{m}_e^g = 0,\ p_w^g = p_e^g,\ u_n^g = 0,$$

$$\frac{\rho_P^{m+1} - \rho_P^m}{\Delta t}\Delta V = \dot{m}_w^l - \dot{m}_e^l + \dot{m}_s^l - \dot{m}_n^g$$

$$\dot{m}_n^g = -j_r\left[\max\left(\dot{m}_e^l, 0\right) + \max\left(-\dot{m}_w^l, 0\right) + \max\left(-\dot{m}_s^l, 0\right)\right], \tag{11.20}$$

$$\left(\frac{\partial T}{\partial y}\right)_n = \left(\frac{\partial c}{\partial y}\right)_n = 0$$

$$T_P^m \Delta\dot{m} + \sum_{i=e,w,n,s} T_i^m \dot{m}_i = \frac{C}{\left(c_p\right)_P^m}\dot{m}_n^g$$

$$c_P^m \Delta\dot{m} + \sum_{i=e,w,n,s} c_i^m \dot{m}_i = \left(c_n^g - c_P^m\right)\dot{m}_n^g$$

is the interface gas mass flow rate calculated according to TPT in relation with the reduced absorbed/generated mass flow rate j_r (kg gas/kg absorbent mixture) (Staicovici 2000a, b, c; Staicovici and Isvoranu 2004), and the flow rates traversing faces *e, w* and *s*. Also, in equations above we neglected the diffuzive terms for the north face in case of T and c, and forces equilibrium at interface results in:

$$\tau_{xy}^s = \tau_{xy}^n + \frac{d\sigma}{dx}. \tag{11.21}$$

Unlike all previous works (e.g. Rie 1991), considering the gas–liquid equilibrium conditions at interface, here for the first time the whole system is supposed to evolve in the cell in a non-equilibrium state, between an initial state, chosen at random (vapor concentration and temperature are $c_n = 0.987$ and $T_n = 293.15$ K, respectively, solution initial conditions are $c = 0.3$, $T = 293.15$ K and the pressure is $p_g = p = 1$ bar) and a final one, of an ideal state, established by the equilibrium points value of sources, chosen again at random and given by $p_g = p = 1$ bar; $c_g = 0.987$; $c = 0.34$; $T_g = T = 293.15$ K. Using the reduced form, the mass current absorbed or generated depends, according to TPT, on the natural force $(-X_r)$ and phenomenological coefficient $L_{11,r}$ by:

$$j_r = L_{11,r}(-X_r)$$

The reduced natural force, governing the system evolution towards the ideal point and emphasizing the *i.p.a.* effect, is plotted in Fig. 9.23 with respect to solution mass fraction. The thermo-physical and dynamical properties were adjusted every step of time and grid. Next we shall give some information regarding these properties. The surface tension of the ammonia-water mixture was considered linear with respect to the liquid phase mass fraction of pure components (NH_3–H_2O, IIR 1994), according to

$$\sigma_{NH_3/H_2O} = c\sigma_{NH_3} + (1 - c)\sigma_{H_2O} \tag{11.22}$$

where pure component surface tension are functions of temperature, *t*(*K*) (Geana et al. 1993), given by

$$\sigma_{NH_3(H_2O)} = \sigma_{1,NH_3(H_2O)} \left(\frac{t_{c,NH_3(H_2O)} - t}{t_{c,NH_3(H_2O)} - t_1} \right)^{n_{1,NH_3(H_2O)}} \tag{11.23}$$

In Eq. (11.23), $\sigma_{1,NH_3(H_2O)}$ is a certain surface tension value, calculated for the temperature $t_{1,NH_3(H_2O)}$ (K), and $t_{c,NH_3(H_2O)}$ (K) and $n_{1,NH_3(H_2O)}$ hold for the critical temperatures and a certain corresponding exponent value of the pure components, respectively. The thermal conductivity of the liquid phase mixture was assessed according to Geana et al. (1993)

$$\lambda_{NH_3/H_2O} = c\lambda_{NH_3} + (1-c)\lambda_{H_2O} - 0.72c(1-c)\left(\lambda_{H_2O} - \lambda_{NH_3}\right) \tag{11.24}$$

where pure components thermal conductivity is function of liquid temperature, *t* (°C), given by

$$\lambda_{NH_3(H_2O)} = l_{1,NH_3(H_2O)} + l_{2,NH_3(H_2O)}t + l_{3,NH_3(H_2O)}t^2 \tag{11.25}$$

and $l_{i,NH_3(H_2O)}, i = 1, 2, 3$, are coefficients given in Geana et al. (1993). The dynamic viscosity was assessed in relation with the liquid mass fraction *c* and temperature, *t* (°C), determining the coefficients $B_{i,j}, i = 0, \ldots, 6, j = 0, \ldots, 3$ of a double polynomial fit of the known viscosity plot (Chiriac 1972):

$$\mu_{NH_3/H_2O} = \sum_{j=0}^{3} t^j \sum_{i=0}^{6} B_{i,j}c^i \tag{11.26}$$

Additionally, the other thermo-physical and dynamical properties of the ammonia-water mixture (density(ρ), enthalpy, specific heat capacity and expansion coefficients (β, γ)) were calculated the from mixture equation of state (Ziegler and Trepp 1984). The mass diffusivity coefficient *D* was considered a function of the liquid mass fraction *c*, and has been calculated according to the following equation:

$$D = (3c + 2)^* 10^{-9} \tag{11.27}$$

Equation (11.27) is valid within the range $0.0 \leq c \leq 0.7$ and has been obtained by the linear interpolation of a previous work results (Kojima and Kashiwagi 1995).

11.4 Pseudo-Marangoni Ammonia–Water Cell Modeling Results

The results were obtained for cell parameters $l = h = 0.05$ and $d = 0.0125$, Fig. 11.3. A selection is given in Figs. 11.5, 11.6, 11.7, 11.8, 11.9 and 11.10. Figure 11.5 shows the film of the modeled Marangoni convection for velocity vectors $(\vec{u}(x, y) + \vec{v}(x, y))$, from the first absorption moments, $t \cong 0.23$ s, till the end of the process, $t \geq 240$ s. The cell mean ammonia mass fraction evolution is

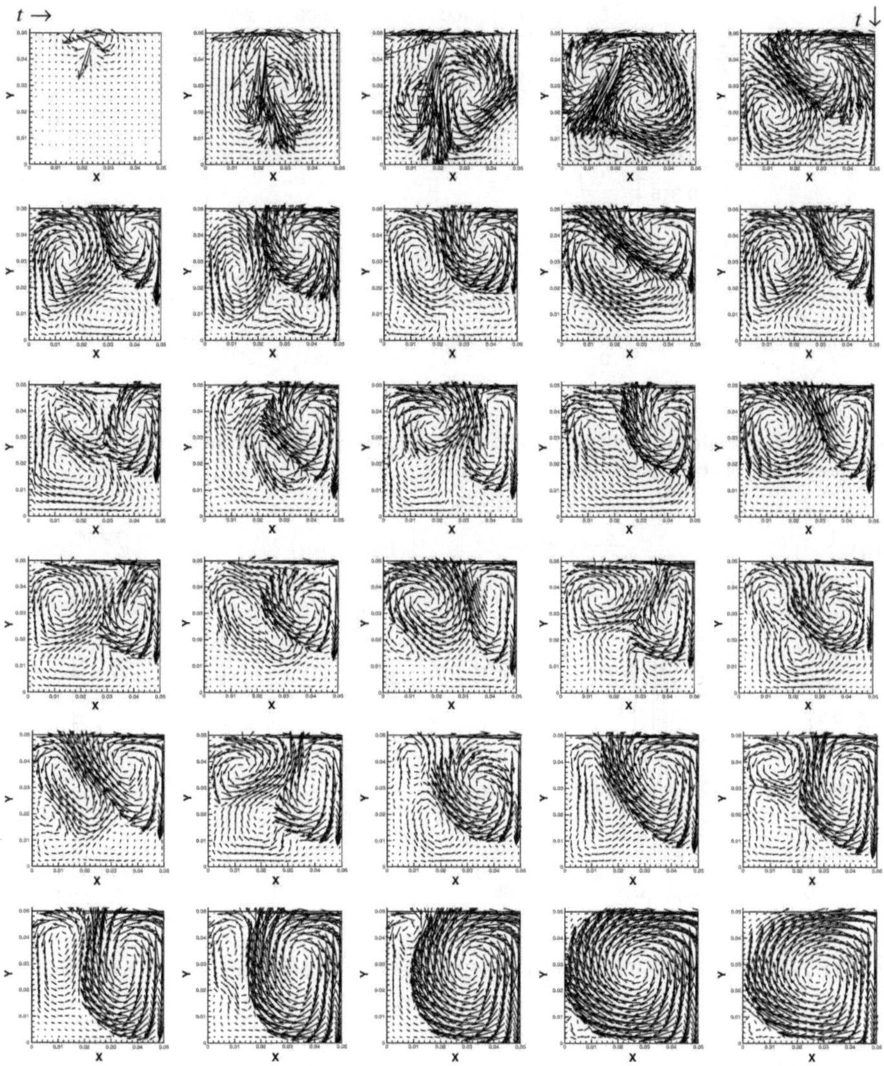

Fig. 11.5 Marangoni convection film for $(\vec{u}(x,y)+\vec{v}(x,y))$. The *top arrows* $t \rightarrow$ and $t \downarrow$ indicate the time increase in every row and from row to row, respectively. The time can be calculated for each figure in seconds from equations $t_{I-V} = 0.23 + (n-1)^{*}0.24$, $n = 1,\ldots,25$ and $t_{VI} = 6.0 + 10^{*}0.24^{*}m$, $m = 1,\ldots,4$ for the first five rows and for the sixth, respectively. The two last figures characterize fluid movement for the rest of process, $t \approx 15.6 - 240$ s

plotted in Fig. 11.6. In the first minute, the absorption rate is high, and the smaller whirlpools, formed in the beginning, are gradually replaced by a big central one, driving almost all cell fluid. Then, absorption increases little, the ammonia mass fraction asymptotically approaches its ideal point value ($c = 0.34$), and the big central whirlpool dominates the whole fluid movement. When absorption starts, an ammonia mass fraction and temperature difference appears between fluid surfaces

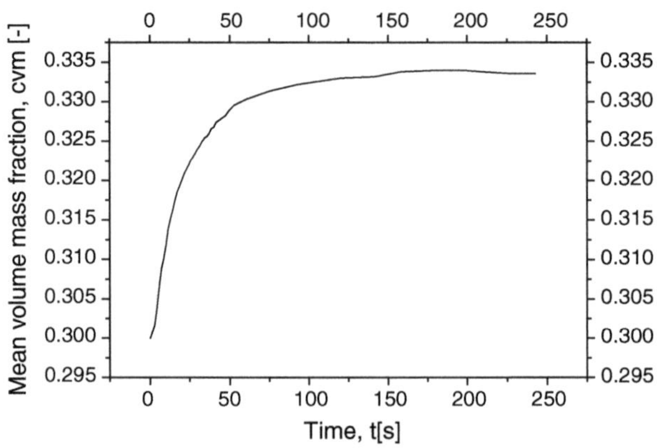

Fig. 11.6 Ammonia mass fraction evolution during absorption in the modeled ammonia-water pseudo-Marangoni cell convection

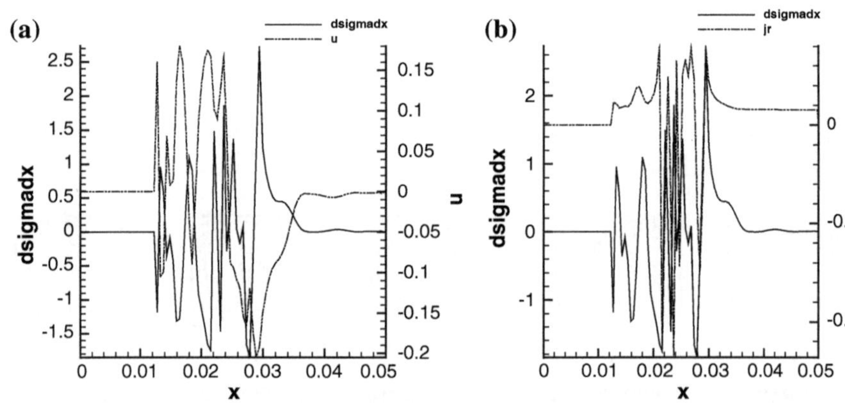

Fig. 11.7 First moments of absorption modeling (t = 0.23 s): **a** $\frac{\partial \sigma(x)}{\partial x}$ and $u(x, h)$; **b** $\frac{\partial \sigma(x)}{\partial x}$ and $j_r(x)$

located under the "drop" and the adjacent absorbing free surface (see pseudo-Marangoni cell explanation, given in Sect. 11.3), causing unbalanced surface tension gradients and top fluid movement release. First moments of absorption and fluid movement are described by Fig. 11.7. The velocity $u(x, h)$ rigorously obeys inequalities (11.12) with respect to $\frac{\partial \sigma(x)}{\partial x}$ variation, Fig. 11.7a, strictly determined by the $j_r(x)$ variation, Fig. 11.7b. Figures 11.8, 11.9 and 11.10, obtained for three different absorption times (corresponding to absorption beginning, middle and end of process) and noted each with (a) to (f), represent functions $\frac{\partial \sigma(x)}{\partial x}$ and $u(x,h)$, $\frac{\partial \sigma(x)}{\partial x}$ and $j_r(x)$, $(\vec{u}(x,y) + \vec{v}(x,y))$, $p(x,y)$, $c\ (x,y)$ and $T(x, y)$, respectively. An interpretation of the plotted results follows, with emphasis of Marangoni convection explanation, given in Sect. 11.2:

1. $j_r(x)$ plots reveal the occurrence of *i.p.a.* effects, characterized by figures of merit having magnitude order of $f_{abs,gen} \approx 10^1 - 10^2$;

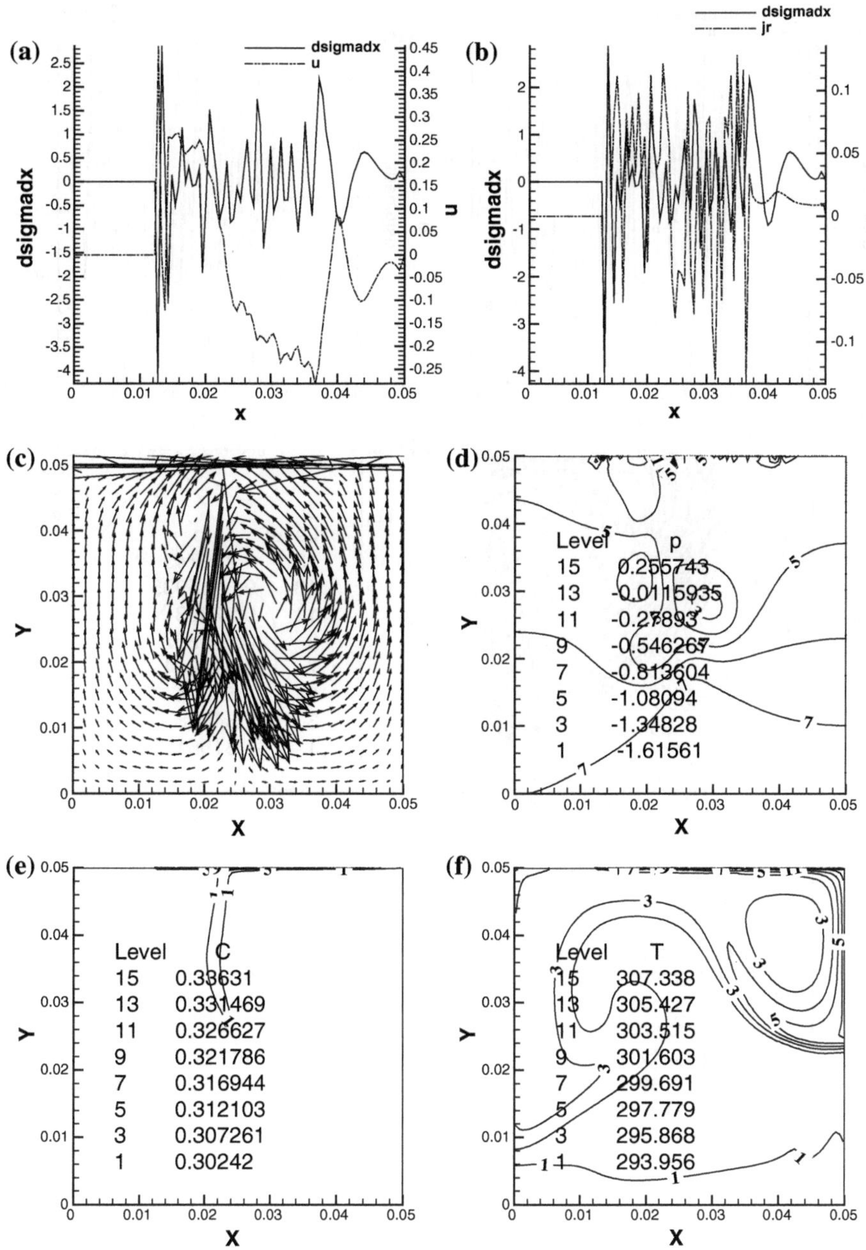

Fig. 11.8 Modeling results of the absorption ammonia-water pseudo-Marangoni cell (time 0.47 s): **a** $\frac{\partial \sigma(x)}{\partial x}$ and $u(x, h)$; **b** $\frac{\partial \sigma(x)}{\partial x}$ and $j_r(x)$; **c** $(\vec{u}(x, y) + \vec{v}(x, y))$; **d** $p(x, y)$; **e** $c(x, y)$; **f** $T(x, y)$

2. Surprisingly for an absorption cell, its free surface suffers not only absorption processes, $j_r > 0$, but generation ones also, $j_r < 0$; both processes occur either simultaneously in adjacent places, or in same place but different times; the

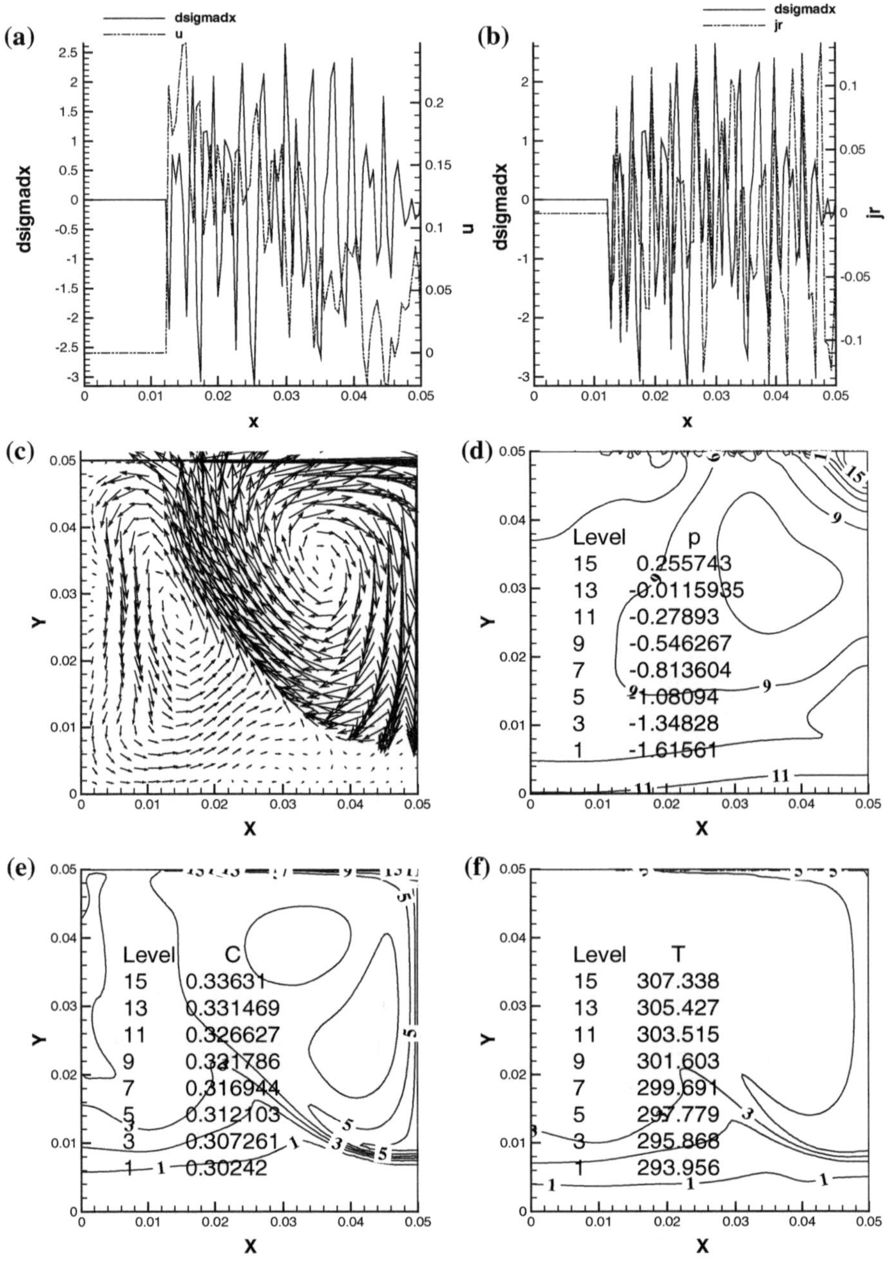

Fig. 11.9 Modeling results of the absorption ammonia-water pseudo-Marangoni cell (time 5.75 s): **a** $\frac{\partial\sigma(x)}{\partial x}$ and $u(x, h)$; **b** $\frac{\partial\sigma(x)}{\partial x}$ and $j_r(x)$; **c** $(\vec{u}(x, y) + \vec{v}(x, y))$; **d** $p(x, y)$; **e** $c(x, y)$; **f** $T(x, y)$

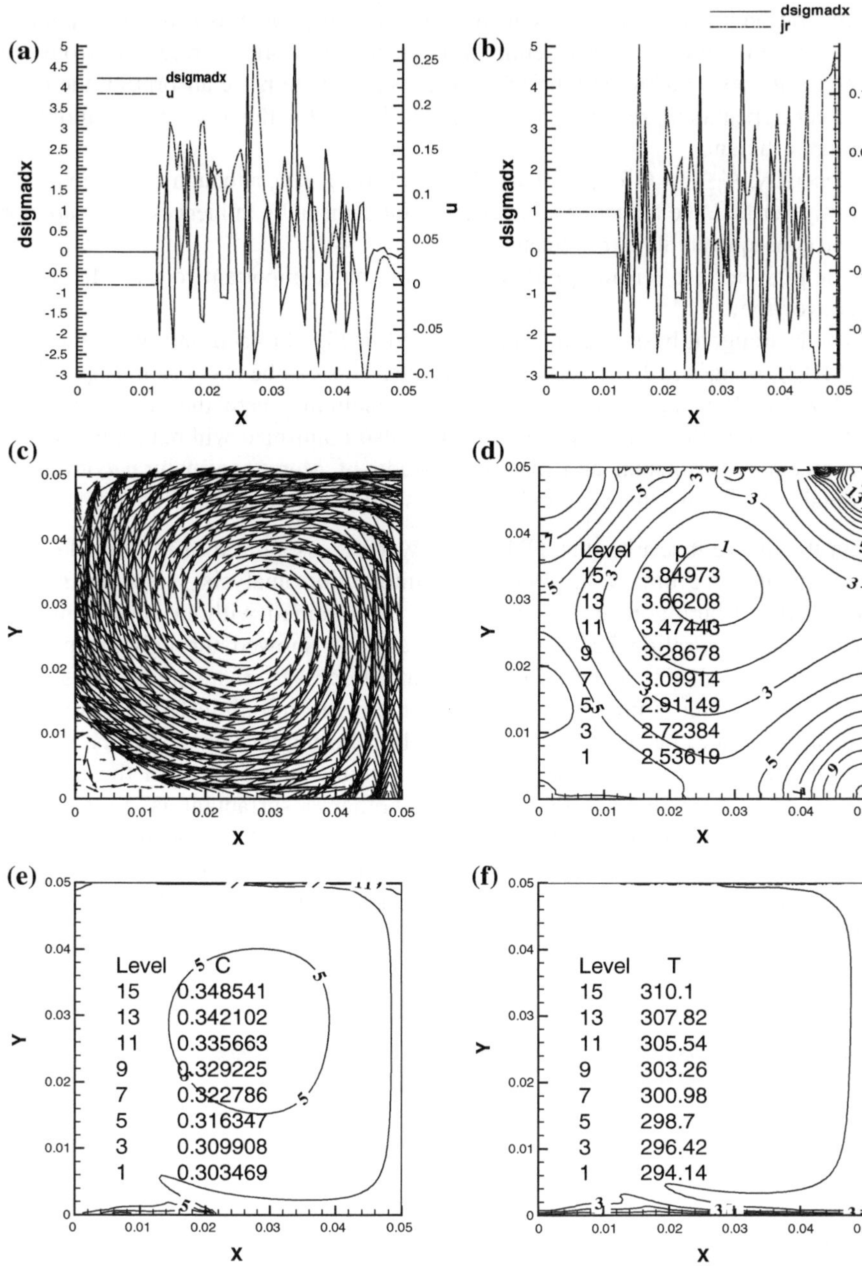

Fig. 11.10 Modeling results of the absorption ammonia-water pseudo-Marangoni cell (time 15.6 s): **a** $\frac{\partial\sigma(x)}{\partial x}$ and $u(x,h)$; **b** $\frac{\partial\sigma(x)}{\partial x}$ and $j_r(x)$; **c** $(\vec{u}(x,y)+\vec{v}(x,y))$; **d** $p(x,y)$; **e** $c(x,y)$; **f** $T(x,y)$

generation process diminishes absorption efficiency, but this is the prize paid by the Marangoni cell for producing fluid mechanical work (convection); from this point of view, bubble flow based absorbers stimulate more absorption the *i.p.a.* effect appearance, receive mechanical work and for this reason they might be more efficient;

3. Generally, $\frac{\partial\sigma(x)}{\partial x} > 0$ for $j_r > 0$ and $\frac{\partial\sigma(x)}{\partial x} < 0$ for $j_r < 0$; although this is not rigorously verified, the reduced mass current j_r fundamentally influences $\frac{\partial\sigma(x)}{\partial x}$ variation and consequently fluid movement;
4. The high surface tension gradients play the most important role in Marangoni convection;
5. The Marangoni basic mechanism, outlined in Fig. 11.4a–d and by inequalities (11.12), is rigorously verified for all results. This can be checked in Figs. 11.8a, 11.9a, 11.10a, and 11.8c, 11.9c, 11.10c, including here the eddy formation process. Additionally, this mechanism is also confirmed without ambiguity by a recent experimental and numerical study of Marangoni-natural convection obtained only by "thermo-capillarity" in shallow liquid layers (Tan 2004).

From the above it results that TPT is a powerful, essential tool in refined qualitative/quantitative binary bi-phase local interaction analysis. Also, the results suggest that Marangoni convection could be obtained without surfactant use, not only in absorption processes, as we already showed through the present modeling, but also in generation of binary processes, if adequate cells were constructed, which stimulate by their geometry an initial surface tension gradient appearance, because, once started, the Marangoni convection is capable to evolve by itself (to be self maintained). Being independent of the refrigerant/absorbent working combination, of whether a surfactant is dissolved or not in an absorbent, of whether the processes are of thermo-capillarity or of distill-capillarity, the Marangoni convection basic mechanism explanation, outlined in the previous papers (Staicovici and Isvoranu 2003, 2004; Isvoranu and Staicovici 2004), is general.

11.5 Pseudo-Marangoni Water–Lithiumbromide Cell Modeling and Modeling Results

11.5.1 TPT Application to the Water–Lithiumbromide Case

11.5.1.1 Natural Force

The reduced natural force ($-X_r$) of a binary two-phase mixture, given in Chap. 9, has been expressed in our case by means of the *p*, *T*, *x*, *s* and *h* known, verified, gas–liquid phase equilibrium data, which are used by the most part of H_2O–LiBr absorption community (McNeely Lowell 1974).

11.5.1.2 Phenomenological Coefficient

The reduced phenomenological coefficient $L_{11,r}$ preserves the value considered for the ammonia-water system:

$$L_{11,r} = |R|^{-1} \tag{11.28}$$

Indeed, let us consider the phenomenological equation, giving the gas interaction mass current:

$$\mathrm{j}'_{\mathrm{ig}} = L_{11}(-X) \tag{11.29}$$

and remember the reduced values of the mass current, Eq. (9.16), phenomenological coefficient and of the natural force, Eq. (9.26), respectively, as:

$$j_r = \frac{j'_{ig}}{j''_{il}}; \quad L_{11,r} = \frac{L_{11}}{(j''_{il}/R)} \text{ and } (-X_r) = \frac{(-X)}{R} \tag{11.30}$$

From Eq. (11.29), with Eqs. (11.30), (9.28) results:

$$\mathrm{j}_{\mathrm{r}} = L_{11,r}(-X_r) \tag{11.31}$$

Equation (11.28) results from considering Eq. (11.31) and bearing in mind that far from equilibrium the natural force equals the lost exergy during gas–liquid interaction, therefore the quantity:

$$\frac{(-X_r)}{j_r} \sim R \tag{11.32}$$

is a specific interaction energy, proportional to the gas constant. This result has been verified in several particular cases, analyzing force and phenomenological coefficient dependence, and choosing the domain where this would be week. Figure (11.11) plots this dependence for two case studies, of water-lithium bromide and ammonia-water systems, respectively. For natural force computation we used mass fraction values c not too close to that of the ideal point. As it may be remarked, in both cases the phenomenological coefficient becomes quasi-independent with respect to the force variation for values close to 1.0×10^{-4}, which roughly is $|R|^{-1}$ (Staicovici 2005b).

11.5.1.3 Ideal Point Approaching (i.p.a.) Effect

For our model, we have chosen as ideal point, a point from the water-lithiumbromide usual absorption domain, having the following values of state parameters:

$$\begin{aligned} p_l &= p_g = 1.275\,\mathrm{kPa} \\ T_l &= T_g = 314.15\,\mathrm{K} \\ c_e &= 45.0;\ y = 100.0 \end{aligned} \tag{11.33}$$

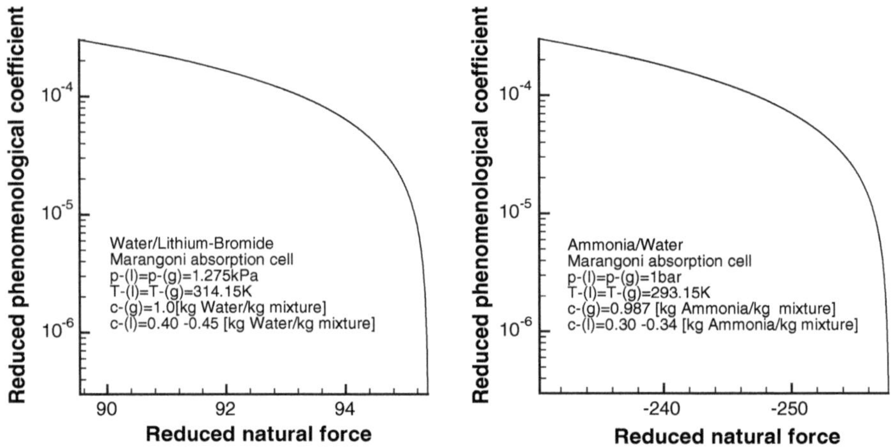

Fig. 11.11 Verification of the phenomenological coefficient value, $L_{11,r}$, in case of the H_2O–LiBr and NH_3–H_2O, respectively

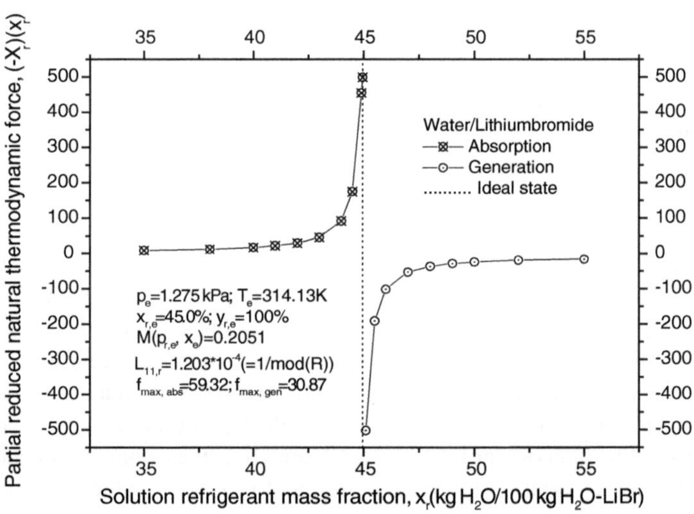

Fig. 11.12 Case of the H_2O–LiBr system: natural force partial function plotted against liquid mass fraction, showing an *i.p.a.* effect behaviour

Figures 11.12, 11.13 and 11.14 plot the reduced values of natural force, Eq. (9.34), mass current, Eq. (9.28) and entropy source, Eq. (9.27), as partial functions of liquid phase refrigerant mass fraction, x. As we can remark, the force and mass current respect the natural force postulate (see Sect. 9.3.), in case of water-lithium bromide too, showing an *i.p.a.* effect behavior when the liquid mass fraction gets closer to the ideal value, $x_e = 45.0$.

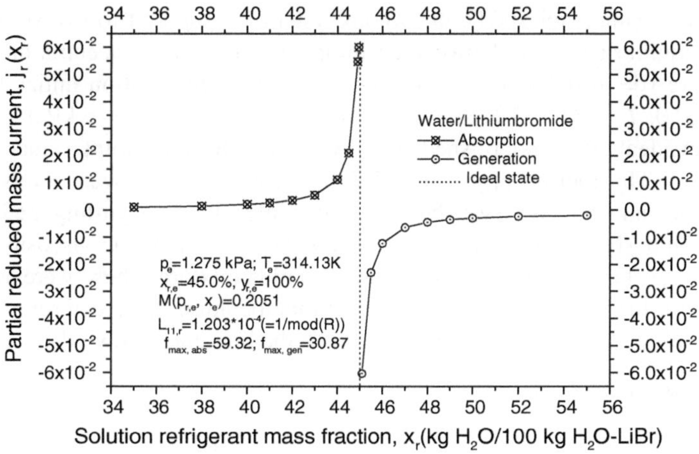

Fig. 11.13 Case of the H_2O–LiBr system: mass current partial function plotted against liquid mass fraction, showing an *i.p.a.* effect behavior

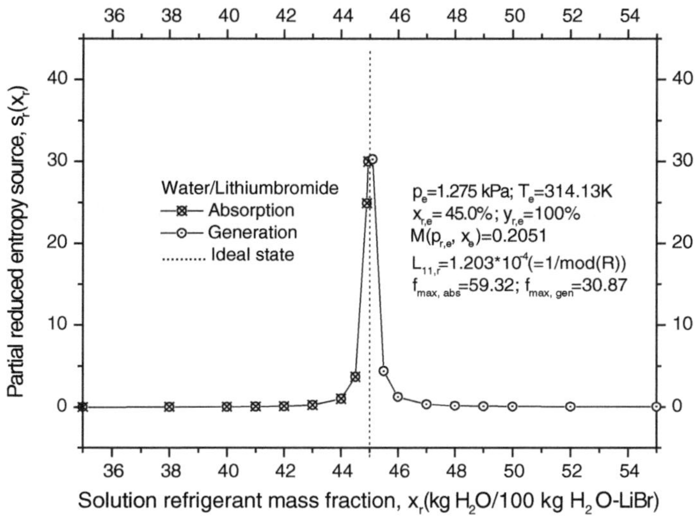

Fig. 11.14 Case of the H_2O–LiBr system: entropy source partial function plotted against liquid mass fraction

11.5.2 Pseudo-Marangoni Water–Lithiumbromide Cell Modeling

The induced convection in a pseudo-Marangoni cell, Fig. 11.3, was modelled with the water-lithium bromide system. The model is similar to that used to the ammonia-water cell, Sect. 11.3. Similarly to Sect. 11.3 (Staicovici and Isvoranu 2003, 2004;

Isvoranu and Staicovici 2004), the whole system is supposed to evolve in the cell in a non-equilibrium state, between an initial state (vapour concentration and temperature are $y = 100.0$ and $T_g = 314.15$ K, respectively, solution initial conditions are $c = 40.0$, $T = 314.15$ K and the pressure is $p_g = p = 1.275$ kPa) and a final one, of an ideal state, established here by the equilibrium point values given by Eq. (11.33). The thermo-physical and dynamical properties were adjusted every step of time and grid. Next, we shall give some information regarding the liquid surface tension σ (N/m), thermal conductivity λ (W/mK), dynamic viscosity μ (Pa*s) and density ρ (kg/m^3) of the water-lithiumbromide mixture. These were assessed in relation with the liquid mass fraction, c, and temperature, t, by determining coefficients $a_{i,j}$ of the double polynomials fitted with the literature known data, as follows:

Surface tension

$$\sigma_{H_2O/LiBr} = 1.0^*10^{-2}\sum_{i=0}^{5} c^i \sum_{j=0}^{4} a_{i,j}t^j \tag{11.34}$$

where c (kg LiBr/kg mixture) and t (°C) (Chiriac and Cartas 1972).

Thermal conductivity

$$\lambda_{H_2O/LiBr} = \sum_{i=0}^{1} c^i \sum_{j=0}^{2} a_{i,j}t^j \tag{11.35}$$

where c (kg LiBr/kg mixture) and t (°C) (Klein and Alvarado 1992).

Dynamic viscosity

$$\mu_{H_2O/LiBr} = exp\left(\sum_{i=0}^{2} t^i \sum_{j=0}^{2} a_{i,j}c^j\right) \tag{11.36}$$

where c (kg LiBr/kg mixture) and t (°C) (Klein and Alvarado 1992).

Density

$$\rho_{H_2O/LiBr} = 1.0^*10^{3}\sum_{i=0}^{1} c^i \sum_{j=0}^{3} a_{i,j}t^j \tag{11.37}$$

where c (kg water/kg mixture) and t (K) (Klein and Alvarado 1992). Additionally, the other necessary thermo-physical and dynamical properties of the water/lithium bromide mixture (entropy, specific heat capacity, expansion coefficients (β, γ) etc.) were calculated from mixture equation of state (McNeely Lowell 1974). The mass diffusivity coefficient is almost constant for $c = 0.40$–0.45, that is $D \cong 1.0 \times 10^{-9}\text{m}^2/\text{s}$ (Kashiwagi et al. 2012).

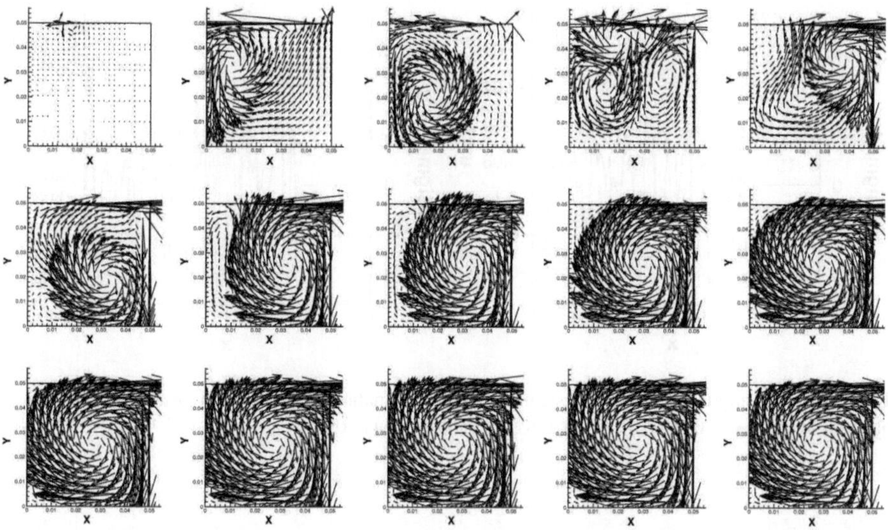

Fig. 11.15 Flow patterns in a *square* Marangoni cell with absorption in the Water-Lithiumbromide medium, $AR = 1.0$ ($h = 5.0/l = 5.0/b = 1.25$) (cm/cm/cm). Absorption times are given sequentially from the *left* to the *right* in each row, for the first, second and third row respectively, as: 1.275; 3.095; 5.37; 7.645; 9.92; 12.2; 11.47; 16.75; 19.02; 21.3; 23.57; 28.12; 46.32; 91.82; 228.3 – 453.7 s

11.5.3 Pseudo-Marangoni Water–Lithiumbromide Cell Modeling Results

The results were obtained for cell parameters $h/l = 2.5/10.0$, $h/l = 5.0/5.0$, $h/l = 10.0/2.5$, and $d/l = 0.25$ (cm/cm). Figure 11.15 shows the flow patterns for velocity vectors $(\vec{u}(x, y) + \vec{v}(x, y))$, in a square Marangoni cell with absorption in the Water-Lithiumbromide medium $(h = 5.0/l = 5.0/d = 1.25)$ (cm/cm/cm), from the first moments, $t \cong 1.275$ s, till the end of the process, $t \geq 453.7$ s. Similar to the ammonia-water case, in the first period, the absorption rate is higher, and the smaller whirlpools, formed in the beginning, are gradually replaced by a big central one, driving almost all cell fluid. Then, absorption increases little, the water mass fraction asymptotically approaches its ideal point value ($c = 45.0$), and the big central whirlpool dominates the whole fluid movement. The cell mean water mass fraction evolution is plotted in Fig. 11.20, dashed line.

Besides ammonia-water, the Marangoni convection basic mechanism explanation has been verified in case of the water-lithiumbromide medium, too. For example, Fig. 11.16 shows $\frac{\partial \sigma(x)}{\partial x}$, $u(x, h)$ and $j_r(x)$ variation, plotted against the mixture mass fraction of the square cell. The velocity $u(x, h)$ rigorously obeys inequalities (11.12), with respect to $\frac{\partial \sigma(x)}{\partial x}$ variation, left Fig. 11.16, strictly determined by the $j_r(x)$ variation, right Fig. 11.16. A same explanation, synthesized by inequalities (11.12), can be found in an excellent, clear-cut plot obtained for the long water-lithiumbromide Marangoni cell, $h/l = 2.5/10.0$, given in Fig. 11.17.

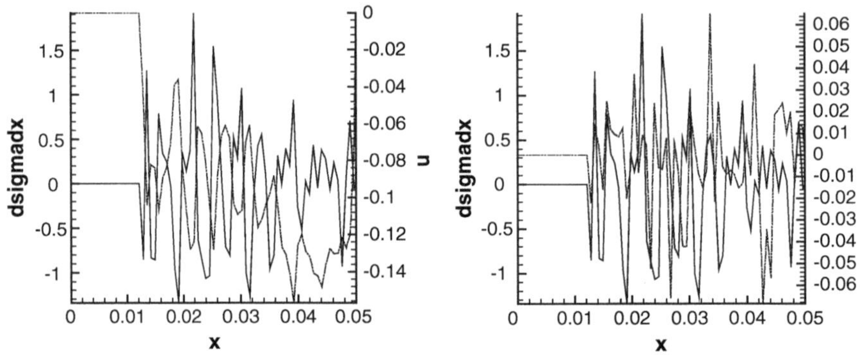

Fig. 11.16 Marangoni convection basic mechanism explanation, verified for the water-lithiumbromide absorption cell, $t = 3.095$ s, $AR = 1.0$. Legend: *solid* $\frac{\partial\sigma}{\partial x}$, *dot* u, or j_r

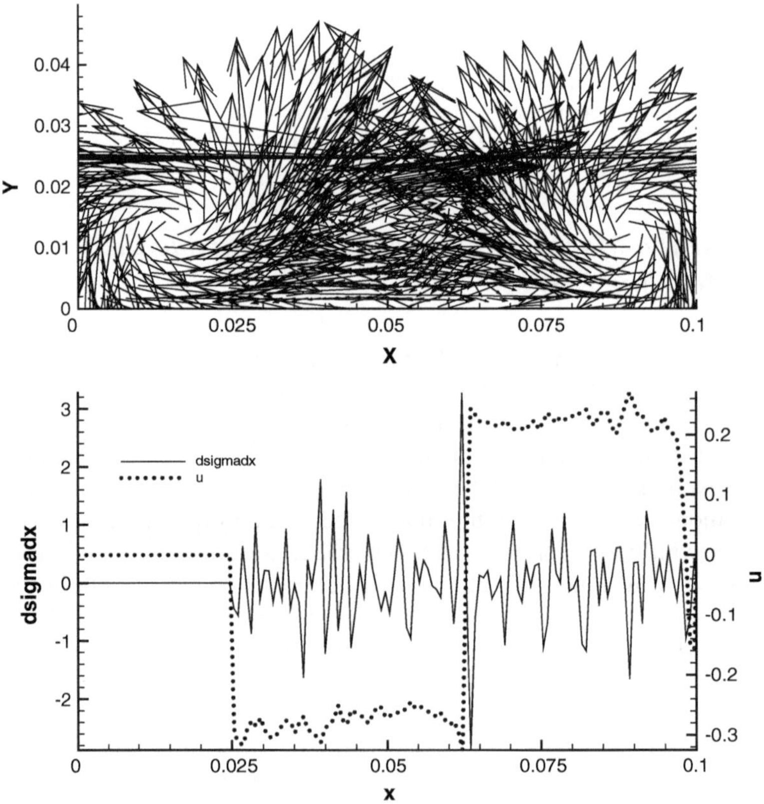

Fig. 11.17 Marangoni convection basic mechanism explanation, verified for the water-lithiumbromide absorption cell, $t = 107.4$ s, $h/l = 2.5/10.0$, $d/l = 0.25$

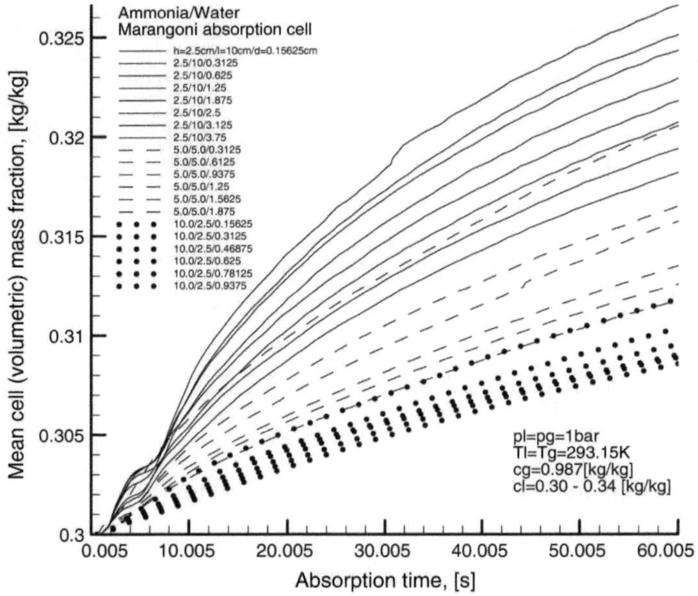

Fig. 11.18 Pseudo-Marangoni ammonia-water absorption cell, one-minute absorption: mean volumetric mass fraction against absorption time, for different *h*/*l* and *d*/*l* ratios

The absorption efficiency has been analysed with TPT model tool for the ammonia-water and water-lithiumbromide systems. Concerning the ammonia-water, Fig. 11.18 shows the cell mean ammonia mass fraction with respect to absorption time, for various *h*/*l* and *b*/*l* ratios. For the same absorption type, here materialized by the Marangoni convection cell, Fig. 11.18 proves that, globally speaking, the absorption is a surface phenomenon, because the higher the absorbing surface is (increasing with aspect ratio *h*/*l* decrease), the higher the absorption is, that is the higher the cell mean mass fraction is for same absorption time. However, the local absorption type is most important, and as it was emphasized to Chap. 9, the *i.p.a.* effect appearance stimulation is the most effective way to improve an absorption process. If we defined a quantity in order to measure the absorption efficiency by:

$$\varepsilon = \frac{c - c_i}{c_e - c_i} \tag{11.38}$$

where c, c_i and c_e hold for refrigerant mass fractions corresponding to a certain period of absorption, initial and final state, respectively, then we can compare different Marangoni cell geometries, for a same absorption time. Figure 11.19 plots the absorption efficiency against the *h*/*l* ratio, for the same *d*/*l* values. Our computation followed also to determine a certain drop dimension, or *d*/*l* value for each *h*/*l* configuration, corresponding to a maximal absorption efficiency, which normally should exist between two quasi-similar states, of near zero absorption (pure diffusion absorption, without Marangoni convection stimulation, when $d = 0.0$), and no

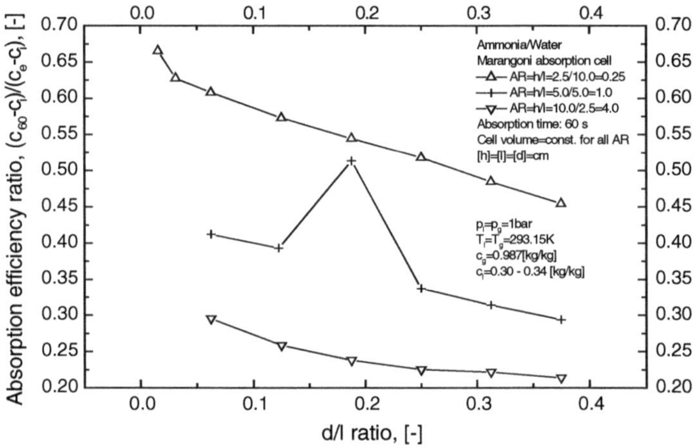

Fig. 11.19 Absorption efficiency against *d/l* ratio and different *h/l* ratios, ammonia-water system

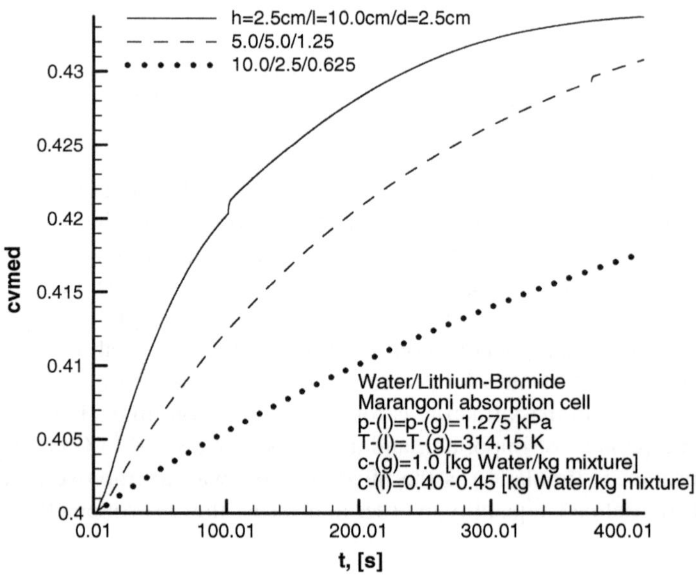

Fig. 11.20 Pseudo-Marangoni absorption cell, Water/Lithium-Bromide system: mean volumetric mass fraction against absorption time, for different *h/l* and *d/l* ratios

absorption at all, for $d = l$, respectively. This *d/l* optimal value was found in case of the square cell configuration, only, as Fig. 11.19 shows $(d/l)_{optimal} \cong 0.18$.

The results of absorption efficiency analysis in case of the water-lithiumbromide system are given in Fig. 11.20. This plotts the cell mean water mass fraction against the absorption time for three *h/l* values and same *b/l* ratio. The absorption efficiency is plotted in Fig. 11.21 against the aspect ratio, for same *b/l* ratio and

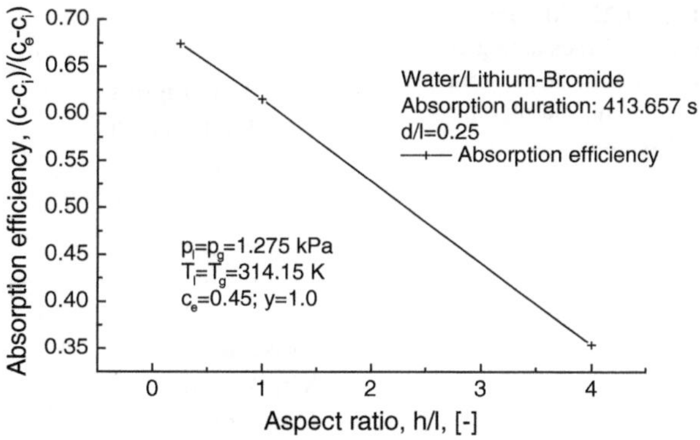

Fig. 11.21 Water-lithiumbromide system: absorption efficiency plotted against cell aspect ratio

absorption time. The absorption increase with absorption surface increase is confirmed in case of the water-lithiumbromide system too. As we can remark, comparing Figs. 11.19 and 11.21 for approximately the same efficiencies and cell configuration, the water-lithiumbromide absorption time is roughly 6–7 times higher than that of the ammonia-water system.

11.6 Inclined Surface Marangoni Convection Cell Evaluation. Ammonia-Water Absorption–Desorption Mass and Heat Exchangers TPT Model Application

11.6.1 Introduction

The absorption technology feasibility depends to a good extent on efficiency of absorption/desorption mass and heat transfer processes. For this reason the improvement of these processes has been its continuous concern. One practical way to improve the transfer is, as already mentioned, the Marangoni convection stimulation. By doing this, an increase of mass and heat transfer by a factor of 2 to 6 is possible for instance in absorption devices. The technical literature is very rich in presenting this topic (see Sect. 11.1). The absorbent flow is more often than not of falling film type with or without surfactant on plane or tubular surfaces in horizontal or vertical mounting (multi-tube & shell). Usually, the mass and heat transfer takes place on external part of exchange surfaces, while the heat/sink sources flow on internal parts. The falling film based devices have global heat transfer coefficients with values ranging from 400 to 600 W/m^2 K in horizontal mounting, up to 3,000–4,000 W/m^2 K in vertical mounting (Niebergall 1959; Fujita 1993; Kang et al. 1999).

Besides qualities, a falling film is confronted, however, with a few problems: (a) its transfer properties are sensitive to mounting position deviation; (b) dry-out phenomena may occur during operation, when the transfer falls down suddenly

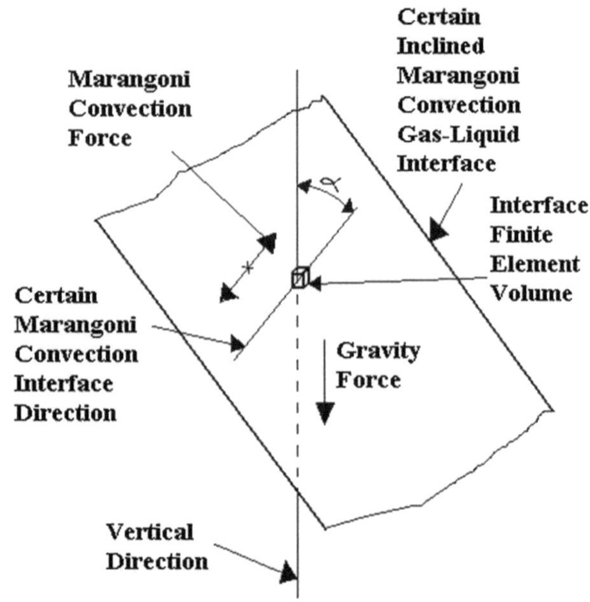

Fig. 11.22 Marangoni-gravity forces acting an interface finite element volume in Marangoni convection

much; (c) the flow plane position has an important influence on Marangoni convection. This section aims at giving a possible solution to these problems, still keeping the thin films advantages (Staicovici 2008c). The main tools used for investigating the mass and heat transfer is in our case the hydrodynamics applied to an ammonia-water Marangoni cell and the TPT.

11.6.2 Marangoni-Gravity Forces Dimensionless Criterion

The Marangoni convection appearance in a certain volume of fluid is conditioned by the interaction of Marangoni forces with other different forces acting on this volume of fluid, like those of viscosity, gravity, surface tension etc. Out of these, the Marangoni-gravity forces interaction is independent of the others ones and can be treated separately. To this extent, let us consider a certain inclined Marangoni convection gas–liquid interface, Fig. 11.22. Here, an interface finite element of volume is acted by the gravity force upon the vertical direction and by the Marangoni convection force upon a certain direction belonging to the interface. The two directions make the ungle α (*rad*). A dimensionless local criterion can be introduced:

$$C_{Mg/mg} = -\left|\frac{Volume\_Local\_Marangoni\_Force}{Volume\_Gravity\_Force}\right| \tag{11.39}$$

It shows the relative magnitude of the two specific (volume) forces (N/m^3). In Eq. (11.39) we consider that the gravity force is projected on the Marangoni force direction and has always an opposite direction to this force:

$$Volume\_Gravity\_Force = \gamma cos\alpha \tag{11.40}$$

where γ (N/m^3) is the interface fluid (thin film) volume element specific weight. Concerning the Marangoni force, this can be expressed making use of the Laplace Eq. (10.12), which in this case has the form:

$$\Delta p = div\vec{\sigma} \tag{11.41}$$

where Δp (Pa)] and σ $\left(\mathrm{J/m^2}\right)$ hold for pressure difference acting the element of volume opposite faces and surface tension, expressed as interface free energy variation, respectively, both considered along the x coordinate. Applying *grad* in Eq. (11.41) along the Marangoni force coordinate x direction, we get, Eq. (11.6):

$$Volume\_Local\_Marangoni\_Force = \frac{\partial}{\partial x}(\Delta p) = \frac{\partial^2 \sigma}{\partial x^2} \tag{11.42}$$

Introducing Eqs. (11.40) and (11.42) in Eq. (11.39), we obtain (Staicovici 2008c):

$$C_{Mg/mg} = -\left|\frac{\frac{\partial^2 \sigma}{\partial x^2}}{\gamma cos\alpha}\right| \tag{11.43}$$

The thin film flows having criterion $\left|C_{Mg/mg}\right| > 1.0$ favour Marangoni convection. On the contrary, those with $\left|C_{Mg/mg}\right| \leq 1.0$ hinder it. The assessment of this criterion in case of the ammonia-water and water-lithiumbromide media, using results obtained in Staicovici and Isvoranu (2004), Staicovici (2005b) and Isvoranu and Staicovici (2004), show clearly that for thin film vertical flow, $\alpha \cong 0.0$, $\left|C_{Mg/mg}\right| \cong 0.2 - 0.5 < 1.0$ holds true, therefore the gravity force hinders much Marangoni convection full development. Starting from this result, we remark that flows must be horizontalized in case of the above two mentioned working combinations, in order to have an efficient Marangoni convection in absorption/desorption processes (Staicovici 2008c).

11.6.3 Proposed Mass and Heat Exchanger

Bearing in mind the result obtained in the last paragraph, we propose here a new mass and heat exchanger (Staicovici 2008c). The key exchange element, Fig. 11.23, is of tube & fin type. The Heat/sink sources flow in tube and the absorbent thin film flows on the fin upper part, provided with small vertical edge walls which establish the film thickness. The whole mass and heat exchanger can be

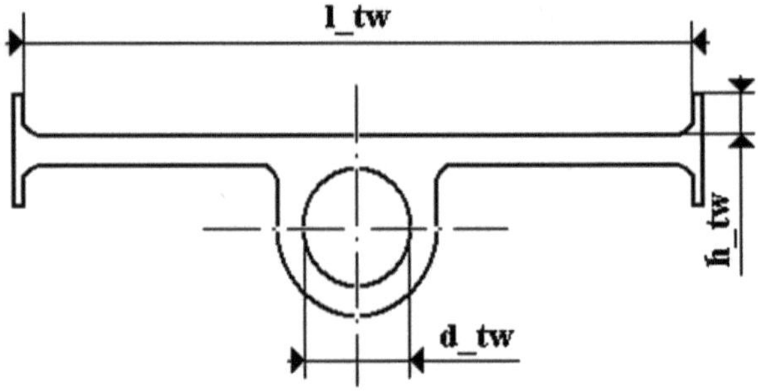

Fig. 11.23 Tube & fin key exchange element

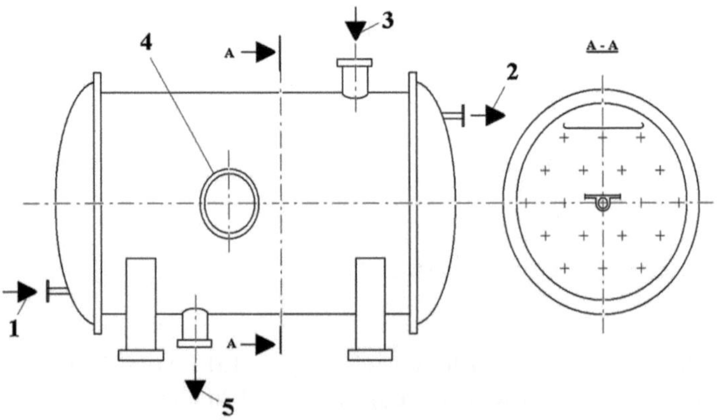

1-Sink Source Inlet; 2-Sink Source Outlet
3-Poor Solution Inlet; 4-Refrigerant Gas
Inlet; 5-Rich Solution Outlet

Fig. 11.24 Proposed multi-tube & fin & shell mass and heat exchanger

build with this element, mounted horizontally, resulting for example in a multi-tube & fin & shell type, Fig. 11.24. Its manufacture is within actual common technological limits.

11.6.4 Mass and Heat Exchange Model

For our model, the exchange element in Fig. 11.23 was considered the place where many Marangoni convection cells are formed simultaneously during absorbent flow with random values of surfactant floating drop length and cell length. The schematic of a single Marangoni cell, which was statically modeled (cell doesn't move on fin) for absorption and desorption processes and which it is showing surfactant floating drop semi-length d and cell semi-length l and height h, is depicted in Fig. 11.3. The model governing equations and numerical solving details are included in Sect. 11.3. Amongst other functions, the model enables the heat exchange calculation of cell side between cell fluid and cell bottom and of the sink or heat source side between source and element tube wall for different Re numbers.

11.6.5 Model Results

The cell side model results are given in Fig. 11.25 for absorption (a–d) and desorption (e–h) (generation) processes respectively. They represent a temporal (*t*) evolution of the ammonia mean mass fraction (*cvmed*, red colour), (kg/kg), in the whole cell, fluid-cell bottom (cell side) global heat transfer (*alfa*, green colour), (W/m^2 K), and the heat exchange through cell bottom (*heat*, blue colour), (W). The following input values have been used: $l = 0.012$ m; $l = 0.006$ m; $h = 0.003$ m; $d = l/4$ m. Also, the thermodynamic parameters were $T_{cooling} = 291.15$ K; $T_{cooling} = 293.15$ K; $x_{initial} = 0.3$; $y = 0.989$; $p = 1$ bar, for absorption, and $T_{heating} = 293.15$ K; $T_{heating} = 295.15$ K; $x_{initial} = 0.38$; $p = 0.88$, for desorption. The ideal points which the system evolves to in absorption and generation are characterized by the following parameters: $T_{ip,abs} = 295$ K, $x_{ip,abs} = 0.34$, $p_{ip,abs} = 1$ bar ($y_{ip,abs} = 0.9865$) and $T_{ip,gen} = 291$ K, $x_{ip,gen} = 0.345$, $p_{ip,gen} = 0.88$ bar, respectively. The fin efficiency has been considered in all applications equal to unit. From the diagrams it results that cell semi-length "*l*" decrease sensibly increases "*alfa*", but cell bottom temperature "*T*" variation doesn't alter it. In this respect, it is worth mentioning that, under the assumptions made, the systems are capable to operate both in absorption and generation with low temperature difference between the sources and the ideal points, i.e. (2–4) °C. This explains why in Chap. 6, the air conditioning applications have been considered $dT_{DI}/dT_{DO} = 2.0\,°C$ (see e.g. Table 6.2). Also, it is worthwile to mention that time "*t*" for absorbent cell to reach the ideal points conditions in absorption or desorption is less 1 s, the parameter responsible for that being cell height "*h*". The cell side (bottom) global heat transfer coefficients "*alfa*" are higher in absorption (2,500–3,500 W/m^2 K), and lower in desorption (1,700–2,500 W/m^2 K). Improvement of these figures is possible.

The global heat transfer of the sink/heat source sides has been modeled in cojuncture with the hybrid coabsorbent heat pump, having a capacity of 24 kW (see Sect. 3.2.5.3). The global heat transfer (Marangoni cell+fin+tube source fluid) has been assessed next for a turbulent flow regime in tube. The results are given in Tables 11.1 and 11.2, for absorption and desorption respectively. In these Tables, "*alfa Mg*" is the cell side global heat transfer, "*alfa global*" is the exchanger device global heat transfer, N_{tv} is the number of fin & tube elements and A_{tot} is the total heat transfer area, (m^2). The calculus has considered the following input data: (a) absorber load $Q_a = 24$ kW; (b) desorber load, $Q_d = 25{,}46$ kW; (c) tube inner diameter, $d_{tv} = 0.004$ m; (d) breadth/length of the Marangoni convection active element, 0.020/1.2 m/m and 0.025/1.2 m/m; (e) absorbent absorber

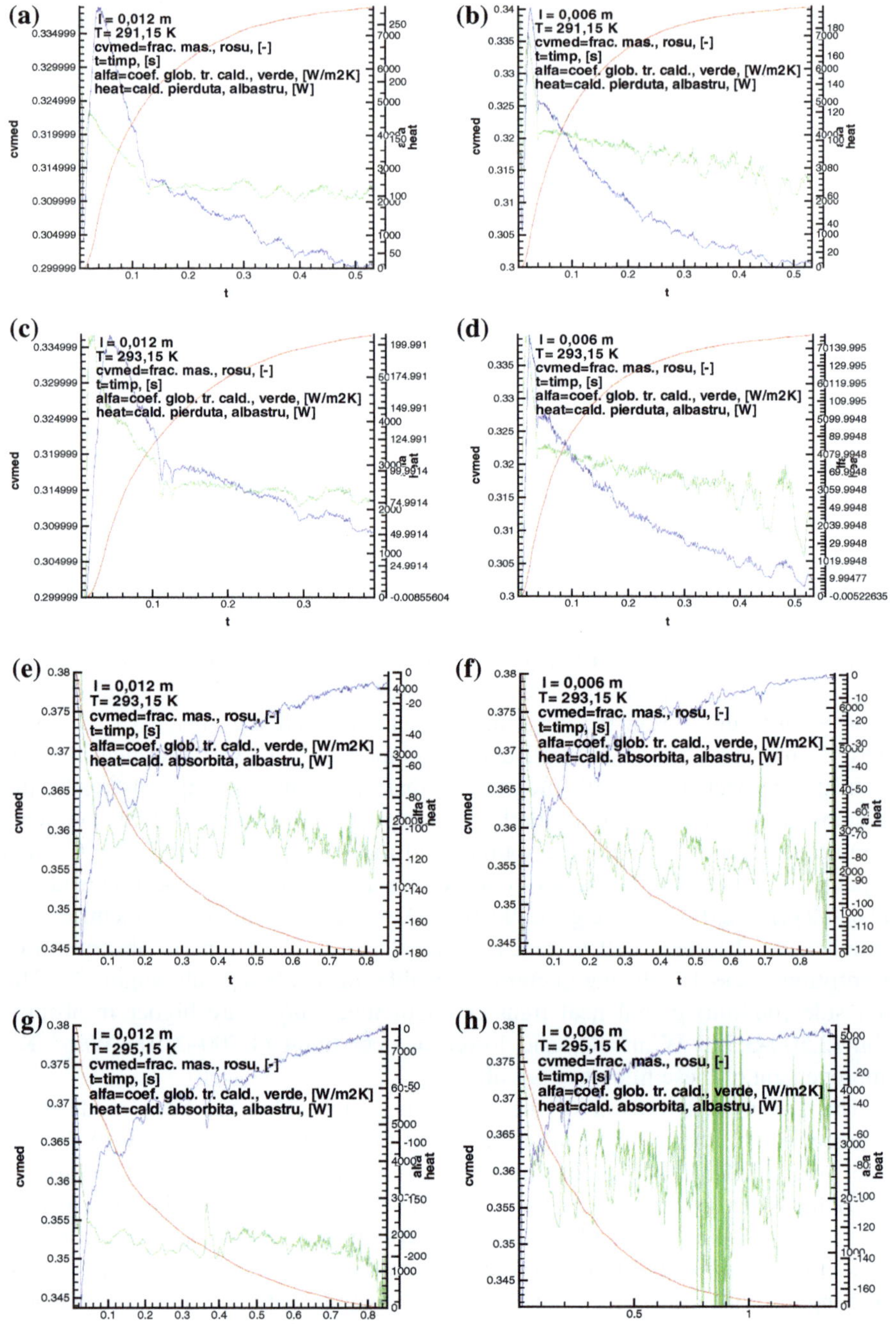

Fig. 11.25 Temporal evolution of absorption (**a**–**d**) and desorption (**e**–**f**) Marangoni cell several parameters

Table 11.1 Absorption global heat transfer results simulation for a Marangoni convection multi-tube & fin & shell mass and heat exchanger

	Absorbtion			
"*alfa Mg*" (W/m^2K)	3,450	3,400	2,663	2,646
"*alfa global*" (W/m^2K)	2,783	2,744	2,173	2,160
N_{tv}	120	121	153	154
A_{tot} (m^2)	2.88	2.92	3.68	3.70

Table 11.2 Desorption global heat transfer results simulation for a Marangoni convection multi-tube & fin & shell mass and heat exchanger

	Desorbtion			
"*alfa Mg*" (W/m^2K)	2,525	2,396	1,842	1,752
"*alfa global*" (W/m^2K)	1,997	2,185	1,681	1,601
N_{tv}	138	140	147	177
A_{tot} (m^2)	3.8	3.98	5.05	5.3

flow rate 0.1146 kg/s; (f) idem (e) desorber, 0.112 kg/s; (g) mean logarithmic temperature difference, 3 °C; (h) source temperature variation for a single pass, 1 °C; (i) maximum pressure decrease for source single pass, max 1,7 bar.

Comparing our results with older literature for ammonia-water without surfactants (e.g. Niebergall 1959), our proposal could benefit of higher global heat transfer coefficients by 3–6 times as in tower falling film absorbers and by 2–3 times as in tower or horizontal falling film desorbers. Concerning recent results for ammonia-water, measured heat transfer coefficients range from 1,000 to 2,200 W/m^2 K (Kang et al. 1999), in absorption experiments of falling films on horizontal tube bundles, which are below those presented here. Also, Kang et al. (1999), absorption tests with vertical surfaces falling films show experimental heat transfer coefficients ranging from 500 to 4,500 W/m^2 K, for various working conditions, which proof experimentally the existence of the ideal point approaching (*i.p.a.*) effect, theoretically discovered in 1997 (Staicovici 2000a, b, c, 2002).

Concluding, the results obtained here encourage a second phase of a laboratory model achievement and test. This conclusion is even more motivated, bearing in mind that our model considered a static Marangoni convection cell only, while including cell movement on fin it is very likely that heat transfer be increased comparatively.

References

Beutler, A., et al. (1996). Surfactants and fluid properties. *International Journal of Refrigeration, 19*(5), 342–346.

Bahloul, A., et al. (2003). Effect on surface tension on convection in a binary fluid layer under a zero gravity environment. *International Journal of Heat and Mass Transfer, 46*(10), 1759–1771.

Boyadjiev, C. H., & Halatchev, I. (1998). Nonlinear mass transfer and Marangoni effect in gas-liquid systems. *International Journal of Heat and Mass Transfer, 41*(1), 197–202.
Chiriac, F., & Cartas, V. (1972). *Cooling Plants, 1972, project design guide, annex for cooling agents and thermodynamic and physical properties*. Bucharest: Civil Engineering Institute. (in Romanian).
Daiguji, H., Hihara, E., & Saito, T. (1997). Mechanism of absorption enhancement by surfactant. *International Journal of Heat and Mass Transfer, 40*(8), 1743–1752.
Fujita, T. (1993). Falling liquid films in absorption machines. *International Journal of Refrigeration, 16*(4), 282–294.
Ganzevles, F. L. A., & van der Geld, C. W. M. (1998). Marangoni convection in binary drops in air cooled from below. *International Journal of Heat and Mass Transfer, 41*(10), 1293–1312.
Geana, M., Ionescu, P., Vais, A., & Ivanus, G. (1993). *Fluid physical properties*. Bucharest: Technical Publishing House. (in Romanian).
Hihara, E., & Saito, T. (1994). Effect of surfactant on falling film absorption. *International Journal of Refrigeration, 16*(5), 339–346.
Hoffmann, L., et al. (1996). Experimental investigation of heat transfer in a horizontal tube falling film absorber with aqueous solutions of LiBr with and without surfactants. *International Journal of Refrigeration, 19*(5), 331–341.
Hozawa, M., et al. (1991). Marangoni convection during steam absorption into aqueous LiBr solution with surfactant. *Journal of Chemical Engineering of Japan, 24*(2), 209–214.
Isvoranu, D., & Staicovici, M. D. (2004). Marangoni convection mechanism explanation using the two-point theory (TPT) of mass and heat transfer and the ammonia-water medium. *International Journal of Heat and Mass Transfer, 47*(2004), 3769–3782.
Kim, K. J., Berman, N. S., & Word, B. D. (1996). The interfacial turbulence in falling film absorption: Effects of additives. *International Journal of Refrigeration, 19*(5), 322–330.
Kang, Y. T., Akisawa, A., & Kashiwagi, T. (1999). Visualization and model development on Marangoni convection in NH_3–H_2O system. *International Journal of Refrigeration, 22*, 640–649.
Ha, V.-M. (2002). Onset of Marangoni instability of a two-component evaporating droplet. *International Journal of Heat and Mass Transfer, 45*(26), 5143–5158.
Kashiwagi, T. (1988). Basic mechanism of absorption heat and mass transfer enhancement by the Marangoni effect. *Newsletter, IEA Heat Pump Center, 6*(4), 2–6.
Kashiwagi, T., et al. (2012). Heat and mass diffusions in the absorption of water vapour by aqueous solution of lithium bromide. *Transactions of the Japan Society of Refrigerating and Air Conditioning Engineers, 1*(1), 89–98.
Klein, S. A., & Alvarado, F. L. (1992–2002). Engineering Equations Solver, commercial code.
Kumar, V., et al. (2003). Effect of thermo-capillary convection in an industrial Czochralsky crucible: numerical simulation. *International Journal of Heat and Mass Transfer, 46*(9), 1641–1652.
Kojima, M., & Kashiwagi, T. (1995). Mass diffusivity measurements for ammonia vapor absorption processes. In: *Proceedings of the19th International Congress of Refrigeration* (Vol. IV(a), pp 353–360). Hague.
Li, M., Zang, D., & Zhu, T. (2002). Instability of the Marangoni convection in a liquid bridge with liquid encapsulation under microgravity conditions. *International Journal of Heat and Mass Transfer, 45*(1), 157–164.
McNeely Lowell, A. (1974). Thermodynamic properties of aqueous solutions of lithium bromide. *ASHRAE Transactions, 85*(T1), 413–434 (PH-79-3, No. 3).
Möller, R., & Knoche, K. F. (1996). Surfactants with NH_3–H_2O. *International Journal of Refrigeration, 19*(5), 317–321.
Niebergall, W. (1959). *Sorptionskältemaschinen*. Berlin: Springer.
Nordgren, M., & Setterwall, F. (1996). An experimental study of the effects of surfactant on a falling liquid film. *International Journal of Refrigeration, 19*(5), 310–316.
Rie, D.-H., & Kashiwagi, T. (1991). Computer simulation of vapor absorption enhancement into H_2O/LiBr absorbent by Marangoni convection. *International Journal of Japan Society Mechanical Engineering: Series III, 34*(3), 355–361.

Staicovici, M. D., & Isvoranu, D. (2003). The Marangoni convection: A two-point theory of mass and heat transfer and Laplace equation new wording approach. In: *Proceedings of 21th International Congress of Refrigeration*. Washington, 17–22 Aug 2003.

Staicovici, M. D., & Isvoranu, D. (2004). The Marangoni convection explanation: a two-point theory (TPT) of mass and heat transfer and new Laplace equation approach. In: *Proceedings of CHT-04, April 2004*. Norway.

Staicovici, M. D. (1998). Variational numerical and analytical approach of the capillary rise effect. *Revue Roumaine des Sciences Techniques-Mecanique Appliquee, 43*(1), 85–99.

Staicovici, M. D. (2000a). A phenomenological theory of polycomponent interactions in non- ideal mixture. Applications to the ammonia/water and other working pairs. *International Journal of Refrigeration, 23*, 153–167.

Staicovici, M. D. (2000b). A non-equilibrium phenomenological theory of the mass and heat transfer in physical and chemical interactions. Part I: Theory and its application to NH_3/H_2O and other working systems. *International Journal of Heat and Mass Transfer, 43*(22), 4153–4173.

Staicovici, M. D. (2000c). A non-equilibrium phenomenological theory of the mass and heat transfer in physical and chemical interactions. Part II: Modeling of the NH_3/H_2O bubble absorption, analytical study of absorption and experiments. *International Journal of Heat and Mass Transfer, 43*(22), 4175–4188.

Staicovici, M. D. (2002). A non-equilibrium phenomenological (two-point) theory of mass and heat transfer. Forces, system-source interactions and thermodynamic cycle application. *International Journal of Thermal Sciences, 41*, 737–752.

Staicovici, M. D., Isvoranu, D. (2005b). Model of the water/lithium bromide absorption process in a Marangoni convection cell, using the Two-Point Theory (TPT) of mass and heat transfer. In: *Proceedings of HEFAT2005 Conference*. 19–22 Sept 2005.

Staicovici, M. D. (2008c). Inclined surface Marangoni convection cells evaluation and ammonia / water absorption-desorption mass and heat exchangers TPT model application. In: *Proceedings of the 19th International Symposium on Transport Phenomena*. Rejkjavik, ICELAND, 17–20 Aug 2008.

IIR. (1994). Tables and diagrams for the refrigeration industry. Thermodynamic and physical properties, ammonia-water. *International Institute of Refrigeration.*

Tan, L. H., Leonardi, E., Barber, T. J., & Leong, S. S. (2004). Experimental and numerical study of Marangoni-natural convection in shallow layers. In: *Proceedings of CHT-04*. Norway, Apr 2004.

Velde, O., et al. (2001). Numerical investigation of the Lorentz force influenced Marangoni convection relevant to aluminium surface alloying. *International Journal of Heat and Mass Transfer, 44*(14), 2751–2762.

Yao, Y. L., Liu, F., & Hu, W. R. (1996). How to determine critical Marangoni number in half floating zone convection. *International Journal of Heat and Mass Transfer, 39*(12), 2539–2544.

Ziegler, B., & Trepp, C. (1984). Equation of state for ammonia-water mixtures. *International Journal of Refrigeration, 7*, 101–106.

Errata to: Coabsorbent and Thermal Recovery Compression Heat Pumping Technologies

Mihail-Dan Staicovici

Errata to:
M.-D. Staicovici, *Coabsorbent and Thermal Recovery Compression Heat Pumping Technologies*, DOI 10.1007/978-3-642-54684-6

The online version of the original book can be found under DOI 10.1007/978-3-642-54684-6

M.-D. Staicovici
Clean Energy Research, S.C. Varia Energia S.R.L. & S.C.,
Incorporate Power-absorption Engineering S.R.L.,
Bucharest, Romania

M.-D. Staicovici, *Coabsorbent and Thermal Recovery Compression Heat Pumping Technologies*, Heat and Mass Transfer, DOI: 10.1007/978-3-642-54684-6_12,

Corrections to be made to the book "Coabsorbent and Thermal Recovery Compression Heat Pumping Technologies, author Mihail-Dan Staicovici

Page	Present text	Corrections to be made = the text which the "Present text" must be replaced with
viii, row 12 from top	The discharge gas superheat recovery is converted…	The recovered discharge gas superheat is converted…
17, row 14 from bottom	elaborated it 1824.	elaborated it in 1824.
21, row 10 from bottom	Than,	Then,
22, row 2 from bottom	in eqn. (1.98)	inequality (1.98)
22, row 5 from bottom	In in eqn. (1.98)	In inequality (1.98)
23, 1st row from top	In eqn. (1.100)	inequality (1.100)
23, row 8 from top	in eqn. (1.101)	inequality (1.101)
23, row 10 from top	In eqn. (1.102)	Inequality (1.102)
23, row 2 from bottom	in eqn. (1.108)	inequality (1.108)
23, row 3 from bottom	ineqn. (1.105)	Inequality (1.105)
23, row 12 from bottom	In in eqn. (1.105)	In inequality (1.105)
23, row 15 from bottom	In eqn. (1.103)	Inequality (1.103)
27, row 3 from top	calculated with the help the arithmetical mean	calculated with the help of the arithmetical mean

(Continued)

(Continued)

Page	Present text	Corrections to be made = the text which the "Present text" must be replaced with
29, row 18 from bottom	where $-(dE)_{irrev}$ is the exergy dissipation	where $-\frac{\partial E}{\partial t}dt$ is the exergy dissipation
36, row 12 from bottom	free energy relationship as $U = F + TS$, Eq. (1.155),	free energy relationship as $U = F + TS$, Eq. (1.164),
45, row 7 from top	Eq. (1.208) partial derivatives are given by Eqs. (1.160) and (1.161) of Table 1.1, …	Eq. (1.208) partial derivatives are given by Eqs. (1.169) and (1.170) of Table 1.1, …
61, row 6 from top	Introducing Eq. (1.244) in Eq. (1.243), it is obtained:	Introducing Eqs. (1.244) and (1.243) in Eq. (1.242), it is obtained:
61, Eq. (1.248)	$\left[\frac{\partial q_{mix}}{\partial(1-y)}\right]_{m_1}\left[\frac{\partial(1-y)}{\partial m_2}\right]_{m_1} = -y\left[\frac{\partial q_{mix}}{\partial y}\right]_{m_2}$	$\left[\frac{\partial q_{mix}}{\partial(1-y)}\right]_{m_1}\left[\frac{\partial(1-y)}{\partial m_2}\right]_{m_1} = -y\left[\frac{\partial q_{mix}}{\partial y}\right]_{m_1}$
61, Eq. (1.249)	$q_{d2} = q_{mix} - y\left[\frac{\partial q_{mix}}{\partial y}\right]_{m_2}$	$q_{d2} = q_{mix} - y\left[\frac{\partial q_{mix}}{\partial y}\right]_{m_1}$
61, rows 4 and 5 from bottom	Eqs. (1.243) and (1.246)	Eqs. (1.242) and (1.249)
Fig.1.21	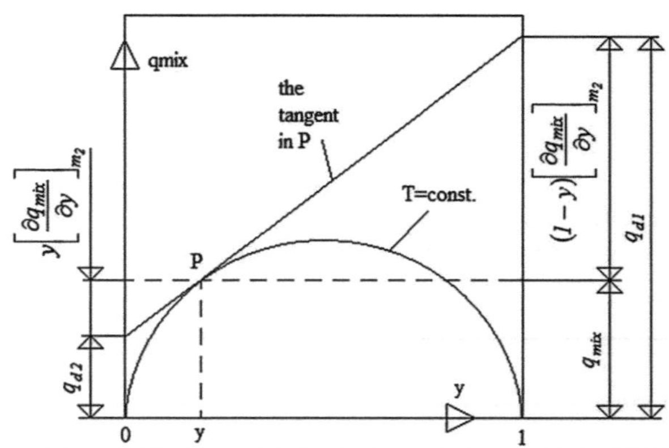	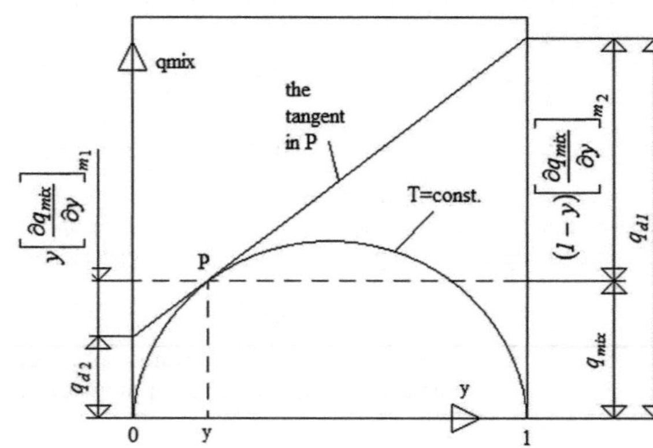

(Continued)

(Continued)

Page	Present text	Corrections to be made = the text which the "Present text" must be replaced with
62, Eq. (1.252)	$(1-y)\left[\frac{\partial q_{d_1}}{\partial y}\right]_{m_2}+y\left[\frac{\partial q_{d_2}}{\partial y}\right]_{m_1}=0$	$(1-y)\left[\frac{\partial q_{d_1}}{\partial y}\right]_{m_1}+y\left[\frac{\partial q_{d_2}}{\partial y}\right]_{m_2}=0$
62, rows 9 and 10 from bottom	Eqs. (1.245) and (1.249) of q_{d1} and q_{d2} are further partially derived with respect to y for $m_2 = const.$ and with respect to $(1-y)$ for $m_1 = const.$, respectively.	Eqs. (1.245) and (1.249) of q_{d1} and q_{d2} are further partially derived with respect to y for $m_1 = const.$ and with respect to $(1-y)$ for $m_2 = const.$, respectively.
62, Eq. (1.253)	$\left[\frac{\partial q_{d_1}}{\partial y}\right]_{m_2}+\left[\frac{\partial q_{d_2}}{\partial y}\right]_{m_1}=0$	$\left[\frac{\partial q_{d_1}}{\partial y}\right]_{m_1}+\left[\frac{\partial q_{d_2}}{\partial y}\right]_{m_2}=0$
62, Eq. (1.254)	$\left[\frac{\partial q_{d_1}}{\partial y}\right]_{m_2}-\left[\frac{\partial q_{d_2}}{\partial y}\right]_{m_1}=-2\left[\left[\frac{\partial q_{mix}}{\partial y}\right]_{m_1}-\left[\frac{\partial q_{mix}}{\partial y}\right]_{m_2}\right]$	$\left[\frac{\partial q_{d_1}}{\partial y}\right]_{m_1}-\left[\frac{\partial q_{d_2}}{\partial y}\right]_{m_2}=2\left[\left[\frac{\partial q_{mix}}{\partial y}\right]_{m_1}-\left[\frac{\partial q_{mix}}{\partial y}\right]_{m_2}\right]$
62, row 3 from bottom	Equations (1.252) and (1.253) are solved together for $\left[\frac{\partial q_{d_1}}{\partial y}\right]_{m_2}$ and $\left[\frac{\partial q_{d_2}}{\partial y}\right]_{m_1}$, obtaining:	Equations (1.252) and (1.253) are solved together for $\left[\frac{\partial q_{d_1}}{\partial y}\right]_{m_1}$ and $\left[\frac{\partial q_{d_2}}{\partial y}\right]_{m_2}$, obtaining:
62, Eq. (1.255)	$\left[\frac{\partial q_{d_1}}{\partial y}\right]_{m_2}=\left[\frac{\partial q_{d_2}}{\partial y}\right]_{m_1}=0$	$\left[\frac{\partial q_{d_1}}{\partial y}\right]_{m_1}=\left[\frac{\partial q_{d_2}}{\partial y}\right]_{m_2}=0$
70, row 7 from bottom	The specific Gibbs free enthalpy is at the same time the chemical potential of the system, $\varphi=\left(\frac{\partial\Phi}{\partial G}\right)_{T,p}$, according to eqn. (1.163).	The specific Gibbs free enthalpy is at the same time the chemical potential of the system, $\varphi=\left(\frac{\partial\Phi}{\partial G}\right)_{T,p}$, according to eqn. (1.172).
70, row 9 from bottom	specific Gibbs free enthalpy assessment, φ, $\varphi=\frac{\Phi}{G}$, ($\Phi(T,p)=H-TS$, Table 1.1, eqn. (1.159))…	specific Gibbs free enthalpy assessment, φ, $\varphi=\frac{\Phi}{G}$, ($\Phi(T,p)=H-TS$, Table 1.1, eqn. (1.168))…
73, row 12 from bottom	In eqn. (1.283), the last bracket of the right member can be calculated from eqns. (1.276) and (1.277)…	In eqn. (1.283), the last bracket of the right member can be calculated from gas phase eqns. (1.276) and (1.277)…
75, row 3 from bottom	*Gas phase*: $c_p^l(T,p_0)=b_1+b_2T+b_3T^2$ (1.294) $v^g(T,p)=\frac{RT}{p}+c_1+\frac{c_2}{T^3}+\frac{c_3}{T^{11}}+\frac{c_4p^2}{T^{11}}$ (1.295)	$c_p^l(T,p_0)=b_1+b_2T+b_3T^2$ (1.294) *Gas phase*: $v^g(T,p)=\frac{RT}{p}+c_1+\frac{c_2}{T^3}+\frac{c_3}{T^{11}}+\frac{c_4p^2}{T^{11}}$ (1.295)

(Continued)

(Continued)

Page	Present text	Corrections to be made = the text which the "Present text" must be replaced with
75, row 5 from top	($y = 0$; $Y = 0$)	($y = 0$; $Y = 1$)
78, 1st row from top	Introducing eqns. (1.295), (1.296) and (1.306) in eqn. (1.304), the analytical expression of the gas phase free enthalpy results:	Introducing eqns. (1.295), (1.296) and (1.306) in eqn. (1.307), the analytical expression of the gas phase free enthalpy results:
128, row 8 from the bottom	This time, temperature is the internal heating temperature….	This time, T_M temperature is the internal heating temperature….
186, row 6 from top	for seen	foreseen
208, Caption of Fig. 417	Fig. 4.44	Fig. 4.45
208, row 2 from top	(see Fig. 4.44 of Sect. 4.3.2)	(see Fig. 4.45 of Sect. 4.3.1)
209, row 2 from the top	Fig. 4.44	Fig. 4.43
210, row 5 from bottom	$y_{GO,1}$	$\Delta T_{gax,R}$
211, row 6 from top	$\Delta T_{R,i,gax(\Delta T_{gax,R,\max})}$	$y_{R,i,gax} = y_{R,i,gax}\left(\Delta T_{gax,R,\max}\right)$
212, row 3 from the bottom	$\Delta T_{gax,R}\max \leq \Delta T_{gax,R} \leq \Delta T_{gax,R}\max$	$\Delta T_{gax,R}\min \leq \Delta T_{gax,R} \leq \Delta T_{gax,R}\max$
214, 4th row down from Eq. (4.122)	y_{G,j,gax^e}	$y^e_{G,j,gax}$
215, 7th row down from Eq. (4.129)	Fig. 4.37a of Sect. 4.2.3	Fig. 4.38a of Sect. 4.2.3
215,	***GHE-Gax problem study cases***	*GHE-Gax problem study cases*

(Continued)

(Continued)

Page	Present text	Corrections to be made = the text which the "Present text" must be replaced with
215, row 3 from the bottom	Resorber Heat Excess (RHE) Gax Operation Model	Generator Heat Excess (GHE) Gax Operation Model
215, row 12 from the bottom	Results of the RHE-Gax Model Run with $\Delta T_{gax,R,\min}$- Infinite Equivalent Solutions to A GHE-Gax Problem	Results of the GHE-Gax Model Run with an intermediate $\Delta T_{gax,G}$- Infinite Equivalent Solutions to a GHE-Gax Problem
216, row 12 down from the top	$q_{G,1,gax^e} = 574.6$	$q^e_{G,1,gax} = 574.6$
218, row 5 from the bottom	According to the 6th study case, running the 4.2.1.3.2. sub-sub-paragraph model for the configuration…	According to the 6th study case, running the Sect. Generator Heat Excess (GHE) Gax Operation Model for the configuration…
250, row 13 from top	Equations (5.6) and (1.214)…	Equations (5.6) and (1.223)…
251, row 2 from bottom	Eq. (1.217)…	Eq. (1.226)…
265, row 3 from top	Important properties of the cascades at hand is emphasized next	Important properties of the cascades at hand are emphasized next
266, row 2 from top	Eq. (1.81)…	Eq. (1.65)…
266, row 6 from top	Eqs. (5.81) (5.82), (5.76) and (5.77)	Eqs. (5.81) and (5.82), Eqs. (5.76) and (5.77)
267, row 7 from top	Eqs. (1.208) and (1.209)…	Eqs. (1.217) and (1.218)…
267, row 6 from bottom	Eqs. (1.208) and (1.209)…	Eqs. (1.217) and (1.218)…

(Continued)

(Continued)

Page	Present text	Corrections to be made = the text which the "Present text" must be replaced with
268, 1st row from bottom	(last column, the CO_2-NH_3 known cascade). Table 5.3.	(last column, the CO_2-NH_3 known cascade).
300, row 18 from bottom	Appendix 1	Appendix 7
309, row 12 from bottom	, it results that in Eq. …	, it results that inequality …
309, row 13 from bottom	Indeed, from the obvious in equations	Indeed, from the obvious inequalities
315, 1st row from bottom	Appendix 2	Appendix 7
324, row 7 from top	**Appendix 1**	**Appendix 7**
324, row 8 from top	Equation (7.34) …	i) Equation (7.34) …
324, row 11 from top	Considering in Eqs. (A7.1, 7.20, 7.28) …	Considering in Eq. (A7.1), Eqs. (7.19, 7.28) …
324, row 13 from top	**Appendix 2**	
324, row 14 from top	A simple, yet not simplistic, explanation …	ii) A simple, yet not simplistic, explanation …
331, row 12 from top	Appendix 1	Appendix 8
334, row 2 from top	(see Appendix 1 of this chapter)	(see Appendix 8 of this chapter)
338, row 13 from bottom	Appendix 1	Appendix 8
338, row 16 from bottom	Appendix 1	Appendix 8

(Continued)

(Continued)

Page	Present text	Corrections to be made = the text which the "Present text" must be replaced with
369, 1st row from bottom	Appendix 1	Appendix 8
369, row 7 from bottom	$1 > \frac{n+1}{n+2}x < 1$	$\frac{n+1}{n+2}x < 1$
373, row 4 from bottom	**Appendix 1**	**Appendix 8**
375, 1st row		$\eta_C(\xi) = 1 - \frac{\ln \frac{T_{2a}}{T_1}}{\frac{T_{2a}}{T_1} - 1} \equiv \eta_{C,TFC} = 1 - \frac{2T_1}{T_{2a} - T_1}\frac{T_{2a} - T_1}{T_{2a} + T_1}$ (A8.4)
378, row 4 from top	$T_{ax} - T_{T_{2a}}$	$T_{ax} - T_{2a}$
391, row 6 from top	(see Appendix of this chapter)	(see Appendix 9 of this chapter)
403, row 2 from top	Eqs. (9.57) and (9.70)	Eqs. (9.56) and (9.70)
430, row 2 from bottom	(see Appendix)	(see Appendix 9)
431, row 10 from top	(see Appendix)	(see Appendix 9)
449, Eq. (9.180)	$\frac{h_1 - h_2}{T_{f,w}} + \frac{h_3 - h_4}{T_{f,s}} + \left(h_{ep,w,2} - h_{ep,w,1}\right)\left(\frac{1}{T_{f,w}} - \frac{1}{T_{ep,w}}\right) + \left(h_{f,s,4} - h_{f,s,3}\right)\left(\frac{1}{T_{f,s}} - \frac{1}{T_{ep,s}}\right)$	$\frac{h_1 - h_2}{T_{f,w}} + \frac{h_3 - h_4}{T_{f,s}} + \left(h_{ep,w,2} - h_{ep,w,1}\right)\left(\frac{1}{T_{f,w}} - \frac{1}{T_{ep,w}}\right) + \left(h_{f,s,4} - h_{f,s,3}\right)\left(\frac{1}{T_{f,s}} - \frac{1}{T_{ep,s}}\right) = 0$

(Continued)

(Continued)

Page	Present text	Corrections to be made = the text which the "Present text" must be replaced with
451, row 12 from top	**Appendix 1**	**Appendix 9**
451, row 13 from top	The natural …	i) The natural …
452, row 3 from top	**Appendix 1**	
452, row 3 from top	Using the vector …	ii) Using the vector …
463, row 2 from top	Appendix	Appendix 10
465, row 14 from top	**Appendix**	**Appendix 10**
495, row 3 from bottom	The Heat/sink sources…	The heat/sink sources…

MIX
Papier aus verantwortungsvollen Quellen
Paper from responsible sources
FSC® C105338

If you have any concerns about our products,
you can contact us on
ProductSafety@springernature.com

In case Publisher is established outside the EU,
the EU authorized representative is:
Springer Nature Customer Service Center GmbH
Europaplatz 3, 69115 Heidelberg, Germany

Printed by Libri Plureos GmbH
in Hamburg, Germany